Plants as a Promising Biofactory for Bioactive Compounds

Plants as a Promising Biofactory for Bioactive Compounds

Editor

Jianfeng Xu

MDPI

Basel • Beijing • Wuhan • Barcelona • Belgrade • Novi Sad • Cluj • Manchester

Editor
Jianfeng Xu
Arkansas Biosciences Institute
Arkansas State University
Jonesboro
United States

Editorial Office
MDPI
St. Alban-Anlage 66
4052 Basel, Switzerland

This is a reprint of articles from the Special Issue published online in the open access journal *Life* (ISSN 2075-1729) (available at: www.mdpi.com/journal/life/special_issues/G5Y906W57L).

For citation purposes, cite each article independently as indicated on the article page online and as indicated below:

Lastname, A.A.; Lastname, B.B. Article Title. *Journal Name* **Year**, *Volume Number*, Page Range.

ISBN 978-3-0365-9465-1 (Hbk)
ISBN 978-3-0365-9464-4 (PDF)
doi.org/10.3390/books978-3-0365-9464-4

Contents

About the Editor vii

Preface ix

Jianfeng Xu
Harnessing the Power of Plants: A Green Factory for Bioactive Compounds
Reprinted from: *Life* **2023**, *13*, 2041, doi:10.3390/life13102041 1

Dayanand Dalawai, Hosakatte Niranjana Murthy, Yaser Hassan Dewir, Joseph Kadanthottu Sebastian and Anish Nag
Phytochemical Composition, Bioactive Compounds, and Antioxidant Properties of Different Parts of *Andrographis macrobotrys* Nees
Reprinted from: *Life* **2023**, *13*, 1166, doi:10.3390/life13051166 5

Hasan Karagecili, Ebubekir İzol, Ekrem Kirecci and İlhami Gulcin
Determination of Antioxidant, Anti-Alzheimer, Antidiabetic, Antiglaucoma and Antimicrobial Effects of Zivzik Pomegranate (*Punica granatum*)—A Chemical Profiling by LC-MS/MS
Reprinted from: *Life* **2023**, *13*, 735, doi:10.3390/life13030735 18

Nancy E. Rodríguez-Garza, Ramiro Quintanilla-Licea, César I. Romo-Sáenz, Joel H. Elizondo-Luevano, Patricia Tamez-Guerra and Cristina Rodríguez-Padilla et al.
In Vitro Biological Activity and Lymphoma Cell Growth Inhibition by Selected Mexican Medicinal Plants
Reprinted from: *Life* **2023**, *13*, 958, doi:10.3390/life13040958 45

Elham Amin, Mohamed Sadek Abdel-Bakky, Hamdoon A. Mohammed and Marwa H. A. Hassan
Chemical Profiling and Molecular Docking Study of *Agathophora alopecuroides*
Reprinted from: *Life* **2022**, *12*, 1852, doi:10.3390/life12111852 59

Muzaffer Mutlu, Zeynebe Bingol, Eda Mehtap Uc, Ekrem Köksal, Ahmet C. Goren and Saleh H. Alwasel et al.
Comprehensive Metabolite Profiling of Cinnamon (*Cinnamomum zeylanicum*) Leaf Oil Using LC-HR/MS, GC/MS, and GC-FID: Determination of Antiglaucoma, Antioxidant, Anticholinergic, and Antidiabetic Profiles
Reprinted from: *Life* **2023**, *13*, 136, doi:10.3390/life13010136 74

Reham F. El-Kased and Dina M. El-Kersh
GC–MS Profiling of Naturally Extracted Essential Oils: Antimicrobial and Beverage Preservative Actions
Reprinted from: *Life* **2022**, *12*, 1587, doi:10.3390/life12101587 92

Vaida Sirgedaitė-Šėžienė, Ieva Čėsnienė, Gabija Leleikaitė, Virgilijus Baliuckas and Dorotėja Vaitiekūnaitė
Phenolic and Antioxidant Compound Accumulation of *Quercus robur* Bark Diverges Based on Tree Genotype, Phenology and Extraction Method
Reprinted from: *Life* **2023**, *13*, 710, doi:10.3390/life13030710 106

Arnold A. Shamilov, Daniil N. Olennikov, Dmitryi I. Pozdnyakov, Valentina N. Bubenchikova, Ekaterina R. Garsiya and Mikhail V. Larskii
Caucasian Blueberry: Comparative Study of Phenolic Compounds and Neuroprotective and Antioxidant Potential of *Vaccinium myrtillus* and *Vaccinium arctostaphylos* Leaves
Reprinted from: *Life* **2022**, *12*, 2079, doi:10.3390/life12122079 . . . . **118**

Arifa Khanam, Ashfaq Ahmad, Neelam Iftikhar, Qasim Ali, Tabinda Fatima and Farhan Khashim Alswailmi et al.
Variation in Phenolic Profile, Antioxidant, and Anti-Inflammatory Activities of *Salvadora oleoides* Decene. and *Salvadora persica* L. Fruits and Aerial Part Extracts
Reprinted from: *Life* **2022**, *12*, 1446, doi:10.3390/life12091446 . . . . **133**

K. Thomas Klasson, Yunci Qi, Gillian O. Bruni, Tristan T. Watson, Bretlyn T. Pancio and Evan Terrell
Recovery of Aconitic Acid from Sweet Sorghum Plant Extract Using a Solvent Mixture, and Its Potential Use as a Nematicide
Reprinted from: *Life* **2023**, *13*, 724, doi:10.3390/life13030724 . . . . **148**

Xuan Wang, Zhuoyu He, Huan Yang, Cong He, Changyi Wang and Aliya Fazal et al.
Genome-Wide Identification of *LeBAHDs* in *Lithospermum erythrorhizon* and In Vivo Transgenic Studies Confirm the Critical Roles of *LeBAHD1/LeSAT1* in the Conversion of Shikonin to Acetylshikonin
Reprinted from: *Life* **2022**, *12*, 1775, doi:10.3390/life12111775 . . . . **160**

Emily Leggatt, Alistair Griffiths, Simon Budge, Anthony D. Stead, Alan C. Gange and Paul F. Devlin
Addition of Arbuscular Mycorrhizal Fungi Enhances Terpene Synthase Expression in *Salvia rosmarinus* Cultivars
Reprinted from: *Life* **2023**, *13*, 315, doi:10.3390/life13020315 . . . . **178**

François Guerineau
Properties of Human Gastric Lipase Produced by Plant Roots
Reprinted from: *Life* **2022**, *12*, 1249, doi:10.3390/life12081249 . . . . **195**

Talha Bin Emran, Fahadul Islam, Nikhil Nath, Hriday Sutradhar, Rajib Das and Saikat Mitra et al.
Naringin and Naringenin Polyphenols in Neurological Diseases: Understandings from a Therapeutic Viewpoint
Reprinted from: *Life* **2022**, *13*, 99, doi:10.3390/life13010099 . . . . **204**

Salvador Manzur-Valdespino, José Arias-Rico, Esther Ramírez-Moreno, María de Cortes Sánchez-Mata, Osmar Antonio Jaramillo-Morales and Julieta Angel-García et al.
Applications and Pharmacological Properties of Cactus Pear (*Opuntia* spp.) Peel: A Review
Reprinted from: *Life* **2022**, *12*, 1903, doi:10.3390/life12111903 . . . . **225**

Michaela Havrlentová, Václav Dvořáček, Lucie Jurkaninová and Veronika Gregusová
Unraveling the Potential of β-D-Glucans in *Poales*: From Characterization to Biosynthesis and Factors Affecting the Content
Reprinted from: *Life* **2023**, *13*, 1387, doi:10.3390/life13061387 . . . . **240**

Corbin England, Jonathan TrejoMartinez, Paula PerezSanchez, Uddhab Karki and Jianfeng Xu
Plants as Biofactories for Therapeutic Proteins and Antiviral Compounds to Combat COVID-19
Reprinted from: *Life* **2023**, *13*, 617, doi:10.3390/life13030617 . . . . **258**

About the Editor

Jianfeng Xu

Dr. Jianfeng Xu is a research professor at the Arkansas Biosciences Institute (ABI). He was tenured in the College of Agriculture at Arkansas State University (A-State). He earned his PhD degree in Biochemical Engineering from Dalian University of Technology in China and his Bachelor's degree in Environmental Engineering from the same university. Before joining the faculty at Arkansas State University (A-State) in 2008, he worked as a Senior Research Associate at Cornell University and a Research Scientist/Postdoctoral Fellow at Ohio University and the State Key Laboratory of Biochemical Engineering in the Chinese Academy of Sciences (Beijing). In addition, he also worked at Beijing Kaizheng Biotech Development Co Ltd for a short period of time.

Preface

It is with great pleasure that I introduce the reprint of the Special Issue entitled "Plants as a Promising Biofactory for Bioactive Compounds". Originally published in *Life*, this collection of research articles showcases the extraordinary potential of nature's green laboratories for the production of bioactive compounds.

Plants, with their remarkable ability to synthesize an astonishing array of bioactive compounds, have always been a source of inspiration and wonder. These bioactive compounds, often referred to as phytochemicals, encompass a wide spectrum of chemical classes and are known for their beneficial effects on human health, as well as their role in plant defense and ecological interactions. In addition to natural compounds, plants can also be genetically engineered to produce valuable recombinant proteins and enzymes for therapeutic and industrial applications. This Special Issue explores the diverse ways in which plants serve as remarkable biofactories for producing active compounds and enzymes with potential applications in medicine, agriculture, and industry.

As the global demand for sustainable sources of bioactive compounds continues to grow, the need for innovative solutions becomes ever more urgent. This reprint aims to serve as a valuable resource for researchers, academicians, students, and industry professionals who are interested in the fascinating world of plant-based bioactive compounds. The potential of plants as biofactories is vast and ever-evolving, and I am confident that the insights provided in this Special Issue will inspire further research and innovation in the field.

Finally, I want to extend my gratitude to the authors for their exceptional contributions to this Special Issue and to the reviewers for their dedication to maintaining the highest standards of research. I hope that this compilation inspires further exploration and collaboration, driving us closer to a future in which plants, as biofactories, are harnessed to their full potential for the betterment of society and the planet.

Jianfeng Xu
Editor

life

Editorial

Harnessing the Power of Plants: A Green Factory for Bioactive Compounds

Jianfeng Xu [1,2]

1 Arkansas Biosciences Institute, Arkansas State University, Jonesboro, AR 72401, USA; jxu@astate.edu; Tel.: +1-870-680-4812
2 College of Agriculture, Arkansas State University, Jonesboro, AR 72401, USA

The plant kingdom has long been revered for its complex biochemical pathways, which give rise to an incredible array of bioactive compounds. These small-molecule compounds can be generally classified as alkaloids, terpenoids/terpenes, and phenolic compounds. Many of these compounds have proven to be of great value to human beings as pharmaceuticals or nutraceuticals, as they exhibit antioxidant, antitumor, anti-inflammatory, and antibacterial effects and also show potential functions in immunomodulation, neuroprotection, and antiallergy. Moreover, the agricultural sector has benefited significantly from the use of bioactive compounds in the form of agrochemicals. Pesticides derived from plant-based terpenoids, for instance, help safeguard crops from pests and diseases while minimizing environmental harm. In addition to native compounds, plants can also be genetically engineered to produce valuable recombinant proteins (termed "molecular pharming") for therapeutic and industrial applications (e.g., cytokines, blood proteins, antibodies, vaccines, and industrial enzymes). The driving forces behind the rapid growth of plant-based biofactories include their relatively low production cost, product safety, and easy scale-up process.

This Special Issue of Life, entitled "*Plants as a Promising Biofactory for Bioactive Compounds*," showcases the latest research on the utilization of plant biofactories for producing compounds with multiple biological effects, including antioxidant, antitumor, antidiabetic, anti-inflammatory, antimicrobial, antialzheimer, antihemolytic, neuroprotective, and pesticide activities. The 17 articles contained within this Special Issue, of which 13 are research papers and 4 are reviews, encompass a broad spectrum of research, ranging from fundamental studies elucidating the biosynthetic pathways of bioactive compounds to applied research harnessing the biotechnological potential of plant-based production systems. They collectively highlight the incredible potential of plants as green factories for producing bioactive compounds.

One recurring theme in these papers is the extraction, isolation, and structural characterization of existing or new bioactive compounds with nutraceutical and therapeutic potential from medicine plants. Advanced analytic tools, such as LC-HR/MS, GC/MS, and GC-FID, were commonly exploited to determine the bioactive compound profile or phytochemical composition in plant extracts. The pharmaceutical or nutraceutical activities of the separated compounds or crude plant extracts were analyzed in vitro or in vivo. The paper by Murthy and Dewir's group determined the phytochemical composition of different parts of *Andrographis macrobotrys* Nees and demonstrated the importance of *A. macrobotrys* as a source of medicine and antioxidants [1]. Karagecili and colleagues analyzed the chemical profiling of peel and seed extracts of Zivzik Pomegranate (*Punica granatum*) via LC-MS/MS and determined the antioxidant, antialzheimer, antidiabetic, antiglaucoma, and antimicrobial effects of different phenolic and flavonoid compounds [2]. Inspired by the remarkable antitumor potential of many plant secondary metabolites, Gomez-Flores' team conducted an assay on of the effects of methanol extracts from 15 selected Mexican

check for updates

Citation: Xu, J. Harnessing the Power of Plants: A Green Factory for Bioactive Compounds. *Life* **2023**, *13*, 2041. https://doi.org/10.3390/life13102041

Received: 30 September 2023
Accepted: 10 October 2023
Published: 11 October 2023

medicinal plants on the growth inhibition of mouse lymphoma cells, the toxicity and proliferation of human peripheral blood mononuclear cells, and their antioxidant, hemolytic, and antihemolytic activities [3]. Their results validate the possibility of identifying bioactive compounds showing antitumor, antioxidant, and antihemolytic activity from medicinal plants. The paper by Hassan and colleagues presented the first metabolomic profiling of the halophyte *Agathophora alopecuroides*, a potential antidiabetic plant, using the LC-HRMS/MS technique [4]. It is worth mentioning that a molecular docking technique was adopted for this study to highlight the bioactive compounds, of which two alkaloids and four flavonoids were identified as being responsible for the observed antidiabetic activity.

In addition to analyzing the broad spectrum of compounds in medicinal plants, five papers focus on two specific types of bioactive compounds: essential oils and phenolic compounds. In a study conducted by Gulcin's group, the antioxidant and antidiabetic properties of essential oil from cinnamon (*Cinnamomum zeylanicum*) leaves were evaluated and investigated for the first time. Furthermore, the inhibitory effects of cinnamon oil on enzymes associated with various metabolic diseases were also determined [5]. In another study, El-Kased and EI-Kersh analyzed the GC–MS profiling of eight natural essential oils and demonstrated their antimicrobial effects and their beverage preservative actions [6]. For phenolic compounds, Vaitiekūnaitė group's research indicated that the accumulation of phenolic and antioxidant compounds in *Quercus robur* Bark diverges based on tree genotype, phenology, and the extraction method used [7]. Olennikov's team profiled and quantified the phenolic compounds of leaf extracts of two Caucasian blueberries, namely *Vaccinium myrtillus* L. and *Vaccinium arctostaphylos* L., via HPLC–PDA–ESI–tQ–MS and showed their neuroprotective and antioxidant potential [8]. Similarly, the study by Hussain and colleagues demonstrated that polyphenols extracted from *Salvadora oleoides* Decene. and *Salvadora persica* L. fruit and aerial parts could potentially be used as antioxidant, analgesic, and anti-inflammatory agents [9]. Other than having applications as nutraceuticals and therapeutics, plant-derived compounds can also be used as pesticides. In their study, Klasson and colleagues extracted aconitic acid from sweet sorghum syrup and used it as a nematicide [10]. The extract was found to effectively reduce the motility of the parasitic *Meloidogyne incognita* and cause a high mortality of the nematode.

The utilization of advanced biotechnological methods such as metabolic engineering, genome editing, and plant tissue culture have allowed researchers to fine-tune the biosynthetic pathways of plants, thereby improving the yields and quality of bioactive compounds. There are two research papers in this Special Issue that tackle this research area. An excellent paper by Qi's group explored the characterization, evolution, expression patterns, and gene function of BAHDs in *Lithospermum erythrorhizon* (LeBAHDs) via bioinformatics and transgenic analysis [11]. The plant BAHD acyltransferase family has been well known to acylate plant metabolites and participate in plant secondary metabolic processes. The results from the study by Qi's group revealed that the overexpression of LeBAHD1 in *L. erythrorhizon* hairy roots significantly increased the content of acetylshikonin and the conversion rate of shikonin to acetylshikonin, whereas the CRISPR/Cas9-based knockout of LeBAHD1 in hairy roots displayed the opposite trend. This study not only confirmed the in vivo function of LeBAHD1 in the biosynthesis of acetylshikonin but also provided new insights into the biosynthetic pathway of shikonin and its derivatives. In another study on improving essential oil production in aromatic plants, Devlin's group found that the addition of Arbuscular Mycorrhizal Fungi (AMF) profoundly influenced terpene synthase expression in six rosemary cultivars without impacting plant growth [12]. Their findings demonstrated the potential for the use of AMF in the improvement of aroma in culinary herbs within a commercial setting.

Apart from the extensive body of research dedicated to exploring bioactive small molecules derived from plants, a notable study by Dr. Guerineau reported the successful production of recombinant human gastric lipase by *Arabidopsis thaliana* root culture while also providing a comprehensive characterization of the enzyme's properties [13]. This investigation showcases the application of molecular pharming in plants as a means to synthesize recombinant proteins, which are big molecules.

In addition to these noteworthy research articles, this Special Issue also contains four review papers that discuss the pharmacological applications of plant-derived bioactive compounds. Emran and colleagues discussed the potential of naringin and naringenin polyphenols as therapeutics for neurodegenerative diseases such as Alzheimer's disease and Parkinson's disease, as well as other neurological conditions such as anxiety, depression, and chronic hyperglycemic peripheral neuropathy [14]. Cruz-Cansino and colleagues focused on the exploitation of by-products from fruits and vegetables. Specifically, the applications and pharmacological properties of bioactive compounds from cactus pear (*Opuntia* spp.) peel were discussed [15]. In addition to bioactive secondary metabolites, a review paper by Havrlentová's group focused on other types of bioactive molecules from plants, namely β-D-glucan, a cell wall polysaccharide [16]. The comprehensive introduction to this bioactive biomolecule from Poales expands our understanding of the characteristics, functions, and applications of this cell wall polysaccharide and opens new avenues for future research and advancements. Finally, in light of the recent outbreak of COVID-19, caused by severe acute respiratory syndrome coronavirus 2 (SARS-CoV-2), which had a profound impact on public health and the global economy, Xu's group presented a comprehensive review on the latest advancements in plant-derived antiviral medicines, including both small-molecule compounds and recombinant therapeutics used to combat COVID-19 [17]. Particularly, the unprecedented opportunity of molecular pharming in plants for developing vaccines, antibodies, and other biologics against COVID-19 was demonstrated. This review sheds light on the future development of plant sources of antiviral medicines for treating COVID-19 and other viral infections.

In a world where the exploration of sustainable sources for bioactive compounds has become increasingly significant, the potential of plants as biofactories has garnered substantial attention. This collection of 17 papers represents a pivotal step forward in our understanding of how plants can serve as potent sources of bioactive molecules with diverse applications. The interdisciplinary nature of these contributions underscores the collaborative efforts of botanists, chemists, biotechnologists, and pharmacologists, all working towards a common goal: harnessing the power of plants to produce valuable compounds.

Funding: This work was supported by the National Institute of Health (Grant No. R15DK128757) and the Arkansas Biosciences Institute, the major research component of the Arkansas Tobacco Settlement Proceeds Act of 2000.

Acknowledgments: We thank all the contributors to this Special Issue, as well as the kind reviewers that contributed to improving the manuscripts.

Conflicts of Interest: The author declares no conflict of interest.

References

1. Dalawai, D.; Murthy, H.N.; Dewir, Y.H.; Sebastian, J.K.; Nag, A. Phytochemical Composition, Bioactive Compounds, and *Antioxidant Properties* of Different Parts of Andrographis macrobotrys Nees. *Life* **2023**, *13*, 1166. [CrossRef]
2. Karagecili, H.; Izol, E.; Kirecci, E.; Gulcin, I. Determination of Antioxidant, Anti-Alzheimer, Antidiabetic, Antiglaucoma and Antimicrobial Effects of Zivzik Pomegranate (*Punica granatum*)—A Chemical Profiling by LC-MS/MS. *Life* **2023**, *13*, 735. [CrossRef]
3. Rodriguez-Garza, N.E.; Quintanilla-Licea, R.; Romo-Saenz, C.I.; Elizondo-Luevano, J.H.; Tamez-Guerra, P.; Rodriguez-Padilla, C.; Gomez-Flores, R. In Vitro Biological Activity and Lymphoma Cell Growth Inhibition by Selected Mexican Medicinal Plants. *Life* **2023**, *13*, 958. [CrossRef] [PubMed]
4. Amin, E.; Abdel-Bakky, M.S.; Mohammed, H.A.; Hassan, M.H.A. Chemical Profiling and Molecular Docking Study of *Agathophora alopecuroides*. *Life* **2022**, *12*, 1852. [CrossRef] [PubMed]
5. Mutlu, M.; Bingol, Z.; Uc, E.M.; Koksal, E.; Goren, A.C.; Alwasel, S.H.; Gulcin, I. Comprehensive Metabolite Profiling of Cinnamon (*Cinnamomum zeylanicum*) Leaf Oil Using LC-HR/MS, GC/MS, and GC-FID: Determination of Antiglaucoma, Antioxidant, Anticholinergic, and Antidiabetic Profiles. *Life* **2023**, *13*, 136. [CrossRef] [PubMed]
6. El-Kased, R.F.; El-Kersh, D.M. GC-MS Profiling of Naturally Extracted Essential Oils: Antimicrobial and Beverage Preservative Actions. *Life* **2022**, *12*, 1587. [CrossRef]

7. Sirgedaite-Seziene, V.; Cesniene, I.; Leleikaite, G.; Baliuckas, V.; Vaitiekunaite, D. Phenolic and Antioxidant Compound Accumulation of Quercus robur Bark Diverges Based on Tree Genotype, Phenology and Extraction Method. *Life* **2023**, *13*, 710. [CrossRef] [PubMed]
8. Shamilov, A.A.; Olennikov, D.N.; Pozdnyakov, D.I.; Bubenchikova, V.N.; Garsiya, E.R.; Larskii, M.V. Caucasian Blueberry: Comparative Study of Phenolic Compounds and Neuroprotective and Antioxidant Potential of *Vaccinium myrtillus* and *Vaccinium arctostaphylos* Leaves. *Life* **2022**, *12*, 2079. [CrossRef] [PubMed]
9. Khanam, A.; Ahmad, A.; Iftikhar, N.; Ali, Q.; Fatima, T.; Alswailmi, F.K.; Hussain, A.I.; Alnasser, S.M.A.; Akhtar, J. Variation in Phenolic Profile, Antioxidant, and Anti-Inflammatory Activities of *Salvadora oleoides* Decene. and *Salvadora persica* L. Fruits and Aerial Part Extracts. *Life* **2022**, *12*, 1446. [CrossRef] [PubMed]
10. Klasson, K.T.; Qi, Y.; Bruni, G.O.; Watson, T.T.; Pancio, B.T.; Terrell, E. Recovery of Aconitic Acid from Sweet Sorghum Plant Extract Using a Solvent Mixture, and Its Potential Use as a Nematicide. *Life* **2023**, *13*, 724. [CrossRef] [PubMed]
11. Wang, X.; He, Z.; Yang, H.; He, C.; Wang, C.; Fazal, A.; Lai, X.; Yang, L.; Wen, Z.; Yang, M.; et al. Genome-Wide Identification of LeBAHDs in *Lithospermum erythrorhizon* and In Vivo Transgenic Studies Confirm the Critical Roles of LeBAHD1/LeSAT1 in the Conversion of Shikonin to Acetylshikonin. *Life* **2022**, *12*, 1775. [CrossRef] [PubMed]
12. Leggatt, E.; Griffiths, A.; Budge, S.; Stead, A.D.; Gange, A.C.; Devlin, P.F. Addition of Arbuscular Mycorrhizal Fungi Enhances Terpene Synthase Expression in *Salvia rosmarinus* Cultivars. *Life* **2023**, *13*, 315. [CrossRef] [PubMed]
13. Guerineau, F. Properties of Human Gastric Lipase Produced by Plant Roots. *Life* **2022**, *12*, 1249. [CrossRef] [PubMed]
14. Emran, T.B.; Islam, F.; Nath, N.; Sutradhar, H.; Das, R.; Mitra, S.; Alshahrani, M.M.; Alhasaniah, A.H.; Sharma, R. Naringin and Naringenin Polyphenols in Neurological Diseases: Understandings from a Therapeutic Viewpoint. *Life* **2022**, *13*, 99. [CrossRef] [PubMed]
15. Manzur-Valdespino, S.; Arias-Rico, J.; Ramirez-Moreno, E.; Sanchez-Mata, M.C.; Jaramillo-Morales, O.A.; Angel-Garcia, J.; Zafra-Rojas, Q.Y.; Barrera-Galvez, R.; Cruz-Cansino, N.D.S. Applications and Pharmacological Properties of Cactus Pear (*Opuntia* spp.) Peel: A Review. *Life* **2022**, *12*, 1903. [CrossRef] [PubMed]
16. Havrlentova, M.; Dvoracek, V.; Jurkaninova, L.; Gregusova, V. Unraveling the Potential of beta-D-Glucans in Poales: From Characterization to Biosynthesis and Factors Affecting the Content. *Life* **2023**, *13*, 1387. [CrossRef] [PubMed]
17. England, C.; TrejoMartinez, J.; PerezSanchez, P.; Karki, U.; Xu, J.F. Plants as Biofactories for Therapeutic Proteins and Antiviral Compounds to Combat COVID-19. *Life* **2023**, *13*, 617. [CrossRef] [PubMed]

life

MDPI

Article

Phytochemical Composition, Bioactive Compounds, and Antioxidant Properties of Different Parts of *Andrographis macrobotrys* Nees

Dayanand Dalawai [1], Hosakatte Niranjana Murthy [1,2,*], Yaser Hassan Dewir [3,*], Joseph Kadanthottu Sebastian [4] and Anish Nag [4]

1 Department of Botany, Karnatak University, Dharwad 580003, India
2 Department of Horticultural Science, Chungbuk National University, Cheongju 28644, Republic of Korea
3 Plant Production Department, College of Food & Agriculture Sciences, King Saud University, Riyadh 11451, Saudi Arabia
4 Department of Life Sciences, Christ University, Bangalore 560029, India
* Correspondence: hnmurthy60@gmail.com (H.N.M.); ydewir@ksu.edu.sa (Y.H.D.)

Abstract: *Andrographis macrobotrys* Nees is an ethnomedicinal plant belonging to the family Acanthaceae, distributed in the moist deciduous and semi-evergreen forests of the southern Western Ghats of India. The objective of this research was to determine the phytochemical composition and bioactive chemical components using gas chromatography and mass spectrometry (GC-MS) and to check the antioxidant potential of the plant part extracts. *A. macrobotrys* roots, stems, and leaves were obtained from the species' natural habitat in the Western Ghats, India. The bioactive compounds were extracted using a Soxhlet extractor at 55–60 °C for 8 h in methanol. Identification analysis of *A. macrobotrys* bioactive compound was performed using GC-MS. Quantitative estimation of phytochemicals was carried out, and the antioxidant capacity of the plant extracts was determined by 2,2′-diphenyl-1-picrylhydrazyl radical scavenging (DPPH) and ferric reducing assays (FRAP). *A. macrobotrys* has a higher concentration of phenolics in its stem extract than in its root or leaf extracts (124.28 mg and 73.01 mg, respectively), according to spectrophotometric measurements. GC-MS analysis revealed the presence of phytochemicals such as azulene, 2,4-di-tert-butylphenol, benzoic acid, 4-ethoxy-ethyl ester, eicosane, 3-heptadecanol, isopropyl myristate, hexadecanoic acid methyl ester, hexadecanoic acid, 1-butyl-cyclohexanol, 9,12-octadecadienoic acid, alpha-monostearin, and 5-hydroxy-7,8-dimethoxyflavone belonging to various classes of flavonoids, terpenoids, phenolics, fatty acids, and aromatic compounds. Significant bioactive phytochemicals include 2,4-di-tert-butylphenol, 2-methoxy-4-vinylphenol, 5-hydroxy-7,8-dimethoxyflavone, azulene, salvigenin, squalene, and tetrapentacontane. In addition, the antioxidant capability of each of the three extracts was assessed. The stem extract demonstrated impressive DPPH scavenging and ferric reduction activities, with EC_{50} values of 79 mg/mL and 0.537 ± 0.02 OD at 0.2 mg/mL, respectively. The results demonstrated the importance of *A. macrobotrys* as a source of medicine and antioxidants.

Keywords: *Andrographis*; antioxidants; GS-MS analysis; bioactive compounds

Citation: Dalawai, D.; Murthy, H.N.; Dewir, Y.H.; Sebastian, J.K.; Nag, A. Phytochemical Composition, Bioactive Compounds, and Antioxidant Properties of Different Parts of *Andrographis macrobotrys* Nees. *Life* **2023**, *13*, 1166. https://doi.org/10.3390/life13051166

Academic Editor: Jianfeng Xu

Received: 25 March 2023
Revised: 5 May 2023
Accepted: 7 May 2023
Published: 11 May 2023

1. Introduction

Many secondary metabolites that plants produce have developed over time as protection against pathogens and herbivores. The use of secondary metabolites in pharmaceuticals, nutraceuticals, herbal cosmetics, and food supplements is widespread. Due to their advantages over synthetic pharmaceuticals, such as fewer side effects, better compatibility with human physiology, and lower prices, the demand for plant-based chemical compounds is rising globally [1]. The concept of linking a plant's phytochemicals to its pharmacological activity is becoming increasingly popular. Hence, after identifying plants that are significant from an ethnopharmacological perspective, chemical compounds from those plants are extracted, isolated, and characterized employing highly sophisticated chromatography techniques, including gas chromatography and mass spectroscopy (GC-MS).

Acanthaceae is a family of plants that includes the genus *Andrographis* Wall. ex. Nees, which is composed of therapeutic plants. *Andrographis* is represented by 48 species that are annual and perennial herbaceous plants distributed in tropical regions of Asia. The majority of *Andrographis* species are endemic to India, and some of them are lesser known but significant medicinal plants utilized by native and tribal people in the Deccan Plateau and Western Ghats, particularly *A. paniculata*, often known as "King of Bitter," which has many uses in modern medicine. It has a wide range of pharmacological actions, including anticancer, antidiabetic, hepatoprotective, anti-inflammatory, antiviral, antioxidative, and antibacterial activities [2]. Diterpenoids, including andrographolide, neoandrographolide, deoxyandrographolide, and 14-deoxy-11,12-didehydroandrographolide, as well as flavonoids, are thought to be responsible for its pharmacological effects [3,4]. Since ancient times, local communities and practitioners of traditional medicine have used a number of *Andrographis* species known for their ethnomedicinal properties to treat a variety of conditions, including wounds, fever, snake bites, constipation, jaundice, diabetes, skin diseases, and a number of other disorders [3,5]. It follows that members of the *Andrographis* genus have a great deal of potential for medicinal uses, and there is a lot of room for isolating and characterizing phytochemicals from lesser known species for use in medicine.

Due to its extensive therapeutic benefits, there is a significant demand for "*Andrographis*" raw material on a global scale. To accommodate this demand, *A. paniculata* has been cultivated in Asian nations. The overexploitation of *A. paniculata* natural populations, however, is one of the main causes of the species' natural distribution becoming depleted. *A. macrobotrys* was identified by Dalawai et al. [4] as a potential substitute for the biosynthesis of neoandrographolide, a significant diterpenoid molecule. The neoandrographolide level in *A. macrobotrys* was 102.03 mg/g DW, while it was 11.72 mg/g DW in *A. paniculata* [4]. This suggests that *A. macrobotrys* has the potential to be the most effective source of neoandrorapholide and that it may also lighten the load on *A. paniculata*. In addition, only a few species, including *A. paniculata*, *A. echioides*, and *A. producta*, have been successfully identified as phytochemicals and their compositions using the GC-MS technique [5,6]. Other species, such as *A. macrobotrys*, have not yet had their phytochemical constituents thoroughly investigated.

In order to fully appreciate *A. macrobotrys* potential as a medicine, it is important to investigate and identify its chemical makeup. After this is accomplished, researchers can shift their focus to species conservation and sustainable use. Hence, using GC-MS techniques, the current study was aimed at investigating specific chemical components of the root, stem, and leaf of *A. macrobotrys*. Furthermore, in vitro techniques have been used to study the antioxidative qualities of methanolic extracts of roots, stems, and leaves.

2. Materials and Methods

2.1. Plant Materials

A. macrobotrys roots stems, and leaves were obtained from the species' natural habitat near Hebri, Udupi, Karnataka, India (13.41789928, 74.94768698). An identification of the plant specimen was made using the Flora of the Presidency of Madras [7] and a voucher specimen (DSD-03) was deposited in the Shivaji University Herbarium in Kolhapur, India.

2.2. Chemicals and Reagents

Analytical-grade chemicals were employed in all of the tests. Methanol, sodium carbonate, sodium nitrate, ferric chloride, aluminium chloride, hydrochloric acid, 2,4,6-tripyridyl-s-triazine, 2,2-diphenyl-1-picrylhydrazyl (DPPH), butylated hydroxyl anisole, ascorbic acid, and Folin-Coicalteu (FC) reagents were procured from Sigma Aldrich Chemical Co. (St. Louis, MO, USA).

2.3. Methanolic Extract Preparation

The plant's root, stem, and leaves were shade-dried before being separately blended to a fine powder. They were left for 24 h in a hot air oven to dry at 35 °C. Each part's powder (100 g) was extracted with 100 mL of methanol using a Soxhlet extractor at 55–60 °C for 8 h. Each part's extract was dried in a rotary evaporator under reduced pressure at 40 °C to remove any excess methanol before being used in further phytochemical and antioxidant studies.

2.4. Determination of Phytochemical Composition

In accordance with Folin and Ciocalteau's [8] description of the spectrophotometric analysis, the amount of total phenolics in the methanolic extracts of the root, stem, and leaf were measured. The Folin–Ciocalteau (FC) reagent (1 mL) and 0.5 mL (1 g/mL) of the methanolic extracts were added to test tubes containing 2.5 mL of deionized distilled water. The mixture was given six minutes to stand before the addition of 0.5 mL 20% sodium carbonate solution. After 30 min of incubation at room temperature, the produced color's absorbance was measured at 760 nm on a UV–visible spectrophotometer (UV-1601, Shimadzu, Kyoto, Japan). By comparing the results to the gallic acid standard curve, the amount of total phenolic content was determined and reported as mg gallic acid equivalent (GAE) per g dry sample.

Spectrophotometric analysis was used to assess the flavonoid content in the extracts of the root, stem, and leaf [9]. Each sample's 0.5 mL of methanolic extract was combined with 2.5 mL of distilled water and 0.15 mL of sodium nitrite (5%) solution, and permitted to sit for 6 min. The reaction was then continued for another 5 min with the addition of 0.3 mL of 10% aluminium chloride. After mixing 2 mL of 1M sodium hydroxide, a spectrophotometer (UV-1601, Shimadzu) was used to read the absorbance at 510 nm. The amount of flavonoid content was calculated using the standard calibration curve of quercetin and reported as mg of quercetin equivalent (QE) per g of dry samples.

The methanolic extracts were used in the determination of tannins present in the samples spectrophotometrically according to the method developed by Schanderi [10]. Each extract (0.5 mL) was mixed with 2.5 mL of distilled water and 0.25 mL of Folin–Denis reagent, followed by 0.5 mL of 30% sodium carbonate. Then, all the reagents were mixed well to complete the reaction and incubated for 30 min at room temperature. The absorbance of the developed color was measured at 700 nm using a spectrophotometer (UV-1601, Shimadzu). The known amount of tannic acid was used to draw the calibration curve, and the amount of tannin in samples was determined and expressed as mg of tannic acid equivalent (TAE) per gram of dry samples.

2.5. Gas Chromatography and Mass Spectroscopy (GC-MS) Analysis

The chemical compounds present in the methanolic extracts of the samples were separated using gas chromatography and mass spectrometry (Model: QP2010S; Shimadzu Corporation, Kyoto, Japan), an advanced analytical device outfitted with Rxi-5Sil MS capillary column (length 30 m 0.25 mm ID, 0.25 m film thickness). The carrier gas, with a constant flow rate of 1 mL/min, was helium (99.9995%). The column oven temperature was originally maintained at 60 °C before being raised to 260 °C by 5 °C/min and maintained for 5 min. The split injection was used to inject the 1 L diluted samples, with a 4 min solvent delay. The show lasted 30 min in total. Temperatures of 200 °C and 280 °C, respectively, were selected for the interface line and ion source [11]. By contrasting the retention durations of real compounds with the mass spectra from the NIST 11 and WILEY 8 mass spectral libraries, the separated components were determined. Kovats retention index was determined using the structural formulae of the compounds as previously described [12]. For quality control of the gas chromatography analysis, we estimated a resolution between two peak separations using the formula: $[\{1.18(Rt2 - Rt1)\}/(W2 + W1)]$, while Rt2 and Rt1 are the retention times of peak 2 and 1, respectively; W is the peak width.

In additio, we estimated the signal-to-noise ratio for each of the sample runs, based on corresponding blank runs.

2.6. *In Vitro Antioxidant Activity*

The extracts' capacity to scavenge DPPH radicals was assessed [13]. To make the stock solution, 24 mg of DPPH was dissolved in 100 mL of methanol. The working solution was created by combining methanol with DPPH stock solution to achieve an absorbance of around 0.99 $\pm$ 0.02 at 515 nm using a spectrophotometer (UV-1601, Shimadzu). A sample with a range of quantities (0.2–1.0 mg/mL) was combined with aliquots (3 mL) of this working solution (DPPH). After thoroughly shaking, the reaction mixture was allowed to sit at room temperature for 30 min in the dark. At 515 nm, the absorbance was then measured. The control was prepared concurrently without any sample. As standards, butylated hydroxyl anisole (BHA) and ascorbic acid were utilized. The following equation was used to obtain the DPPH scavenging activity percentage. [(Control OD − Sample OD)/Control OD] $\times$ 100 equals the percentage of DPPH scavenging activity.

The Benzie and Strain method was employed while performing the FRAP assay [14]. Plant extracts were added to 3 mL of newly made FRAP reagent (300 mM acetate buffer at pH 3.6, 10 mM 2,4,6-tripyridyl-s-triazine (TPTZ) in 40 mM HCl, and 20 mM $FeCl_3$ in the ratio 10:1:1) at varied concentrations (0.2–1.0 mg/mL) and then heated to 37 °C in a hot water bath for 10 min. Distilled water was used to adjust the final volume to 4 mL, and it was incubated for 10 min at room temperature in the dark. At 593 nm, optical density was observed. As standards, butylated hydroxyl anisole and ascorbic acid were utilized. More antioxidant power was shown by an increase in optical density (OD).

2.7. *Statistical Analysis*

All experiments were conducted in triplicate using three different lots. The mean, percentage, EC50 values, and standard deviation were calculated using the statistical software SPSS, Version 17.

3. Results

3.1. *Phytochemical Composition*

In the root, stem, and leaf of *A. macrobotrys*, phytochemicals, including total phenolics, flavonoids, and tannins, were quantified and expressed as gallic acid equivalent (GAE), quercetin equivalent (QE), and tannic acid equivalent (TAE), respectively. The total phenolics, flavonoids, and tannins of different plant parts are shown in Table 1. The concentration of phenolics is the highest among all the plant parts.

Table 1. Phytochemical composition of *Andrographis macrobotrys* plant extracts.

Plant Part	Phenolics (mg GAE/g DW)	Flavonoids mg QE/g DW	Tannins (mg TAE/g DW)
Root	124.28 ± 0.24^{b}	9.06 ± 0.79^{c}	59.49 ± 0.22^{b}
Stem	180.98 ± 0.76^{a}	18.04 ± 0.08^{b}	86.39 ± 0.06^{a}
Leaf	73.01 ± 0.35^{c}	57.04 ± 0.43^{a}	57.33 ± 0.14^{b}

Values are mean $\pm$ standard deviation (n = 3). Mean values followed by different superscript in a column are significantly different ($p < 0.05$) according to Duncan's multiple range test.

3.2. *Gas Chromatography and Mass Spectrometry (GC-MS) Analysis*

Extracts of the root, stem, and leaf of *A. macrobotrys* were subjected to gas chromatography and mass spectrometry (GC-MS) profiling to identify bioactive chemical compounds. Average resolution between two peaks was found to be ~1.0 for all GC runs reflecting 98% separation. We also noted a signal-to-noise ratio of around 5:1 for all the runs. These two parameters were well within the limit as stipulated by USP and ICH. The GC-MS analysis revealed the presence of various kinds of metabolites, including phenolics, sesquiterpenoids, isoprenoids, fatty acids, benzofurans, and aromatic compounds. The total number of compounds found in all the extracts was 104, including 24 from the root, 36 from the stem, and 44 from the leaf. The compounds from the root (Figure 1A), stem (Figure 1B), and leaf (Figure 1C) are presented in Tables 2–4, respectively. Additionally, a number of the phytochemicals in the extracts have been reported to be biologically active substances. Table 5 lists key bioactive substances that are present in the roots, stem, and leaves of *A. macrobotrys* and have been proven to exhibit biological activity.

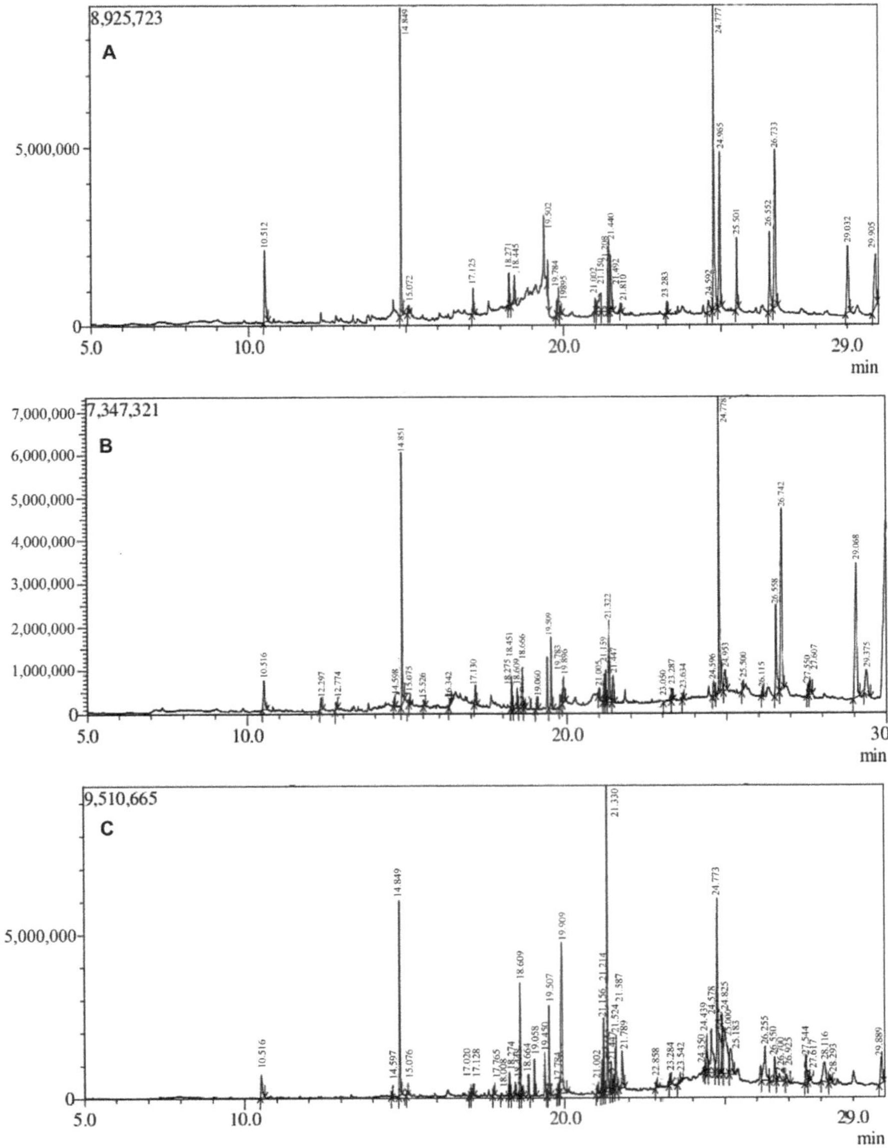

Figure 1. GCMS profile of the (**A**) root, (**B**) stem, and (**C**) leaves of *Andrographis macrobotrys*.

Table 2. GC-MS profile of *Andrographis macrobotrys'* root.

Sl. No.	Compound Name	Retention Time	Concentration (%)
1	Azulene	10.512	4.71
2	2,4-Di-tert-butylphenol	14.849	11.46
3	4-Ethoxy-ethyl esterbenzoic acid	15.072	0.28
4	Eicosane	17.125	1.02
5	3-Heptadecanol	18.271	1.28
6	Isopropyl myristate	18.445	1.26
7	Hexadecanoic acid, methyl ester	19.502	0.93
8	2,3-Dimethyl-3-hexanol	19.784	0.57
9	Hexadecanoic acid	19.895	0.36
10	1-Butyl-cyclohexanol	21.002	0.59
11	Methyl octadeca-9,12-dienoate	21.150	2.73
12	6-Octadecenoic acid, methyl ester, (Z)-	21.208	10.02
13	Tetratriacontane	21.440	5.64
14	4,4′-Thiobis[2-(1,1-dimethylethyl)-5-methyl-phenol	21.492	3.79
15	2,6,10,14-Tetramethyl-hexadecane	21.810	0.34
16	Dotriacontane	23.283	0.55
17	2-Ethylbutyric acid, eicosyl ester	24.592	1.05
18	Hexadecanoic acid, 2-hydroxy-1-(hydroxymethyl) ethyl ester	24.777	15.47
19	Asaraldehyde	24.965	7.99
20	Acetosyringone	25.501	3.23
21	alpha-Monostearin	26.552	5.04
22	5-Hydroxy-7,8-dimethoxyflavone	26.733	10.65
23	5-Hydroxy-6,7,4′-trimethoxyflavone/salvigenin	29.032	5.55
24	3,5-Dihydroxy-6,7,8-trimethoxyflavone	29.905	5.50

Table 3. GC-MS profile of *Andrographis macrobotrys'* stem.

Sl. No.	Compound Name	Retention Time	Concentration (%)
1	Azulene	10.516	2.48
2	2-Methoxy-4-vinylphenol	12.297	0.66
3	2,6-Dimethoxy-phenol or syringol	12.774	0.45
4	Nonadecane	14.598	0.29
5	2,4-Di-tert-butylphenol	14.851	11.24
6	Ethyl 4-ethoxybenzoate	15.075	0.28
7	4-Methyl-2,5-dimethoxybenzaldehyde	15.526	0.26
8	N-Phenyl aniline	16.342	0.44
9	Eicosane	17.130	0.94
10	3-Heptadecanol	18.275	1.29
11	Isopropyl myristate	18.451	0.69
12	Neophytadiene	18.609	1.52
13	6,10,14-Trimethyl-2-pentadecanone	18.666	0.30
14	3,7,11,15-Tetramethyl-2-hexadecen-1-ol	19.060	0.55
15	Hexadecanoic acid, methyl ester	19.509	3.09
16	3-Ethyl-3-pentanol	19.783	1.03
17	Hexadecanoic acid	19.896	1.80
18	1-Butyl-cyclohexanol	21.005	0.47
19	9,12-Octadecadienoic acid, methyl ester	21.159	1.28
20	8,11,14-Docosatrienoic acid, methyl ester	21.216	1.52
21	Phytol	21.322	4.13
22	Octadecanoic acid, methyl ester	21.447	1.66
23	2-Mono-myristin	23.050	0.61
24	Tetrapentacontane	23.287	0.53
25	Octadecane	23.634	0.24
26	Ethyl 3-hydroxytridecanoate	24.596	1.33
27	Hexadecanoic acid, 2-hydroxy-1-(Hydroxymethyl) ethyl ester	24.778	16.03
28	Asaraldehyde	24.953	1.95

Table 3. *Cont.*

Sl. No.	Compound Name	Retention Time	Concentration (%)
29	3-Acetyl biphenyl	25.500	0.42
30	N-{4-[2-(1,1-Dimethylethyl)-5-oxo-1,3-dioxolan-4-Yl]butyl}formamide	26.115	0.41
31	alpha-Monostearin	26.558	6.51
32	5-Hydroxy-7,8-dimethoxyflavone	26.742	15.35
33	Squalene	27.550	0.81
34	Stigmasta-5,22-dien-3-ol	27.607	1.39
35	5-Hydroxy-6,7,4′-trimethoxyflavone/Salvigenin	29.068	14.53
36	(3 beta,24S)-Stigmast-5-en-3-ol	29.375	3.53

Table 4. GC-MS profile of *Andrographis macrobotrys′* leaf.

Sl. No.	Compound Name	Retention Time	Concentration (%)
1	Azulene	10.516	1.62
2	Nonadecane	14.597	0.19
3	2,4-Di-tert-butylphenol	14.849	6.92
4	Ethyl 4-ethoxybenzoate	15.076	0.17
5	1-{2-[3-(2-Acetyloxiran-2-yl)-1,1-dimethylpropyl]cycloprop-2-enyl}ethenone	17.020	0.14
6	Eicosane	17.128	0.46
7	Tetradecanoic acid	17.765	0.36
8	Loliolide	18.008	0.24
9	3-Heptadecanol	18.274	0.86
10	Isopropyl myristate	18.448	0.46
11	Neophytadiene	18.609	3.91
12	1-Dodecanol, 3,7,11-trimethyl-	18.664	0.40
13	3,7,11,15-Tetramethyl-2-hexadecen-1-ol	19.058	1.18
14	Methyl palmitoleate	19.450	0.26
15	Hexadecanoic acid, methyl ester	19.507	3.16
16	3-Ethyl-3-pentanol	19.784	0.32
17	Phytane	19.817	0.28
18	Hexadecanoic acid	19.909	10.19
19	1-Butyl-cyclohexanol	21.002	0.36
20	9,12-Octadecadienoic acid, methyl ester	21.156	1.22
21	9,12,15-Octadecatrienoic acid, methyl ester	21.214	2.76
22	Phytol	21.330	12.21
23	Octadecanoic acid, methyl ester	21.447	1.40
24	9,12-Octadecadienoic acid	21.524	0.81
25	cis,cis,cis-7,10,13-Hexadecatrienal	21.587	2.65
26	Octadecanoic acid	21.789	1.83
27	3-Cyclopentylpropionic acid, 2-dimethylaminoethyl ester	22.858	0.26
28	Tetrapentacontane	23.284	0.44
29	Eicosanoic acid	23.542	0.18
30	3-Cyclopentylpropionic acid, 2-dimethylaminoethyl ester	24.350	0.17
31	3,4-Dihydro-2(1h)-isoquinolinecarboxamidine	24.439	2.07
32	2-Ethylbutyric acid, eicosyl ester	24.578	4.99
33	Hexadecanoic acid, 2-hydroxy-1-(hydroxymethyl) ethyl ester	24.773	8.99
34	3,7,11-Trimethyl-2,6,10-dodecatrien-1-ol,	24.825	3.93
35	Geranyl linalool isomer	25.000	7.44
36	2,6,10,14,18-Pentamethyl-2,6,10,14,18-icosapentaene	25.183	3.08
37	4,22-Stigmastadiene-3-one	26.255	3.79
38	alpha-Monostearin	26.550	1.29
39	5-Hydroxy-7,8-dimethoxyflavone	26.700	0.89
40	(3 beta)-Cholest-5-en-3-ol	26.925	0.57
41	Squalene	27.544	1.64
42	(3beta)-Stigmast-5-en-3-ol	28.116	2.76
43	Aromadendrene	28.293	0.58
44	3,5-Dihydroxy-6,7,8-trimethoxyflavone	29.889	2.33

Table 5. Important bioactive compounds detected by GCMS analysis in the root, stem, and leaves of *A. macrobotrys* having potential biological activities.

Bioactive Compounds	Root	Stem	Leaves	RT	KI	Biological Activity	References
Azulene	+	+	+	10.516	518.92	Anti-inflammatory	[15]
2-Methoxy-4-vinylphenol	−	+	−	12.297	905.88	Anticancer, anti-inflammatory, and antioxidant	[16–18]
2,6-Dimethoxy-phenol or Syringol	−	+	−	12.774	1000	Antioxidant	[19]
2,4-Di-tert-butylphenol	+	+	+	14.849	1372.27	Anticancer, anti-inflammatory, and antioxidant	[20]
Eicosane	+	+	+	17.128	1725.4	anti-inflammatory, analgesic, and antipyretic	[21]
Loliolide	−	−	+	18.008	1849.32	Herbivore resistance	[22]
Neophytadiene	−	+	+	18.609	1930.51	Anti-inflammatory	[23]
Hexadecanoic acid, methyl ester	+	+	+	19.507	2047.07	Anti-Inflammatory	[24]
Phytol	−	+	+	21.322	2267.1	Anti-inflammatory and immunomodulating	[25,26]
2-Mono-myristin	−	+	−	23.05	2459.83	Antimicrobial	[27]
Tetrapentacontane	−	+	−	23.287	2485.13	Antimicrobial, antioxidant	[28]
Asaraldehyde	+	−	−	24.965	2657.22	Anti-obesity	[29]
5-Hydroxy-7,8-dimethoxyflavone	+	+	+	26.733	2826.44	Neuroprotective and lipid lowering	[30,31]
Squalene	−	+	+	27.544	2900.36	Antioxidant, antitumor, and colon cancer	[32,33]
Stigmasta-5,22-dien-3-ol	−	+	−	27.607	2906.28	Antimicrobial	[34]
Aromadendrene	−	−	+	28.293	2966.72	Antibacterial activity	[35]
5-Hydroxy-6,7,4′-trimethoxyflavone/Salvigenin	+	−	−	29.032	3030.49	Neuroprotective and lipid lowering	[30,31]
3,5-Dihydroxy-6,7,8-trimethoxyflavone	+	−	−	29.905	3103.76	Antitumor	[36]

+ or − represents presence or absence of the bioactive compound; RT = Retention time, KI = Kovats retention Index.

3.3. *In Vitro Antioxidant Potential*

The dark purple color scheme of the DPPH free radicals is brought on by their single electron pair. Samples containing antioxidants decolorize the DPPH solution by scavenging the electrons. Thus, the reduction in the purple color of the DPPH solution is directly linked to the antioxidant capacities of the methanolic extracts of the samples as they scavenge the pair of electrons. The DPPH free radical scavenging activities of the extracts are expressed as their mg/mL EC_{50} value. The lower value of EC_{50} indicates higher radical scavenging activity. Ascorbic acid and BHA had EC_{50} values of 0.44 ± 0.03 mg/mL and 0.37 ± 0.02 mg/mL, respectively, whereas the root, stem, and leaf methanolic extracts had DPPH radical scavenging activities of 1.16 ± 0.2 mg/mL, 0.79 ± 0.15 mg/mL, and 1.92 ± 0.25 mg/mL, respectively (Figure 2).

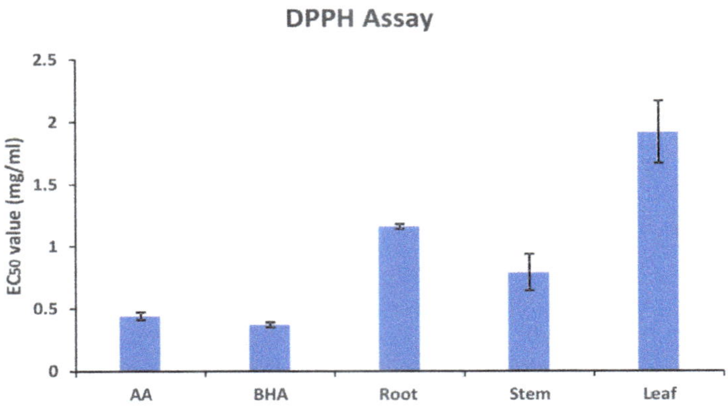

Figure 2. DPPH analysis of the root, stem, and leaves of *Andrographis macrobotrys.*

The colorless ferric-tripyridyltriazine (Fe3+-TPTZ) complex reduced to ferrous-tripyridyltriazine (Fe2+-TPTZ), an intense, blue-colored complex, by the activity of antioxidants present in the sample, due to the dose–response relationship. The sample containing higher levels of antioxidants may result in more intense color development. The methanolic extracts of the stem demonstrated the highest antioxidant activity, with ODs of 0.537 $\pm$ 0.02 and 1.367 $\pm$ 0.03, at concentrations of 0.2 and 1.0 mg/mL, respectively. The antioxidant capacity of various samples was in the following order: leaf, root, and stem (Figure 3). The activity of the extracts is comparable to that of ascorbic acid and BHA. The leaves and stem contributed to the highest phenolic and flavonoid content, which can be correlated with antioxidant activities.

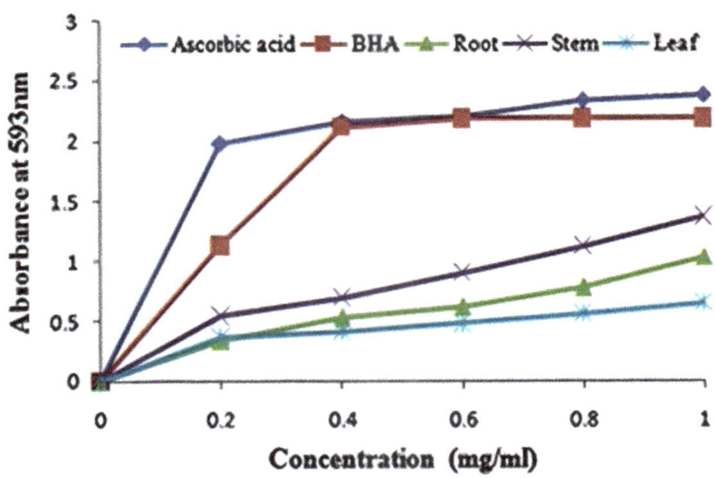

Figure 3. FRAP analysis of the root, stem, and leaves of *Andrographis macrobotrys.*

4. Discussion

A. paniculata, the most popular species in the genus, is well known for its exceptional medicinal qualities [2,37]. It is therefore highly sought after in the pharmaceutical and herbal industries. Asian countries have used plant medications as an alternative to Western medicine. The plant is also well known for its incredible ability to combat viral infections. Because of this, demand for the raw materials and goods produced by *A. paniculata* increased dramatically during the COVID-19 outbreak [38]. The quest for alternatives with potential on par with *A. paniculata* has accelerated [39,40] in an effort to minimize the strain. The current phytochemical investigation was carried out on the closely similar species *A. macrobotrys* in order to confirm that, and it identified a variety of phytochemicals.

The investigation's findings unambiguously show that *A. macrobotrys*'s stem extract (124.28 mg GAE/g) contains a higher amount of phenolics than the plant's root or leaf extracts (73.01 mg GAE/g, respectively). Similar findings also applied to *A. paniculata*, where stem extracts revealed a higher phenolic content than leaves [41]. The amount of flavonoids in leaf extract was six times greater than that in root extract (9.06 mg QE/g), or 57.04 mg QE/gm. The best source of tannins was the stem (86.39 mg TAE/g), followed by the root (59.49 mg TAE/g) and leaf extract (57.33 mg TAE/g), which had roughly identical amounts. Our estimation shows that the stem is the best source of phytochemicals that are soluble in methanol when compared to the leaf and root. The root, stem, and leaf of *A. producta* all had varying amounts of phenolic, flavonoid, and tannin content in the methanolic extract [6], with the stem having the greatest levels (163.61 mg GAE/g) and TAE (84.52 mg/g) of phenolics and tannins, respectively. This further demonstrates the stem's excellent status as a source of total phenolic chemicals. Furthermore, *A. macrobotrys* possessed the highest level of phenolic compounds (180.98 mg GAE/g), compared to the stems of *A. producta* [6], *A. paniculata* [42], and *A. echioides* [43].

A. macrobotrys methanolic extracts were chemically profiled by GC-MS, and the results showed that each extract included a variety of high- and low-molecular-weight metabolites in varied amounts. Twelve substances were included in all the extracts: azulene, 2,4-bis(1,1-dimethylethyl)-phenol, benzoic acid, 4-ethoxy-ethyl ester, eicosane, 3-heptadecanol, isopropyl myristate, hexadecanoic acid, methyl ester, hexadecanoic acid, 1-butyl-cyclohexanol; 9,12-octadecadienoic acid, methyl ester; alpha-monostearin; 5-hydroxy-7,8-dimethoxyflavone. The GC-MS profiling's primary objective was to identify the biologically active compounds present in this crucial plant for medicine. The discovered bioactive substances exhibited similarities to the known *Andrographis* species [44]. *A macrobotrys* contains a lot of chemicals that are biologically active, as would be expected. In our previous studies [4], the plant was found to be a rich source of the bioactive compound Neoandrographolide (102.03 mg/g DW) through HPLC analysis, when compared to other *Andrographis* species.

The plant extracts showed significant antioxidant potential in comparison to the other *Andrographis* species [45]. The antioxidant and other biological activities of these plants were demonstrated by both crude and identified compounds. A dimer produced by the oxidation of 2,6-dimethoxy phenol (syringol) by laccase, which increased antioxidant activity for FRAP, TEAC, and DPPH by 119.32, 53.15, and 93.25%, respectively, in comparison to the substrate [19]. A flavoring ingredient known as 2-methyl-4-vinylphenol was shown to be a powerful anticancer agent that inhibited the migration of Panc-1 and SNU-213 pancreatic cancer cells as well as their viability by preventing the expression of their nuclear antigens [16]. A monoacylglycerol derivative known as 2-Monomyristin has shown antibacterial efficacy against an *Escherichia coli* bacterial stain [27]. Trimethoxy flavone and salvigenin at a dose of 25 M inhibited the oxidative stress-induced apoptosis in neuroblastoma SH-SY5Y cells by activating antioxidant factors in the neuroprotective assessment [30] (Table 1). According to Serino et al. [31], salvigenin exhibits biological activity, lowering lipid levels (−22.5% palmitic acid biosynthesis at 30 M) while increasing mitochondrial functionality (+15.4% at 30 M). With a MIC value of 125 g/mL, Aromadendrene, a terpenoid found in *Eucalyptus globulus*, was discovered to be particularly efficient against methicillin-resistant *Staphylococcus aureus* [35] (Table 1). According to Murata et al. [22], the carotenoid metabolite loliolide functions as a strong endogenous inducer and mediates the host's defense reaction against three herbivores. In studies using 7,12-dimethylbenzanthracene to produce skin tumors, squalene (5%) was found to have an anticancer effect and to have reduced 26.67% of tumors in the test group [32] (Table 1). However, several chemical compounds such as 4,22-stigmastadiene-3-one, 3,4-dihydro-2(1h)-isoquinoline carboxamidine,

4,22-stigmastadiene-3-one, acetosyringone, alpha-monostearin, eicosane, geranyl linalool isomer, and tetrapentacontane belonging to various classes of phytochemicals need to be examined in detail for their biological activities.

Due to their protective properties, various plant species and their parts have been employed to cure chronic ailments. These medicinal herbs' protective function is related to their chemical components, which also have antioxidant activities [46]. Polyphenols from plant sources, including phenolic acids and flavonoids, demonstrate effective antioxidant action [47]. With an EC_{50} value of 0.79 mg/mL, the methanolic extract from *A. macrobotrys* stem displayed the highest DPPH radical scavenging activity and is comparable to the EC_{50} values of the reference substances evaluated, such as ascorbic acid (0.44 mg/mL) and BHA (0.37 mg/mL) (Figure 2). This could be attributed to the higher phenolic content in the stem compared to other parts. As compared to ascorbic acid (3.22 mg/mL EC_{50}), the stem of *A. producta* showed good DPPH radical scavenging action with a value of 3.58 mg/mL [6]. The stem of *A. macrobotrys* performed better than *A. producta* at scavenging DPPH radicals. Antioxidants from stems were shown to have comparable reducing activity to BHA (1.129 ± 0.02 OD at 0.2 mg/mL) and the highest Fe3+-TPTZ reducing potential (1.367 ± 0.03 OD at 1.0 mg/mL concentration). However, the decrease in Fe3+-TPTZ complex was substantially less active with root and leaf extracts (Figure 3). At 1.0 mg/mL concentration, *A. producta*'s ferric-reducing activity was determined to be 1.742 ± 0.02 OD (stem), 1.139 ± 0.03 OD (root), and 0.866 ± 0.016 OD (leaf) [6]. As of right now, the stem of *A. macrobotrys* (1.367 ± 0.03 OD) reduces ferric oxide more effectively than the root (1.139 ± 0.03 OD) and less effectively than the stem of *A. producta* (1.742 ± 0.02 OD). The root, stem, and leaf of *A. macrobotrys* exhibit the same antioxidant potential according to both DPPH and FRAP experiments. Due to its phenolic concentration and bioactive chemicals, such as 2,4-di-tert-butylphenol, syringe, squalene, and tetrapentacontane, the stem may have a high antioxidant potential. Since *A. macrobotrys* possesses valuable elements that account for antioxidant activity, it could be exploited in food and pharmaceutical preparations. The antioxidant potential of phytochemicals plays an essential role in bioactive food and pharmaceutical resources [48].

5. Conclusions

Various parts of *Andrographis macrobotrys* were analyzed quantitatively to determine the distribution and variety of phytochemicals. From the various plant components, large amounts of phenolics, flavonoids, and tannins were recovered. Due to the presence of phenolic chemicals, this study demonstrated that plant components have high antioxidative potential. According to GC-MS analysis, there were several key bioactive phytochemicals present, including 2,4-di-tert-butylphenol, 2-methoxy-4-vinylphenol, 5-hydroxy-7,8-dimethoxyflavone, azulene, salvigenin, squalene, and tetrapentacontane. Numerous other substances have been found for biological evaluation to understand their importance, including 4,22-stigmastadiene-3-one, 3,4-dihydro-2(1h)-isoquinoline carboxamidine, 4,22-stigmastadiene-3-one, acetosyringone, and alpha-monostearin, among others. Therefore, *A. macrobotrys* is a source of practical phytochemicals, as the present investigation showed. The plants can be utilized as an alternative to *A. paniculata*, preventing their overuse and extinction.

Author Contributions: Conceptualization, D.D. and H.N.M.; methodology, D.D. and H.N.M.; investigation, D.D. and H.N.M.; formal analysis and data curation, D.D. and H.N.M.; writing—original draft preparation; D.D., J.K.S. and H.N.M.; writing—review and editing, D.D., H.N.M. and Y.H.D.; validation, data analysis, and interpretation; J.K.S., A.N. and Y.H.D.; visualization, Y.H.D. All authors have read and agreed to the published version of the manuscript.

Funding: The authors acknowledge Researchers Supporting Project number (RSP2023R375), King Saud University, Riyadh, Saudi Arabia.

Institutional Review Board Statement: Not applicable.

Informed Consent Statement: Not applicable.

Data Availability Statement: All data are presented in the article.

Acknowledgments: The authors acknowledge the Researchers Supporting Project number (RSP2023R375), King Saud University, Riyadh, Saudi Arabia.

Conflicts of Interest: The authors declare no conflict of interest.

References

1. Sen, S.; Chakraborty, R.; De, B. Challenges and opportunities in the advancement of herbal medicine: India's position and role in a global context. *J. Herb. Med* **2011**, *1*, 67–75. [CrossRef]
2. Subramanian, R.; Zaini Asmawi, M.; Sadikun, A. A Bitter plant with a sweet future? A comprehensive review of an oriental medicinal plant: *Andrographis paniculata. Phytochem. Rev.* **2012**, *11*, 39–75. [CrossRef]
3. Ignacimuthu, S.; Ayyanar, M.; Sivaraman, K.S. Ethnobotanical investigations among tribes in Madurai district of Tamil Nadu (India). *J. Ethnobiol. Ethnomed.* **2006**, *2*, 25. [CrossRef] [PubMed]
4. Dalawai, D.; Aware, C.; Jadhav, J.P.; Murthy, H.N. RP-HPLC analysis of diterpene lactones in leaves and stem of different species of *Andrographis. Nat. Prod. Res.* **2021**, *35*, 2239–2242. [CrossRef] [PubMed]
5. Kshirsagar, R.D.; Singh, N.P. Some less known ethnomedicinal uses from Mysore and Coorg districts, Karnataka State, India. *J. Ethnopharmacol.* **2001**, *75*, 231–238. [CrossRef]
6. Dalawai, D.; Murthy, H.N. Chemical profile and antioxidant properties of *Andrographis producta* (C. B. Clarke) Gamble. *Pharmacogn. J.* **2020**, *13*, 475–485. [CrossRef]
7. Gamble, J.S. *Flora of the Presidency of Madras, Part VI, (Scrophulariaceae to Plantaginaceae)*; Adlard and Son: London, UK, 1924.
8. Folin, O.; Ciocalteu, V. On Tyrosine and Tryptophane determinations in proteins. *J. Biol. Chem.* **1927**, *73*, 627–650. [CrossRef]
9. Prior, R.L.; Wu, X.; Schaich, K. Standardized methods for the determination of antioxidant capacity and phenolics in foods and dietary supplements. *J. Agric. Food Chem.* **2005**, *53*, 4290–4302. [CrossRef]
10. Schanderi, S.H. *Method in Food Analysis*; Academic Press: New York, NY, USA, 1970.
11. Roy, S.; Rao, K.; Bhuvaneswari, C.; Giri, A.; Mangamoori, L.N. Phytochemical analysis of *Andrographis paniculata* extract and its antimicrobial activity. *World J. Microbiol. Biotechnol.* **2010**, *26*, 85–91. [CrossRef]
12. Spivakovskii, G.I.; Tishchenko, A.I.; Zaslavskii, I.I.; Wulfson, N.S. Calculation of retention indices of compounds from their structural formulae for combined identification by Gas Chromatography-Mass Spectrometry. *J. Chromatogr. A* **1977**, *144*, 1–16. [CrossRef]
13. Brand-Williams, W.; Cuvelier, M.E.; Berset, C. Use of a free radical method to evaluate antioxidant activity. *LWT-Food Sci. Technol.* **1995**, *28*, 25–30. [CrossRef]
14. Benzie, I.F.F.; Strain, J.J. The ferric reducing ability of plasma (FRAP) as a measure of "Antioxidant Power": The FRAP Assay. *Anal. Biochem.* **1996**, *239*, 70–76. [CrossRef] [PubMed]
15. Guarrera, M.; Turbino, L.; Rebora, A. The anti-inflammatory activity of azulene. *J. Eur. Acad. Dermatol. Venerol.* **2001**, *15*, 486–487. [CrossRef] [PubMed]
16. Kim, S.H.; Jeong, K.S.; Ryu, S.Y.; Kim, T.H. *Panax ginseng* prevents apoptosis in hair follicles and accelerates recovery of hair medullary cells in irradiated mice. *In Vivo* **1998**, *12*, 219–222. [PubMed]
17. Jeong, J.B.; Hong, S.C.; Jeong, H.J.; Koo, J.S. Anti-inflammatory effect of 2-methoxy-4-vinylphenol via the suppression of NF-KB and MAPK activation, and acetylation of histone H3. *Arch. Pharm. Res.* **2011**, *34*, 2109–2116. [CrossRef]
18. Rubab, M.; Chelliah, R.; Saravanakumar, K.; Barathikannan, K.; Wei, S.; Kim, J.-R.; Yoo, D.; Wang, M.-H.; Oh, D.-H. Bioactive potential of 2-methoxy-4-vinylphenol and benzofuran from *Brassica oleracea* L. var. *capitata* f, *rubra* (Red Cabbage) on oxidative and microbiological stability of beef meat. *Foods* **2020**, *9*, 568. [CrossRef]
19. Adelakun, O.E.; Kudanga, T.; Green, I.R.; le Roes-Hill, M.; Burton, S.G. Enzymatic modification of 2,6-dimethoxyphenol for the synthesis of dimers with high antioxidant capacity. *Process Biochem.* **2012**, *47*, 1926–1932. [CrossRef]
20. Zhao, F.; Wang, P.; Lucardi, R.; Su, Z.; Li, S. Natural sources and bioactivities of 2,4-di-tert-butylphenol and its analogs. *Toxins* **2020**, *12*, 35. [CrossRef]
21. Chuah, X.; Okechukwu, P.; Amini, F.; Teo, S. Eicosane, pentadecane and palmitic acid: The effects in in vitro wound healing studies. *Asian Pac. J. Trop. Biomed.* **2018**, *8*, 490. [CrossRef]
22. Murata, M.; Nakai, Y.; Kawazu, K.; Ishizaka, M.; Kajiwara, H.; Abe, H.; Takeuchi, K.; Ichinose, Y.; Mitsuhara, I.; Mochizuki, A.; et al. Loliolide, a carotenoid metabolite, is a potential endogenous inducer of herbivore resistance. *Plant Physiol.* **2019**, *179*, 1822–1833. [CrossRef]
23. Bhardwaj, M.; Sali, V.K.; Mani, S.; Vasanthi, H.R. Neophytadiene from *Turbinaria ornata* suppresses LPS-induced inflammatory response in RAW 264.7 macrophages and Sprague dawley rats. *Inflammation* **2020**, *43*, 937–950. [CrossRef] [PubMed]
24. Aparna, V.; Dileep, K.V.; Mandal, P.K.; Karthe, P.; Sadasivan, C.; Haridas, M. Anti-inflammatory property of n-hexadecanoic acid: Structural evidence and kinetic assessment. *Chem. Biol. Drug Des.* **2012**, *80*, 434–439. [CrossRef] [PubMed]
25. Silva, R.O.; Sousa, F.B.M.; Damasceno, S.R.B.; Carvalho, N.S.; Silva, V.G.; Oliveira, F.R.M.A.; Sousa, D.P.; Aragão, K.S.; Barbosa, A.L.R.; Freitas, R.M.; et al. Phytol, a diterpene alcohol, inhibits the inflammatory response by reducing cytokine production and oxidative stress. *Fundam. Clin. Pharmacol.* **2014**, *28*, 455–464. [CrossRef]

26. Nakanishi, T.; Anraku, M.; Suzuki, R.; Kono, T.; Erickson, L.; Kawahara, S. Novel immunomodulatory effects of phytanic acid and its related substances in mice. *J. Funct. Foods* **2016**, *21*, 283–289. [CrossRef]
27. Jumina; Nurmala, A.; Fitria, A.; Pranowo, D.; Sholikhah, E.; Kurniawan, Y.; Kuswandi, B. Monomyristin and monopalmitin derivatives: Synthesis and evaluation as potential antibacterial and antifungal agents. *Molecules* **2018**, *23*, 3141. [CrossRef] [PubMed]
28. Xuanji, X.; Zengjun, G.; Hui, Z.; Xia, L.; Jun, L.; Dandan, L.; Jun, L. Chemical composition, in vitro antioxidant activity and α-glucosidase inhibitory effects of the essential oil and methanolic extract of *Elsholtzia densa* Benth. *Nat. Prod. Res.* **2016**, *30*, 2707–2711. [CrossRef]
29. Wu, M.-R.; Hou, M.-H.; Lin, Y.-L.; Kuo, C.-F. 2,4,5-TMBA, a Natural inhibitor of cyclooxygenase-2, suppresses adipogenesis and promotes lipolysis in 3T3-L1 adipocytes. *J. Agric. Food Chem.* **2012**, *60*, 7262–7269. [CrossRef]
30. Rafatian, G.; Khodagholi, F.; Farimani, M.M.; Abraki, S.B.; Gardaneh, M. Increase of autophagy and attenuation of apoptosis by salvigenin promote survival of SH-SY5Y cells following treatment with H_2O_2. *Mol. Cell. Biochem.* **2012**, *371*, 9–22. [CrossRef]
31. Serino, E.; Chahardoli, A.; Badolati, N.; Sirignano, C.; Jalilian, F.; Mojarrab, M.; Farhangi, Z.; Rigano, D.; Stornaiuolo, M.; Shokoohinia, Y.; et al. Salvigenin, a trimethoxylated flavone from *Achillea wilhelmsii* C. Koch, exerts combined lipid-lowering and mitochondrial stimulatory effects. *Antioxidants* **2021**, *10*, 1042. [CrossRef]
32. Huang, F.; Long, Y.; Liang, Q.; Purushotham, B.; Swamy, M.K.; Duan, Y. Safed Musli (*Chlorophytum borivilianum* L.) Callus-mediated biosynthesis of silver nanoparticles and evaluation of their antimicrobial activity and cytotoxicity against human colon cancer cells. *J. Nanomater.* **2019**, *2019*, 2418785. [CrossRef]
33. Rao, C. Chemopreventive effect of squalene on colon cancer. *Carcinogenesis* **1998**, *19*, 287–290. [CrossRef] [PubMed]
34. Achika, J.; Ndukwe, G.; Ayo, R. Isolation, Characterization and antimicrobial activity of 3β, 22E-stigmasta-5, 22-dien-3-ol from the aerial part of *Aeschynomene uniflora* E. Mey. *J. Pharm. Res. Int.* **2016**, *11*, 1–8. [CrossRef]
35. Mulyaningsih, S.; Sporer, F.; Reichling, J.; Wink, M. Antibacterial activity of essential oils from *Eucalyptus* and selected components against multidrug-resistant bacterial pathogens. *Pharm. Biol.* **2011**, *49*, 893–899. [CrossRef] [PubMed]
36. Thomas, C.M.; Wood, R.C.; Wyatt, J.E.; Pendleton, M.H.; Torrenegra, R.D.; Rodriguez, O.E.; Harirforoosh, S.; Ballester, M.; Lightner, J.; Krishnan, K.; et al. Anti-neoplastic activity of two flavone somers Derived from *Gnaphalium elegans* and *Achyrocline bogotensis*. *PLoS ONE* **2012**, *7*, e39806. [CrossRef] [PubMed]
37. Kumar, S.; Singh, B.; Bajpai, V. *Andrographis paniculata* (Burm.f.) Nees: Traditional uses, phytochemistry, pharmacological properties and quality control/quality assurance. *J. Ethnopharmacol.* **2021**, *275*, 114054. [CrossRef] [PubMed]
38. Lim, X.Y.; Chan, J.S.W.; Tan, T.Y.C.; Teh, B.P.; Mohd Abd Razak, M.R.; Mohamad, S.; Syed Mohamed, A.F. *Andrographis paniculata* (Burm. F.) Wall. Ex Nees, Andrographolide, and andrographolide analogues as SARS-CoV-2 antivirals? A rapid review. *Nat. Prod. Commun.* **2021**, *16*, 1934578X2110166. [CrossRef]
39. Parlapally, S.; Cherukupalli, N.; Bhumireddy, S.R.; Sripadi, P.; Anisetti, R.; Giri, C.C.; Khareedu, V.R.; Reddy Vudem, D. Chemical profiling and anti-psoriatic activity of methanolic extract of *Andrographis nallamalayana* J.L.Ellis. *Nat. Prod. Res.* **2016**, *30*, 1256–1261. [CrossRef]
40. Zhang, H.; Li, S.; Si, Y.; Xu, H. Andrographolide and its derivatives: Current achievements and future perspectives. *Eur. J. Med. Chem.* **2021**, *224*, 113710. [CrossRef]
41. Adiguna, S.P.; Panggabean, J.A.; Swasono, R.T.; Rahmawati, S.I.; Izzati, F.; Bayu, A.; Putra, M.Y.; Formisano, C.; Giuseppina, C. Evaluations of Andrographolide-rich fractions of *Andrographis paniculata* with enhanced potential antioxidant, anticancer, antihypertensive, and anti-Inflammatory activities. *Plants* **2023**, *12*, 1220. [CrossRef]
42. Kurzawa, M.; Filipiak-Szok, A.; Kłodzińska, E.; Szłyk, E. Determination of phytochemicals, antioxidant activity and total phenolic content in *Andrographis paniculata* using chromatographic methods. *J. Chromatogr. B* **2015**, *995–996*, 101–106. [CrossRef]
43. Gurupriya, S.; Cathrine, L. Qualitative and quantitative phytochemical analysis of *Andrographis echioides* leaves. *Int. J. Life Sci. Pharm. Res.* **2022**, *11*, P148–P158. [CrossRef]
44. Jiang, M.; Sheng, F.; Zhang, Z.; Ma, X.; Gao, T.; Fu, C.; Li, P. *Andrographis paniculata* (Burm.f.) Nees and its major constituent andrographolide as potential antiviral agents. *J. Ethnopharmcol.* **2021**, *272*, 113954. [CrossRef] [PubMed]
45. Low, M.; Khoo, C.S.; Münch, G.; Govindaraghavan, S.; Sucher, N.J. An in vitro study of anti-inflammatory activity of standardised *Andrographis paniculata* extracts and pure andrographolide. *BMC Complement. Altern. Med.* **2015**, *15*, 18. [CrossRef] [PubMed]
46. Zhang, Y.-J.; Gan, R.-Y.; Li, S.; Zhou, Y.; Li, A.-N.; Xu, D.-P.; Li, H.-B. Antioxidant phytochemicals for the prevention and treatment of chronic diseases. *Molecules* **2015**, *20*, 21138–21156. [CrossRef] [PubMed]
47. Senguttuvan, J.; Paulsamy, S.; Karthika, K. Phytochemical analysis and evaluation of leaf and root parts of the medicinal herb, *Hypochaeris radicata* L. for in vitro antioxidant activities. *Asian Pac. J. Trop. Biomed.* **2014**, *4*, S359–S367. [CrossRef]
48. Yu, M.; Gouvinhas, I.; Rocha, J.; Barros, A.I.R.N.A. Phytochemical and antioxidant analysis of medicinal and food plants towards bioactive food and pharmaceutical resources. *Sci. Rep.* **2021**, *11*, 10041. [CrossRef]

life

Article

Determination of Antioxidant, Anti-Alzheimer, Antidiabetic, Antiglaucoma and Antimicrobial Effects of Zivzik Pomegranate (*Punica granatum*)—A Chemical Profiling by LC-MS/MS

Hasan Karagecili [1,*], Ebubekir İzol [2,3], Ekrem Kirecci [4] and İlhami Gulcin [3,*]

1 Department of Nursing, Faculty of Health Sciences, Siirt University, 56100 Siirt, Turkey
2 Bee and Natural Products R & D and P & D Application and Research Center, Bingöl University, 12000 Bingol, Turkey
3 Department of Chemistry, Faculty of Science, Ataturk University, 25240 Erzurum, Turkey
4 Department of Basic Medical Sciences, Faculty of Medicine, Microbiology, Kahramanmaraş Sütçü İmam University, 46050 Kahramanmaras, Turkey
* Correspondence: hasankaragecili@siirt.edu.tr (H.K.); igulcin@atauni.edu.tr (İ.G.); Tel.: +90-4422314375 (İ.G.)

Abstract: Zivzik pomegranate (*Punica granatum*) has recently sparked considerable interest due to its nutritional and antioxidant properties. To evaluate the antioxidant capacities of *P. granatum* juice, ethanol (EEZP), and water (WEZP) extracts from peel and seed, the antioxidant methods of 2,2′-azino-bis-3-ethylbenzthiazoline-6-sulphonic acid radical ($ABTS^{•+}$) scavenging, 1,1-diphenyl-2-picrylhydrazyl free radical ($DPPH^{•}$) scavenging, Fe^{3+}-2,4,6-tris(2-pyridyl)-S-triazine (TPTZ) reducing, Fe^{3+} reducing, and Cu^{2+} reducing methods were used. The antioxidant capacities of samples were compared with the most commonly used synthetic antioxidants, i.e., BHA, BHT, α-tocopherol, and Trolox. In terms of setting an example, the IC_{50} values of EEZP for $ABTS^{•+}$ and $DPPH^{•}$ scavenging activities were found to be lower than standards, at 5.9 and 16.1 μg/mL, respectively. The phenolic and flavonoid contents in EEZP peel were 59.7 mg GAE/g and 88.0 mg QE/g, respectively. Inhibition of α-glycosidase, α-amylase, acetylcholinesterase, and human carbonic anhydrase II (hCA II) enzymes was also investigated. EEZP demonstrated IC_{50} values of 7.3 μg/mL against α-glycosidase, 317.7 μg/mL against α-amylase, 19.7 μg/mL against acetylcholinesterase (AChE), and 106.3 μg/mL against CA II enzymes. A total of 53 phenolic compounds were scanned, and 30 compounds were determined using LC-MS/MS. *E. coli* and *S. aureus* bacteria were resistant to all four antibiotics used as standards in hospitals.

Keywords: Zivzik pomegranate; enzyme inhibition; *Punica granatum*; antioxidant; α-glycosidase; acetylcholinesterase; carbonic anhydrase; LC-MS/MS analysis

check for updates

Citation: Karagecili, H.; İzol, E.; Kirecci, E.; Gulcin, İ. Determination of Antioxidant, Anti-Alzheimer, Antidiabetic, Antiglaucoma and Antimicrobial Effects of Zivzik Pomegranate (*Punica granatum*)—A Chemical Profiling by LC-MS/MS. *Life* **2023**, *13*, 735. https://doi.org/10.3390/life13030735

Academic Editors: Balazs Barna and Jianfeng Xu

Received: 19 January 2023
Revised: 23 February 2023
Accepted: 7 March 2023
Published: 9 March 2023

1. Introduction

Pomegranate (*Punica granatum* L.) is an antiquity fruit that is primarily grown in western Asia, although it is also grown in other parts of the world, including the Mediterranean region. Its utilization has been linked to a variety of health advantages since ancient times [1,2]. Pomegranates are members of the Punicaceae family and have distinctive characteristics. Unsaturated–polyunsaturated fatty acids, vitamins, sugar, polysaccharides, polyphenols, and minerals can all be found in pomegranate seeds. Pomegranate seed oil in particular contains significant amounts of phenolic compounds, fatty acids, linoleic acid, gallic acid, and ellagic acid [3]. Pomegranate is one of the fruits that contain significant amounts of bioactive phenolic compounds, which are frequently used as botanical components in dietary supplements and herbal medicines [4]. Anthocyanins, anthocynidins, proanthocyanidins, flavonoids, vitamins, sterols, lignans, saccharides, fatty acids, organic acids, terpenes, and terpenoids are just a few of the bioactive components of pomegranates consumed in the human diet. In particular, proanthocyanidins, which have a considerable impact on human health, were investigated, focusing on their systemic lipid-lowering effects, as well as their hypoglycemic and anti-inflammatory abilities in the intestinal epithelium [5]. Other bioactive components include ellagic acid and its derivative, gallic acid. In addition to ellagic acid and its derivatives, ellagitannins and gallotannins are important bioactive components of *P. granatum* [6]. Likewise, because of their well-known potential biological and pharmaceutical properties, secondary metabolites of plants have been widely used in traditional medicine. These metabolites play protective roles in plants and exhibit various biological and pharmaceutical traits with positive effects on health [7].

Antioxidant defenses such as enzymatic antioxidants and antioxidant food ingredients are present in all aerobic organisms and are used to either remove or repair damaged molecules [8–10]. By delaying the lipid peroxidation process, which is one of the main causes of the degradation of pharmaceutical and food items during processing and storage, antioxidants can remove free radicals and lengthen shelf life. The effects of free radicals and ROS can be prevented by antioxidants [11]. Enzymes and antioxidant components make up the antioxidant defense system. They have the ability to replace or fix broken biomolecules in living things such as lipids, carbohydrates, nucleic acids, and proteins. The oxidation of these biomolecules is postponed, avoided, and inhibited by antioxidants. They consist of phenols and polyphenols as potent substances that lessen or neutralize the harmful and undesirable effects of ROS [12,13]. However, the human body can be assisted in reducing oxidative damage caused by free radicals and ROS by antioxidant supplements or foods [14]. As free radicals or active oxygen scavengers, many antioxidant compounds found naturally in plant sources have been identified. The search for natural antioxidants for use in food or medicine has recently attracted increasing attention because synthetic antioxidants, the use of which is restricted due to their side effects such as carcinogenicity, are becoming harder to find [15]. Recently, propyl gallate, *tert*-butyl hydroquinone, butylated hydroxyanisole (BHA), and butylated hydroxytoluene (BHT) have become the most commonly used antioxidants. However, BHA and BHT have been subject to legislative restrictions because of concerns about their carcinogenic and toxic effects [16]. As a result, there is an increase in consumer preference for natural antioxidants, as well as an increased interest in natural and safer antioxidants for food applications, both of which have sparked efforts to investigate natural sources of antioxidants [17,18]. The leaves, roots, seeds, and fruits of most plants contain natural antioxidants. Numerous fruits and vegetables are rich in substances such as polyphenols, ascorbic acid, carotenoids, and tocopherols that have positive effects on health. Consuming fruits and vegetables can lower the risk of developing chronic diseases such as cardiovascular disease and cancer [19]. Medicinal plants, which have been the subject of many studies to date, are among the most significant natural antioxidant sources. Medicinal plants contain a high number of phenols. Cereals, plants, and fruits are the main sources of natural antioxidants in the human diet [20]. Phenolic compounds are plant secondary metabolites that prevent degenerative disorders such as cataracts, cardiovascular disease, cancer, hypercholesterolemia, rheumatoid arthritis, diabetes, and arteriosclerosis [21,22].

Alzheimer's disease (AD) is regarded as one of the most pressing global health concerns of our time. Numerous AD treatment strategies have been developed. The inhibition of acetylcholinesterase (AChE) and butyrylcholinesterase (BChE) is one of the most important strategies [23,24]. In particular, in several medical conditions, such as carcinogenesis, coronary atherosclerosis, AD, and age-related disorders, lipid peroxidative destruction has been linked to biologically active substances with antioxidant effects. AChE and BChE are activity parameters that are still considered to be a part of prophylaxis to treat neurological disorders associated with AD [25]. Alzheimer's disease (AD) is initially distinguished by the development of memory loss and other cognitive disorders and is believed to be connected to acetylcholine (ACh) deficiency, inflammation, and oxidative stress. Consuming plants with antioxidant capabilities can therefore stop the progression of AD and neurodegeneration [26].

It is thought that these disorders can be cured by inhibiting the essential enzymes linked to them. However, the synthetic drugs that are used to inhibit these important enzymes have a lot of negative side effects [27]. Researchers were encouraged to find an alternative natural product with fewer or no negative effects in an effort to solve this issue [28]. Natural substances such as AChE inhibitors (AChEIs) have frequently been used in clinical trials, particularly for the treatment of AD. Phenolic substances served as the first drugs for the treatment of AD and were also discovered to be AChEIs [29].

Inhibiting enzymes that hydrolyze carbohydrates, such as α-glucosidase and α-amylase, is one of the current methods for treating T2DM. By delaying glucose absorption, postprandial plasma glucose levels can be lowered, and hyperglycemia can be suppressed [30]. The hydrolysis of both oligosaccharides and polysaccharides into monosaccharide components such as glucose and fructose is accomplished by the enzyme glycosidase, which is released from intestine cells. As a result, the control of type 2 diabetes mellitus and hyperglycemia requires the use of α-glycosidase inhibitors (α-GIs) [31].

Carbonic anhydrase enzymes (CAs) are Zn^{2+}-including metalloenzymes, which catalyze the reversible hydration of carbon dioxide (CO_2) to proton and bicarbonate (HCO_3^-) [32]. Numerous biochemical functions such as ureagenesis, lipogenesis, and gluconeogenesis are carried out by CAs [33]. CA inhibition has therapeutic uses in the treatment of infection, convulsions, glaucoma, and cancer [34]. They also maintain fluid balance throughout the body, especially in the eyes, kidneys, and stomach. High intraocular pressure (IOP) due to glaucoma (IOP) can be relieved with the use of inhibitors of carbonic anhydrase (CAIs) [35].

With the current study, we sought to investigate the chemical components and biological activities of *P. granatum* obtained from the Zivzik village of Siirt province in the southeastern Anatolian region of Turkey. The following steps were taken to accomplish this: (a) The phenolic and flavonoid profiles of *P. granatum* were determined using LC-MS/MS analysis; (b) the antioxidant abilities of *P. granatum* were measured using DPPH, DMPD, ABTS, Cu^{2+} reducing (CUPRAC), Fe^{3+} reducing, Fe^{3+}-TPTZ reducing (FRAP), and Folin–Ciocalteu techniques; (c) the inhibitory effect of *P. granatum* on some metabolic enzymes, including AChE, hCA II, and α-glycosidase, was investigated for determination of possible relationships with AD, glaucoma, and diabetes mellitus; and (d) Gram-positive (*Staphylococcus aureus*) and Gram-negative (*Escherichia coli*) microorganisms were used to test the antimicrobial activity.

2. Materials and Methods

2.1. Chemicals

Acetylcholinesterase, acetylcholine iodide, α-glycosidase, p-nitrophenyl-D-glycopyranoside, DPPH (1,1-diphenyl-2-picryl-hydrazyl), ABTS (2,2-azino-bis 3-ethylbenzthiazoline-6-sulfonic acid), neocuproine (2,9-dimethyl-1,10-phenanthroline), BHT (butylated hydroxytoluene), BHA (butylated hydroxyanisole), α-tocopherol, Trolox, (Ferrozine) 3-(2-pyridyl)-5,6-bis(4-phenyl-sulfonic acid)-1,2,4-triazine, (TCA) trichloroacetic acid, and standard phenolic compounds for LC-MS/MS were purchased from Sigma (Sigma-Aldrich GmbH, Steinheim, Germany). The other materials were procured from Sigma-Aldrich or Merck.

2.2. Plant Materials

Zivzik pomegranate (*Punica granatum*) was defined as the Siirt ecotype by Assoc. Prof. Dr. Mehmet Fidan from the Siirt University Department of Biology. Zivzik pomegranates were obtained from Dişlipinar village (Zivzik) in the Şirvan district of Siirt province (altitude: 764 m (2506 ft)). *P. granatum* ethanol and water extract were dissolved in ethanol to determine their antioxidant activities and in DMSO for tests of enzyme inhibition due to the potential inhibitory effects of ethanol.

2.3. Preparation of Zivzik Pomegranate (*Punica granatum*)'s Extracts

The extraction procedure was performed as previously described [36]. Water extracts of *P. granatum* (WEZP) were prepared using 100 mL of distilled water and 25 g of dried

P. granatum peel and seeds that had been ground in a mill. This mixture was boiled for 20 min in a magnetic stirrer. The filtrates of the extracts were frozen and lyophilized in a lyophilizator at −50 °C under a pressure of 5 mmHg (Labconco, Freezone).

For ethanol extracts of *P. granatum* (WEZP), 25 g of dried *P. granatum* peel and seeds were milled before being combined with 100 mL of ethanol and stirred in a magnetic stirrer for 1 h. Filtrates were collected after the extracts had been filtered. A rotary evaporator (RE 100 Bibby, Stone Staffordshire, England) operating at 50 °C was used to remove the ethanol. Before being used in experimental studies, all of the extracts were kept in a dark plastic bottle at a temperature of 20 °C [37]. The yield of *P. granatum* extraction was calculated using the following equation:

$$\text{Yield} = \text{Weight of } P.\ granatum \text{ extract (g)}/\text{weight of raw extract (g)} \times 100\%$$

The yield of *P. granatum* extracts were calculated as follows: WEZP peel = $9.4/15 \times 100 = 62.7\%$; WEZP seed = $8.6/15 \times 100 = 57.3\%$; EEZP peel = $4/15 \times 100 = 26.7\%$; EEZP seed = $2.92/15 \times 100 = 19.5\%$. In order to obtain *P. granatum* juice, first, *P. granatum* were peeled, and pomegranate seeds were obtained. Then, *P. granatum* juice was obtained by pressing the *P. granatum* arils through a cheesecloth.

2.4. Total Phenolic Contents

The method described by Singleton and Rossi [38] was used to quantify the phenolics in the WEZP and EEZP peel and seed and *P. granatum* juice with a few minor modifications [39,40]. First 0.5 mL of each extracted sample was transferred to Folin–Ciocalteu reagent (FCR, 1.0 mL). The solution was then thoroughly blended and neutralized with carbonate (0.5 mL, 1%). After two hours of incubation in the dark at room temperature, the absorbances were measured at 760 nm in comparison to a blank sample, which included water. The phenolic content was expressed as milligrams of gallic acid equivalents (GAE) per gram of WEZP, EEZP, and *P. granatum* juice. The standard curve of gallic acid for total phenolic contents (r^2: 0.9408) is presented in Figure 1.

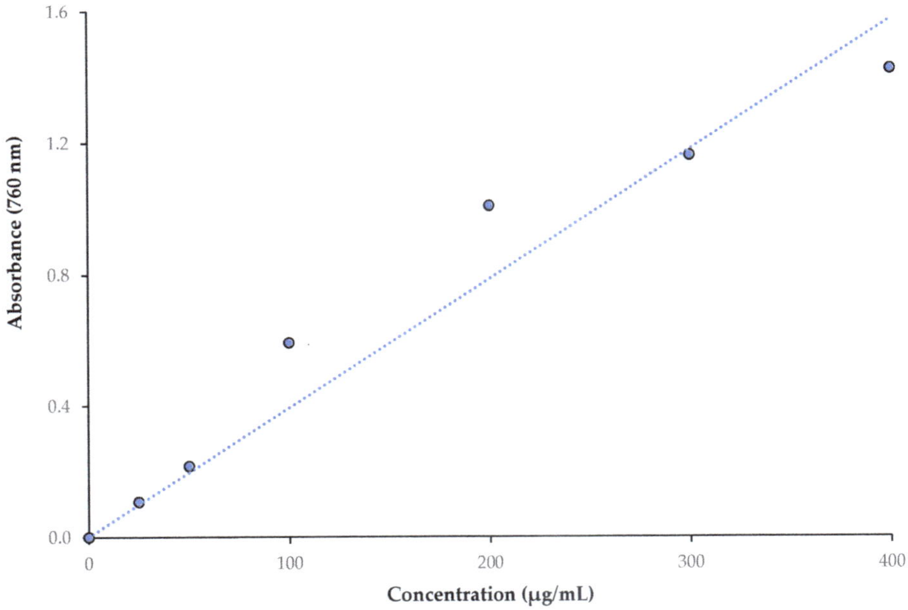

Figure 1. The standard curve of gallic acid for total phenolic contents (r^2: 0.9408).

2.5. Total Flavonoid Contents

A class of polyphenolic substances known as flavonoids is widely distributed in plants and frequently found in the human diet. Based on a previously described method [41], a colorimetric assay was used to estimate the total flavonoid contents in WEZP, EEZP, and *P. granatum* juice. To this end, 0.5 mL of sample was combined with 1.5 mL of 95% methanol. Then, 0.5 mL CH_3COOK (1.0 M) and 2.3 mL of deionized water were combined with 1.5 mL of 10% $Al(NO_3)$, and the samples were vortexed. Then, the vortexed samples were kept at 25 °C for 40 min in the dark. Absorbance measurements were taken at a wavelength of 415 nm. Quercetin equivalents (QE) are reported as mg per gram of extract in this study. The standard curve of total flavonoid contents is obtained from Figure 2.

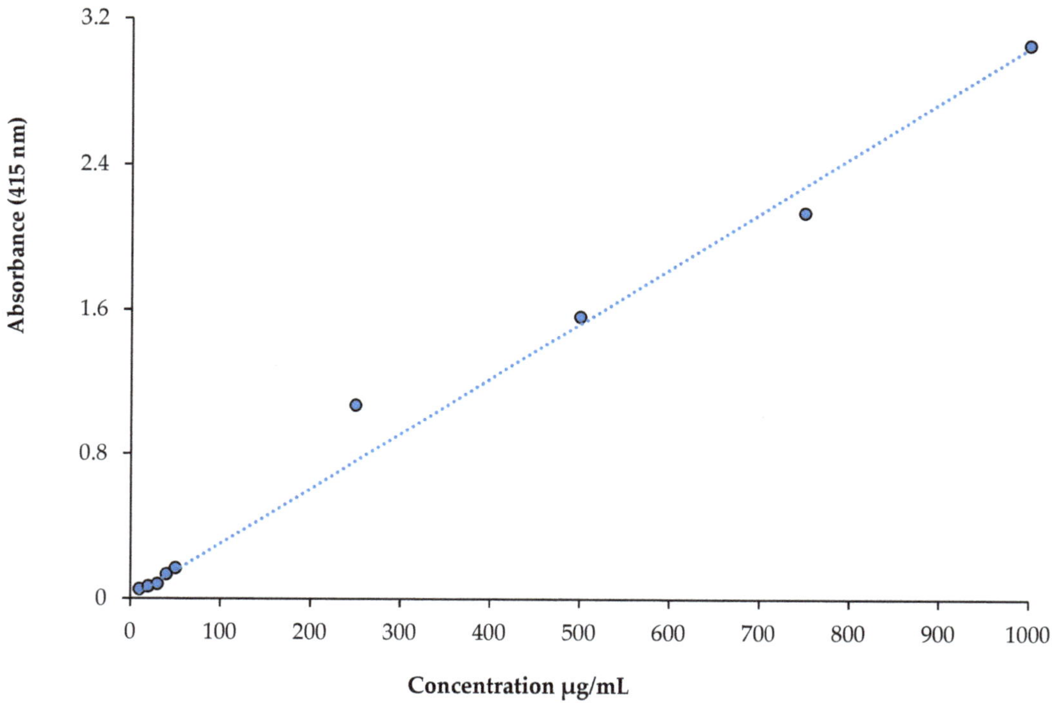

Figure 2. The standard curve of quercetin for total flavonoid contents (r^2: 0.9877).

2.6. Analaysis of Polyphenolic Composition by LC-MS/MS

2.6.1. Sample Preparation

First, 100 mg of each WEZP and EEZP was dissolved in 5 mL of water–ethanol (50:50 *v/v*) in a volumetric flask, and 1 mL of this solution was added to another volumetric flask with a capacity of 5 mL. Then, 100 μL of *P. granatum* extracts were added and diluted to the volume with water–ethanol (50:50 *v/v*). An aliquot of 1.5 mL from the final solution was transferred into a vial with a cap, and 10 μL of the sample was injected into the LC-MS/MS. Throughout the experiment, the samples in the autosampler were kept at 15 °C [42].

2.6.2. Method Validation Parameters and LC-MS/MS Analysis

The analytical approach utilized in this investigation was in accordance with the latest studies. The LC-MS/MS study was carried out by the Dicle University Central Research Laboratory. This chromatographic method was successfully carried out by Yılmaz [43] and adapted for *P. granatum* ethanol extracts. A total of 53 phytochemical standards were obtained as reference from Sigma-Aldrich (Steinheim, Germany). They were used to analyze phytochemicals in EEZP and WEZP.

2.7. *Fe^{3+} Reducing Capacity*

The Fe^{3+} reducing capacities of *P. granatum*, EEZP, WEZP, and *P. granatum* juice were assessed on the basis of the method proposed by Oyaizu [44], as also previously described in [45]. In a summary, various concentrations of samples in 0.75 mL of distilled water (10–30 µg/mL) were added into the same volume of buffer solution (1.25 mL, pH 6.6; 0.2 M) and 1.25 mL of $K_3Fe(CN)_6$ (1%, *w*/*w*). Trichloroacetic acid (TCA) (1.25 mL, 10%) was used to acidify the mixture after it has been incubated at 50 °C for 30 min. The absorbances of the fruit extracts were recorded at 700 nm after an aliquot of 0.1%, 0.25 mL, and $FeCl_3$ solution had been added to the mixture. Phosphate buffer solution was used as a blank sample. Activity measurements for the Fe^{3+} reducing ability at each concentration were conducted in triplicate.

2.8. *Cu^{2+} Reducing Capacity*

The Cu^{2+} reducing abilities of EEZP, WEZP, and *P. granatum* juice were measured according to the method used by Apak et al. [46], which was thoroughly described in [47], To this end, the same volumes of 0.25 mL of $CuCl_2$ solution (10 mM), 0.25 mL of neocuproine solution (7.5 mM), and 0.25 mL of acetate buffer (1.0 M) were added to the EEZP and WEZP solutions (10–30 µg/mL) in a test tube. The total volumes of mixtures were adjusted to 2 mL with distilled water and vigorously mixed. Then, the glass tubes were closed and retained at 25 °C until use in experiments. Finally, after 30 min, the absorbances were spectrophotometrically recorded at 450 nm. Acetate buffer solution was used as a blank sample. Increased reaction mixture absorbance suggests increased reduction capacity. Activity measurements for Cu^{2+} reducing ability at each concentration conducted in triplicate [48].

2.9. *Fe^{3+}-TPTZ Reducing Capacity*

The Fe^{3+}-TPTZ reducing capacity of EEZP, WEZP, and *P. granatum* juice in acidic solution were measured at 593 nm [49]. TPTZ solution (10 mM, 2.25 mL) with $FeCl_3$ (20 mM, 2.25 mL) in acetate buffer made up the FRAP reagent solution (2.5 mL, pH 3.6, 0.3 M). The mixture was then incubated at 37 °C in the dark for 30 min after EEZP, WEZP, and *P. granatum* juice (10–30 g/mL) were dissolved in buffer solution (5 mL). The absorbance of the samples was then measured. Activity measurements for Fe^{3+}-TPTZ reducing ability at each concentration were conducted in triplicate [50].

2.10. *DPPH$^\bullet$ Scavenging Activity*

The bleaching of a purple DPPH solution in methanol allows for the presence of certain pure antioxidant compounds with hydrogen-atom- or electron-donating properties to be determined. Stable DPPH$^\bullet$ is the reagent used in this spectrophotometric assay [51]. The method described by Blois [52], as previously applied by Gulcin [53], was used with minor modification to estimate the DPPH$^\bullet$ free radical scavenging capacity of EEZP, WEZP, and *P. granatum* juice; a stable free radical called DPPH was monitored for bleaching at a specific wavelength while the sample was present. The DPPH$^\bullet$ solution was prepared daily. Aluminum foil was used to cover the solution flask, which was stirred for 16 h at 4 °C while being kept in the dark. Shortly after preparing a 0.1 mM DPPH$^\bullet$ solution in ethanol, 0.5 mL of this solution was combined with 2 mL of EEZP, WEZP, and *P. granatum* juice at various concentrations (10–30 g/mL). After being vortexed, the samples were incubated at 30 °C in the dark for 30 min. Absorbance was measured at 517 nm in comparison to blank samples. The scavenging of DPPH free radicals is indicated by a decrease in absorbance [54]. When DPPH is reduced by an antioxidant or another radical species, its absorption falls below that of the radical form, which absorbs at 517 nm. The absorbance at 517 nm decreased proportionately to an increase in DPPH's non-radical forms when a hydrogen atom or electron was transferred to the odd electron [55]. Absorbance decreases indicate that DPPH is actively scavenging free radicals. Activity measurements for DPPH radical scavenging activity at each concentration were conducted in triplicate [56].

2.11. *ABTS$^{•+}$ Scavenging Activity*

A relatively stable free radical, ABTS, also decolorizes in its non-radical state. The method of Re et al. [57] was used to determine ABTS$^{•+}$ scavenging activity. This technique involves adding an antioxidant to a prepared ABTS radical solution, and after a set amount of time, the remaining ABTS$^{•+}$ is measured spectrophotometrically at 734 nm [58]. Then, 2 mM ABTS in water was combined with 2.45 mM potassium persulfate ($K_2S_2O_8$) to create ABTS$^{•+}$, which was then left to sit for 6 h at room temperature in the dark. The ABTS started to oxidize right away, but it took over 6 h for the absorbance to reach its peak and stabilize. Under storage conditions at room temperature in the dark, the radical cation is stable in this form for longer than two days. In order to perform the assay, the solution was diluted in phosphate buffer (pH 7.4), providing an absorbance of 0.700 ± 0.025 at 734 nm and equilibrated to 30 °C, the temperature at which all assays were carried out. Then, 3 mL of EEZP, WEZP, and *P. granatum* juice in ethanol at 10–30 µg/mL were combined with 1 mL of the ABTS$^{•+}$ solution. After mixing for 30 min, the absorbance was measured, and the radical scavenging percentage was computed for each concentration in comparison to a blank containing no scavenger. The percentage reduction in absorbance was used to determine the degree of decolorization. Activity measurements for ABTS radical scavenging activity at each concentration were conducted in triplicate [59].

2.12. *Enzyme Inhibition Studies*

2.12.1. Acetylcholinesterase Inhibition Study

The cholinergic enzyme-inhibitory abilities of EEZP, WEZP, and *P. granatum* juice were determined using Ellman's methodology [60] as described in a previous study [61]. This was accomplished using AChE serum from electric eels. Briefly, a specific *P. granatum* concentration (10–30 µg/mL) in buffer (1.0 M Tris/HCl, 100 µL, pH 8.0) was transferred to the enzyme solution (50 µL, 5.32 10^{-3} EU). The mixtures were kept at 20 °C for 10 min. Then, 50 µL of mixtures containing DTNB (5,5′-dithio-bis(2-nitro-benzoic acid) (0.5 mM)) and acetylthiocholine iodide (AChI) was added. The reaction medium was then started, and the mixture's absorbances were measured spectrophotometrically at 412 nm [62].

2.12.2. α-Glycosidase Inhibition Study

The inhibitory abilities of the WEZP, EEZP, and *P. granatum* juice on α-glycosidase were determined based on the method of Tao et al. [63], as described in detail in [64]. Various amounts of WEZP, EEZP, and *P. granatum* juice were transferred to phosphate buffer (75 µL, pH 7.4) for this purpose. Then, 20 µL of α-glycosidase solution was transferred to the same buffer and incubated for 10 min. The final mixture was mixed with 50 µL of p-nitrophenyl-D-glycopyranoside (p-NPG) dissolved in the same buffer. The mixture was then incubated again at room temperature (37 °C), and the absorbances were measured at 405 nm against a blank sample made up of phosphate buffer.

2.12.3. α-Amylase Inhibition Study

The inhibitory effects of WEZP, EEZP, and *P. granatum* juice on α-amylase were measured according to Xiao et al [65]. Briefly, 40 mL of 0.4 M alkaline solution was used to dissolve 1 g of starch, which was then heated for 30 min at 80 °C. The pH of the mixture was adjusted to 6.9, and the total volume was adjusted to 100 mL with deionized water. Then, different amounts of WEZP, EEZP, and *P. granatum* juice and 35 µL of starch prepared in buffer solution (pH 6.9) were mixed. Then, 20 µL of enzyme was added to the mixture and incubated at 40 °C for 60 min. Finally, 50 µL of HCl (0.1 M) was added to the mixture, and the reaction was stopped. The absorbance of samples was measured at 580 nm. The blank sample contained buffer solution (pH 6.9).

2.12.4. hCA Inhibition Study

The sepharose-4B-L-tyrosine sulfanilamide affinity technique was used to separate and purify CA II isoenzymes using human blood samples, as previously reported [66]. The protein levels were determined at 595 nm using the Bradford method after the enzymes had been purified [67]. The spectrophotometric Verpoorte's method (Shimadzu, UVmini-1240 UV–VIS) was used to perform CA activity [68]. Acetazolamide (AZA) was utilized as a reference standard [69].

2.13. Antimicrobial Studies

2.13.1. Microorganisms to Be Used in the Study

Microorganisms that can be potentially harmful to humans were used in this study. Gram-positive bacteria (*S. aureus* ATCC 25923) and Gram-negative bacteria (*E. coli* clinical isolate) were used for the assessment of antibacterial activity [70]. Bacterial strains were derived from stock cultures (clinical isolates and standard strains) of Kahramanmaras Sutcu İmam University Faculty of Medicine, Department of Medical Microbiology, Microbiology Laboratory.

2.13.2. Identification of *E. coli* Clinical Isolates

The identification of *E. coli* clinical isolates was realized according to the method of Deniz et al. [71]. Pathogen bacterial isolations from various clinical samples collected from patients and delivered to the laboratory under sterile conditions were inoculated on blood agar and EMB agar, and the media were incubated at 37 °C for 48 h. Colonies of *E. coli* bacteria grown in culture media were identified as species by Gram staining, biochemical tests, and the BD Phoenix 100 identification system.

2.13.3. Antimicrobial Activity Determination

The antimicrobial activity of the WEZP, EEZP, and *P. granatum* juice was determined by a disk diffusion method [72]. The test microorganism agar cultures were prepared in accordance with the procedure described by Gulcin et al. [73]. Bacterial strains were grown on blood agar medium (Oxoid CM55, Basingstoke, Hampshire, UK). In the study, pathogens to be evaluated were inoculated into Tryptone soy broth (Oxoid CM129, Basingstoke, Hampshire, UK). Facultative anaerobes and aerobes, including some fungi, were cultivated using tryptone soy broth, a highly nutritive and versatile medium that is recommended for general laboratory use. Prepared cultures were incubated for 24 h at 37 °C. For the antimicrobial test, 50 µL of WEZP, EEZP, and *P. granatum* juice was added to sterile 6 mm diameter filter paper discs, and susceptibility measurements were conducted on Mueller Hinton agar (Oxoid CM337, Basingstoke, Hampshire, UK) medium with the diffusion technique prescribed in Clinical and Laboratory Standards (CLSI 2018).

The growth inhibition zones around the discs containing antibiotics and WEZP, EEZP, and *P. granatum* juice were measured and recorded. The presence of antimicrobial activity was shown by clear zones of inhibition surrounding the discs [74]. Plant extracts, amoxicillin-clavulanic acid (20/10 µg/disc), gentamicin (10 µg/disc), ampicillin-sulbactam (10/10 µg/disc), and ciprofloxacin (5 µg/disc, BD BBL™ Sensi-Disc™) were compared with standard antimicrobial discs. Antimicrobial test results were analyzed according to the references suggested by the Clinical and Laboratory Standards [75].

2.14. Statistical Analysis

All experiments are repeated three times for each sample. The results are reported as the mean ± SD. (n = 3) and were evaluated using one-way ANOVA followed by Tukey's post hoc test; $p < 0.05$ was considered statistically significant.

3. Results

3.1. Total Phenolics, Total Flavonoids, and LC-MS/MS Analysis Results

The phenolic and flavonoid contents in EEZP peel were measured as 59.7 mg GAE/g and 88.0 mg QE/g, respectively, in this study. Between 6.36 and 1.78 mg GAE/100 mL of total phenolics were present in five different pomegranate cultivars. The total flavonoid content varied from 4.93 to 2.24 mg GAE/100 mL [2]. "Wonderful" pomegranate fruit mineral concentration, bioactivity, and internal quality were improved using foliar nutrient applications. Total phenolic content in *P. granatum* juice ranged from 2091 to 3735 mg/L GAE [76]. The polyphenol and flavonoid contents of pomegranate peel acetone extract (338 ± 20 mg/g GAE and 60.8 ± 9.3 mg/g QE, respectively) were significantly ($p < 0.05$) higher than those of water and ethanol extracts. Additionally, it was discovered that the polyphenol and flavonoid levels of acetone extract were higher than those found in methanol, ethanol, and ethyl acetate extracts of the fruit peels of various Pakistani pomegranate varieties, including "Desi", "Kandhari", and "Badana" [77]. In this study, *P. granatum* extracts were shown to have comparable effective amounts of polyphenolics.

Using fifty-three phenolics as standard compounds, the LC-MS/MS method was utilized to identify the major phenolic components in *P. granatum* extracts. The elucidation of phenolic compounds was accomplished by comparison of their chromatographic behavior, UV spectra, and MS information with references, and thirty compounds were measured (Table 1 and Figure 3). Table 1 shows the mean values of each chemical based on the LC-MS/MS tests. The major compounds detected in ethanol extract of *P. granatum* were ellagic acid (199.967 mg/g), catechin (27.664 mg/g), epigallocatechin gallate (25.600 mg/g), epicatechin (24.210 mg/g), nicotiflorin (23.535 mg/g), astragalin (20.551 mg/g), gallic acid (20.021 mg/g), epigallocatechin (19.148 mg/g), quinic acid (17.460 mg/g), tannic acid (12.300 mg/g), aconitic acid (8.190 mg/g), hesperidin (6.136 mg/g), isoquercitrin (4.056 mg/g), rutin (2.732 mg/g), fumaric acid (2.128 mg/g), cosmosiin (2.036 mg/g), luteolin (1.126 mg/g), and epicatechin gallate (1.060 mg/g). Also, the compounds include protocatechuic acid, protocatechuic aldehyde, caffeic acid, vanillin, piceid, p-cumaric acid, cynaroside, quercetin, naringenin, kaempferol, apigenin, amentoflavone, gentisic acid, and chlorogenic acid were detected. Only quinic acid was in found a higher amount (44.662 mg/g) in water extract of *P. granatum*, whereas 1,5-dicaffeoylquinic acid, 4-OH-benzoic acid, vanillic acid, syringic acid, daidzin, syringic aldehyde, ferulic acid-D3-IS, ferulic acid, coumarin, sinapic acid, salicylic acid, miquelianin rutin D3-IS, *O*-coumaric acid, rosmarinic acid, genistin, quercitrin, fisetin, daidzein, quercetin-D3-IS, hesperetin, genistein, chrysin, and acacetin were not recorded in EEZP and WEZP. The chemical structures of the most plentiful phenolics in *P. granatum* are presented in Figure 4.

3.2. Reducing Ability Results

As summarized in Table 2 and Figure 5A, *P. granatum* extracts showed a potent Fe^{3+} reducing profile. However, the Fe^{3+} reducing ability of a 30 μg/mL concentration of *P. granatum* extracts, phenolic compounds, and standards decreased in the following order: α-tocopherol (2.778 ± 0.248, r^2: 0.9999) > Trolox (2.334 ± 0.167, r^2: 0.9997) > BHA (2.319 ± 0.041, r^2: 0.9629) > BHT (1.873 ± 0.152, r^2: 0.9918) > *P. granatum* juice (1.810 ± 0.149, r^2: 0.7433) > WEZP peel (1.278 ± 0.143, r^2: 0.9995) > EEZP peel (1.219 ± 0.028, r^2: 0.9253) > EEZP seed (0.258 ± 0.005, r^2:0.9712) > WEZP seed (0.229 ± 0.033, r^2: 0.9252). All analyses were carried out in triplicate. Depending on the reducing antioxidant capacity of *P. granatum* extracts, the test solution's yellow color in this assay shifted to various shades of green and blue.

Table 1. Analytical method validation parameters and chemical profiles of WEZP and EEZP by LC-MS/MS analysis.

No	Analyte	RT [a]	M.I. (m/z) [b]	F.I. (m/z) [c]	Ion. Mode	Equation	r^2 [d]	RSD% [e] Interday	RSD% [e] Intraday	Linearity Range (mg/L)	LOD/LOQ (μg/L) [f]	Recovery (%) Interday	Recovery (%) Intraday	U [g]	Gr. No [i]	Phenolics WEZP	Phenolics EEZP
1	Quinic acid	3.0	190.8	93.0	Neg	$y = -0.0129989 + 2.97989x$	0.996	0.69	0.51	0.1–5	25.7/33.3	1.0011	1.0083	0.0372	1	44.662	17.460
2	Fumaric aid	3.9	115.2	40.9	Neg	$y = -0.0817862 + 1.03467x$	0.995	1.05	1.02	1–50	135.7/167.9	0.9963	1.0016	0.0091	1	N.D.	2.128
3	Aconitic acid	4.0	172.8	129.0	Neg	$y = -0.7014530 + 32.9994x$	0.971	2.07	0.93	0.1–5	16.4/31.4	0.9968	1.0068	0.0247	1	N.D.	8.190
4	Gallic acid	4.4	168.8	79.0	Neg	$y = 0.0547697 + 20.8152x$	0.999	1.60	0.81	0.1–5	13.2/17.0	1.0010	0.9947	0.0112	1	0.846	20.021
5	Epigallocatechin	6.7	304.8	219.0	Neg	$y = -0.00494986 + 0.0483704x$	0.998	1.22	0.73	1–50	237.5/265.9	0.9969	1.0040	0.0184	3	N.D.	19.148
6	Protocatechuic acid	6.8	152.8	108.0	Neg	$y = 0.211373 + 12.8622x$	0.957	1.43	0.76	0.1–5	21.9/38.6	0.9972	1.0055	0.0350	1	N.D.	0.371
7	Catechin	7.4	288.8	203.1	Neg	$y = -0.00370053 + 0.431369x$	0.999	2.14	1.08	0.2–10	55.0/78.0	1.0024	1.0045	0.0221	3	0.195	27.664
8	Gentisic acid	8.3	152.8	109.0	Neg	$y = -0.0238983 + 12.1494x$	0.997	1.81	1.22	0.1–5	18.5/28.2	0.9963	1.0077	0.0167	1	N.D.	N.D.
9	Chlorogenic acid	8.4	353.0	85.0	Neg	$y = 0.289983 + 36.3926x$	0.995	2.15	1.52	0.1–5	13.1/17.6	1.0000	1.0023	0.0213	1	N.D.	N.D.
10	Protocatechuic aldehyde	8.5	137.2	92.0	Neg	$y = 0.257085 + 25.4657x$	0.996	2.08	0.57	0.1–5	15.4/22.2	1.0002	0.9988	0.0396	1	N.D.	0.154
11	Tannic acid	9.2	182.8	78.0	Neg	$y = 0.0126307 + 26.9263x$	0.999	2.40	1.16	0.05–2.5	15.3/22.7	0.9970	0.9950	0.0190	1	1.694	12.287
12	Epigallocatechin gallate	9.4	457.0	305.1	Neg	$y = -0.0380744 + 1.61233x$	0.999	1.30	0.63	0.2–10	61.0/86.0	0.9981	1.0079	0.0147	3	0.090	25.600
13	1,5-Dicaffeoylquinic acid	9.8	515.0	191.0	Neg	$y = -0.0164044 + 16.6535x$	0.999	2.42	1.48	0.1–5	5.8/9.4	0.9983	0.9997	0.0306	1	N.D.	N.D.
14	4-OH Benzoic acid	10.5	137.2	65.0	Neg	$y = -0.0240747 + 5.06492x$	0.999	1.24	0.97	0.2–10	68.4/88.1	1.0032	1.0068	0.0237	1	N.D.	N.D.
15	Epicatechin	11.6	289.0	203.0	Neg	$y = -0.0172078 + 0.0833424x$	0.996	1.47	0.62	1–50	139.6/161.6	1.0013	1.0012	0.0221	3	N.D.	24.210
16	Vanillic acid	11.8	166.8	108.0	Neg	$y = -0.0480183 + 0.779564x$	0.999	1.92	0.76	1–50	141.9/164.9	1.0022	0.9998	0.0145	1	N.D.	N.D.
17	Caffeic acid	12.1	179.0	134.0	Neg	$y = 0.120319 + 95.4610x$	0.999	1.11	1.25	0.05–2.5	7.7/9.5	1.0015	1.0042	0.0152	1	N.D.	0.096
18	Syringic acid	12.6	196.8	166.9	Neg	$y = -0.0458599 + 0.663948x$	0.998	1.18	1.09	1–50	82.3/104.5	1.0006	1.0072	0.0129	1	N.D.	N.D.

Table 1. *Cont.*

No	Analyte	RT [a]	M.I. (m/z) [b]	F.I. (m/z) [c]	Ion. Mode	Equation	r^2 [d]	RSD% [e] Interday	RSD% [e] Intraday	Linearity Range (mg/L)	LOD/LOQ (μg/L) [f]	Recovery (%) Interday	Recovery (%) Intraday	U [g]	Gr. No [i]	Phenolics WEZP	Phenolics EEZP
19	Vanillin	13.9	153.1	125.0	Poz	$y = 0.00185898 + 20.7382x$	0.996	1.10	0.85	0.1–5	24.5/30.4	1.0009	0.9967	0.0122	1	N.D.	0.201
20	Syringic aldehyde	14.6	181.0	151.1	Neg	$y = -0.0128684 + 7.90153x$	0.999	2.51	0.77	0.4–20	19.7/28.0	1.0001	0.9964	0.0215	1	N.D.	N.D.
21	Daidzin	15.2	417.1	199.0	Poz	$y = 9.45747 + 152.338x$	0.996	2.25	1.32	0.05–2.5	7.0/9.5	0.9955	1.0017	0.0202	2	N.D.	N.D.
22	Epicatechin gallate	15.5	441.0	289.0	Neg	$y = -0.0142216 + 1.06768x$	0.997	1.63	1.28	0.1–5	19.5/28.5	0.9984	0.9946	0.0229	3	N.D.	1.060
23	Piceid	17.2	391.0	135/106.9	Poz	$y = 0.00772525 + 25.4181x$	0.999	1.94	1.16	0.05–2.5	13.8/17.8	1.0042	0.9979	0.0199	1	N.D.	0.23
24	*p*-Coumaric acid	17.8	163.0	93.0	Neg	$y = 0.0249034 + 18.5180x$	0.999	1.92	1.43	0.1–5	25.9/34.9	1.0049	1.0001	0.0194	1	N.D.	0.874
25	Ferulic acid-D3-IS [h]	18.8	196.2	152.1	Neg	N.A.	N.A.	N.A.	N.A.	N.A.	N.A.	N.A.	N.A.	0.0170	1	N.A.	N.A.
26	Ferulic acid	18.8	192.8	149.0	Neg	$y = -0.0735254 + 1.34476x$	0.999	1.44	0.53	1–50	11.8/15.6	0.9951	0.9976	0.0181	1	N.D.	N.D.
27	Sinapic acid	18.9	222.8	193.0	Neg	$y = -0.0929932 + 0.836324x$	0.999	1.45	0.52	0.2–10	65.2/82.3	1.0031	1.0037	0.0317	1	N.D.	N.D.
28	Coumarin	20.9	146.9	103.1	Poz	$y = 0.0633397 + 136.508x$	0.999	2.11	1.54	0.05–2.5	214.2/247.3	0.9950	0.9958	0.0383	1	N.D.	N.D.
29	Salicylic acid	21.8	137.2	65.0	Neg	$y = 0.239287 + 153.659x$	0.999	1.48	1.18	0.05–2.5	6.0/8.3	0.9950	0.9998	0.0158	1	N.D.	N.D.
30	Cynaroside	23.7	447.0	284.0	Neg	$y = 0.280246 + 6.13360x$	0.997	1.56	1.12	0.05–2.5	12.1/16.0	1.0072	1.0002	0.0366	2	N.D.	0.926
31	Miquelianin	24.1	477.0	150.9	Neg	$y = -0.00991585 + 5.50334x$	0.999	1.31	0.95	0.1–5	10.6/14.7	0.9934	0.9965	0.0220	2	N.D.	N.D.
32	Rutin-D3-IS [h]	25.5	612.2	304.1	Neg	N.A.	N.A.	N.A.	N.A.	N.A.	N.A.	N.A.	N.A.	N.A.	2	N.A.	N.A.
33	Rutin	25.6	608.9	301.0	Neg	$y = -0.0771907 + 2.89868x$	0.999	1.38	1.09	0.1–5	15.7/22.7	0.9977	1.0033	0.0247	2	0.024	2.732
34	Isoquercitrin	25.6	463.0	271.0	Neg	$y = -0.111120 + 4.10546x$	0.998	2.13	0.78	0.1–5	8.7/13.5	1.0057	0.9963	0.0220	2	0.038	4.056
35	Hesperidin	25.8	611.2	449.0	Poz	$y = 0.139055 + 13.2785x$	0.999	1.84	1.35	0.1–5	19.0/26.0	0.9967	1.0043	0.0335	2	0.063	6.136
36	*o*-Coumaric acid	26.1	162.8	93.0	Neg	$y = 0.00837193 + 11.2147x$	0.999	2.11	1.46	0.1–5	31.8/40.4	1.0044	0.9986	0.0147	1	N.D.	N.D.
37	Genistin	26.3	431.0	239.0	Neg	$y = 1.65808 + 7.57459x$	0.991	2.01	1.28	0.1–5	14.9/21.7	1.0062	1.0047	0.0083	2	N.D.	N.D.
38	Rosmarinic acid	26.6	359.0	197.0	Neg	$y = -0.0117238 + 8.04377x$	0.999	1.24	0.86	0.1–5	16.2/21.2	1.0056	1.0002	0.0130	1	N.D.	N.D.

Table 1. *Cont.*

No	Analyte	RT [a]	M.I. (m/z) [b]	F.I. (m/z) [c]	Ion. Mode	Equation	r^2 [d]	RSD% [e] Interday	RSD% [e] Intraday	Linearity Range (mg/L)	LOD/LOQ (µg/L) [f]	Recovery (%) Interday	Recovery (%) Intraday	U [g]	Gr. No [i]	Phenolics WEZP	Phenolics EEZP
39	Ellagic acid	27.6	301.0	284.0	Neg	$y = 0.00877034 + 0.663741x$	0.999	1.57	1.23	0.4–20	56.9/71.0	1.0005	1.0048	0.0364	1	1.518	199.967
40	Cosmosiin	28.2	431.0	269.0	Neg	$y = -0.708662 + 8.62498x$	0.998	1.65	1.30	0.1–5	6.3/9.2	0.9940	0.9973	0.0083	2	0.019	2.036
41	Quercitrin	29.8	447.0	301.0	Neg	$y = -0.00153274 + 3.20368x$	0.999	2.24	1.16	0.1–5	4.8/6.4	0.9960	0.9978	0.0268	2	N.D.	N.D.
42	Astragalin	30.4	447.0	255.0	Neg	$y = 0.00825333 + 3.51189x$	0.999	2.08	1.72	0.1–5	6.6/8.2	0.9968	0.9957	0.0114	2	0.243	20.551
43	Nicotiflorin	30.6	592.9	255.0/284.0	Neg	$y = 0.00499333 + 2.62351x$	0.999	1.48	1.23	0.05–2.5	11.9/16.7	0.9954	1.0044	0.0108	2	0.273	23.535
44	Fisetin	30.6	285.0	163.0	Neg	$y = 0.0365705 + 8.09472x$	0.999	1.75	1.19	0.1–5	10.1/12.7	0.9980	1.0042	0.0231	3	N.D.	N.D.
45	Daidzein	34.0	253.0	223.0	Neg	$y = -0.0329252 + 6.23004x$	0.999	2.18	1.73	0.1–5	9.8/11.6	0.9926	0.9963	0.0370	3	N.D.	N.D.
46	Quercetin-D3-IS [h]	35.6	304.0	275.9	Neg	N.A.	N.A.	N.A.	N.A.	N.A.	N.A.	N.A.	N.A.	N.A.	3	N.A.	N.A.
47	Quercetin	35.7	301.0	272.9	Neg	$y = +0.00597342 + 3.39417x$	0.999	1.89	1.38	0.1–5	15.5/19.0	0.9967	0.9971	0.0175	3	N.D.	0.136
48	Naringenin	35.9	270.9	119.0	Neg	$y = -0.00393403 + 14.6424x$	0.999	2.34	1.69	0.1–5	2.6/3.9	1.0062	1.0020	0.0392	3	0.004	0.234
49	Hesperetin	36.7	301.0	136.0/286.0	Neg	$y = +0.0442350 + 6.07160x$	0.999	2.47	2.13	0.1–5	7.1/9.1	0.9998	0.9963	0.0321	3	N.D.	N.D.
50	Luteolin	36.7	284.8	151.0/175.0	Neg	$y = -0.0541723 + 30.7422x$	0.999	1.67	1.28	0.05–2.5	2.6/4.1	0.9952	1.0029	0.0313	3	0.005	1.126
51	Genistein	36.9	269.0	135.0	Neg	$y = -0.00507501 + 12.1933x$	0.999	1.48	1.19	0.05–2.5	3.7/5.3	1.0069	1.0012	0.0337	3	N.D.	N.D.
52	Kaempferol	37.9	285.0	239.0	Neg	$y = 0.00459557 + 3.13754x$	0.999	1.49	1.26	0.05–2.5	10.2/15.4	0.9992	0.9990	0.0212	3	N.D.	0.357
53	Apigenin	38.2	268.8	151.0/149.0	Neg	$y = 0.119018 + 34.8730x$	0.998	1.17	0.96	0.05–2.5	1.3/2.0	0.9985	1.0003	0.0178	3	N.D.	0.338
54	Amentoflavone	39.7	537.0	417.0	Neg	$y = 0.727280 + 33.3658x$	0.992	1.35	1.12	0.05–2.5	2.8/5.1	0.9991	1.0044	0.0340	3	0.013	0.009
55	Chrysin	40.5	252.8	145.0/119.0	Neg	$y = -0.0777300 + 18.8873x$	0.999	1.46	1.21	0.05–2.5	1.5/2.8	0.9922	1.0050	0.0323	3	N.D.	N.D.
56	Acacetin	40.7	283.0	239.0	Neg	$y = -0.559818 + 163.062x$	0.997	1.67	1.28	0.02–1	1.5/2.5	0.9949	1.0011	0.0363	3	N.D.	N.D.

[a] RT: retention time; [b] MI (*m/z*): molecular ions of the standard analytes (*m/z* ratio); [c] FI (*m/z*): fragment ions; [d] r^2: coefficient of determination; [e] *RSD*: relative standard deviation; [f] *LOD/LOQ* (µg/L): limit of detection/quantification; [g] *U* (%): percent relative uncertainty at 95% confidence level ($k = 2$); [h] IS: internal standard; [i] Gr. No: grouping of internal standards (these numbers indicate which IS stands for which phenolic compound); N.D.: not detected; N.A.: not applicable.

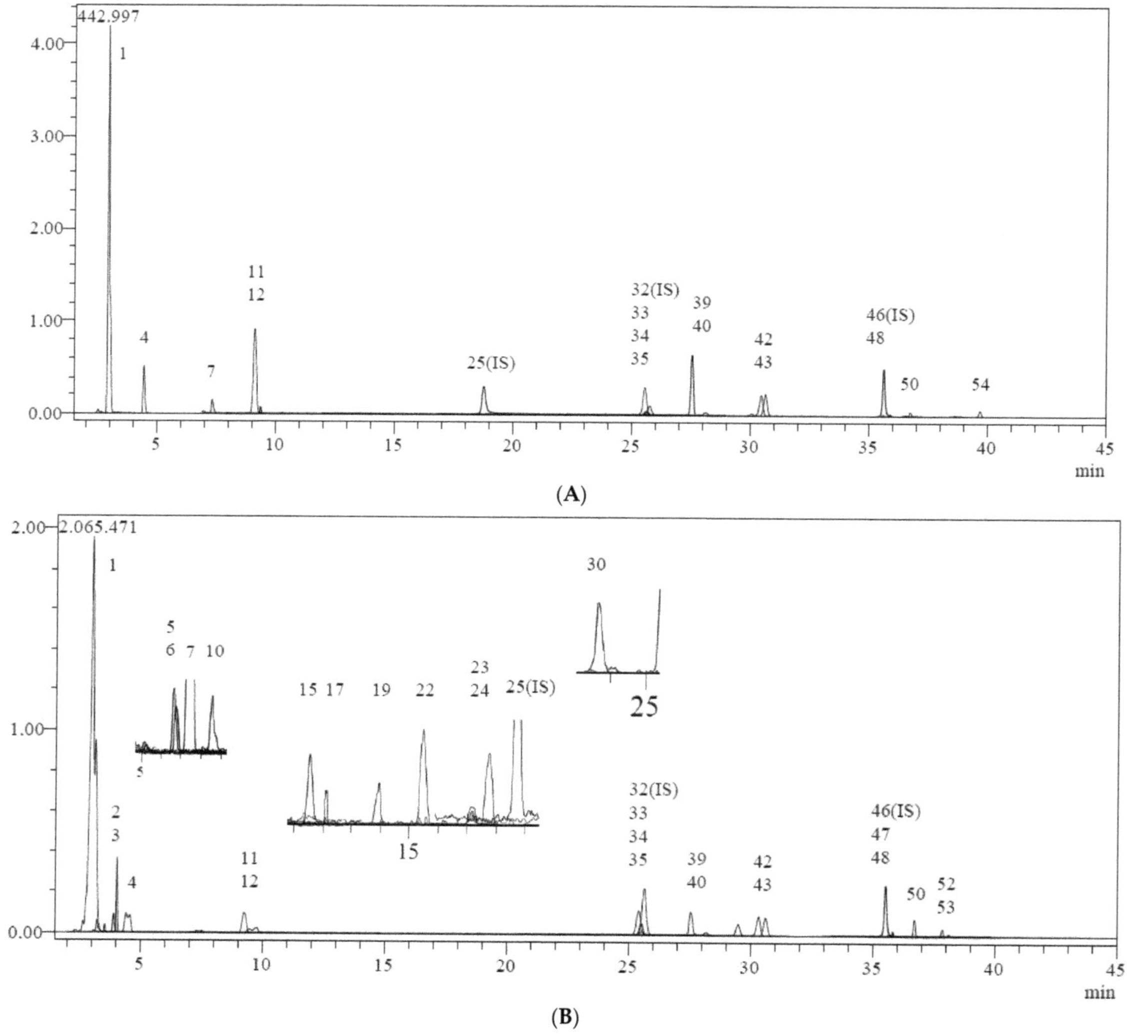

Figure 3. (**A**) Chromatogram of *P. granatum* water extract compounds. (**B**) Chromatogram of *P. granatum* ethanol extract compounds (1. quinic acid, 2. fumaric acid, 3. aconitic acid, 4. gallic acid, 5. epigallocatechin 7. catechin, 11. tannic acid, 12. epigallocatechin gallate, 15. epicatechin, 22. epicatechin gallate, 33. rutin, 34. isoquercitrin, 35. hesperidin, 39. ellagic acid, 40. cosmosiin, 42. astragalin, 43. nicotiflorin, 50. luteolin).

Cu^{2+} reducing abilities of the phenolic composition in *P. granatum* extracts and juice are presented in Table 2 and Figure 5B. It was determined that there was a strong relationship between the Cu^{2+} reducing impact and different concentrations of phenolics in *P. granatum* extracts. However, at a concentration of 30 μg/mL, the significant absorbance of reducing ability was demonstrated by phenolics in *P. granatum* extracts. On the other hand, the Cu^{2+} reducing abilities of *P. granatum* extracts and standards were found as follows: BHT (2.865 ± 0.038, r^2: 0.9991) > BHA (2.849 ± 0.020, r^2: 0.9999) > *P. granatum* juice (2.790 ± 0.045, r^2: 0.9999) > Trolox (2.555 ± 0.022, r^2: 0.9987) > α-tocopherol (2.185 ± 0.110, r^2: 0.9986) > WEZP peel (0.927 ± 0.022, r^2: 0.9965) > EEZP peel (0.878 ± 0.017, r^2: 0.9967) > EEZP seed (0.194 ± 0.008, r^2: 0.9974) > WEZP seed (0.114 ± 0.034, r^2: 0.8485).

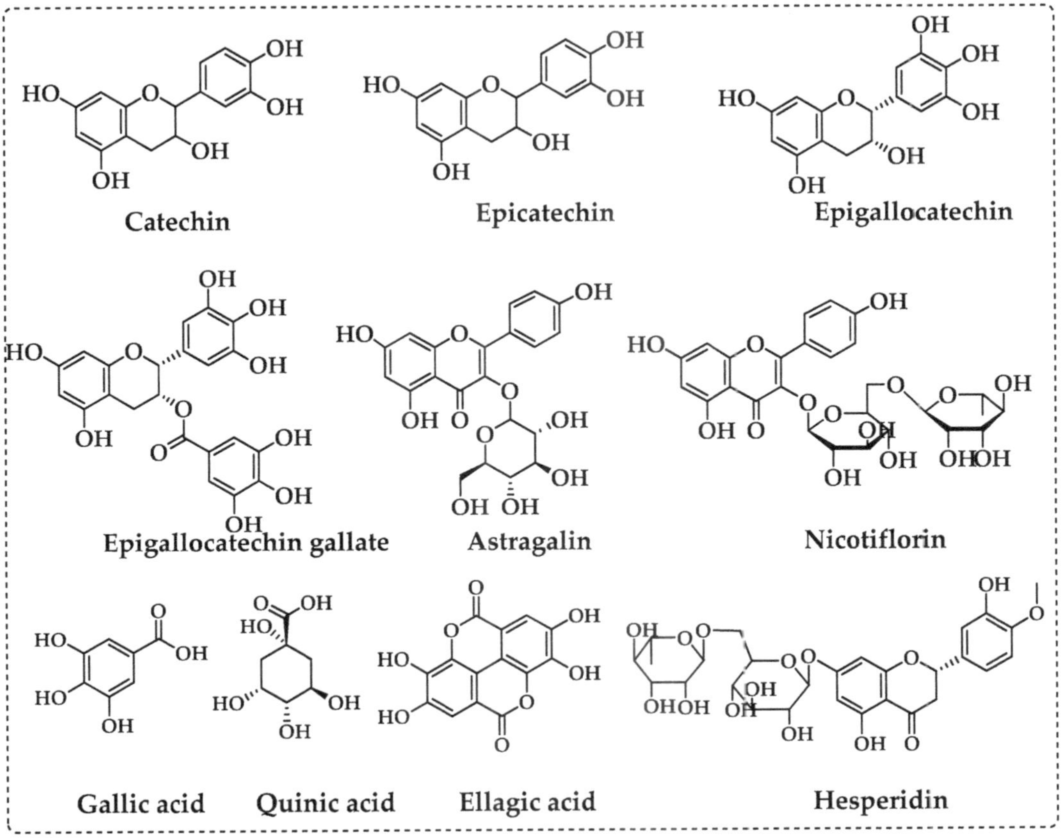

Figure 4. The structures of the ten most abundant phenolic compounds in *P. granatum*.

Table 2. Fe^{3+}, Cu^{2+}, and Fe^{3+}-TPTZ reducing capabilities of *P. granatum* extracts, juice, and positive controls at 30 μg/mL (BHA: butylated hydroxyanisole; BHT: butylated hydroxytoluene).

Antioxidant	Fe^{3+} Reducing *		Cu^{2+} Reducing *		Fe^{3+}-TPTZ Reducing *	
	λ_{700}	r^2	λ_{450}	r^2	λ_{593}	r^2
BHA	2.319 ± 0.041 [a]	0.9629	2.849 ± 0.020 [a]	0.9994	2.151 ± 0.020 [b]	0.9367
BHT	1.873 ± 0.152 [b]	0.9918	2.865 ± 0.038 [a]	0.9991	2.031 ± 0.190 [b]	0.9670
Trolox	2.334 ± 0.167 [a]	0.9997	2.555 ± 0.022 [a]	0.9987	2.108 ± 0.026 [b]	0.9291
α-Tocopherol	2.778 ± 0.248 [a]	0.9999	2.185 ± 0.110 [b]	0.9986	2.434 ± 0.103 [a]	0.8714
WEZP peel	1.278 ± 0.143 [c]	0.9995	0.927 ± 0.022 [c]	0.9965	1.903 ± 0.052 [b]	0.9875
WEZP seed	0.229 ± 0.033 [d]	0.9252	0.114 ± 0.034 [d]	0.8485	0.483 ± 0.023 [c]	0.9124
EEZP peel	1.219 ± 0.028 [c]	0.9253	0.878 ± 0.017 [c]	0.9967	2.086 ± 0.080 [b]	0.9866
EEZP seed	0.258 ± 0.005 [d]	0.9712	0.194 ± 0.008 [d]	0.9974	0.606 ± 0.011 [c]	0.9471
P. granatum juice	1.810 ± 0.149 [b]	0.4020	2.790 ± 0.045 [a]	0.9999	2.230 ± 0.010 [b]	0.9056

* All values are averages of three parallel measurements (n = 3) and presented as mean ± SD. Different letters in the same column indicate a significant difference between the means ($p < 0.05$ regarded as significant).

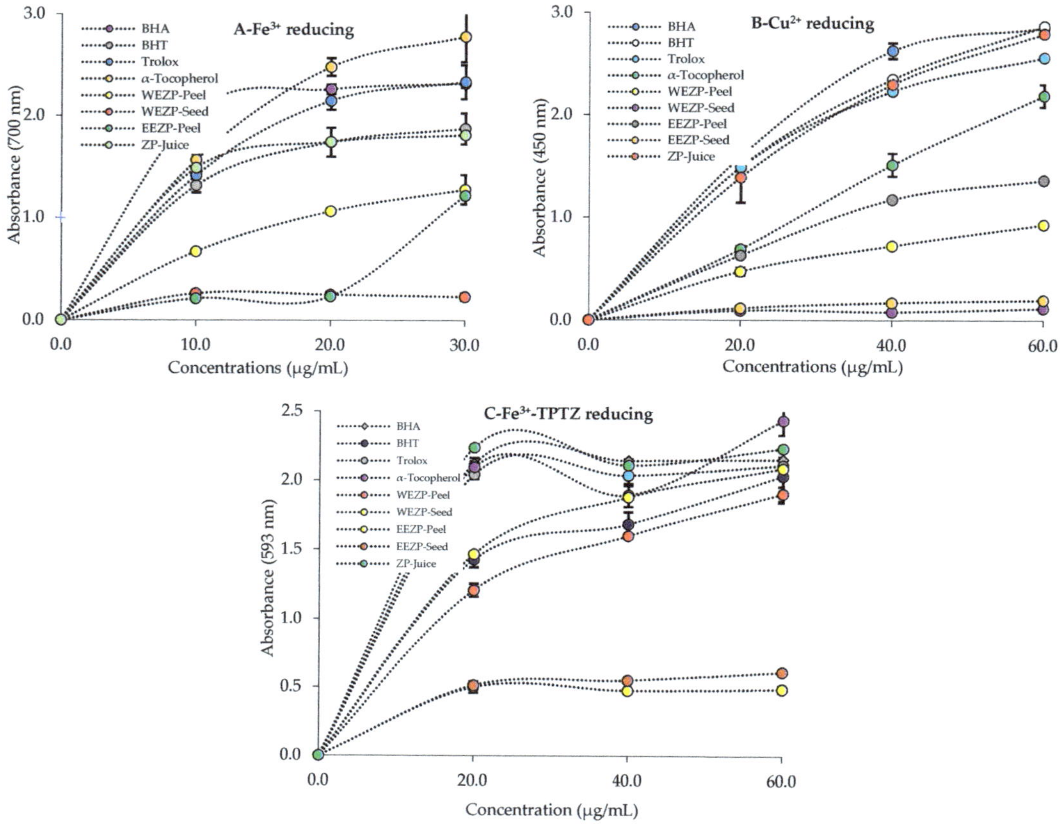

Figure 5. Fe^{3+} (**A**), Cu^{2+} (**B**), and Fe^{3+}-TPTZ (**C**) reducing abilities of *P. granatum* and standards.

An Fe^{3+}-TPTZ reducing assay was used to determine the reducing abilities of *P. granatum* extracts and standards. The reduction powers of samples dropped in the following sequence according to the results provided in Table 2 and Figure 5C: α-tocopherol (2.434 ± 0.103, r^2: 0.8714) > ZP juice (2.230 ± 0.010, r^2: 0.9056) > BHA (2.151 ± 0.020, r^2: 0.9367) > Trolox (2.108 ± 0.026, r^2: 0.9291) > EEZP peel (2.086 ± 0.080, r^2: 0.9866) > BHT (2.031 ± 0.190, r^2: 0.9670) > WEZP peel (1.903 ± 0.052, r^2: 0.9875) > EEZP seed (0.606 ± 0.011, r^2: 0.9471) > WEZP seed (0.483 ± 0.023, r^2: 0.9124). In this method, the higher the absorbance readings, the better the test samples' ability to reduce.

3.3. Radical Scavenging Results

P. granatum extracts and juice are thought to have natural antioxidant potential if they have DPPH$^{\bullet}$ scavenging ability. The DPPH$^{\bullet}$ scavenging activity of *P. granatum* extracts was measured, and the IC_{50} value was derived (Table 3, Figure 6A). *P. granatum* extracts and juice demonstrated concentration-dependent radical scavenging activity (Figure 6A). The DPPH$^{\bullet}$ scavenging capability of *P. granatum* extracts, juice, and standards was decreased as follows: ascorbic acid (IC_{50}: 5.82 μg/mL) > Trolox (IC_{50}: 6.03 μg/mL) > BHA (IC_{50}: 6.86 μg/mL) > α-Tocopherol (7.70 μg/mL) > EEZP peel (IC_{50}: 16.10 μg/mL) WEZP peel (IC_{50}: 31.50 μg/mL) > BHT (IC_{50}: 49.50 μg/mL). DPPH$^{\bullet}$ scavenging abilities of WEZP seed, EEZP seed, and *P. granatum* juice could not be measured due to the color blur that occurred during measurements.

Table 3. IC_{50} values (μg/mL) of DPPH• and ABTS•+ scavenging activities of *P. granatum* and standards.

Antioxidant	DPPH• Scavenging		ABTS•+ Scavenging	
	IC_{50}	r^2	IC_{50}	r^2
BHA	6.86	0.9949	6.35	0.9746
BHT	49.50	0.9957	12.60	0.9995
Trolox	6.03	0.9925	16.50	0.9775
α-Tocopherol	7.70	0.9961	18.72	0.9347
Ascorbic acid	5.82	0.9668	11.74	0.9983
WEZP peel	31.50	0.9995	8.80	0.9178
WEZP seed	-	-	-	-
EEZP peel	16.10	0.9310	5.90	0.9669
EEZP seed	-	-	-	-
P. granatum Juice	-	-	-	-

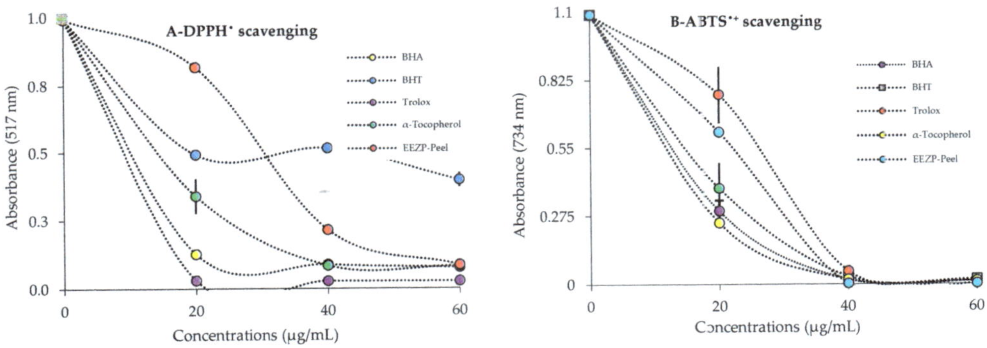

Figure 6. Radical scavenging effects of *P. granatum* and positive controls. (**A**) DPPH• scavenging ability; (**B**) ABTS•+ scavenging ability.

IC_{50} values of ABTS•+ scavenging of *P. granatum* extracts and juice and reference radical scavenger agents such as Trolox, α-tocopherol, BHT, and BHA were detected in the following range: EEZP peel (IC_{50}: 5.90 μg/mL, r^2: 0.9669) > BHA (IC_{50}: 6.35 μg/mL, r^2: 0.9746) > WEZP peel (IC_{50}: 8.80 μg/mL, r^2: 0.9178) > ascorbic acid (IC_{50}: 11.74 μg/mL, r^2: 0.9983) > BHT (IC_{50}: 12.60 μg/mL, r^2: 0.9995) Trolox (IC_{50}:1 6.50 μg/mL, r^2: 0.9775) > α-Tocopherol (18.72 μg/mL, r^2: 0.9347) (Table 3 and Figure 6B). As with DPPH• scavenging measurements, ABTS•+ scavenging abilities of WEZP seed, EEZP seed, and *P. granatum* juice could not be measured due to the color blur that occurred during measurements.

3.4. Enzyme Inhibition Results

For α-glycosidase enzyme, *P. granatum* extracts and juice exhibited effective inhibition effects. From this perspective, EEZP peel demonstrated an IC_{50} value of 7.3 μg/mL (r^2: 0.9941), WEZP seed exhibited an IC_{50} value of 7.3 μg/mL (r^2: 0.8819), *P. granatum* juice displayed an IC_{50} value of 27.1 μg/mL (r^2: 0.9665), and WEZP peel showed an IC_{50} value of 28.8 (r^2: 0.9420) (Table 4). However, the IC_{50} value could not be determined for EEZP seed. On the other hand, acarbose, as a standard for α-glycosidase and α-amylase, showed a value of 22,800 μM against α-glycosidase [63].

In inhibition studies conducted using similar methods, *P. granatum* extracts and juice were assayed for α-amylase inhibition ability, the results of which are presented in Table 4. For α-amylase enzyme, EEZP peel demonstrated an IC_{50} value of 317.7 μg/mL (r^2: 0.7778), WEZP seed exhibited an IC_{50} value of 375.8 μg/mL (r^2: 0.8193), WEZP peel showed an IC_{50} value of 494.3 μg/mL (r^2: 0.7705), and *P. granatum* juice displayed an IC_{50} value of 70.1 μg/mL (r^2: 0.9999) (Table 4).

Table 4. The half-maximal inhibition concentration (IC_{50}; μg/mL) of *P. granatum* towards acetylcholinesterase, α-glycosidase, α-amylase, and carbonic anhydrase II enzymes.

Enzyme	AChE		hCA II		α-Glycosidase		α-Amylase	
	IC_{50}	r^2	IC_{50}	r^2	IC_{50}	r^2	IC_{50}	r^2
WEZP peel	20.0	0.9976	36.4	0.9957	28.8	0.9420	494.3	0.7705
WEZP seed	20.4	0.9851	144.5	0.9906	6.4	0.8819	375.8	0.8193
EEZP peel	19.7	0.9869	106.3	0.9941	7.3	0.9399	317.7	0.7778
EEZP seed	17.8	0.9976	30.4	0.8881	-	-		
P. granatum juice	22.6	0.9951	94.0	0.9909	27.1	0.9665	70.1	0.9999
Standards	5.97 [1]	0.9706	8.4 [2]	0.9825	22,800 [3]	-		

[1] Acetazolamide (AZA) was used as a standard inhibitor for carbonic anhydrase II isoenzyme. [2] Tacrine was used as a standard inhibitor for acetylcholinesterase. [3] Acarbose was used as a standard inhibitor for α-glycosidase and α-amylase enzymes [63].

In addition, dominant cytosolic CA II isoform is frequently linked to a number of illnesses, including osteoporosis, glaucoma, and renal tubular acidosis. The CA inhibitory effects of *P. granatum* extracts and juice were decreased in the following order (Table 4): WEZP seed (IC_{50}: 144.5 μg/mL; r^2: 0.9906) > EEZP peel (IC_{50}: 106.3 μg/mL; r^2: 0.9941) > *P. granatum* juice (IC_{50}: 94.0 μg/mL; r^2: 0.9909) > WEZP peel (IC_{50}: 36.4 μg/mL; r^2: 0.9957) > EEZP seed (IC_{50}: 30.4 μg/mL; r^2: 0.9999) > acetazolamide (IC_{50}: 8.4 μg/mL; r^2: 0.9825). AZA was employed as a control for the inhibition of CA isoenzymes [78].

AChE was the first FDA-approved therapeutic target for the AD treatment, and many drugs are currently produced and marketed for this purpose. The AChE-inhibitory capacity of *P. granatum* extracts and juice was enhanced in the following order (Table 4): *P. granatum* juice (IC_{50}, 22.6 μg/mL; r^2: 0.9951) > WEZP seed (IC_{50}: 20.4 μg/mL; r^2: 0.9851) > WEZP peel (IC_{50}: 20.0 μg/mL; r^2: 0.9976) > EEZP peel (IC_{50}: 19.7 μg/mL; r^2: 0.9869) > EEZP seed (IC_{50}: 17.8 μg/mL; r^2: 0.9976) > tacrine (IC_{50}: 5.97 μg/mL; r^2: 0.9706; as a positive control for the inhibition of cholinergic enzymes) [79].

Urinary tract infections, respiratory pneumonia, surgical site infections, bacteremia, gastrointestinal disorders, and skin infections are among the most common nosocomial infections. *Staphylococcus aureus*, as a Gram-positive microorganism, and *E. coli*, as a Gram-negative microorganism, are the most prevalent pathogens that cause these infections according to the Center for Disease Control and Prevention (Atlanta, USA) [80]. We chose to test the effectiveness of *P. granatum* extracts and juice against these microorganisms since they are notoriously difficult to eradicate due to their resistance to most antimicrobial agents. Antimicrobial results are shown in Table 5.

Table 5. Antimicrobial activities of *P. granatum* extracts (50 μg/disk). Amc 30: amoxycillin/clavulanic acid antimicrobial susceptibility disks (30 μg/disk); Sxt 25: trimethoprim/sulfamethoxazole (25 μg/disk); Cip 5: ciprofloxacin (5 μg/disk); Gnt 10: gentamicin (10 μg/disk).

Sample	Antimicrobial Zone (mm)	
	Escherichia coli ATCC 39628	*Staphylococcuc aureus* ATCC 25923
WEZP peel	8	8
WEZP seed	R, N.D.	9
EEZP peel	R, N.D.	R, N.D.
EEZP seed	R, N.D.	R, N.D.
P. granatum juice	10	R, N.D.
Amc/Clav-30	10, R	R, N.D.
Sxt-25	R, N.D.	R, N.D.
Cip-5	R, N.D.	10, R
Gnt-10	11, R	R, N.D.

N.D.: activity not detected at this concentration; S: sensitivity; R: resistant.

4. Discussion

A vital and important component of the human diet is phenolic chemicals, which are present in all plants. Their biological activity, which includes antioxidant properties, has attracted considerable attention [81]. Ellagic acid, a phenolic compound found in large amounts in dicotyledonous plants, has been shown in numerous studies to possess potent anti-inflammation and antioxidant properties. Furthermore, research shows that ellagic acid can lessen damage in neurodegenerative conditions such as AD, Parkinson's disease, and cerebral ischemia by enhancing neuronal viability, reducing neuronal defects, and preventing neuronal damage [82]. A brand-new diabetes medication was made with plant flavonoids including epicatechin, catechin, and rutin, which have strong anti-inflammatory and antioxidant properties. Their combination can be improved through a mixture design experiment to produce a novel, safe, multitarget antidiabetic formulation, making it an effective combination for the management of diabetes and the associated complications. Rutin, catechin, and epicatechin all have strong antihyperglycemic properties; their synergistic combination assures a novel formulation that might actually be a viable alternative to current medications [83]. Accounting for roughly 59% of the total catechins, epigallocatechin gallate (EGCG) is the most prevalent flavanol. The beneficial effects of EGCG include its impact on metabolism, which lowers the risk of type 2 diabetes; its ability to block antimicrobial activity; and its antioxidant properties against neurodegenerative diseases such as AD [84]. In multi-infarct dementia model rats, nicotiflorin has protective effects such as energy metabolism failure, lowering memory dysfunction, and oxidative stress [85]. Astragalin has a wide spectrum of medicinal effects, including anti-inflammatory, antioxidant, neurological, cardioprotective, antidiabetic, and anticancer effects [86]. Resveratrol, quercetin, catechin, and gallic acid are examples of polyphenols that have antioxidant properties that prevent oxidative damage to DNA and inhibit LDL oxidation in vitro [87]. Antioxidant quinic acid has demonstrated anticancer activity by inducing apoptosis-mediated cytotoxicity in breast cancer cells. Additionally, it has shown a potent affinity for selectins, angiogenesis factors that are elevated in breast cancer tissue [88]. Tannic acid has antimutagenic and anticancer properties. Microorganisms can be killed by tannic acid (bacteria and viruses). Additionally, it functions as a homeostatic agent and an antioxidant. Tannic acid also has the ability to reduce the development of free radicals, which are responsible for a number of diseases, including those that affect the cardiovascular system, Parkinson's disease, diabetes, and AD. Tannic acid also has demonstrated anticancer properties. Tannic acid is currently being researched as an organic polymer additive owing to its bioactive characteristics and its ability to improve the capabilities of materials for biomedical applications [89]. By restoring the normal expression levels of the genes related to insulin signaling and glucose metabolism that were disturbed in the liver of high-fat-diet-induced obese mice, hesperidin has the potential to have an antidiabetic effect [90].

An important indication of a compound's potential antioxidant activity may be found in the reduction capacity of that substance. ROS and free radicals are capable of receiving electron donations from antioxidant compounds, which converts them into more stable and unreactive species [91]. The diversity, high amount of ingredients, and rich phenolic contents might contribute to the antioxidant potential of *P. granatum*. The reduction potentials of phenolic compounds in *P. granatum* were determined with reduction systems, including Cu^{2+}, Fe^{3+}, and Fe^{3+}-TPTZ reducing abilities. The radical scavenging properties of *P. granatum* ethanol extracts was examined by DPPH and ABTS radical scavenging assays. *P. granatum* possesses reducing properties, which may neutralize oxidants and ROS.

The reduction of $Fe^{3+}(CN^{-})_6$ to $Fe^{2+}(CN^{-})_6$ and the absorbance resulting formation of Perl's Prussian Blue complex after the addition of excess ferric ions (Fe^{3+}) were used to measure the ability of *P. granatum* extracts to reduce Fe^{3+}. The reducing power assay described by Oyaizu [44], with a minor modification, was applied to assess the reducing ability of *P. granatum* extracts [92]. In this assay, Fe^{3+} was converted to Fe^{2+} in the presence of reductants or plant extracts [93]. The addition of Fe^{3+} to compounds caused an $Fe_4[Fe(CN^-)_6]_3$ complex, with maximum absorption at 700 nm [94].

The chromogenic oxidant of neocuproine (Nc) was used in the CUPRAC method. Antioxidants reduce the cupric neocuproine complex [Cu(II)-Nc] to the cuprous neocuproine complex [Cu(I)-Nc], which exhibits maximum absorbance at 450 nm [95]. The CUPRAC method is a convenient, inexpensive, selective, stable of antioxidants [96,97]. The reducing capacity of pure compounds or plant extracts can be determined using the FRAP test. A ferric salt is utilized as an oxidant in the electron transfer process, which is the basis of the FRAP test. Due to its colored combination with TPTZ, which exhibits maximum absorbance at 593 nm, Fe^{2+} may be recorded spectrophotometrically [98]. The reducing capacity can be effectively ascertained using this method. First, in a redox-linked colorimetric reaction, the FRAP assay uses the sample's antioxidants as reductants. Second, the FRAP assay procedure is fairly straightforward and is simple to standardize. The FRAP assay was created to assess the ability of biological fluids and aqueous solutions of pure compounds to reduce ferric ions. It has also been used to assess the antioxidant capacity of polyphenols [99]. In this study, we determined the Fe^{3+}, Cu^{2+}, and Fe^{3+}-TPTZ reducing abilities of aqueous extract of *P. granatum* peel as concentration-dependent (10–30 μg/mL). In this test, the test solution's color changed from yellow to various shades of green and blue depending on the antioxidant samples' reducing power. A compound's reducing capacity might be a good predictor of its potential antioxidant action.

In terms of the harm caused to living organisms by free radicals and ROS, radical scavenging is very important [100]. Due to its quick analysis time compared to other techniques, DPPH's scavenging ability for free radicals has been commonly used to assess antioxidant activity [101]. For example, the $DPPH^{\bullet}$ test, which is based on scavenging of DPPH radicals to the non-radical form of DPPH-H, is commonly used to determine antioxidant activity [102,103]. A freshly made DPPH solution displays a deep purple hue with an absorption peak at 517 nm. When an antioxidant is present in the medium, this purple color typically vanishes. An indicator of the amount of free DPPH that has been reduced by the antioxidant is a decrease in absorbance [104]. As observed in this and previous studies, *P. granatum* has a comparable or better antioxidant potential relative to standard antioxidants. In another study, the IC_{50} values of acetone and ethanol extracts of *P. granatum* peel for DPPH scavenging activity were found to be 1.56 and 7.09 μg/mL, respectively [77]. The IC_{50} of methanolic extract of *P. granatum* for DPPH radical scavenging was reported to be 0.16 ± 0.07 mg/mL [105]. The IC_{50} values of aqueous and ethanolic extracts from *P. granatum* fruit peel for DPPH radical scavenging were found to be 471.7 and 509.16 g/mL, respectively, [106]. All analyses were performed in triplicate.

The ABTS radicals were produced in an ABTS/$K_2S_2O_8$ system. The test is a decolorization approach in which the ABTS radical is created directly in a stable state prior to treatment with suspected antioxidants. The improved approach for producing $ABTS^{\bullet+}$ reported here involves the direct creation of a blue/green $ABTS^{\bullet+}$ chromophore via a reaction between ABTS and $K_2S_2O_8$ [107]. One spectrophotometric technique used to assess the overall antioxidant ability of pure materials, mixtures, and beverages is based on the generation of an ABTS radical cation [108]. When compared to positive controls, the data clearly reveal that *P. granatum* approximated an effective $ABTS^{\bullet+}$ scavenging ability. *P. granatum* samples showed a radical scavenging effect higher than that of reference standard antioxidants. A lower IC_{50} value, as in DPPH free radical scavenging activity, suggests more $ABTS^{\bullet+}$ scavenging ability.

α-Glycosidase plays a crucial role in the metabolism of carbohydrates and is associated with diabetes, cancer, and viral infections. Because of its numerous biological functions, α-Glycosidase is regarded as a promising drug target [109]. Several α-glycosidase inhibitors have recently been found and are currently being researched. Acarbose and miglitol, two commonly prescribed diabetes medications, competitively inhibit α-glycosidase in the brush border of the small intestine. This prevents the hydrolysis of carbohydrates and reduces postprandial hyperglycemia [110]. α-Glycosidase inhibitors may play a significant role in the therapeutic approach to type 2 diabetes mellitus. Postprandial hyperglycemia is a notable and early defect in diabetic diseases, and lowering blood glucose levels can slow the progression of secondary complications related to diabetic diseases [111]. The results reveal that ethanol extract of *P. granatum* has less inhibitory effects than that of acarbose (IC_{50}: 22,800 nM) [63]. According to various subsequent studies, IC_{50} value of *P. granatum* peel extract for inhibition of α-glycosidase activity was 5.56 2.23 μg/mL. Punicalagins may be responsible for this activity [112]. The ethanolic extract of *P. granatum* fruit peel demonstrated concentration-dependent inhibition of α-glucosidase, with activity ranging from 53.34 2.0 to 15.18 1.4 U/L. Aqueous extract, on the other hand, showed activity ranging from 65.48 1.8 to 20.2 1.3 U/L at the different tested concentrations [105]. The results of α-glycosidase inhibition of *P. granatum* extract are quite significant and indicate potential use of *P. granatum* for DM disease.

In order to properly digest carbohydrates, digestive enzymes such as α-amylase and α-glycosidase are essential glycoside hydrolases. Both of these enzymes are found on the cells that line the intestine, where they hydrolyze polysaccharides into monosaccharide units that can be absorbed. Certain inhibitors can block the actions of both digestive enzymes to reduce body weight and regulate blood glucose levels. A relatively safe source of inhibitors is plant-based food [113]. Because α-amylase plays a significant role in the digestion of dietary starches, its inhibition helps to prevent and control postprandial hyperglycemia. As a result, numerous studies have looked into and discovered the inhibition of α-amylase by natural products, such as plant extracts, in recent years [114]. *P. granatum* peel extracts in both aqueous and methanolic form were found to have no effect on the enzyme α-amylase in earlier research [115]. The acetone extract of *P. granatum* peel demonstrated excellent α-amylase inhibitory glycemic control potential, as well as dose-dependent but moderate antiglycation activity (IC_{50}: 16.2 5.6 μg/mL), with 61% inhibition at 80 g/mL [77]. Measurements of the in vitro inhibition of α-glucosidase and α-amylase by *P. granatum* bark extracts were performed at two different concentrations (166 and 332 μg/mL) [115].

The most common and primary cause of dementia in the elderly is AD, a common neurodegenerative disease. The most significant biochemical change associated with AD is a decrease in AChE levels in the brain [116]. According to studies, the decline in acetyltransferase activity and choline (Ch) causes acetylcholine (ACh) to decrease as a neurotransmitter. As a result, cholinesterase (ChE) inhibitors have been the focus of research studies on the treatment of this illness as a symptomatic intervention [117]. AChE-inhibitory medicines are utilized in the treatment of AD. However, these medications have several undesired side effects. Therefore, research on use of novel AChE inhibitors with antioxidant ability is greatly needed [118]. It is known that the predominant AChE inhibitory effects are related to aromatic chemicals and, to a lesser extent, aliphatic molecules [119]. Although AChE inhibitors are used to treat AD, they can only bring about short-term relief. Medicinal herbs have long been known to rich in cholinesterase inhibitors. Phenolic chemicals are primarily responsible for medicinal plants' suppression of cholinergic enzymes [120]. The in vitro cholinesterase-inhibitory effect of *P. granatum* peel extract is noteworthy, and its methanol extract was found to be more effective than its ethanol extract. The higher AChE activity of methanolic (IC_{50}: 32 μg/mL) and ethanolic (IC_{50}: 42 μg/mL) extract was correlated with the bioactive metabolite content of the extracts [121]. The inhibition level of *P. granatum* ethanol extract was slightly lower compared to that of tacrine.

Numerous diseases, including glaucoma, epilepsy, edema, and altitude sickness, are caused by the ubiquitous, physiologically dominant cytosolic isoform CA II [122]. CA isoform activation and inhibition are important therapeutic targets to treat a variety of diseases, including glaucoma, cancer, edema, obesity, epilepsy, hypertension, and osteoporosis [123]. CA II suppression reduces HCO_3^- generation and, as a result, aqueous humor secretion, resulting in reduced ocular pressure [124]. Among them, glaucoma is a multifactorial optical disease that is mostly associated with high intraocular pressure (IOP), which can result in blindness. Therefore, hCA inhibitor medications such as acetazolamide, brinzolamide, and dorzolamide can reduce IOP after topical treatment [125].

One of the most prevalent Gram-positive bacteria that causes food poisoning among is *Staphylococcus aureus*, which is derived people who consumed contaminate food [126]. A Gram-negative bacterium called *Escherichia coli* is a part of the typical human flora. Preservatives are required to stop its growth because an enterohemorrhagic strain of *E. coli* has been implicated in severe cases of food poisoning [14]. Bacteria (*E. coli* and *S. aureus*) were resistant to all four antibiotics used as standards. Although some of them had a zone diameter of 10 mm, etc., they were considered resistant because they could not reach the standard sensitivity diameter according to the CLSI criteria.

Some of the extracts were resistant (R) because they did not form any zone diameter (N.D.). However, in some extracts, zones with 7–10 mm intervals, that is, areas in which the extract had an antimicrobial effect and the bacteria were destroyed, were observed. Discs with diameters of 7–10 mm formed at 50 μg (concentration-adjusted) ratios on each extraction disc are very good when compared to standard antibiotics, as observed in extracts that were completely zoneless, that is, resistant.

The benefits of *P. granatum* can be increased by drinking smoothies made from minor Mediterranean crop purées and *P. granatum* juice as a good way to increase the consumption of these healthy but underutilized fruits. The effects of an ethanol extract of *P. granatum* seeds on the central nervous system (CNS) in mice were studied. The results showed that *P. granatum* extract exhibit anxiolytic activity at all doses and induced increased sleeping latency and decreased sleeping time.

The effects of an ethanolic extract of *P. granatum* seeds on the CNS of mice were studied. The results showed that *P. granatum* extract exhibited anxiolytic activity at all dose levels and induced increased sleeping latency and decreased sleeping time [127]. The flavonoids in *P. granatum* vary greatly. For instance, flavonoids in plants can be found either in free form (aglycones) or linked to sugars. Glycosylated flavonoids are the most common, and glycosylated anthocyanidins, for example, are recognized as an essential flavonoid class known as anthocyanins. Anthocyanidins are light-sensitive and have been linked to sugars. *O*-glycosides are the most common type of flavonoid glycoside, but *C*-glycosides are also present. The benefits of P. granatum can be increased by drinking smoothies made from minor Mediterranean crop purées and *P. granatum* juice as a good way to increase consumption of these healthy but underutilized fruits [128,129].

5. Conclusions

Zivzik pomegranate (*Punica granatum*) has various qualities and contains quantities of bioactive secondary metabolites, phenolics, and flavonoids. This product, which is rich, nutritious, and contributes to human health, has been used since prehistoric times. In LC-MS/MS analysis, the major components detected in *P. granatum* extracts were ellagic acid, catechin, epigallocatechin gallate, epicatechin, nicotiflorin, astragalin, gallic acid, epigallocatechin, quinic acid, tannic acid, aconitic acid, hesperidin, isoquercitrin, rutin, fumaric acid, cosmosiin, luteolin, and epicatechin gallate. Furthermore, the *P. granatum* ethanol extract was found to be rich in phenolic contents, antioxidant ability, reducing power, AChE, α-glycosidase, α-amylase, and hCA II inhibition. *P. granatum* can also be used as a natural remedy to treat severe T2DM, AD, and glaucoma disease, as well as in food and pharmaceutical applications. From this perspective, inhibition studies on the AChE enzyme are planned to determine the anti-Alzheimer effects of WEZP and EEZP. In addition, the inhibition of CA II enzyme was analyzed to determine the link with glaucoma. Similarly, some studies have been carried out to identify the antidiabetic potential of *P. granatum* extracts on α-amylase and α-glycosidase. Additionally, Fe^{2+}, Cu^{2+}, and Fe^{3+}-TPTZ reduction, as well as DPPH and ABTS scavenging, tests were performed to understand the antioxidant potential of *P. granatum*. Furthermore, total phenolic and flavonoid contents in *P. granatum* were established for both extracts. Finally, an analysis of the phenolic compounds was performed via LC-MS/MS to define the biological activity of the chemical profile of *P. granatum*. However, the possible cytotoxic or other undesirable effects of *P. granatum* should be more comprehensively detailed in the future.

Author Contributions: Conceptualization, H.K., E.İ., İ.G. and E.K.; methodology and investigation, H.K., E.İ., İ.G. and E.K; software, validation, and visualization, H.K., E.İ., İ.G. and E.K; data curation, writing—original draft preparation, writing—review and editing, supervision, funding, and acquisition, H.K., E.İ., İ.G. and E.K. All authors have read and agreed to the published version of the manuscript.

Funding: This research received no external funding.

Institutional Review Board Statement: Not applicable.

Informed Consent Statement: Not applicable.

Data Availability Statement: Data are publicly available in an accessible repository.

Acknowledgments: Scientific Research Projects Coordination at Siirt University (2021-SÜSBF-040) provided funding for this study.

Conflicts of Interest: The authors declare no conflict of interest.

References

1. Kandylis, P.; Kokkinomagoulos, E. Food applications and potential health benefits of pomegranate and its derivatives. *Foods* **2020**, *9*, 122. [CrossRef] [PubMed]
2. Nikdel, K.; Seifi, E.; Babaie, H.; Sharifani, M.; Hemmati, K. Physicochemical properties and antioxidant activities of five Iranian pomegranate cultivars (*Punica granatum* L.) in maturation stage. *Acta Agric. Slov.* **2016**, *107*, 277–286. [CrossRef]
3. Esther, L.D.; Khusro, A.; Immanuel, P.; Esmail, G.A.; Al-Dhabi, N.A.; Arasu, M.V. Photo-activated synthesis and characterization of gold nanoparticles from *Punica granatum* L. seed oil: An assessment on antioxidant and anticancer properties for functional yoghurt nutraceuticals. *J. Photochem. Photobiol.* **2020**, *206*, 111868. [CrossRef] [PubMed]
4. Qu, W.; Li, P.; Hong, J.; Hong, J.; Liu, Z.; Chen, Y.; Breksa, A.P.; Pan, Z. Thermal stability of liquid antioxidative extracts from pomegranate peel. *J. Sci. Food Agric.* **2014**, *94*, 1005–1012. [CrossRef]
5. Mannino, G.; Chinigo, G.; Serio, G.; Genova, T.; Gentile, C.; Munaron, L.; Bertea, C.M. Proanthocyanidins and where to find them: A meta-analytic approach to investigate their chemistry, biosynthesis, distribution, and effect on human health. *Antioxidants* **2021**, *10*, 1229. [CrossRef]
6. Borochov, N.H.; Judeinstein, S.; Tripler, E.; Harari, M.; Greenberg, A.; Shomer, I.; Holland, D. Seasonal and cultivar variations in antioxidant and sensory quality of pomegranate (*Punica granatum* L.) fruit. *J. Food Comp. Anal.* **2009**, *22*, 189–195. [CrossRef]
7. Gulcin, I.; Kaya, R.; Goren, A.C.; Akincioglu, H.; Topal, M.; Bingol, Z.; Cakmak, C.K.: Sarikaya, B.O.S.; Durmaz, L.; Alwasel, S. Anticholinergic, antidiabetic and antioxidant activities of cinnamon (*Cinnamomum verum*) bark extracts: Polyphenol contents analysis by LC-MS/MS. *Int. J. Food Prop.* **2019**, *22*, 1511–1526. [CrossRef]
8. Gulcin, I.; Bursal, E.; Sehitoglu, M.H.; Bilsel, M.; Goren, A.C. Polyphenol contents and antioxidant activity of lyophilized aqueous extract of propolis from Erzurum, Turkey. *Food Chem. Toxicol.* **2010**, *48*, 2227–2238. [CrossRef]
9. Gulcin, I.; Berashvili, D.; Gepdiremen, A. Antiradical and antioxidant activity of total anthocyanins from *Perilla pankinensis* decne. *J. Ethnopharmacol.* **2005**, *101*, 287–293. [CrossRef]
10. Gulcin, I.; Elmastas, M.; Aboul-Enein, H.Y. Antioxidant activity of clove oil-A powerful antioxidant source. *Arab. J. Chem.* **2012**, *5*, 489–499. [CrossRef]
11. Ak, T.; Gulcin, I. Antioxidant and radical scavenging properties of curcumin. *Chem. Biol. Interact.* **2008**, *174*, 27–37. [CrossRef] [PubMed]
12. Gulcin, I.; Goren, A.C.; Taslimi, P.; Alwasel, S.H.; Kilic, O.; Bursal, E. Anticholinergic, antidiabetic and antioxidant activities of Anatolian pennyroyal (*Mentha pulegium*)-Analysis of its polyphenol contents by LC-MS/MS. *Biocat. Agric. Biotechnol.* **2020**, *23*, 101441. [CrossRef]
13. Gulcin, I. Antioxidant activity of caffeic acid (3,4-dihydroxycinnamic acid). *Toxicology* **2006**, *217*, 213–220. [CrossRef] [PubMed]
14. Gulcin, I.; Kufrevioglu, O.I.; Oktay, M.; Buyukokuroglu, M.E. Antioxidant, antimicrobial, antiulcer and analgesic activities of nettle (*Urtica dioica* L.). *J. Ethnopharmacol.* **2004**, *90*, 205–215. [CrossRef]
15. Gulcin, I.; Oktay, M.; Kireçci, E.; Küfrevioğlu, Ö.I. Screening of antioxidant and antimicrobial activities of anise (*Pimpinella anisum* L.) seed extracts. *Food Chem.* **2003**, *83*, 371–382. [CrossRef]
16. Koksal, E.; Bursal, E.; Gulcin, I.; Korkmaz, M.; Çaglayan, C.; Goren, A.C.; Alwasel, S.H. Antioxidant activity and polyphenol content of Turkish thyme (*Thymus vulgaris*) monitored by LC-MS/MS. *Int. J. Food Prop.* **2017**, *20*, 514–525. [CrossRef]
17. Gulcin, I.; Tel, A.Z.; Kirecci, E. Antioxidant, antimicrobial, antifungal and antiradical activities of *Cyclotrichium niveum* (Boiss.) Manden and Scheng. *Int. J. Food Prop.* **2008**, *11*, 450–471. [CrossRef]
18. Gulcin, I. Antioxidant properties of resveratrol: A structure-activity insight. *Innov. Food Sci. Emerg.* **2010**, *11*, 210–218. [CrossRef]
19. Bursal, E.; Gulcin, I. Polyphenol contents and in vitro antioxidant activities of lyophilised aqueous extract of kiwifruit (*Actinidia deliciosa*). *Food Res. Int.* **2011**, *44*, 1482–1489. [CrossRef]
20. Bursal, E.; Koksal, E.; Gulcin, I.; Bilsel, G.; Goren, A.C. Antioxidant activity and polyphenol content of cherry stem (*Cerasus avium* L.) determined by LC-MS/MS. *Food Res. Int.* **2013**, *51*, 66–74. [CrossRef]

21. Cetin, C.K.; Gulcin, I. Anticholinergic and antioxidant activities of usnic acid-an activity-structure insight. *Toxicol. Rep.* **2019**, *6*, 1273–1280. [CrossRef] [PubMed]
22. Yemis, O.; Bakkalbasi, E.; Artik, N. Antioxidative activities of grape (*Vitis vinifera*) seed extracts obtained from different varieties grown in Turkey. *Int. J. Food Sci. Technol.* **2008**, *43*, 154–159. [CrossRef]
23. Ceylan, R.; Zengin, G.; Guler, G.O.; Aktumsek, A. Bioactive constituents of *Lathyrus czeczottianus* and ethyl acetate and water extracts and their biological activities: An endemic plant to Turkey. *S. Afr. J. Bot.* **2020**, *143*, 306–311. [CrossRef]
24. Zengin, R.; Gok, Y.; Demir, Y.; Sen, B.; Taskin-Tok, T.; Aktas, A.; Demirci, O.; Gulcin, I.; Aygun, M. Fluorinated benzimidazolium salts: Synthesis, characterization, molecular docking studies and inhibitory properties against some metabolic enzymes. *J. Fluor. Chem.* **2023**, *267*, 110094. [CrossRef]
25. Karakaya, S.; Bingol, Z.; Koca, M.; Dagoglu, S.; Pınar, N.M.; Demirci, B.; Gulcin, I.; Brestic, M.; Sytar, O. Identification of non-alkaloid natural compounds of *Angelica purpurascens* (Avé-Lall.) Gilli. (Apiaceae) with cholinesterase and carbonic anhydrase inhibition potential. *Saudi Pharm. J.* **2020**, *28*, 1–14. [CrossRef] [PubMed]
26. Mutlu, M.; Bingöl, Z.; Uc, E.M.; Koksal, E.; Goren, A.C.; Alwasel, S.H.; Gulcin, I. Comprehensive metabolite profiling of cinnamon (*Cinnamomum zeylanicum*) leaf oil using LC-HR/MS, GC/MS, and GC-FID: Determination of antiglaucoma, antioxidant, anticholinergic, and antidiabetic profiles. *Life* **2023**, *13*, 136. [CrossRef]
27. Bulut, Z.; Abul, N.; Halıc Poslu, A.; Gulcin, I.; Ece, A.; Erçag, E.; Koz, O.; Koz, G. Structural characterization and biological evaluation of uracil-appended benzylic amines as acetylcholinesterase and carbonic anhydrase I and II inhibitors. *J. Mol. Struct.* **2023**, *1280*, 135047. [CrossRef]
28. Ozcan, K. Antibacterial, antioxidant and enzyme inhibition activity capacities of *Doronicum macrolepis* (FREYN&SINT): An endemic plant from Turkey. *Saudi Pharm. J.* **2020**, *28*, 95–100.
29. Yigit, B.; Taslimi, P.; Barut Celepci, D.; Taskin-Tok, T.; Yigit, M.; Aygun, M.; Ozdemir, I.; Gulcin, I. Novel PEPPSI-type N-heterocyclic carbene palladium(II) complexes: Synthesis, characterization, in silico studies and enzyme inhibitory properties against some metabolic enzymes. *Inorg. Chim. Acta* **2023**, *544*, 121239. [CrossRef]
30. Gulcin, I.; Taslimi, P.; Aygun, A.; Sadeghian, N.; Bastem, E.; Kufrevioglu, O.I.; Turkan, F.; Şen, F. Antidiabetic and antiparasitic potentials: Inhibition effects of some natural antioxidant compounds on α-glycosidase, α-amylase and human glutathione *S*-transferase enzymes. *Int. J. Biol. Macromol.* **2018**, *119*, 741–746. [CrossRef]
31. Taslimi, P.; Koksal, E.; Goren, A.C.; Bursal, E.; Aras, A.; Kilic, O.; Alwasel, S.; Gulcin, I. Anti-Alzheimer, antidiabetic and antioxidant potential of *Satureja cuneifolia* and analysis of its phenolic contents by LC-MS/MS. *Arab. J. Chem.* **2020**, *13*, 4528–4537. [CrossRef]
32. Durmaz, L.; Erturk, A.; Akyüz, M.; Polat, K.L.; Uc, E.M.; Bingol, Z.; Saglamtas, R.; Alwasel, S.; Gulcin, I. Screening of carbonic anhydrase, acetylcholinesterase, butyrylcholinesterase, and α-glycosidase enzyme inhibition effects and antioxidant activity of coumestrol. *Molecules* **2022**, *27*, 3091. [CrossRef]
33. Karakaya, S.; Bingol, Z.; Koca, M.; Demirci, B.; Gulcin, I.; Baser, K.H.C. Screening of non-alkaloid acetylcholinesterase and carbonic anhydrase isoenzymes inhibitors of *Leiotulus dasyanthus* (K. Koch) Pimenov & Ostr. (Apiaceae). *J. Essent. Oil Res.* **2020**, *32*, 227–241.
34. Yildirim, A.; Atmaca, U.; Keskin, A.; Topal, M.; Celik, M.; Gulcin, I. N-Acylsulfonamides strongly inhibit human carbonic anhydrase isoenzymes I and II. *Bioorg. Med. Chem.* **2015**, *23*, 2598–2605. [CrossRef]
35. Durmaz, L.; Kiziltas, H.; Guven, L.; Karagecili, H.; Alwasel, S.; Gulcin, I. Antioxidant, antidiabetic, anticholinergic, and antiglaucoma effects of magnofluorine. *Molecules* **2020**, *27*, 5902. [CrossRef] [PubMed]
36. Oktay, M.; Gülçin, I.; Kufrevioglu, O.I. Determination of in vitro antioxidant activity of fennel (*Foeniculum vulgare*) seed extracts. *Lebensm. Wissen. Technol.* **2003**, *36*, 263–271. [CrossRef]
37. Gulcin, I.; Sat, I.G.; Beydemir, S.; Elmastas, M.; Kufrevioglu, O.I. Comparison of antioxidant activity of clove (*Eugenia caryophylata* Thunb) buds and lavender (*Lavandula stoechas* L.). *Food Chem.* **2004**, *87*, 393–400. [CrossRef]
38. Singleton, V.L.; Rossi, J.A. Colorimetry of total phenolics with phosphomolybdic-phosphotungstic acid reagents. *Am. J. Enol. Vitic.* **1965**, *16*, 144–158.
39. Elmastas, M.; Türkekul, I.; Ozturk, L.; Gulcin, I.; Isıldak, O.; Aboul-Enein, H.Y. The antioxidant activity of two wild edible mushrooms (*Morchella vulgaris* and *Morchella esculanta*). *Comb. Chem. High Throughput Screen.* **2006**, *9*, 443–448. [CrossRef]
40. Bursal, E.; Aras, A.; Kilic, O.; Taslimi, P.; Goren, A.C.; Gulcin, I. Phytochemical content, antioxidant activity and enzyme inhibition effect of *Salvia eriophora* Boiss. & Kotschy against acetylcholinesterase, α-amylase, butyrylcholinesterase and α-glycosidase enzymes. *J. Food Biochem.* **2019**, *43*, e12776.
41. Gulcin, I.; Topal, F.; Cakmakci, R.; Bilsel, M.; Goren, A.C.; Erdogan, U. Pomological features, nutritional quality, polyphenol content analysis, and antioxidant properties of domesticated and 3 wild ecotype forms of raspberries (*Rubus idaeus* L.). *J. Food Sci.* **2011**, *76*, 585–593. [CrossRef] [PubMed]
42. Karagecili, H.; Yilmaz, M.A.; Erturk, A.; Kiziltas, H.; Guven, L.; Alwasel, S.H.; Gulcin, I. Comprehensive metabolite profiling of Berdav propolis using LC-MS/MS: Determination of antioxidant, anticholinergic, antiglaucoma, and antidiabetic effects. *Molecules* **2023**, *28*, 1739. [CrossRef]
43. Yilmaz, M.A. Simultaneous quantitative screening of 53 phytochemicals in 33 species of medicinal and aromatic plants: A detailed, robust and comprehensive LC–MS/MS method validation. *Ind. Crops Prod.* **2020**, *149*, 112347. [CrossRef]

44. Oyaizu, M. Studies on products of browning reaction. Antioxidative activities of products of browning reaction prepared from glucosamine. *Jpn. J. Nutr. Dietet.* **1986**, *44*, 307–315. [CrossRef]
45. Gocer, H.; Gulcin, I. Caffeic acid phenethyl ester (CAPE): Correlation of structure and antioxidant properties. *Int. J. Food Sci. Nutr.* **2011**, *62*, 821–825. [CrossRef]
46. Apak, R.; Guclu, K.; Ozyurek, M.; Esin, K.S.; Ercag, E. The cupric ion reducing antioxidant capacity and polyphenolic content of some herbal teas. *Int. J. Food Sci. Nutr.* **2006**, *57*, 292–304. [CrossRef]
47. Koksal, E.; Gulcin, I. Antioxidant activity of cauliflower (*Brassica oleracea* L.). *Turk. J. Agric. For.* **2008**, *32*, 65–78.
48. Talaz, O.; Gulcin, I.; Goksu, S.; Saracoglu, N. Antioxidant activity of 5,10-dihydroindeno[1,2-b]indoles containing substituents on dihydroindeno part. *Bioorg. Med. Chem.* **2009**, *17*, 6583–6589. [CrossRef]
49. Topal, M.; Gocer, H.; Topal, F.; Kalin, P.; Polat Kose, P.; Gulcin, I.; Cetin Cakmak, K.C.; Kucuk, M.; Durmaz, L.; Goren, A.C.; et al. Antioxidant, antiradical and anticholinergic properties of cynarin purified from the illyrian thistle (*Onopordum illyricum* L.). *J. Enzyme Inhib. Med. Chem.* **2006**, *31*, 266–275. [CrossRef]
50. Cetinkaya, Y.; Gocer, H.; Menzek, A.; Gulcin, I. Synthesis and antioxidant properties of (3,4-dihydroxyphenyl)(2,3,4- trihydroxyphenyl)methanone and its derivatives. *Arch. Pharm.* **2012**, *345*, 323–334. [CrossRef]
51. Gulcin, I.; Beydemir, S.; Sat, I.G.; Kufrevioglu, O.I. Evaluation of antioxidant activity of cornelian cherry (*Cornus mas* L.). *Acta Aliment. Hung.* **2005**, *34*, 193–202. [CrossRef]
52. Blois, M.S. Antioxidant determinations by the use of a stable free radical. *Nature* **1958**, *181*, 1199–1200. [CrossRef]
53. Gulcin, I. The antioxidant and radical scavenging activities of black pepper (*Piper nigrum*) seeds. *Int. J. Food Sci. Nutr.* **2005**, *56*, 491–499. [CrossRef] [PubMed]
54. Gulcin, I.; Dastan, A. Synthesis of dimeric phenol derivatives and determination of in vitro antioxidant and radical scavenging activities. *J. Enzyme Inhib. Med. Chem.* **2007**, *22*, 685–695. [CrossRef] [PubMed]
55. Ancerewicz, J.; Migliavacca, E.; Carrupt, P.A.; Testa, B.; Brée, F.; Zini, R.; Tillement, J.P.; Labidalle, S.; Guyot, D.; Chauvet-Monges, A.M.; et al. Structure-property relationships of trimetazidine derivatives and model compounds as potential antioxidants. *Free Radic. Biol. Med.* **1998**, *25*, 113–120. [CrossRef] [PubMed]
56. Gulcin, I.; Sat, I.G.; Beydemir, S.; Kufrevioglu, O.I. Evaluation of the in vitro antioxidant properties of extracts of broccoli (*Brassica oleracea* L.). *Ital. J. Food Sci.* **2004**, *16*, 17–30.
57. Re, R.; Pellegrini, N.; Proteggente, A.; Pannala, A.; Yang, M.; Rice-Evans, C. Antioxidant activity applying an improved ABTS radical cation decolorization assay. *Free Radic. Biol. Med.* **1999**, *26*, 1231–1237. [CrossRef]
58. Gulcin, I.; Mshvildadze, V.; Gepdiremen, A.; Elias, R. Screening of antiradical and antioxidant activity of monodesmosides and crude extract from *Leontice smirnowii* tuber. *Phytomedicine* **2006**, *13*, 343–351. [CrossRef]
59. Cakmakci, S.; Topdas, E.F.; Kalın, P.; Han, H.; Sekerci, P.; Polat Kose, L.; Gulcin, I. Antioxidant capacity and functionality of oleaster (*Elaeagnus angustifolia* L.) flour and crust in a new kind of fruity ice cream. *Int. J. Food Sci. Technol.* **2015**, *50*, 472–481. [CrossRef]
60. Ellman, G.L.; Courtney, K.D.; Andres, V.; Featherstone, R.M. A new and rapid colorimetric determination of acetylcholinesterase activity. *Biochem. Pharmacol.* **1961**, *7*, 2952. [CrossRef]
61. Ozbey, F.; Taslimi, P.; Gulcin, I.; Maras, A.; Goksu, S.; Supuran, C.T. Synthesis of diaryl ethers with acetylcholinesterase, butyrylcholinesterase and carbonic anhydrase inhibitory actions. *J. Enzyme Inhib. Med. Chem.* **2016**, *31*, 79–85. [CrossRef] [PubMed]
62. Savikin, K.; Zivkovic, J.; Alimpic, A.; Zdunic, G.; Jankovic, T.; Duletic-Lausevic, S.; Menkovic, N. Activity guided fractionation of pomegranate extract and its antioxidant, antidiabetic and antineurodegenerative properties. *Ind. Crops Prod.* **2018**, *113*, 142–149. [CrossRef]
63. Tao, Y.; Zhang, Y.; Cheng, Y.; Wang, Y. Rapid screening and identification of α-glucosidase inhibitors from mulberry leaves using enzyme-immobilized magnetic beads coupled with HPLC/MS and NMR. *Biomed. Chromatogr.* **2013**, *27*, 148–155. [CrossRef] [PubMed]
64. Gondolova, G.; Taslimi, P.; Medjidov, A.; Farzaliyev, V.; Sujayev, A.; Huseynova, M.; Sahin, O.; Yalcın, B.; Turkan, F.; Gulcin, I. Synthesis, crystal structure and biological evaluation of spectroscopic characterization of Ni(II) and Co(II) complexes with N-salicyloil-N′-maleoil-hydrazine as anticholinergic and antidiabetic agents. *J. Biochem. Mol. Toxicol.* **2018**, *32*, e22197. [CrossRef] [PubMed]
65. Xiao, Z.; Storms, R.; Tsang, A. A quantitative starch-iodine method for measuring alpha-amylase and glucoamylase activities. *Anal. Biochem.* **2006**, *351*, 146–148. [CrossRef] [PubMed]
66. Kocyigit, U.M.; Budak, Y.; Gurdere, M.B.; Erturk, F.; Yencilek, B.; Taslimi, P.; Gulcin, I.; Ceylan, M. Synthesis, characterization, anticancer, antimicrobial and carbonic anhydrase inhibition profiles of novel (3aR,4S,7R,7aS)-2-(4-((E)-3-(3-aryl)acryloyl) phenyl)-3a,4,7,7a-tetrahydro-1H-4,7-methanoisoindole-1,3(2H)-dione derivatives. *Bioorg. Chem.* **2017**, *70*, 118–125. [CrossRef]
67. Bradford, M.M. A rapid and sensitive method for the quantitation of microgram quantities of protein utilizing the principle of protein-dye binding. *Anal. Biochem.* **1976**, *67*, 248–254. [CrossRef]
68. Verpoorte, J.A.; Mehta, S.; Edsall, J.T. Esterase activities of human carbonic anhydrases B and C. *J. Biol. Chem.* **1967**, *242*, 4221–4229. [CrossRef]

69. Huseynova, M.; Taslimi, P.; Medjidov, A.; Farzaliyev, V.; Aliyeva, M.; Gondolova, G.; Sahin, O.; Yalcın, B.; Sujayev, A.; Orman, E.B.; et al. Synthesis, characterization, crystal structure, electrochemical studies and biological evaluation of metal complexes with thiosemicarbazone of glyoxylic acid. *Polyhedron* **2018**, *155*, 25–33. [CrossRef]
70. Tohma, H.; Koksal, E.; Kılıc, O.; Alan, Y.; Yılmaz, M.A.; Gulcin, I.; Bursal, E.; Alwasel, S.H. RP-HPLC/MS/MS analysis of the phenolic compounds, antioxidant and antimicrobial activities of Salvia L. species. *Antioxidants* **2016**, *5*, 38. [CrossRef]
71. Deniz, S.; Büyük, F.; Murat, K. Investigation of CTX-M, TEM and SHV type extended spectrum beta-lactamase activity with automated system and molecular methods in *Escherichia coli* and *Klebsiella pneumonia* strains isolated from individuals with urinary track infections. *Kafkas J. Med. Sci.* **2021**, *11*, 307–317. [CrossRef]
72. Kocyigit, U.M.; Budak, Y.; Gurdere, M.B.; Erturk, F.; Yencilek, B.; Taslimi, P.; Gulcin, I.; Ceylan, M. Synthesis of chalcone-imide derivatives and investigation of their anticancer and antimicrobial activities, carbonic anhydrase and acetylcholinesterase enzymes inhibition profiles. *Arch. Physiol. Biochem.* **2018**, *124*, 61–68. [CrossRef]
73. Kocyigit, U.M.; Okten, S.; Cakmak, O.; Burhan, G.; Atas, M.; Taslimi, P.; Gulcin, I. Arylated quinoline and tetrahydroquinolines: Synthesis, characterization and their metabolic enzyme inhibitory and antimicrobial activities. *ChemistrySelect* **2022**, *7*, 202203469. [CrossRef]
74. Koksal, E.; Tohma, S.H.; Kılıc, O.; Alan, Y.; Aras, A.; Gulcin, I.; Bursal, E. Assessment of antimicrobial and antioxidant activities of *Nepeta trachonitica*-Analysis of its phenolic compounds using HPLC-MS/MS. *Sci. Pharm.* **2017**, *15*, 24. [CrossRef]
75. Limbago, B. M100-S11, Performance standards for antimicrobial susceptibility testing. *Clin. Microbiol. Newslett.* **2001**, *23*, 49.
76. Chater, J.M.; Garner, L.C. Foliar nutrient applications to 'Wonderful' pomegranate (*Punica granatum* L.). I. Effects on fruit mineral nutrient concentrations and internal quality. *Sci. Horticul.* **2019**, *244*, 421–427. [CrossRef]
77. Chukwuma, C.I.; Mashele, S.S.; Akuru, E.A. Evaluation of the in vitro α-amylase inhibitory, antiglycation, and antioxidant properties of *Punica granatum* L. (pomegranate) fruit peel acetone extract and its effect on glucose uptake and oxidative stress in hepatocytes. *J. Food Biochem.* **2020**, *44*, 1–14. [CrossRef]
78. Kucukoglu, K.; Gul, H.I.; Taslimi, P.; Gulcin, I.; Supuran, C.T. Investigation of inhibitory properties of some hydrazone compounds on hCA I, hCA II and AChE enzymes. *Bioorg. Chem.* **2019**, *86*, 316–321. [CrossRef]
79. Bicer, A.; Taslimi, P.; Yakali, G.; Gulcin, I.; Gultekin, M.S.; Turgut Cin, G. Synthesis, characterization, crystal structure of novel bis-thiomethylcyclohexanone derivatives and their inhibitory properties against some metabolic enzymes. *Bioorg. Chem.* **2019**, *82*, 393–404. [CrossRef]
80. Silva, J.C.; Rodrigues, S.; Feas, X.; Estevinho, L.M. Antimicrobial activity, phenolic profile and role in the inflammation of propolis. *Food Chem. Toxicol.* **2012**, *50*, 1790–1795. [CrossRef]
81. Kalın, P.; Gulcin, I.; Goren, A.C. Antioxidant activity and polyphenol content of cranberries (*Vaccinium macrocarpon*). *Rec. Nat. Prod.* **2015**, *9*, 496–502.
82. Zhu, H.; Yan, Y.; Jiang, Y.; Meng, X. Ellagic acid and its anti-aging effects on central nervous system. *Int. J. Mol. Sci.* **2022**, *23*, 10937. [CrossRef] [PubMed]
83. Mechchate, H.; Es-safi, I.; Haddad, H.; Bekkari, H.; Grafov, A.; Bousta, D. Combination of catechin, epicatechin, and rutin: Optimization of a novel complete antidiabetic formulation using a mixture design approach. *J. Nutr. Biochem.* **2021**, *88*, 108520. [CrossRef]
84. Andreu-Fernandez, V.; Toledano, L.A.; Pizarro, N.; Navarro-Tapia, E.; Gomez-Roig, M.D.; de la Torre, R.; Garcia-Algar, O. Bioavailability of epigallocatechin gallate administered with different nutritional strategies in healthy volunteers. *Antioxidants* **2020**, *9*, 440. [CrossRef] [PubMed]
85. Huang, J.L.; Fu, S.T.; Jiang, Y.Y.; Cao, Y.B.; Guo, M.L.; Wang, Y.; Xu, Z. Protective effects of nicotiflorin on reducing memory dysfunction, energy metabolism failure and oxidative stress in multi-infarct dementia model rats. *Pharmacol. Biochem. Behav.* **2007**, *86*, 741–748. [CrossRef]
86. Araujo-Padilla, X.; Ramón-Gallegos, E.; Díaz-Cedillo, F.; Silva-Torres, R. Astragalin identification in graviola pericarp indicates a possible participation in the anticancer activity of pericarp crude extracts: In vitro and in silico approaches. *Arab. J. Chem.* **2022**, *15*, 103720. [CrossRef]
87. Cotea, V.V.; Luchian, C.E.; Bilba, N.; Niculaua, M. Mesoporous silica SBA-15, a new adsorbent for bioactive polyphenols from red wine. *Anal. Chim. Acta* **2012**, *732*, 180–185. [CrossRef]
88. Samimi, S.; Ardestani, M.S.; Dorkoosh, F.A. Preparation of carbon quantum dots-quinic acid for drug delivery of gemcitabine to breast cancer cells. *J. Drug Deliv. Sci. Technol.* **2021**, *61*, 102287. [CrossRef]
89. Kaczmarek, B. Tannic acid with antiviral and antibacterial activity as a promising component of biomaterials-A minireview. *Materials* **2020**, *13*, 3224. [CrossRef]
90. Rajan, P.; Natraj, P.; Ranaweera, S.S.; Dayarathne, L.A.; Lee, Y.J.; Han, C.H. Anti-diabetic effect of hesperidin on palmitate (PA)-treated HepG2 cells and high fat diet-induced obese mice. *Food Res. Int.* **2022**, *162*, 112059. [CrossRef]
91. Gulcin, I.; Elias, R.I.; Gepdiremen, A.; Boyer, L.; Koksal, E. A comparative study on the antioxidant activity of fringe tree (*Chionanthus virginicus* L.) extracts. *Afr. J. Biotechnol.* **2007**, *6*, 410–418.
92. Rezai, M.; Bayrak, C.; Taslimi, P.; Gulcin, I.; Menzek, A. The first synthesis, antioxidant and anticholinergic activities of 1-(4,5-dihydroxybenzyl)pyrrolidin-2-one derivative bromophenols including natural products. *Turk. J. Chem.* **2018**, *42*, 808–825.
93. Gulcin, I.; Topal, F.; Ozturk Sarikaya, S.B.; Bursal, E.; Goren, A.C.; Bilsel, M. Polyphenol contents and antioxidant properties of medlar (*Mespilus germanica* L.). *Rec. Nat. Prod.* **2011**, *5*, 158–175.

94. Han, H.; Yılmaz, H.; Gulcin, I. Antioxidant activity of flaxseed (*Linum usitatissimum* L.) and analysis of its polyphenol contents by LC-MS/MS. *Rec. Nat. Prod.* **2018**, *12*, 397–402. [CrossRef]
95. Karaman, Ş.; Tutem, E.; Sozgen, B.K.; Apak, R. Comparison of total antioxidant capacity and phenolic composition of some apple juices with combined HPLC-CUPRAC assay. *Food Chem.* **2010**, *120*, 1201–1209. [CrossRef]
96. Gulcin, I. Measurement of antioxidant ability of melatonin and serotonin by the DMPD and CUPRAC methods as trolox equivalent. *J. Enzyme Inhib. Med. Chem.* **2008**, *23*, 871–876. [CrossRef] [PubMed]
97. Apak, R.; Calokerinos, A.; Gorinstein, S.; Segundo, M.A.; Hibbert, D.B.; Gulcin, I.; Demirci Cekic, S.; Güçlü, K.; Özyürek, M.; Çelik, S.E.; et al. Methods to Evaluate the Scavenging Activity of antioxidants towards reactive oxygen and nitrogen species (IUPAC Technical Report). *Pure Appl. Chem.* **2022**, *94*, 87–144. [CrossRef]
98. Benzie, I.F.F.; Strain, J.J. The ferric reducing ability of plasma (FRAP) as a measure of "antioxidant power": The FRAP assay. *Anal. Biochem.* **1996**, *239*, 70–76. [CrossRef] [PubMed]
99. Pulido, R.; Bravo, L.; Saura-Calixto, F. Antioxidant activity of dietary polyphenols as determined by a modified ferric reducing/antioxidant power assay. *J. Agric. Food Chem.* **2000**, *48*, 3396–3402. [CrossRef]
100. Kiziltas, H.; Goren, A.C.; Bingol, Z.; Alwasel, S.H.; Gulcin, I. Anticholinergic, antidiabetic and antioxidant activities of *Ferula oriantalis* L. determination of its polyphenol contents by LC-HRMS. *Rec. Nat. Prod.* **2021**, *15*, 513–528. [CrossRef]
101. Eruygur, N.; Kocyigit, U.M.; Taslimi, P.; Atas, M.; Tekin, M.; Gulcin, I. Screening the in vitro antioxidant, antimicrobial, anticholinesterase, antidiabetic activities of endemic *Achillea cucullata* (Asteraceae) ethanol extract. *S. Afr. J. Bot.* **2019**, *120*, 141–145. [CrossRef]
102. Mitra, K.; Uddin, N. Total phenolics, flavonoids, proanthrocyanidins, ascorbic acid contents and in-vitro antioxidant activities of newly developed isolated soya protein. *Discour. J. Agric. Food Sci.* **2014**, *2*, 160–168.
103. Sheng, J.; Zhou, J.; Wang, L.; Xu, J.; Hu, Q. Antioxidant activity of ethanol and petroleum ether extracts from Brazilian propolis. *Eur. Food Res. Technol.* **2007**, *225*, 249–253. [CrossRef]
104. Gulcin, I. Antioxidant activity of food constituents: An overview. *Arch. Toxicol.* **2012**, *86*, 345–391. [CrossRef] [PubMed]
105. Shahbazi, Y. Antibacterial and antioxidant properties of methanolic extracts of apple (*Malus pumila*), grape (*Vitis vinifera*), poegranate (*Punica granatum* L.) and common fig (*Ficus carica* L.) fruits. *Pharm. Sci.* **2017**, *24*, 308–315. [CrossRef]
106. Mayasankaravalli, C.; Deepika, K.; Lydia, E.D.; Agada, R.; Thagriki, D.; Govindasamy, C.; Chinnadurai, V.; Gatar, O.M.O.; Khusro, A.; Kim, Y.O.; et al. Profiling the phyto-constituents of *Punica granatum* fruits peel extract and accessing its in-vitro antioxidant, anti-diabetic, anti-obesity, and angiotensin-converting enzyme inhibitory properties. *Saudi J. Biol. Sci.* **2020**, *27*, 3228–3234. [CrossRef] [PubMed]
107. Gulcin, I.; Beydemir, S.; Topal, F.; Gagua, N.; Bakuridze, A.; Bayram, R.; Gepdiremen, A. Apoptotic, antioxidant and antiradical effects of majdine and isomajdine from *Vinca herbacea* Waldst. and kit. *J. Enzyme Inhib. Med. Chem.* **2012**, *27*, 587–594. [CrossRef] [PubMed]
108. Gulcin, I.; Huyut, Z.; Elmastas, M.; Aboul-Enein, H.Y. Radical scavenging and antioxidant activity of tannic acid. *Arab. J. Chem.* **2010**, *3*, 43–53. [CrossRef]
109. Hashmi, S.; Khan, S.; Shafiq, Z.; Taslimi, P.; Ishaq, M.; Sadeghian, N.; Karaman, S.H.; Akhtar, N.; Islam, M.; Ansari, A.; et al. Probing 4-(diethylamino)-salicylaldehyde-based thiosemicarbazones as multi-target directed ligands against cholinesterases, carbonic anhydrases and α-glycosidase enzymes. *Bioorg. Chem.* **2021**, *107*, 104554. [CrossRef]
110. Taslimi, P.; Caglayan, C.; Farzaliyev, F.; Nabiyev, O.; Sujayev, A.; Türkan, F.; Kaya, R.; Gulcin, I. Synthesis and discovery of potent carbonic anhydrase, acetylcholinesterase, butyrylcholinesterase and α-glycosidase enzymes inhibitors: The novel N,N′-bis-cyanomethylamine and alkoxymethylamine derivatives. *J. Biochem. Mol. Toxicol.* **2018**, *32*, e22042. [CrossRef]
111. Taslimi, P.; Gulcin, I. Antidiabetic potential: In vitro inhibition effects of some natural phenolic compounds on α-glycosidase and α-amylase enzymes. *J. Biochem. Mol. Toxicol.* **2017**, *31*, e21956. [CrossRef] [PubMed]
112. Cam, M.; Icyer, N.C. Phenolics of pomegranate peels: Extraction optimization by central composite design and alpha glucosidase inhibition potentials. *J. Food Sci. Technol.* **2015**, *52*, 1489–1497. [CrossRef] [PubMed]
113. Gulcin, I.; Tel, A.Z.; Goren, A.C.; Taslimi, P.; Alwasel, S. Sage (*Salvia pilifera*): Determination its polyphenol contents, anticholinergic, antidiabetic and antioxidant activities. *J. Food Meas. Charact.* **2019**, *13*, 2062–2074. [CrossRef]
114. Kiziltas, H.; Gören, A.C.; Alwasel, S.; Gulcin, I. Sahlep (*Dactylorhiza osmanica*): Phytochemical analyses by LC-HRMS, molecular docking, antioxidant activity and enzyme inhibition profiles. *Molecules* **2022**, *27*, 6907. [CrossRef]
115. Laaraj, N.; Bouhrim, M.; Kharchoufa, L.; Tiji, S.; Bendaha, H.; Addi, M.; Drouet, S.; Hano, C.; Lorenzo, J.M.; Bnouham, M.; et al. Phytochemical analysis, α-glucosidase and α-amylase inhibitory activities and acute toxicity studies of extracts 797 from pomegranate (*Punica granatum*) bark, a valuable agro-industrial by-product. *Foods* **2022**, *11*, 1353. [CrossRef]
116. Akincioglu, A.; Akincioglu, H.; Gulcin, I.; Durdagi, S.; Supuran, C.T.; Goksu, S. Discovery of potent carbonic anhydrase and acetylcholine esterase inhibitors: Novel sulfamoylcarbamates and sulfamides derived from acetophenones. *Bioorg. Med. Chem.* **2015**, *23*, 3592–3602. [CrossRef]
117. Kiziltas, H.; Goren, A.C.; Alwasel, S.; Gulcin, I. Comprehensive metabolic profiling of Acantholimon caryophyllaceum using LC-HRMS and evaluation of antioxidant activities, enzyme inhibition properties and molecular docking studies. *S. Afr. J. Bot.* **2022**, *151*, 743–751. [CrossRef]

118. Gulcin, I.; Scozzafava, A.; Supuran, C.T.; Akıncıoglu, H.; Koksal, Z.; Turkan, F.; Alwasel, S. The effect of caffeic acid phenethyl ester (CAPE) metabolic enzymes including acetylcholinesterase, butyrylcholinesterase, glutathione s-transferase, lactoperoxidase and carbonic anhydrase isoenzymes I, II, IX and XII. *J. Enzyme Inhib. Med. Chem.* **2016**, *31*, 1095–1101. [CrossRef]
119. Holth, T.F.; Tollefsen, K.E. Acetylcholine esterase inhibitors in effluents from oil production platforms in the North Sea. *Aquat. Toxicol.* **2012**, *112–113*, 92–98. [CrossRef]
120. Bursal, E.; Taslimi, P.; Goren, A.; Gulcin, I. Assessments of anticholinergic, antidiabetic, antioxidant activities and phenolic content of *Stachys annua. Biocat. Agric. Biotechnol.* **2020**, *28*, 101711. [CrossRef]
121. Konsoula, Z. A preliminary in vitro investigation of anticholinesterase activity of pomegranate peel extracts. *J. Biotechnol.* **2018**, *280*, S88. [CrossRef]
122. Scozzafava, A.; Kalin, P.; Supuran, C.T.; Gulcin, I.; Alwasel, S.H. The impact of hydroquinone on acetylcholine esterase and certain human carbonic anhydrase isoenzymes (hCA I, II, IX, and XII). *J. Enzyme Inhib. Med. Chem.* **2015**, *30*, 941–946. [CrossRef] [PubMed]
123. Taslimi, P.; Turhan, K.; Turkan, F.; Sedef, K.H.; Turgut, Z.; Gulcin, I. Cholinesterases, α-glycosidase, and carbonic anhydrase inhibition properties of 1H-pyrazolo[1,2-b]phthalazine-5,10-dione derivatives: Synthetic analogues for the treatment of Alzheimer's disease and diabetes mellitus. *Bioorg. Chem.* **2020**, *97*, 103647. [CrossRef] [PubMed]
124. Supuran, C.T. Structure and function of carbonic anhydrases. *Biochem. J.* **2016**, *473*, 22023–22032. [CrossRef]
125. Genc Bilgicli, H.; Kestane, A.; Taslimi, P.; Karabay, O.; Bytyqi-Damoni, A.; Zengin, M.; Gulcin, I. Novel eugenol bearing oxypropanolamines: Synthesis, characterization, antibacterial, antidiabetic, and anticholinergic potentials. *Bioorg. Chem.* **2019**, *88*, 102931. [CrossRef]
126. Rauha, J.P.; Remes, S.; Heinonen, M.; Hopia, A.; Kahkonen, M.; Kujala, T.; Pihlaja, K.; Vuorela, H.; Vuorela, P. Antimicrobial effects of Finnish plant extracts containing flavonoids and other phenolic compounds. *Int. J. Food Microbiol.* **2000**, *56*, 3–12. [CrossRef]
127. Kumar, S.; Maheshwari, K.K.; Singh, V. Central nervous system activity of acute administration of ethanol extract of *Punica granatum* L. seeds in mice. *Ind. J. Exp. Biol.* **2008**, *46*, 811–816.
128. Dias, M.C.; Pinto, D.C.G.A.; Silva, A.M.S. Plant flavonoids: Chemical characteristics and biological activity. *Molecules* **2021**, *26*, 5377. [CrossRef]
129. Cano-Lamadrid, M.; Nowicka, P.; Hernandez, F.; Carbonell-Barrachina, A.A.; Wojdylo, A. Phytochemical composition of smoothies combining pomegranate juice (*Punica granatum* L) and Mediterranean minor crop purees (*Ficus carica, Cydonia oblonga,* and *Ziziphus jujube*). *J. Sci. Food Agric.* **2018**, *98*, 5731–5741. [CrossRef]

life

Article

In Vitro Biological Activity and Lymphoma Cell Growth Inhibition by Selected Mexican Medicinal Plants

Nancy E. Rodríguez-Garza [1,2], Ramiro Quintanilla-Licea [3,*], César I. Romo-Sáenz [1], Joel H. Elizondo-Luevano [2,3], Patricia Tamez-Guerra [1], Cristina Rodríguez-Padilla [1] and Ricardo Gomez-Flores [1,*]

1 Departamento de Microbiología e Inmunología, Facultad de Ciencias Biológicas, Universidad Autónoma de Nuevo León, San Nicolás de los Garza 66455, N.L., Mexico
2 Grupo de Enfermedades Infecciosas y Tropicales (e-INTRO), IBSAL—CIETUS (Instituto de Investigación Biomédica de Salamanca—Centro de Investigación de Enfermedades Tropicales de la Universidad de Salamanca), Facultad de Farmacia, Universidad de Salamanca, 37007 Salamanca, Spain
3 Departamento de Química, Facultad de Ciencias Biológicas, Universidad Autónoma de Nuevo León, San Nicolás de los Garza 66455, N.L., Mexico
* Correspondence: ramiro.quintanillalc@uanl.edu.mx (R.Q.-L.); ricardo.gomezfl@uanl.edu.mx (R.G.-F.); Tel.: +52-81-83763668 (ext. 1476) (R.Q.-L.); +52-81-80207449 (R.G.-F.)

Abstract: Cancer is a major health problem with significant morbidity and mortality. In addition, plants are a source of metabolites with diverse biological properties, including antitumor potential. In this study, we investigated the in vitro murine lymphoma L5178Y-R cell growth inhibition, human peripheral blood mononuclear cells (PBMC) toxicity and proliferation, and antioxidant, hemolytic, and anti-hemolytic activities of methanol extracts from 15 plants of traditional use in Mexico. *Justicia spicigera* caused the highest tumor cell growth inhibition with a half maximal inhibitory concentration (IC_{50}) of 29.10 μg/mL and a selectivity index >34.36 compared with those of PBMC, whereas *Mimosa tenuiflora* showed the highest lymphoproliferative activity from 200 μg/mL compared with that induced by concanavalin A. In addition, *M. tenuiflora* showed an antioxidant effect (IC_{50} = 2.86 μg/mL) higher than that of ascorbic acid. Regarding the hemolytic and anti-hemolytic activity, all extracts presented significant anti-hemolytic activity. The extract of *J. spicigera* is emerging as a possible source of effective antineoplastic compounds.

Keywords: antitumor activity; cancer; ethnobotany; lymphoma; medicinal plants; natural extracts; Mexican plants

Citation: Rodríguez-Garza, N.E.; Quintanilla-Licea, R.; Romo-Sáenz, C.I.; Elizondo-Luevano, J.H.; Tamez-Guerra, P.; Rodríguez-Padilla, C.; Gomez-Flores, R. In Vitro Biological Activity and Lymphoma Cell Growth Inhibition by Selected Mexican Medicinal Plants. *Life* **2023**, *13*, 958. https://doi.org/10.3390/life13040958

Academic Editor: Stefania Lamponi

Received: 15 February 2023
Revised: 20 March 2023
Accepted: 30 March 2023
Published: 6 April 2023

1. Introduction

Cancer is a group of diseases that are characterized by uncontrolled and abnormal cell growth, as well as the potential to invade healthy tissues through metastasis [1]. This is a critical public health problem, causing significant morbidity and mortality [2]. In 2020, there were 19.3 million new cancer cases and 10 million deaths [3]. To date, more than 100 distinct types of cancer are known, which are classified according to the type of cells that were initially affected [1].

Lymphomas derive from T and B lymphocytes or natural killer cells, usually resulting in lymph node enlargement. Therefore, developing new, safe, and more specific biological targets is essential, especially for the most aggressive tumors [3]. In most types of cancers, chemotherapy is the treatment of choice. However, the presence of cancer cells resistant to chemotherapeutic agents [4], as well as the serious side effects they generate in patients, makes it essential to search for new drugs that are more effective and less toxic to patients [5].

Medicinal plants are an important source of metabolites with diverse biological properties that are used as active principles for the treatment of diseases [6]. They have been used for centuries and the World Health Organization recognizes their relevance in public health [7]. In recent years, plant secondary metabolites such as flavonoids, alkaloids, terpenoids, and saponins, among others, have been shown to be potent anticancer agents [8]. Some of the commonly used antineoplastics that have been identified and purified from plants [9] include Paclitaxel (Taxol®), which is a diterpene found in bark extracts from *Taxus brevifolia*, and vincristine, which is an alkaloid extracted from *Vinca rosea*. Another example is strigolactones, which are a group of phytohormones from *Striga* spp. and *Orobanche* spp. parasitic plants [10]. Mexico has a significant plant biodiversity, and more than 15 million people use traditional medicine [11]. In this country, more than 4500 plants have been traditionally used to treat various diseases, including cancer [12].

The aim of the present study was to evaluate the in vitro antitumor potential of methanol extracts from selected Mexican medicinal plants against the murine lymphoma cell line L5178Y-R, as compared with normal human peripheral blood mononuclear cells (PBMC), to determine their selectivity indices (SI). We used 15 plants of traditional medicinal use in Mexico, which belong to the families Acanthaceae, Anacardiaceae, Celastraceae, Compositae, Euphorbiaceae, Leguminosae, Papaveraceae, Poaceae, Rutaceae, Smilacaceae, and Zygophyllaceae, whose selection was based on the amount of reports of plants with anticancer activity [13]. In addition, PBMC proliferation, and antioxidant, hemolytic, and anti-hemolytic activities were evaluated.

2. Materials and Methods

2.1. Plant Material

Plants were purchased from Pacalli® (pacalli.com.mx; Monterrey, Mexico). One specimen of each plant was identified by Professor Dr. Marco Antonio Guzmán-Lucio, curator of the Herbarium of Facultad de Ciencias Biológicas (FCB) at Universidad Autónoma de Nuevo León (UANL), México, where they were assigned voucher numbers (Table 1). Botanical names and families of plant species have been taxonomically validated, using ThePlantList website (http://www.theplantlist.org (accessed 21 January 2023)).

Table 1. Taxonomic identification of medicinal plants used in this study.

Family	Scientific Name	Common Name	Used Part	Voucher Number
Acanthaceae	*Justicia spicigera* Schltdl.	Muicle	Stems, leaves, and flowers	30649
Anacardiaceae	*Amphipterygium adstringens* (Schltdl.) Standl.	Cuachalalate	Bark	30642
Celastraceae	*Semialarium mexicanum* (Miers) Mennega	Cancerina	Bark	30647
Compositae	*Artemisia ludoviciana* Nutt.	Estafiate	Stems, leaves and flowers	30643
Compositae	*Heterotheca inuloides* Cass.	Árnica	Flowers	30646
Compositae	*Psacalium decompositum* (A. Gray) H. Rob. and Brettell	Matarique	Roots	30652
Compositae	*Pseudognaphalium obtusifolium* (L.) Hilliard and B.L. Burtt.	Gordolobo	Flowers	30653
Compositae	*Tagetes lucida* Cav.	Hierbanís or Yerbaniz	Young leaves and stems	30656
Euphorbiaceae	*Jatropha dioica* Sessé	Sangre de Drago	Roots	30648
Leguminosae	*Mimosa tenuiflora* (Willd.) Poir.	Tepezcohuite	Bark	30651
Papaveraceae	*Argemone mexicana* L.	Chicalote	Leaves	29127
Poaceae	*Cymbopogon citratus* (DC.) Stapf.	Zacate limón	Young leaves and stems	30644
Rutaceae	*Ruta chalepensis* L.	Ruda	Young leaves and stems	30654
Smilacaceae	*Smilax aspera* L.	Zarzaparilla	Roots	30655
Zygophyllaceae	*Larrea tridentata* (Sessé and Moc. ex DC.) Coville	Gobernadora	Young leaves and stems	30650

2.2. Plant Extracts Preparation

Plant extraction was performed by placing 25 g of each plant (dried and ground) in a Soxhlet extractor and 500 mL of absolute methanol (CTR Scientific, Monterrey, N.L., México) as extraction solvent. Extraction was maintained for 48 h, after which extracts were filtered and concentrated by vacuum evaporation with a rotary evaporator (Buchi R-3000; Brinkman Instruments, Inc., Westbury, NY, USA). Solvent was removed with a SpeedVac SPD121P concentrator (Thermo Fisher Scientific, San Jose, CA, USA) at 35 °C [14,15]. The extraction yield for each of the extracts was calculated by the following Formula (1):

$$\%\ \text{Yield} = \frac{\text{Final weight of dry extract}}{\text{Initial weight of the plant}} \times 100 \quad (1)$$

Next, 100 mg of each extract was dissolved in one milliliter of dimethyl sulfoxide (DMSO; Sigma-Aldrich, St. Louis, MO, USA), sterilized by filtration, using 0.22 µm pore size-membrane filters (Corning Incorporated, Corning, NY, USA), and stored at −20 °C until use. The final concentration of DMSO used in cell cultures was less than 1% (v/v), which did not affect cell viability [15].

Phytochemical Assay

Each of the crude plant extracts underwent phytochemical screening. The appearance of solids or foam during the reactions allows a semiquantitative evaluation of the presence of secondary metabolites [16]. We evaluated the presence of alkaloids, carbohydrates, coumarins, unsaturations (double bonds), flavonoids, quinones, saponins, sesquiterpene lactones, sterols, and tannins (phenolic groups). These tests were reported as presence (+) or absence (−) of compound groups. Protocols for each test are found as Supplementary Material (Supplementary Material S1: Phytochemical Screening). Solvents and chemicals used in phytochemical screening were purchased from Sigma-Aldrich.

2.3. Cell Lines and Cell Culture Conditions

Murine L5178Y-R lymphoma cells (ATCC CRL-1722) and PBMC were used in this study. PBMC were obtained from a 50 mL to 60 mL blood sample from a healthy donor, using Ficoll-Paque PLUS (GE Healthcare Life Sciences, Pittsburgh, PA, USA) and following supplier's instructions. L5178Y-R cells and PBMC were maintained in RPMI 1640 culture medium (Life Technologies, Grand Island, NY, USA), supplemented with 10% heat-inactivated fetal bovine serum (FBS; Life Technologies) and 1% antibiotic/antifungal solution (Life Technologies) (referred as complete 1640 medium) at 37 °C in an atmosphere of 5% CO_2 in air [17].

2.4. Effect of Extracts on Cell Growth

2.4.1. Antitumor Activity of Plant Methanol Extracts

Cells were incubated in round-bottomed 96-well microplates (Corning Incorporated, Corning, NY, USA) at concentrations of 1×10^4 L5178Y-R cells/well and 1×10^5 PBMC/well in complete RPMI 1640 medium. After 24 h of incubation, cells were treated with methanol extracts at concentrations ranging from 3.9 µg/mL to 1000 µg/mL. The antineoplastic vincristine sulfate (VC) (Hospira, Warwickshire, UK) at 100 µg/mL was used as a positive control and untreated culture medium was used as a negative control [15]. Cells were incubated for 48 h at 37 °C in an atmosphere of 5% CO_2 in air and cell viability was determined using the 3-[4,5-dimethylthiazol-2-yl]-2,5-diphenyltetrazoliumbromide (MTT; Affymetrix, Cleveland, OH, USA) colorimetric method by adding 15 µL of MTT/well (0.5 mg/mL final concentration) and incubating the plate at 37 °C for 3 h [18]. Plates were then decanted, formazan crystals were dissolved with 100 µL of DMSO, and optical densities (OD) were measured at 570 nm in a microplate reader (MULTISKAN GO; Thermo Fisher Scientific, Waltham, MA, USA). Percentage growth inhibition was calculated as follows (2):

$$\%\ \text{Cell growth inhibition} = 100 - \left(\frac{\text{OD}_{570}\ \text{Treated cells}}{\text{OD}_{570}\ \text{Untreated cells}} \times 100\right) \quad (2)$$

2.4.2. Determination of the Selectivity Index

Selectivity indices (SI) were determined to assess cytotoxic potential in tumor cells relative to toxicity in normal cells, where high SI indicates high potency and low cell toxicity [19]. In our study, it was considered that samples with SI values greater than three possess a high selectivity to tumor cells, according to Bezivin et al. (2003) [20]. Plant extract SIs were calculated by dividing the IC_{50} of normal cells (PBMC) by that of tumor cells (L5178Y-R), using the following Formula (3):

$$\text{SI} = \frac{\text{IC}_{50}\ \text{Normal cells}}{\text{IC}_{50}\ \text{Tumor cells}} \quad (3)$$

2.4.3. Effect of Plant Extracts on PBMC Lymphoproliferation

PBMC were incubated in round-bottomed 96-well microplates (Corning Incorporated) at 1×10^5 cells/well in complete RPMI 1640 medium. After 24 h of incubation, cells were treated with methanol extracts at concentrations ranging from 100 μg/mL to 500 μg/mL, using 5 μg/mL concanavalin A (Con A) as a positive control and untreated culture medium as a negative control. Cells were incubated for 48 h at 37 °C in an atmosphere of 5% CO_2 in air, and cell viability was determined using the colorimetric MTT reduction assay, as explained above [18]. Results were expressed as the proliferation index, which was calculated using the following Formula (4) [21]:

$$\text{Proliferation index} = \frac{\text{OD}_{570}\ \text{Treatment} - \text{OD}_{570}\ \text{Negative control}}{\text{OD}_{570}\ \text{Negative control}} \quad (4)$$

In addition, we determined the half maximal effective concentration (EC_{50}) of plant extracts for PBMC proliferation. The percentage of proliferation was calculated by multiplying the proliferation index by 100 [22].

2.4.4. Synergistic Antitumor Activity of Plant Extracts

We evaluated the antitumor potential of combinations of bioactive extracts, as previously reported by Shang et al. (2019) [23]. We selected four plants that showed the highest percentage of tumor cell growth inhibition ($IC_{50} < 50$ μg/mL) against L5178Y-R murine lymphoma cells to determine if the combination of these extracts shows synergistic activity. For the assay, 1×10^4 L5178Y-R lymphoma cells/well were incubated in round-bottomed 96-well plates in complete RPMI 1640 medium at 37 °C in 5% CO_2 in air for 24 h, after which cells were treated with plant methanol extracts, alone or in combination, at concentrations ranging from 7.8 μg/mL to 125 μg/mL, using 100 μg/mL VC as a positive control and untreated culture medium as a negative control. After 48 h of incubation at 37 °C in 5% CO_2 in air, cell viability was determined by the colorimetric MTT reduction assay, as explained above [18]. The type of interaction that occurred between crude methanol extracts was determined using the Synergy Finder 3.0 application (https://synergyfinder.fimm.fi/) (accessed 30 November 2022). Scores obtained in this application indicate the following characteristics: (a) the interaction is likely to be antagonistic, when the score is lower than −10, (b) the interaction is likely to be additive, when the score ranges from −10 to 10, and (c) the interaction is likely to be synergistic, when the score is higher than 10 [24].

2.5. Antioxidant Activity

To determine the antioxidant activity of plant extracts, we incubated 100 μL of the extracts at concentrations ranging from 3.9 μg/mL to 500 μg/mL plus 100 μL of a 0.1 mM

2,2-diphenyl-1-picrylhydrazil methanolic solution (DPPH; Sigma-Aldrich) in flat-bottomed 96-well microplates for 30 min at room temperature in darkness, after which ODs were determined at 517 nm. DMSO was used as a blank and ascorbic acid (J. T. Baker, Phillipsburg, NJ, USA) as a positive control at concentrations ranging from 0.2 µg/mL to 250 µg/mL [25]. Percentage inhibition of DPPH was calculated using the following Formula (5):

$$\%\text{ inhibicion DPPH} = \frac{OD_{517}\text{ Treatment} - OD_{517}\text{ Negative control}}{OD_{517}\text{ Negative control}} \times 100 \quad (5)$$

2.6. Hemolytic and Anti-Hemolytic Activity

We obtained 20 mL of blood from a healthy volunteer in tubes with anticoagulant (BD Vacutainer K2 EDTA; Becton Dickinson & Company, Franklin Lakes, NJ, USA). Red blood cells were washed three times with phosphate buffered saline solution (PBS, pH 7.2) and a 5% erythrocytes suspension was prepared in sterile PBS. For the evaluation of the hemolytic activity in 2 mL tubes, we incubated extracts at concentrations ranging from 200 µg/mL to 1000 µg/mL and 5% erythrocytes suspension for 30 min at 37 °C, after which they were centrifuged at 4 °C for 5 min at 13,000 rpm [26], using distilled water as a positive control for hemolysis and PBS as a negative control. For the evaluation of the antihemolytic effect, we incubated 2,2′-azobis(2-amidinopropane) dihydrochloride (AAPH), which was prepared in PBS at a concentration of 150 mM extract and a erythrocyte suspension for 5 h at 37 °C at 200 rpm in darkness using a rotating incubator (MaxQ, Thermo-Scientific, Waltham, MA, USA), and using PBS with the erythrocyte suspension without AAPH as the negative control and the erythrocyte suspension with 150 mM AAPH as a positive control (C+) [27]. In both cases, 200 µL of the supernatant was taken from each tube and transferred to a flay-bottomed 96-well microplate to measure the OD of the released hemoglobin at 540 nm in a microplate reader [28]. The percentage of hemolysis and anti-hemolytic activity for each sample was calculated using the following Formulas (6) and (7):

$$\%\text{ Hemolysis} = \frac{OD_{540}\text{ Treatment} - OD_{540}\text{ Negative control}}{OD_{540}\text{ Positive control} - OD_{540}\text{ Negative control}} \times 100 \quad (6)$$

$$\%\text{ AAPH Inhibition} = 1 - \left(\frac{OD_{540}\text{ Treatment} - OD\text{ Negative control}}{OD_{540}\text{ Positive control} - OD\text{ Negative control}} \times 100\right) \quad (7)$$

2.7. Ethical Statement

The study with human PBMC and erythrocytes was performed under the approval of the Ethics Committee of FCB-UANL (Registration Number CI-08-2020) and under the consent of a healthy donor (the Letter of Informed Consent for Donors of Human Biological Sample Material and the Institutional Review Board Approval are attached as Supplementary Material), following the provisions of the Official Mexican Technical Standard NOM-253-SSA1-2012. We did not develop studies involving animals.

2.8. Statistical Analysis

Data represent the mean ± SD of triplicate determinations of at least three independent experiments. A one-way analysis of variance was used to determine the significant difference between the tested concentrations. Tukey's post hoc test was used to determine the difference between the treatment means. The Probit test was used to calculate the IC_{50} (half maximal inhibitory concentration) and the EC_{50} (half maximal effective concentration) values. Statistical analyses were performed using the GraphPad Prism 8 statistical package (GraphPad Software Inc., San Diego, CA, USA).

3. Results and Discussion

3.1. Plant Material Identification

Table 1 provides specific information on medicinal plants used in the present study. Of the fifteen plants evaluated, five belong to the Compositae family, and the other plants belong to different families. Plant methanol extracts were analyzed to determine their cytotoxic, lymphoproliferative, antioxidant, hemolytic, and anti-hemolytic activities, and their synergistic antitumor potential of the most active extracts against L5178Y-R tumor cells. These plants have been demonstrated their antitumor properties [13].

3.2. Plant Extract Yields and Phytochemical Analysis

Table 2 shows yields (11.14% to 27.37%) of plant methanol extracts used in this study. All extracts were positive for unsaturation (double bonds) and coumarins. *A. mexicana* and *R. chalepensis* were positive for alkaloids, whereas *J. spicigera*, *S. mexicanum*, *S. aspera*, and *T. lucida* extracts were negative for the sterol test, and only *P. obtusifolium* was positive for saponins. Plant extracts consist of a complex mixture of various compounds such as alkaloids, esters, aldehydes, carbohydrates, terpenes, and polyphenols, among others [29]. In addition, crude extracts, semi-purified fractions, and pure compounds have been used in different approaches testing biological activities [30]. However, it is necessary to search for new sources and specific compounds against cancer [14,31]. In this context, Mexico is an attractive country for its great variety of endemic plants [32].

Table 2. Plant extract yields and phytochemical screening.

Plant Extract	%	Chemical Groups									
		Alk	Carb	Cm	Db	Flv	Qn	Sp	Sl	St	Tn
J. spicigera	27.37	−	+	+	+	−	+	−	−	−	+
A. adstringens	24.82	−	+	+	+	+	+	−	+	+	+
S. mexicanum	11.14	−	+	+	+	−	+	−	+	−	−
A. ludoviciana	17.31	−	+	+	+	+	+	−	+	+	+
H. inuloides	21.21	−	+	+	+	+	−	−	−	+	+
P. decompositum	14.43	−	+	+	+	+	−	−	−	+	+
P. obtusifolium	16.99	−	+	+	+	+	+	+	+	+	+
T. lucida	20.63	−	+	+	+	+	−	−	+	−	+
J. dioica	19.58	−	+	+	+	−	+	−	−	+	−
M. tenuiflora	11.25	−	−	+	+	+	+	−	+	+	+
A. mexicana	11.26	+	−	+	+	+	−	−	+	+	−
C. citratus	23.04	−	+	+	+	+	−	−	−	+	−
R. chalepensis	19.40	+	+	+	+	+	−	−	+	+	+
S. aspera	18.26	−	+	+	+	−	+	−	+	−	−
L. tridentata	26.61	−	−	+	+	+	+	−	+	+	+

%: Extraction yield percentage; Alk: alkaloids, Carb: carbohydrates, Cm: coumarins, Db: Double bonds, Flv: flavonoids, Qn: quinones, Sp: saponins, Sl: sesquiterpene—lactones, St: sterols, Tn: tannins; (+): present, (−): absent.

3.3. Cell Growth Inhibition by Plant Extracts

We evaluated the percentage cell growth inhibition of plant extracts against the tumor cell line L5178Y-R and PBMC, using the colorimetric MTT reduction assay, which allowed us to determine the SI, as explained above [19]. Table 3 shows the effect of plant extracts on cell growth inhibition, where the most active extracts against L5178Y-R cells were *J. spicigera* (IC_{50} = 29.10 μg/mL), *A. ludoviciana* (IC_{50} = 32.39 μg/mL), *J. dioica* (IC_{50} = 39.25 μg/mL), and *M. tenuiflora* (IC_{50} = 47.10 μg/mL). The other extracts showed activities with IC_{50} values higher than that of *A. mexicana* (IC_{50} = 70.73 μg/mL). When the extracts were evaluated on PBMC viability, *J. spicigera*, *A. adstringens*, *S. mexicanum*, *A. ludoviciana*, *H. inuloides*, *P. decompositum*, *M. tenuiflora*, and *S. aspera* did not significantly alter PBMC viability, since they presented IC_{50} > 1000 μg/mL, thus indicating low toxicity for normal cells [31]. VC was also not toxic for PBMC. Previous studies with VC showed its potent anticancer effects on solid tumors, such as hepatocellular carcinoma, breast cancer, and lung cancer [33], in addition to the L5178Y-R cell line [34]. It has been well documented that VC has no cytotoxic effects on PBMC nor on normal L929 and Vero cells, suggesting a selective action of the drug against leukemic cells [14,17,35].

Table 3. Effect of methanol plant extracts on L5178Y-R and PBMC viability.

Plant Extract	IC_{50} (µg/mL)		SI
	L5178Y-R	PBMC	
J. spicigera	29.10 ± 5.23 [a]	>1000 †	>34.36
A. adstringens	197.66 ± 4.07 [e]	>1000 †	>5.07
S. mexicanum	135.00 ± 5.19 [c]	>1000 †	>7.41
A. ludoviciana	32.39 ± 5.59 [a]	>1000 †	>30.87
H. inuloides	140.00 ± 10.03 [cd]	>1000 †	>7.14
P. decompositum	166.50 ± 19.19 [d]	>1000 †	>6.01
P. obtusifolium	710.55 ± 24.33 [g]	745.40 ± 12.90 [d]	1.05
T. lucida	251.17 ± 6.02 [f]	732.05 ± 8.59 [d]	2.91
J. dioica	39.25 ± 3.81 [a]	156.60 ± 10.76 [a]	3.99
M. tenuiflora	47.10 ± 10.19 [ab]	>1000 †	>21.23
A. mexicana	70.73 ± 2.36 [b]	398.45 ± 7.97 [c]	5.63
C. citratus	209.24 ± 6.15 [e]	312.41 ± 2.87 [b]	1.49
R. chalepensis	71.81 ± 1.86 [b]	183.58 ± 11.89 [ab]	2.56
S. aspera	250.00 ± 5.29 [f]	>1000 †	>4.00
L. tridentata	214.64 ± 1.63 [e]	403.05 ± 13.72 [c]	1.88

Data are mean ± SD of the IC_{50} (µg/mL) for each extract against the evaluated cell lines. Different letters within the same column are significantly different, analyzed by the post hoc Tukey test ($p < 0.05$). SI represents IC_{50} for the PBMC cell line divided by that of the L5178Y-R cell line after 72 h, as detailed in the Section 2. VS was used as a positive control at 0.05 µg/mL. † As IC_{50} was above 1000 µg/mL, these values were not considered for Tukey analysis.

In the present study, we found antitumor potential of traditionally used plants in Mexico without affecting human PBMC. Therefore, with the results of cytotoxicity to PBMC, we determined the SI, where a value of 3 or greater was taken as the cut-off point [20]. In the present study, *J. spicigera, A. adstringens, S. mexicanum, A. ludoviciana, H. inuloides, P. decompositum, P. obtusifolium, J. dioica, M. tenuiflora,* and *S. aspera* extracts showed their following SIs: >34.36, >5.07, >7.41, >30.87, >7.14, >6.01, 3.99, >21.23, and >4.00, respectively (Table 3). Extracts with the lowest SI were *P. obtusifolium, T. lucida, C. citratus, R. chalepensis,* and *L. tridentata,* since they showed SI lower than 3. The SI is commonly used to measure the efficacy of drugs, where the reduction of cell survival is evaluated [36]. Thus, we determined the cytotoxic potential of each extract in a tumor cell line (L5178Y-R) in relation to the toxicity in normal cells (PBMC), where a high SI (greater than 3) indicates high potency against L5178Y-R cells and low cell toxicity to PBMCs [19].

3.4. PBMC Proliferation

Figure 1 shows the effect of plant extracts on PBMC proliferation. Extracts that showed the highest and significant cell proliferation activity were *M. tenuiflora* at concentrations ranging from 200 µg/mL to 500 µg/mL, followed by *S. mexicanum, A. adstringens,* and *J. spiciguera*. In addition, *A. ludoviciana, H. inuloides, P. decompositum,* and *S. aspera* extracts showed lower lymphoproliferative activity than that of Con A. The remaining extracts presented null activity at all concentrations tested, and they were not included in Figure 1. Crude plant extracts are complex mixtures that contain a great diversity of secondary metabolites that act synergistically to stimulate a biological response [37], for which the lymphoproliferative activity of PBMC is probably due to the synergistic effects of flavones, phenols, terpenes, sesquiterpenes, and tannins present in the extracts [38].

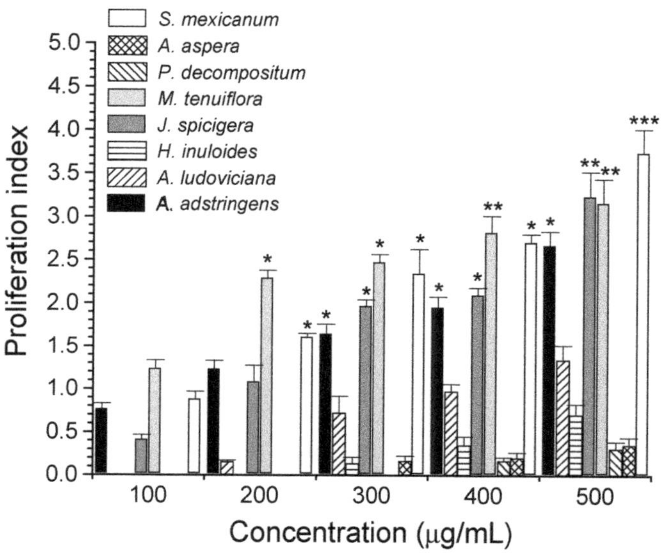

Figure 1. PBMC proliferation by Mexican plant extracts. Values represent the PBMC proliferation index ± SD by plant extracts, using 5 µg/mL Con A as a positive control. * $p < 0.05$, ** $p < 0.01$, *** $p < 0.001$, as compared with Con A (proliferation index = 1.21 ± 0.18).

The proliferative activity of plant extracts on PBMC is shown in Table 4. *M. tenuiflora, A. adstringens,* and *S. mexicanum* extracts showed the highest proliferative effect on PBMC with EC_{50} of 14.93 µg/mL, 18.3 µg/mL, and 21.95 µg/mL, respectively.

Table 4. Lymphoproliferative activity of methanol extracts on PBMC cells.

Plant Extract	EC_{50} (µg/mL)
J. spicigera	154.14 ± 7.3 [c]
A. adstringens	18.30 ± 1.5 [a]
S. mexicanum	21.95 ± 5.6 [ab]
A. ludoviciana	256.99 ± 11.78 [d]
H. inuloides	431.69 ± 19.2 [e]
P. decompositum	560.84 ± 42.01 [f]
P. obtusifolium	ND
T. lucida	ND
J. dioica	ND
M. tenuiflora	14.93 ± 4.3 [a]
A. mexicana	ND
C. citratus	ND
R. chalepensis	ND
S. aspera	595.76 ± 27.9 [g]
L. tridentata	ND

Data are means ± SD of the EC_{50} (µg/mL) for each extract evaluated on PBMC cells. Different letters within the same column are significantly different, analyzed by the post hoc Tukey test ($p < 0.05$). ND: Not determined, because these extracts did not show lymphoproliferative activity at any of the evaluated concentrations.

The proliferative effect of *A. adstringens* may be due to the action of anacardic acids such as 6-pentadecyl salicylic acid (6-PSA), which are the main compounds of this plant. It has been shown that 6-PSA did not reduce PBMC cell viability, as compared with the gastric tumor cell line AGS, where it was determined that it showed cytotoxic and genotoxic activity and induced cell death by apoptosis in a caspase 8-dependent manner, thus evidencing its therapeutic potential [39]. In addition, it has been shown that an extract from *S. mexicanum* bark has antitumor activity against breast cancer cells (MDA-MB-231 and MCF7) but promotes proliferation of non-breast derived cells (MCF 10A) and PBMC, which may be due to the quinone triterpenes pristimerin and tingenone [40], suggesting that *S. mexicanum* cortex has a potential application in cancer treatment [41].

Different saponins isolated from *M. tenuiflora* possess a synergistic antitumor effect against the lymphoma cells RDM 4 and Molt 4, and have a significant mitogenic effect on mouse fibroblast cells LMTK and human fibroblasts [42,43]. This is the first report showing *M. tenuiflora* proliferation activity on PBMC.

3.5. Synergistic Antitumor Activity of Plant Extracts against L5178Y-R Lymphoma Cells

The combination of two or more therapeutic agents is a complementary approach in cancer therapy. Clinical studies have reported the beneficial effects of herbal medicines in the treatment and quality of life of cancer patients when used in combination with conventional therapy [44]. Plants selected for this study were *A. ludoviciana*, *J. dioica*, *J. spicigera*, and *M. tenuiflora* because they showed $IC_{50} < 50$ μg/mL against L5178Y-R cells. Table 5 shows the type of interaction presented by the various combinations of the most active extracts against L5178Y-R cells, where it can be seen that AlJd, AlJs, and JdJs combinations evidenced synergistic activity. In addition, AlMt and JsMt combinations showed additive activity. Only JdMt combination presented antagonistic activity. A series of scientific investigations have demonstrated the synergistic activity of plant extracts, for which the next step was to investigate a synergistic effect between the most active extracts. It is believed that the active compounds of plants modify and inhibit mechanisms of acquired resistance in cells, thus exhibiting a synergistic effect [45].

Table 5. In vitro interaction between bioactive plant extracts on murine lymphoma cell growth.

Extracts Combination	Combination Code	HSA Model Value	Interaction
A. ludoviciana + *J. dioica*	AlJd	13.737	Synergism
A. ludoviciana + *J. spicigera*	AlJs	10.951	Synergism
A. ludoviciana + *M. tenuiflora*	AlMt	−8.663	Additive
J. dioica + *J. spicigera*	JdJs	18.111	Synergism
J. dioica + *M. tenuiflora*	JdMt	−19.151	Antagonism
J. spicigera + *M. tenuiflora*	JsMt	2.985	Additive

The value shown corresponds to that obtained with the HSA model, when evaluating concentrations from 7.8 μg/mL to 125 μg/mL of the extracts.

3.6. Antioxidant, Hemolytic, and Anti-Hemolytic Activities of Plant Extracts

Extensive research indicates that oxidizing agents such as reactive oxygen species (ROS) at low levels have beneficial effects on health. However, excessive accumulation causes various disorders, including carcinogenesis, as ROS play a crucial role in cancer cell survival [46]. In the present study, the antioxidant activity of plant extracts related to the DPPH free radical scavenging effect was compared with that of ascorbic acid, which showed an IC_{50} of 7.23 μg/mL. Extracts with an $IC_{50} \leq 50$ μg/mL were considered to have relevant antioxidant activity [47]. *J. spicigera* (IC_{50} = 15.68 μg/mL), *A. adstringens* (IC_{50} = 17.22 μg/mL), *M. tenuiflora* (IC_{50} = 2.86 μg/mL), and *S. aspera* (IC_{50} = 29.21 μg/mL) extracts showed the highest antioxidant effect among all the extracts evaluated, with *M. tenuiflora* the extract that presented significantly ($p < 0.001$) higher antioxidant activity compared with the positive control, whereas *R. chalepensis* extract (IC_{50} = 859.85 μg/mL) showed the lowest antioxidant effect (Table 6). In general, a natural extract that may be used as a universal antioxidant does not exist. Therefore, it is necessary to screen for specific antioxidant compounds for certain outcomes [30]. Various authors have demonstrated the antioxidant activity of extracts from various plants through the DPPH test and have concluded that polyphenolic components and terpenes are the main source of antioxidant activity in various extracts, such as eucalyptus (*Eucalyptus camaldulensis* Dehnh.), which is probably due to gallic and ellagic acids [48]. Moreover, the resveratrol present in grapes

(*Vitis vinifera* L.) has chemopreventive and therapeutic effects in reducing human breast, uterus, blood, prostate, and ovarian cancers, among others [49]. As mentioned above, *M. tenuiflora* extract maintained a high antioxidant activity, which was higher than that of the positive control. These results showed that one or several antioxidant compounds present in *M. tenuiflora* extract have the potential to act for the free radical scavenging activity (DPPH assay) through a hydrogen transfer reaction [50].

Table 6. Hemolytic and anti-hemolytic antioxidant activity of plant methanol extracts.

Plant Extract	IC_{50} (µg/mL)		
	DPPH	**Hemolysis**	**AAPH**
J. spicigera	15.68 ± 1.79 [c]	642.10 ± 3.30 [d]	16.22 ± 1.18 [d]
A. adstringens	17.22 ± 2.02 [c]	268.40 ± 13.90 [a]	1.07 ± 0.31 [a]
S. mexicanum	581.50 ± 1.09 [j]	>1000 [†]	8.26 ± 1.42 [bc]
A. ludoviciana	89.90 ± 3.92 [f]	>1000 [†]	10.06 ± 0.76 [c]
H. inuloides	61.62 ± 7.95 [e]	427.20 ± 2.81 [b]	1.09 ± 0.08 [a]
P. decompositum	106.10 ± 1.06 [g]	>1000 [†]	1.18 ± 0.24 [a]
P. obtusifolium	528.67 ± 25.78 [i]	>1000 [†]	28.63 ± 2.31 [f]
T. lucida	550.85 ± 16.09 [ij]	843.84 ± 31.93 [f]	12.50 ± 1.14 [c]
J. dioica	116.80 ± 1.70 [h]	565.80 ± 2.07 [c]	14.58 ± 0.48 [d]
M. tenuiflora	2.86 ± 1.26 [a]	>1000 [†]	14.25 ± 1.58 [d]
A. mexicana	565.98 ± 17.60 [ij]	973.88 ± 38.46 [g]	20.32 ± 0.21 [e]
C. citratus	690.40 ± 26.37 [k]	606.82 ± 19.12 [cd]	6.40 ± 1.49 [b]
R. chalepensis	859.85 ± 25.08 [l]	738.73 ± 20.74 [e]	5.66 ± 0.82 [b]
S. aspera	29.21 ± 10.43 [d]	262.80 ± 2.07 [a]	0.88 ± 0.17 [a]
L. tridentata	335.20 ± 1.60 [i]	>1000 [†]	155.57 ± 8.91 [g]
C+	7.23 ± 0.03 [b]	ND	316.14 ± 30.19 [h]

Data are mean ± SD of the IC_{50} (µg/mL) for each evaluated extract. Different letters within the same column are significantly different, analyzed by the post hoc Tukey test ($p < 0.05$). C+: Positive control. ND: Not determined. [†] As IC_{50} was above 1000 µg/mL, these values were not considered for Tukey analysis.

One of the research objectives of studying medicinal plants is to develop and use assays that are easy to use, reproducible, and inexpensive, such as the determination of the toxic activity of plant extracts, fractions, combinations, and/or formulations on human erythrocytes by the hemolysis test [51]. Therefore, we decided to evaluate the hemolytic potential of plant extracts and it was determined that none of the extracts was toxic to human erythrocytes, according to the criteria of López Villarreal et al. (2022) [29], and *A. adstringens* and *S. aspera* are the extracts with the lowest activity, with IC_{50} of 268.40 and 262.80 µg/mL, respectively. *S. mexicanum*, *A. ludoviciana*, *P. decompositum*, *P. obtusifolium*, *M. tenuiflora*, and *L. tridentata* extracts showed IC_{50} > 1000 (Table 6). The median lethal dose (IC_{50}) for an extract to be considered non-toxic is ≥ 1000 µg/mL. Those between 500 µg/mL and 1000 µg/mL are considered slightly toxic, between 100 µg/mL and 500 µg/mL are considered moderately toxic, and between 10 µg/mL and 100 µg/mL are considered highly toxic [47]. In addition, the anti-hemolytic effect of the extracts was determined by the hemolysis test, using the AAPH radical to form peroxyl radicals and induce membrane oxidation in human erythrocytes [28]. We observed a significant anti-hemolytic effect for all extracts evaluated, as compared with the positive control. *A. adstringens*, *H. inuloides*, *P. decompositum*, and *S. aspera* extracts possessed the highest anti-hemolytic activity, with IC_{50} values of 1.07 µg/mL, 1.09 µg/mL, 1.18 µg/mL, and 0.88 µg/mL, respectively, which agrees with a report by Elizondo et al. (2022) [14].

Results on the hemolytic and anti-hemolytic potential of plant extracts emphasize the antioxidant properties of the extracts, since the study of the uptake of azo radicals, such as the AAPH radical, was designed to induce oxidative stress in the lipid and aqueous phases of cells and assess the vulnerability of erythrocytes to oxidative stress [52]. The breakdown of AAPH, which is soluble in water at physiological temperature, generates free radicals that attack the erythrocyte membrane and induce lipid peroxidation that leads to hemolysis [53].

Therefore, these results showed the safe use of medicinal plants and their doses [54]. Taken together, our study demonstrated the antitumor activity of extracts from plants of medicinal use in Mexico as well as their potential to induce PBMC proliferation plus antioxidant and anti-hemolytic activities on human erythrocytes.

4. Conclusions

J. spicigera, *A. ludoviciana*, *J. dioica*, and *M. tenuiflora* methanol extracts significantly inhibited L5178Y-R lymphoma cell growth. Moreover, *J. spicigera*, *A. ludoviciana*, and *M. tenuiflora* presented the highest SI when evaluated against PBMC, with SIs of >34.36, >30.87, and >21.23, respectively, making them promising candidates for further study against other cell lines and for bio-targeted purification of their most active components. It was also demonstrated that the extracts in combination showed in vitro synergistic activity when evaluated against L5178Y-R cells, which opens a new line of research regarding the possible combined use of the different plants. Overall, results of the present study validated the use of medicinal plant extracts and their combinations in cancer. In addition, the evaluation of extracts from plants used in traditional medicine reveals the potential of identifying bioactive compounds showing antitumor, antioxidant, and anti-hemolytic activity.

5. Future Perspectives

In a second stage of this project, we will develop a bio-directed fractionation of the active plant methanol extracts in the L5178Y-R lymphoma model, as well as the purification of the components with biological activity and determination of the mechanisms of molecular action. In addition, in vivo studies of the most promising extracts and active compounds are warranted to validate their potential use in L5178Y-R lymphoma treatment.

Supplementary Materials: The following supporting information can be downloaded at: https://www.mdpi.com/article/10.3390/life13040958/s1, Phytochemical Screening Tests.

Author Contributions: Conceptualization, R.G.-F. and R.Q.-L.; Correspondence R.G.-F. and R.Q.-L.; methodology, R.G.-F. and C.I.R.-S.; software, J.H.E.-L.; validation, J.H.E.-L.; formal analysis, and investigation, N.E.R.-G.; resources, R.G.-F., R.Q.-L. and C.R.-P.; data curation, C.I.R.-S.; writing—original draft preparation, N.E.R.-G.; writing—review and editing, N.E.R.-G. and R.G.-F.; visualization, R.G.-F.; supervision, R.Q.-L.; project administration, P.T.-G. and C.R.-P.; funding acquisition, P.T.-G. All authors have read and agreed to the published version of the manuscript.

Funding: This research was funded by the Consejo Nacional de Ciencia y Tecnología (CONACYT, México) grant 808132 (CVU: 1006989) to N.E.R.-G., grant 935405 (CVU: 418935) to J.H.E.-L., and grant 877783 (CVU: 445572) to C.I.R.-S and the Programa de Apoyo a la Investigación Científica y Tecnológica (PAICYT-UANL 2022) grant CI-09-2022 to R.Q.-L.

Institutional Review Board Statement: The study was conducted according to the guidelines of the Declaration of Helsinki and approved by the Ethics Committee of the UANL, registration number CI-09-2022. The letter of Informed Consent for donors of human biological sample material and the Institutional Review Board Approval is attached as Supplementary Material.

Informed Consent Statement: Informed consent was obtained from all subjects involved in the study.

Data Availability Statement: The datasets generated or analyzed during the present study are available from the corresponding author.

Acknowledgments: Special thanks to the Herbarium, the Phytochemistry Laboratory, and the Immunobiology and Drug Carriers Unit staff of the FCB-UANL.

Conflicts of Interest: The authors declare no conflict of interest.

References

1. National Cancer Institute. Understanding Cancer. Available online: https://www.cancer.gov/about-cancer/understanding/what-is-cancer (accessed on 20 January 2023).
2. World Health Organization. Cancer. Available online: https://www.who.int/news-room/fact-sheets/detail/cancer (accessed on 20 January 2023).
3. Sung, H.; Ferlay, J.; Siegel, R.L.; Laversanne, M.; Soerjomataram, I.; Jemal, A.; Bray, F. Global Cancer Statistics 2020: GLOBOCAN Estimates of Incidence and Mortality Worldwide for 36 Cancers in 185 Countries. *CA Cancer J. Clin.* **2021**, *71*, 209–249. [CrossRef] [PubMed]
4. Cui, C.; Yang, J.; Li, X.; Liu, D.; Fu, L.; Wang, X. Functions and Mechanisms of Circular RNAs in Cancer Radiotherapy and Chemotherapy Resistance. *Mol. Cancer* **2020**, *19*, 58. [CrossRef] [PubMed]
5. Zhang, Q.Y.; Wang, F.X.; Jia, K.K.; Kong, L.D. Natural Product Interventions for Chemotherapy and Radiotherapy-Induced Side Effects. *Front. Pharmacol.* **2018**, *9*, 1253. [CrossRef]
6. Wang, W.; Xu, J.; Fang, H.; Li, Z.; Li, M. Advances and Challenges in Medicinal Plant Breeding. *Plant Sci.* **2020**, *298*, 110573. [CrossRef]
7. World Health Organization. *Estrategia de La OMS Sobre Medicina Tradicional 2014–2023*; Organización Mundial de la Salud: Geneva, Switzerland, 2013; ISBN 9789243506098.
8. Dutt, R.; Garg, V.; Khatri, N.; Madan, A.K. Phytochemicals in Anticancer Drug Development. *Anticancer Agents Med. Chem.* **2019**, *19*, 172–183. [CrossRef]
9. Rosales-Reyes, T.; de la Garza, M.; Arias-Castro, C.; Rodríguez-Mendiola, M.; Fattel-Fazenda, S.; Arce-Popoca, E.; Hernández-García, S.; Villa-Treviño, S. Aqueous Crude Extract of *Rhoeo discolor*, a Mexican Medicinal Plant, Decreases the Formation of Liver Preneoplastic Foci in Rats. *J. Ethnopharmacol.* **2008**, *115*, 381–386. [CrossRef]
10. Fridlender, M.; Kapulnik, Y.; Koltai, H. Plant Derived Substances with Anti-Cancer Activity: From Folklore to Practice. *Front. Plant Sci.* **2015**, *6*, 799. [CrossRef]
11. Rodríguez-Hernández, A.A.; Flores-Soria, F.G.; Patiño-Rodríguez, O.; Escobedo-Moratilla, A. Sanitary Registries and Popular Medicinal Plants Used in Medicines and Herbal Remedies in Mexico (2001–2020): A Review and Potential Perspectives. *Horticulturae* **2022**, *8*, 377. [CrossRef]
12. Lucía, C.P.A.; Jacqueline, B.R.; Alberto, B.R.L.; David, B.A.; Beatriz, R.A. Actualized inventory of medicinal plants used in traditional medicine in Oaxaca, Mexico. *J. Ethnobiol. Ethnomed.* **2021**, *17*, 7. [CrossRef]
13. Alonso-Castro, A.J.; Villarreal, M.L.; Salazar-Olivo, L.A.; Gomez-Sanchez, M.; Dominguez, F.; Garcia-Carranca, A. Mexican Medicinal Plants Used for Cancer Treatment: Pharmacological, Phytochemical and Ethnobotanical Studies. *J. Ethnopharmacol.* **2011**, *133*, 945–972. [CrossRef]
14. Elizondo-Luévano, J.H.; Gomez-Flores, R.; Verde-Star, M.J.; Tamez-Guerra, P.; Romo-Sáenz, C.I.; Chávez-Montes, A.; Rodríguez-Garza, N.E.; Quintanilla-Licea, R. In Vitro Cytotoxic Activity of Methanol Extracts of Selected Medicinal Plants Traditionally Used in Mexico against Human Hepatocellular Carcinoma. *Plants* **2022**, *11*, 2862. [CrossRef] [PubMed]
15. Romero-Arguelles, R.; Romo-Sáenz, C.I.; Morán-Santibáñez, K.; Tamez-Guerra, P.; Quintanilla-Licea, R.; Orozco-Flores, A.A.; Ramírez-Villalobos, J.M.; Tamez-Guerra, R.; Rodríguez-Padilla, C.; Gomez-Flores, R. In Vitro Antitumor Activity of Endophytic and Rhizosphere Gram-Positive Bacteria from *Ibervillea sonorae* (S. Watson) Greene against L5178Y-R Lymphoma Cells. *Int. J. Environ. Res. Public Health* **2022**, *19*, 894. [CrossRef] [PubMed]
16. Guillén-Meléndez, G.A.; Villa-Cedillo, S.A.; Pérez-Hernández, R.A.; Castillo-Velázquez, U.; Salas-Treviño, D.; Saucedo-Cárdenas, O.; Montes-de-Oca-Luna, R.; Gómez-Tristán, C.A.; Garza-Arredondo, A.J.; Zamora-Ávila, D.E.; et al. Cytotoxic Effect In Vitro of *Acalypha monostachya* Extracts over Human Tumor Cell Lines. *Plants* **2021**, *10*, 2326. [CrossRef]
17. Ramírez-Villalobos, J.M.; Romo-Sáenz, C.I.; Morán-Santibañez, K.S.; Tamez-Guerra, P.; Quintanilla-Licea, R.; Orozco-Flores, A.A.; Romero-Arguelles, R.; Tamez-Guerra, R.; Rodríguez-Padilla, C.; Gomez-Flores, R. In Vitro Tumor Cell Growth Inhibition Induced by *Lophocereus marginatus* (DC.) S. Arias and Terrazas Endophytic Fungi Extracts. *Int. J. Environ. Res. Public Health* **2021**, *18*, 9917. [CrossRef]
18. Kury, A.L.T.; Taha, Z.; Talib, W.H. Immunomodulatory and Anticancer Activities of *Hyacinthus orientalis* L.: An in Vitro and in Vivo Study. *Plants* **2021**, *10*, 617. [CrossRef]
19. Mansour, K.A.; El-Neketi, M.; Lahloub, M.-F.; Elbermawi, A. Nanoemulsions of *Jasminum humile* L. and *Jasminum grandiflorum* L. Essential Oils: An Approach to Enhance Their Cytotoxic and Antiviral Effects. *Molecules* **2022**, *27*, 3639. [CrossRef]
20. Bézivin, C.; Tomasi, S.; Lohézic-Le Dévéhat, F.; Boustie, J. Cytotoxic Activity of Some Lichen Extracts on Murine and Human Cancer Cell Lines. *Phytomedicine* **2003**, *10*, 499–503. [CrossRef]
21. Gonzalez Rivas, E.; Ximenez, C.; Nieves-Ramirez, M.E.; Moran Silva, P.; Partida-Rodríguez, O.; Hernandez, E.H.; Rojas Velázquez, L.; Serrano Vázquez, A.; Magaña Nuñez, U. *Entamoeba histolytica* Calreticulin Induces the Expression of Cytokines in Peripheral Blood Mononuclear Cells Isolated From Patients With Amebic Liver Abscess. *Front. Cell. Infect. Microbiol.* **2018**, *8*, 358. [CrossRef]
22. Mavrova, A.T.; Wesselinova, D.; Vassilev, N.; Tsenov, J.A. Design, Synthesis and Antiproliferative Properties of Some New 5-Substituted-2-Iminobenzimidazole Derivatives. *Eur. J. Med. Chem.* **2013**, *63*, 696–701. [CrossRef]
23. Shang, D.; Liu, Y.; Jiang, F.; Ji, F.; Wang, H.; Han, X. Synergistic Antibacterial Activity of Designed Trp-Containing Antibacterial Peptides in Combination With Antibiotics Against Multidrug-Resistant *Staphylococcus epidermidis*. *Front. Microbiol.* **2019**, *10*, 2719. [CrossRef]

24. Ianevski, A.; Giri, A.K.; Aittokallio, T. SynergyFinder 3.0: An Interactive Analysis and Consensus Interpretation of Multi-Drug Synergies across Multiple Samples. *Nucleic Acids Res.* **2022**, *50*, W739–W743. [CrossRef] [PubMed]
25. Elizondo-Luévano, J.H.; Hernández-García, M.E.; Pérez-Narváez, O.A.; Castro-Ríos, R.; Chávez-Montes, A. Berberina, Curcumina y Quercetina Como Potenciales Agentes Con Capacidad Antiparasitaria. *Rev. Biol. Trop.* **2020**, *68*, 1241–1249. [CrossRef]
26. Elizondo-Luévano, J.H.; Castro-Ríos, R.; Vicente, B.; Fernández-Soto, P.; López-Abán, J.; Muro, A.; Chávez-Montes, A. In Vitro Antischistosomal Activity of the *Argemone mexicana* Methanolic Extract and Its Main Component Berberine. *Iran. J. Parasitol.* **2021**, *16*, 91. [CrossRef] [PubMed]
27. Elizondo-Luévano, J.H.; Castro-Ríos, R.; López-Abán, J.; Gorgojo-Galindo, O.; Fernández-Soto, P.; Vicente, B.; Muro, A.; Chávez-Montes, A. Berberine: A Nematocidal Alkaloid from *Argemone mexicana* against *Strongyloides venezuelensis*. *Exp. Parasitol.* **2021**, *220*, 108043. [CrossRef] [PubMed]
28. Elizondo-Luévano, J.H.; Pérez-Narváez, O.A.; Sánchez-García, E.; Castro-Ríos, R.; Hernández-García, M.E.; Chávez-Montes, A. *In-Vitro* Effect of *Kalanchoe daigremontiana* and Its Main Component, Quercetin against *Entamoeba histolytica* and *Trichomonas vaginalis*. *Iran. J. Parasitol.* **2021**, *16*, 394. [CrossRef]
29. López Villarreal, S.M.; Elizondo Luévano, J.H.; Pérez Hernández, R.A.; Sánchez García, E.; Verde Star, M.J.; Castro Ríos, R.; Garza Tapia, M.; Rodríguez Luis, O.E.; Chávez Montes, A. Preliminary Study of the Antimicrobial, Anticoagulant, Antioxidant, Cytotoxic, and Anti-Inflammatory Activity of Five Selected Plants with Therapeutic Application in Dentistry. *Int. J. Environ. Res. Public Health* **2022**, *19*, 7927. [CrossRef]
30. Wong-Paz, J.E.; Contreras-Esquivel, J.C.; Rodríguez-Herrera, R.; Carrillo-Inungaray, M.L.; López, L.I.; Nevárez-Moorillón, G.V.; Aguilar, C.N. Total Phenolic Content, in Vitro Antioxidant Activity and Chemical Composition of Plant Extracts from Semiarid Mexican Region. *Asian Pac. J. Trop. Med.* **2015**, *8*, 104–111. [CrossRef]
31. Popoca, J.; Aguilar, A.; Alonso, D.; Villarreal, M.L. Cytotoxic Activity of Selected Plants Used as Antitumorals in Mexican Traditional Medicine. *J. Ethnopharmacol.* **1998**, *59*, 173–177. [CrossRef]
32. Quiñonez-Bastidas, G.N.; Navarrete, A. Mexican Plants and Derivates Compounds as Alternative for Inflammatory and Neuropathic Pain Treatment—A Review. *Plants* **2021**, *10*, 865. [CrossRef]
33. Moghadam, M.H.; Hajimehdipoor, H.; Saeidnia, S.; Atoofi, A.; Shahrestani, R.; Read, R.W.; Mosaddegh, M. Anti-Proliferative Activity and Apoptotic Potential of Britannin, a Sesquiterpene Lactone from *Inula aucheriana*. *Nat. Prod. Commun.* **2012**, *7*, 1934578X1200700. [CrossRef]
34. Gomez-Flores, R.; Verástegui-Rodríguez, L.; Quintanilla-Licea, R.; Tamez-Guerra, P.; Monreal-Cuevas, E.; Tamez-Guerra, R.; Rodríguez-Padilla, C. Antitumor Properties of *Gymnosperma glutinosum* Leaf Extracts. *Cancer Investig.* **2009**, *27*, 149–155. [CrossRef] [PubMed]
35. Mohammadlou, H.; Hamzeloo-Moghadam, M.; Mohammadi, M.H.; Yami, A.; Gharehbaghian, A. Britannin, a Sesquiterpene Lactone Induces ROS-Dependent Apoptosis in NALM-6, REH, and JURKAT Cell Lines and Produces a Synergistic Effect with Vincristine. *Mol. Biol. Rep.* **2021**, *48*, 6249–6258. [CrossRef]
36. Safaroghli-Azar, A.; Bashash, D.; Sadreazami, P.; Momeny, M.; Ghaffari, S.H. PI3K-δ Inhibition Using CAL-101 Exerts Apoptotic Effects and Increases Doxorubicin-Induced Cell Death in Pre-B-Acute Lymphoblastic Leukemia Cells. *Anticancer Drugs* **2017**, *28*, 436–445. [CrossRef] [PubMed]
37. Coballase-Urrutia, E.; Pedraza-Chaverri, J.; Camacho-Carranza, R.; Cárdenas-Rodríguez, N.; Huerta-Gertrudis, B.; Medina-Campos, O.N.; Mendoza-Cruz, M.; Delgado-Lamas, G.; Espinosa-Aguirre, J.J. Antioxidant Activity of *Heterotheca inuloides* Extracts and of Some of Its Metabolites. *Toxicology* **2010**, *276*, 41–48. [CrossRef] [PubMed]
38. Manzur-Valdespino, S.; Arias-Rico, J.; Ramírez-Moreno, E.; Sánchez-Mata, M.D.C.; Jaramillo-Morales, O.A.; Angel-García, J.; Zafra-Rojas, Q.Y.; Barrera-Gálvez, R.; Cruz-Cansino, N.D.S. Applications and Pharmacological Properties of Cactus Pear (*Opuntia* spp.) Peel: A Review. *Life* **2022**, *12*, 1903. [CrossRef] [PubMed]
39. Alam-Escamilla, D.; Estrada-Muñiz, E.; Solís-Villegas, E.; Elizondo, G.; Vega, L. Genotoxic and Cytostatic Effects of 6-Pentadecyl Salicylic Anacardic Acid in Transformed Cell Lines and Peripheral Blood Mononuclear Cells. *Mutat. Res./Genet. Toxicol. Environ. Mutagen.* **2015**, *777*, 43–53. [CrossRef] [PubMed]
40. Gomes, J.P.M.; Cardoso, C.R.P.; Varanda, E.A.; Molina, J.M.; Fernandez, M.F.; Olea, N.; Carlos, I.Z.; Vilegas, W. Antitumoral, Mutagenic and (Anti)Estrogenic Activities of Tingenone and Pristimerin. *Rev. Bras. Farmacogn.* **2011**, *21*, 963–971. [CrossRef]
41. Maldonado-Cubas, J.; San Martin-Martínez, E.; Quiroz-Reyes, C.N.; Casañas-Pimentel, R.G. Cytotoxic Effect of *Semialarium mexicanum* (Miers) Mennega Root Bark Extracts and Fractions against Breast Cancer Cells. *Physiol. Mol. Biol. Plants* **2018**, *24*, 1185–1201. [CrossRef]
42. Alves, A.S.A.; Santos, G.C.; Albuquerque, U.P. Mimosa *tenuiflora* (Willd.) Poir. BT. In *Medicinal and Aromatic Plants of South America: Brazil*; Albuquerque, U.P., Patil, U., Máthé, Á., Eds.; Springer: Dordrecht, The Netherlands, 2018; pp. 345–353. ISBN 978-94-024-1552-0.
43. Anton, R.; Jiang, Y.; Weniger, B.; Beck, J.P.; Rivier, L. Pharmacognosy of *Mimosa tenuiflora* (Willd.) Poiret. *J. Ethnopharmacol.* **1993**, *38*, 145–152. [CrossRef]
44. Zeinoddini, S.; Nabiuni, M.; Jalali, H. The Synergistic Cytotoxic Effects of Doxorubicin and *Viola odorata* Extract on Human Breast Cancer Cell Line T47-D. *J. Cancer Res. Ther.* **2019**, *15*, 1073. [CrossRef]

45. Buckner, C.A.; Lafrenie, R.M.; Dénommée, J.A.; Caswell, J.M.; Want, D.A.; Gan, G.G.; Leong, Y.C.; Bee, P.C.; Chin, E.; Teh, A.K.H.; et al. We Are IntechOpen, the World's Leading Publisher of Open Access Books Built by Scientists, for Scientists TOP 1 %. *Intech* **2016**, *11*, 13.
46. Acharya, A.; Das, I.; Chandhok, D.; Saha, T. Redox Regulation in Cancer: A Double-Edged Sword with Therapeutic Potential. *Oxidative Med. Cell. Longev.* **2010**, *3*, 23–34. [CrossRef] [PubMed]
47. De La Cruz-Jiménez, L.; Hernández-Torres, M.A.; Monroy-García, I.N.; Rivas-Morales, C.; Verde-Star, M.J.; Gonzalez-Villasana, V.; Viveros-Valdez, E. Biological Activities of Seven Medicinal Plants Used in Chiapas, Mexico. *Plants* **2022**, *11*, 1790. [CrossRef] [PubMed]
48. Amakura, Y.; Umino, Y.; Tsuji, S.; Ito, H.; Hatano, T.; Yoshida, T.; Tonogai, Y. Constituents and Their Antioxidative Effects in Eucalyptus Leaf Extract Used as a Natural Food Additive. *Food Chem.* **2002**, *77*, 47–56. [CrossRef]
49. Rauf, A.; Imran, M.; Butt, M.S.; Nadeem, M.; Peters, D.G.; Mubarak, M.S. Resveratrol as an Anti-Cancer Agent: A Review. *Crit. Rev. Food Sci. Nutr.* **2018**, *58*, 1428–1447. [CrossRef]
50. Mishra, K.; Ojha, H.; Chaudhury, N.K. Estimation of Antiradical Properties of Antioxidants Using DPPH Assay: A Critical Review and Results. *Food Chem.* **2012**, *130*, 1036–1043. [CrossRef]
51. Pacheco-Ordaz, A.; Sánchez-García, E.; Quintanilla-Licea, R.; Bazaldúa-Rodríguez, A.F.; Pérez-Hernández, R.A.; Hernández-García, M.E.; Báez-González, J.G.; Castro-Ríos, R.; Elizondo-Luévano, J.H.; Chávez-Montes, A. Amoebicidal and Trichomonicidal Capacity of Polymeric Nanoparticles Loaded with Extracts of the Plants *Curcuma longa* (Zingiberaceae) and *Berberis vulgaris* (Berberidaceae). *Rev. Biol. Trop.* **2022**, *70*, 319–331. [CrossRef]
52. Shiva Shankar Reddy, C.S.; Subramanyam, M.V.V.; Vani, R.; Asha Devi, S. In Vitro Models of Oxidative Stress in Rat Erythrocytes: Effect of Antioxidant Supplements. *Toxicol. Vitr.* **2007**, *21*, 1355–1364. [CrossRef]
53. Liu, Z.Q.; Han, K.; Lin, Y.J.; Luo, X.Y. Antioxidative or Prooxidative Effect of 4-Hydroxyquinoline Derivatives on Free-Radical-Initiated Hemolysis of Erythrocytes Is Due to Its Distributive Status. *Biochim. Biophys. Acta (BBA)—Gen. Subj.* **2002**, *1570*, 97–103. [CrossRef]
54. Tienda-Vázquez, M.A.; Morreeuw, Z.P.; Sosa-Hernández, J.E.; Cardador-Martínez, A.; Sabath, E.; Melchor-Martínez, E.M.; Iqbal, H.M.N.; Parra-Saldívar, R. Nephroprotective Plants: A Review on the Use in Pre-Renal and Post-Renal Diseases. *Plants* **2022**, *11*, 818. [CrossRef]

life

Article

Chemical Profiling and Molecular Docking Study of *Agathophora alopecuroides*

Elham Amin [1,2], Mohamed Sadek Abdel-Bakky [3,4], Hamdoon A. Mohammed [1,5] and Marwa H. A. Hassan [2,*]

1 Department of Medicinal Chemistry and Pharmacognosy, College of Pharmacy, Qassim University, Buraydah 51452, Saudi Arabia
2 Department of Pharmacognosy, Faculty of Pharmacy, Beni-Suef University, Beni-Suef 62514, Egypt
3 Department of Pharmacology and Toxicology, College of Pharmacy, Qassim University, Buraydah 51452, Saudi Arabia
4 Department of Pharmacology and Toxicology, Faculty of Pharmacy, Al-Azhar University, Cairo 11751, Egypt
5 Department of Pharmacognosy and Medicinal Plants, Faculty of Pharmacy, Al-Azhar University, Cairo 11751, Egypt
* Correspondence: marwa.hassan@pharm.bsu.edu.eg or mh_elsefy@yahoo.com; Tel.: +2-012-7898-2288; Fax: +2-(082)-2317958

Abstract: Natural products continue to provide inspiring chemical moieties that represent a key stone in the drug discovery process. As per our previous research, the halophyte *Agathophora alopecuroides* was noted as a potential antidiabetic plant. However, the chemical profiling and highlighting the metabolite(s) responsible for the observed antidiabetic activity still need to be investigated. Accordingly, the present study presents the chemical profiling of this species using the LC-HRMS/MS technique followed by a study of the ligand–protein interaction using the molecular docking method. LC-HRMS/MS results detected twenty-seven compounds in *A. alopecuroides* extract (AAE) belonging to variable chemical classes. Among the detected compounds, alkaloids, flavonoids, lignans, and iridoids were the most prevailing. In order to highlight the bioactive compounds in AAE, the molecular docking technique was adopted. Results suggested that the two alkaloids (Eburnamonine and Isochondrodendrine) as well as the four flavonoids (Narirutin, Pelargonidin 3-O-rutinoside, Sophora isoflavanone A, and Dracorubin) were responsible for the observed antidiabetic activity. It is worth mentioning that this is the first report for the metabolomic profiling of *A. alopecuroides* as well as the antidiabetic potential of Isochondrodendrine, Sophora isoflavanone A, and Dracorubin that could be a promising target for an antidiabetic drug.

Keywords: *Agathophora alopecuroides*; antidiabetic; LC-HRMS/MS; molecular docking

Citation: Amin, E.; Abdel-Bakky, M.S.; Mohammed, H.A.; Hassan, M.H.A. Chemical Profiling and Molecular Docking Study of *Agathophora alopecuroides*. *Life* **2022**, *12*, 1852. https://doi.org/10.3390/life12111852

Academic Editor: Jianfeng Xu

Received: 18 October 2022
Accepted: 7 November 2022
Published: 11 November 2022

1. Introduction

Diabetes mellitus is a chronic metabolic disease recognized by an increase in blood glucose levels which develops from a deficiency in insulin secretion, action, or both of them [1]. Type 2 diabetes mellitus (T2DM) is the most common health problem and accounts for about 90% of diabetes cases with 4.9 million mortalities throughout the world [2]. Inhibiting the digestion of dietary carbohydrates is one of the effective procedures for the management of postprandial hyperglycemia in T2DM. One of the essential digestive enzymes is pancreatic α-amylase which converts dietary carbohydrates such as starch into smaller oligosaccharides mixture that are further broken down into glucose by α-glucosidase, another important metabolic enzyme. Upon the absorption of glucose, it enters the bloodstream and causes postprandial elevation in blood glucose levels. Therefore, blocking the enzymes α-amylase and α-glucosidase can inhibit the digestion of carbohydrates, postpone glucose uptake, and subsequently lower blood sugar levels [3]. Recently, medicinal plants proved their great therapeutic potential and negligible side effects in the treatment of T2DM.

For instance, several medicinal herbs have been reported to exhibit strong glucosidase and amylase inhibitory properties [4–6].

Halophytes are salt-tolerating plants noted for their ability to produce variable secondary metabolites, such as alkaloids, glycosides, and terpenes. Hence, they could be considered as promising sources for bioactive metabolites that could be used for the treatment of various diseases such as diabetes [7–9]. *A. alopecuroides* is a halophytic species prevalent in the deserts of Saudi Arabia and was reported to exhibit a strong in vitro and in vivo antidiabetic activity [7]; nevertheless, its chemical profile remained to be investigated [10].

Dereplication, defined as the rapid identification of known compounds from natural product extracts, represents an important step in drug discovery programs. This approach combines the benefits of different analytical techniques, modern spectroscopic methods, as well database searches for the prompt characterization of an active compound during the drug discovery process. Recently, advances in technology have provided what is called tandem analytical techniques, such as LC-MS, LC-MS/MS, LC-NMR, HPLC-PDA, and LC-NMR/MS [11]. Tandem mass spectrometry (MS/MS) is a powerful technique for the characterization of target phytoconstituents in complex plant extracts. The high sensitivity, selectivity, and fast screening abilities of the LC-MS/MS technique, compared to other dereplication techniques, rationalized the privilege of this technique for the online identification of secondary metabolites in plant extracts [11].

Furthermore, advances in computational biology have had a great impact on reducing the time, cost, and effort spent while screening the biological activity of natural products. Molecular docking is now widely adopted for predicting the binding mode and binding affinity of a drug-like molecule into the active site of the receptor. A huge number of natural and synthetic compounds could be virtually screened for activity against a wide array of targets, thus reducing the time and effort and giving a rapid expectation for the most promising candidates [12].

Based on this concept and in continuation of our previous research on the halophytes with promising antidiabetic potential [7], which recorded the strong in vitro and in vivo antidiabetic potential of AAE, the current study investigated the metabolic content of AAE using LC-MS/MS followed by the screening of the annonated compounds for the enzyme inhibitory potential against the two carbohydrate-metabolizing enzymes—α-amylase and α-glucosidase—using the molecular docking technique.

2. Materials and Methods

2.1. Plant Materials

Agathophora alopecuroides var. papillosa was collected from the Qassim region in the northcentral Saudi Arabia during October 2021. The taxonomic identity of the plant was confirmed by Ibrahim Aldakhil, botanical expert, Qassim Area, and avoucher sample number QPP-101 was kept at the College of Pharmacy, Qassim University (Buraydah, Qassim, Saudi Arabia). The aerial parts of the plant were carefully cleaned then dried in the shade for two weeks. The dried powdered plant material (500 g) was extracted by maceration in 80% methanol (3 times × 1000 mL) at room temperature with frequent shaking. The methanolic extract was then concentrated by a vacuum rotary evaporator. The dried extract was then kept in an amber-colored vial at 4 °C till further use.

2.2. Metabolites Profiling of the Methanolic Extract of A. alopecuroides

AAE was reconstituted in HPLC grade methanol, filtered through a 0.22 μm PTFE membrane, and then separation was performed adopting Thermo Scientific C_{18} column (AcclaimTM Polar Advantage II, 3 × 150 mm, 3 μm particle size) on an UltiMate 3000 UHPLC system (Dionex). Gradient elution was performed at a flow rate of 0.4 mL/min and a column temperature of 40 °C, using H_2O + 0.1% Formic Acid (A) and 100% Acetonitrile (B) with a 22 min total run time. The injection volume of the sample was 3 μL. The gradient started at 5% B (0–3 min), 80% B (3–10 min), 80% B (10–15 min), and 5% B (15–22 min). High-resolution mass spectrometry was carried out using a MicroTOF QIII (Bruker Daltonic, Bremen, Germany) using an ESI positive ionization and adjusting the following settings: capillary voltage: 4500 V; nebulizer pressure: 2.0 bar; drying gas: 8 L/min at 300 °C. The mass range was 50–1000 *m/z*. The accurate mass data of the molecular ions, provided by the TOF analyzer, were processed by Compass Data Analysis software (Bruker Daltonik GmbH).

2.3. Molecular Docking Study

AutoDock Vina software was used in all molecular docking experiments [13]. All dereplicated compounds were docked against the active sites of both human α-amylase and human α-glucosidase crystal structure (PDB codes: 4W93 and 3L4W, respectively) [14,15]. The binding site was determined according to the enzyme's co-crystallized ligands (Montbretin A and Miglitol, respectively). The co-ordinates of the grid boxes were (x = −9.682; y = 4.274; z = −23.145 and x = 45.424; y = 92.375; z = 34.811). The size of the grid box was set to 20 Å. Exhaustiveness was set to 24. Ten poses were generated for each docking experiment. Docking poses were analyzed and visualized using Pymol software [13]. The full method is provided in the Supplementary Materials.

3. Results

3.1. In Vitro Testing of the Antidibetic Activity of A. alopecuroides (AAE)

This research represents an extension of our previous work on halophytic plants with potential antidiabetic activity. In the previous publication, in vitro testing revealed the strong inhibitory activity of the hydroalcoholic extract (AAE) against α-glucosidase and α-amylase with IC_{50} values 117.9 and 90.9 μg/mL, respectively, compared to 191.4 and 53.3 μg/mL of the standard drug Acarbose [7].

3.2. Metabolomic Profiling of the Methanolic Extract of A. alopecuroides (AAE)

The metabolic profiling using LC-HRMS/MS of AAE led to the annonation of 27 metabolites with variable chemical structures. The annonated metabolites could be classified according to their chemical classes into: nine alkaloids, five flavonoids, four lignans, two iridoid glycosides, two acids, one anthraquinone, one benzisochroman, one furanochomarine, one triterpenoid saponin, and one benzofuran (Table 1, Figures 1 and 2). Using the metabolomic data, the nine alkaloids were annotated from the mass ion peaks at *m/z* 275.152, 294.173, 361.153, 285.290, 328.155, 313.131, 301.131, 594.273, and 356.319, which were in agreement with the molecular formulas $C_{16}H_{21}NO_3$, $C_{19}H_{22}N_2O$, $C_{19}H_{23}NO_6$, $C_{16}H_{15}NO_4$, $C_{19}H_{22}NO_4$, $C_{18}H_{19}NO_4$, $C_{17}H_{19}NO_4$, $C_{36}H_{38}N_2O_6$, and $C_{24}H_{40}N_2$, respectively. These alkaloids were dereplicated as Epinorlycoramine (**1**), Eburnamonine (**2**), 3-Acetylnerbowdine (**3**), Arborinine (**4**), 1,2-Dehydroreticuline (**5**), N-Feruloyltyramine (**16**), Powelline (**19**), Isochondrodendrine (**24**), and Conessine (**26**), respectively. While the five flavonoids were dereplicated from the mass ion peaks at *m/z* 580.179, 579.171, 446.121, 370.142, and 488.162 that matched with the molecular formulas $C_{27}H_{32}O_{14}$, $C_{27}H_{31}O_{14}$, $C_{22}H_{22}O_{10}$, $C_{21}H_{22}O_6$, and $C_{32}H_{24}O_5$, respectively. These metabolites were characterized as Narirutin (**9**), Pelargonidin 3-O-rutinoside (**10**), Biochanin A-β-D-glucoside (**12**), Sophora isoflavanone A (**14**), and Dracorubin (**23**), respectively The four lignans were dereplicated from the molecular ion peaks at *m/z* 372.194, 520.194, 312.121, and 342.131 corresponding to the suggested formulas $C_{22}H_{28}O_5$, $C_{26}H_{32}O_{11}$, $C_{15}H_{20}O_7$, and $C_{16}H_{22}O_3$, respectively, as Veraguensin (**8**), Pinoresinol glucoside (**12**), 4-Hydroxycinnamyl alcohol 4-D-glucoside (**17**), and Coniferin (**18**), respectively. The mass ion peaks at *m/z* 344.147 and 388.137 were in agreement with the molecular formulas $C_{16}H_{24}O_8$ and $C_{17}H_{24}O_{10}$ that were characterized as the two iridoid glycosides Boschnaloside (**11**) and Geniposide (**15**). Additionally, seven mass ion peaks at *m/z* 402.095, 288.136, 300.136, 314.152, 628.304, 278.225, and 634.408, compatible with the molecular formulae $C_{20}H_{18}O_9$, $C_{17}H_{20}O_4$, $C_{18}H_{20}O_4$, $C_{19}H_{22}O_4$, $C_{38}H_{44}O_8$, $C_{18}H_{30}O_2$, and $C_{36}H_{58}O_9$, respectively, were dereplicated as Versiconol acetate (**6**), Karwinaphthol B (**7**), Toxyl angelate (**20**), Heliettin (**21**), Gambogic acid (**22**) Punicic acid (**25**), and Soyasapogenol B 3-*O*-D-glucuronide (**27**), respectively.

Table 1. The LC-HRMS/MS dereplication results of *A. alopecuroides* methanolic extract.

No.	Tentative Identification	Rt	*m/z*	Formula	Ms/Ms-Fragmnets	Biological Source	Nature of Compounds	References
1	Epinorlycoramine	1.2	275.152	$C_{16}H_{21}NO_3$	112.0874-124.0875-154.0978-190.098-202.0982-220.1087	*Narcissus leonensis* plant	Alkaloid	[16]
2	Eburnamonine	1.6	294.173	$C_{19}H_{22}N_2O$	114.1028-145.0519	Leaves of *Kopsia zarutensi*	Alkaloid	[17]
3	3-Acetylnerbowdine	2.0	361.153	$C_{19}H_{23}NO_6$	151.0774	Bulbs of *Nerine bowdenii*	Alkaloid	[18,19]
4	Arborinine	3.3	285.290	$C_{16}H_{15}NO_4$	110.0375-135.0299-153.0546	Leaves of *Glycosmis parva*	Alkaloid	[20]
5	1,2-Dehydroreticuline	4.1	328.155	$C_{19}H_{22}NO_4$	120.0817-132.0821-166.0880-178.0875	Bark and the root of *Xylopia parviflora*	Alkaloid	[21,22]
6	Versiconol acetate	6.7	402.095	$C_{20}H_{18}O_9$	143.0711	Culture of *Aspergillus parasiticus*	Anthraquinone	[23]
7	Karwinaphthol B	7.0	288.136	$C_{17}H_{20}O_4$	112.0879-117.0333-138.0664-145.0289	Roots of *Karwinskia humboldtiana*	Benzisochromans	[24]
8	Veraguensin	7.4	372.194	$C_{22}H_{28}O_5$	177.0550-196.1444-222.1249	Leaves and root bark of *Nectandra turbacensis* (Kunth) Nees	Lignan	[25,26]
9	Narirutin	8.6	580.179	$C_{27}H_{32}O_{14}$	313.0713-415.1029-433.1137	Citrus fruits	Flavanone glycoside	[27]
10	Pelargonidin 3-*O*-rutinoside	8.6	579.171	$C_{27}H_{31}O_{14}$	313.0713-397.0918 415.1029-433.1137	Strawberries	Anthocyanin	[28]
11	Boschnaloside	8.8	344.147	$C_{16}H_{24}O_8$	151.0742-177.0546-186.0527	*Boschniakia rossica* plant	Iridoid glycoside	[29,30]
12	Biochanin A-β-D-glucoside	9.0	446.121	$C_{22}H_{22}O_{10}$	145.0288-175.0619-177.0551	*Trifolium pratense* L. plant	Isoflavone glycoside	[31]
13	Pinoresinol glucoside	9.1	520.194	$C_{26}H_{32}O_{11}$	177.0552-184.0717-186.0523-191.0710	Prunes of *Prunus domestica* L.	Lignan	[32]
14	Sophora isoflavanone A	9.2	370.142	$C_{21}H_{22}O_6$	145.0285-177.0553-284.0695	*Sophora tomentosa* L. plant	Isoflavone	[33]
15	Geniposide	9.2	388.137	$C_{17}H_{24}O_{10}$	149.0614-151.0390-177.0552-186.0541	Fruit of *Gardenia jasminoides* Ellis	Iridoid glycoside	[34]

Table 1. *Cont.*

No.	Tentative Identification	Rt	*m/z*	Formula	Ms/Ms-Fragmnets	Biological Source	Nature of Compounds	References
16	N-Feruloyltyramine	9.3	313.131	$C_{18}H_{19}NO_4$	117.0339-145.0293-149.0607-162.0539	Fruits of *Lycium barbarum* (goji berries) *Bassia indica* and *A. alopecuroides* plants	Alkaloid	[10,35]
17	4-Hydroxycinnamyl alcohol 4-D-glucoside	9.3	312.121	$C_{15}H_{20}O_7$	117.0339-145.0293-149.0607-	*Linum usitatissimum*, Linn. plant	Lignan	[36,37]
18	Coniferin	9.4	342.131	$C_{16}H_{22}O_8$	137.0600-175.0763-177.0546-218.0794	*Paulownia tomentosa* bark	Lignan	[38,39]
19	Powelline	9.7	301.131	$C_{17}H_{19}NO_4$	121.0652-135.0448-163.0395-180.0660	Leaves from *Crinum latifolium* L.	Alkaloid	[40]
20	Toxyl angelate	9.7	300.136	$C_{18}H_{20}O_4$	121.0652-135.0448-145.0286-163.0395-	*Isocoma wrightii* plant	Banzofuran	[41]
21	Heliettin	10.1	314.152	$C_{19}H_{22}O_4$	121.0657-145.0289-177.0553	Stem bark of *Helietta longifoliata* Britt	Furanochomarine	[42,43]
22	Gambogic acid	10.2	628.304	$C_{38}H_{44}O_8$	121.0655-177.0549-201.0549-297.1123	*Garcinia hanburyi* plant	Phenolic acid (Xanthoid derivative)	[44]
23	Dracorubin	10.9	488.162	$C_{32}H_{24}O_5$	121.0654-177.0565-201.0545-323.0921	Resin extracted from the tree *Dracaena draco*	Proanthocyanidine	[45]
24	Isochondrodendrine	12.1	594.273	$C_{36}H_{38}N_2O_6$	565.2679	*Cissampelos mucronate* and *Cissampelos pareira* plants	Alkaloids	[46]
25	Punicic acid	13.0	278.225	$C_{18}H_{30}O_2$	107.0864-121.1017-133.1021-135.1160-149.1334	Pomegranate Seed Oil	Unsaturated fatty acid	[47]
26	Conessine	13.3	356.319	$C_{24}H_{40}N_2$	121.1015-123.1154-135.1162-149.1330	*Holarrhena floribunda* G. Don. plant	Alkaloid	[48]
27	Soyasapogenol B 3-*O*-D-glucuronide	13.5	634.408	$C_{36}H_{58}O_9$	133.0863-177.1134-247.2054-291.2317	Aerial parts of *Lathylus palustris* L.	Triterpenoid saponin	[49,50]

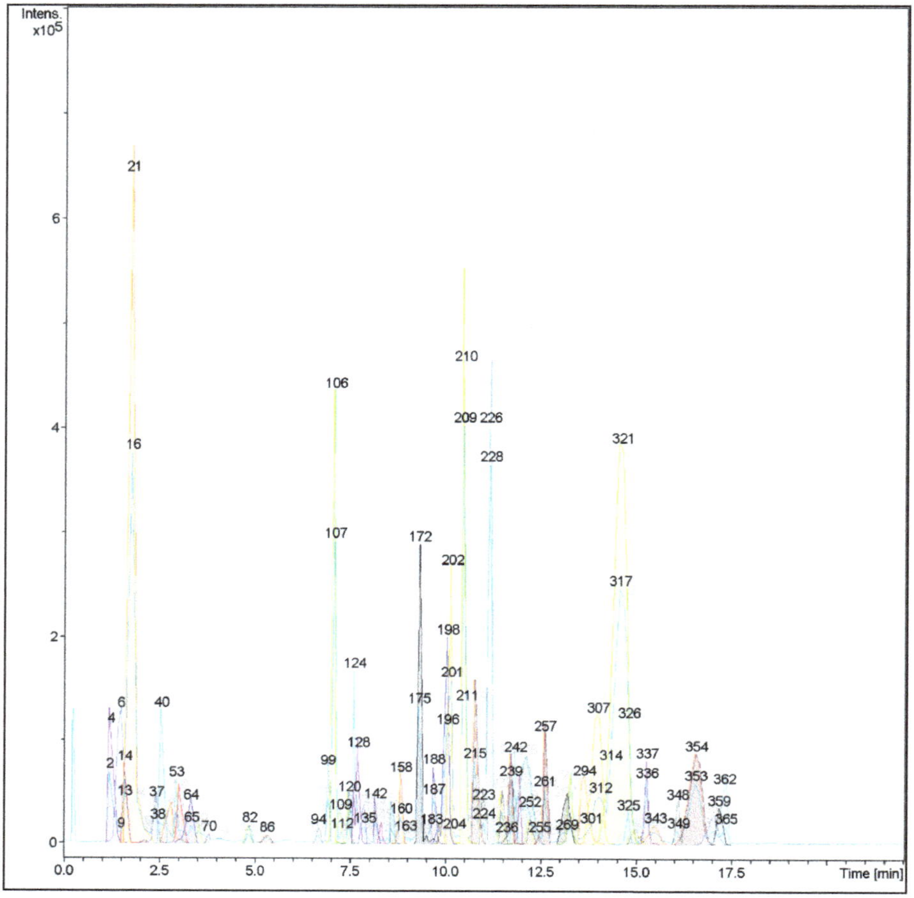

Figure 1. LC-HRMS chromatogram of the dereplicated metabolites of *A. alopecuroides* methanolic extract (positive mode).

3.3. Molecular Docking Analysis

All the dereplicated compounds were subjected to molecular docking against both human α-amylase and human α-glucosidase enzymes (PDB codes: 4W93 and 3L4W, respectively). As shown in (Table 2 and Figure 3) and (Supplementary Materials, Figures S1 and S2), all dereplicated compounds received scores ranging from ~−4.5 to −9.1 kcal/mol against both enzymes. Most of the compounds (21 compounds in the case of α-amylase, and 23 compounds in the case of α-glucosidase) received scores ranging between −5 and −8 kcal/mol. The compounds that achieved the best score (<−8 kcal/mol) against α-amylase enzyme were Eburnamonine (**2**), Narirutin (**9**), Pelargonidin 3-*O*-rutinoside (**10**), and Isochondrodendrine (**24**). Considering α-glucosidase enzyme inhibition, the best scoring compounds (<−8 kcal/mol) were Pelargonidin 3-*O*-rutinoside (**10**), Sophora isoflavanone A (**14**), and Dracorubin (**23**). The two co-crystallized compounds—Montbretin A with amylase and Miglitol with glucosidase—achieved binding scores of −8.1 and −8.0 kcal/mol, respectively, and hence −8.0 kcal/mol was chosen as a cut-off value to select the best scoring compounds with each enzyme.

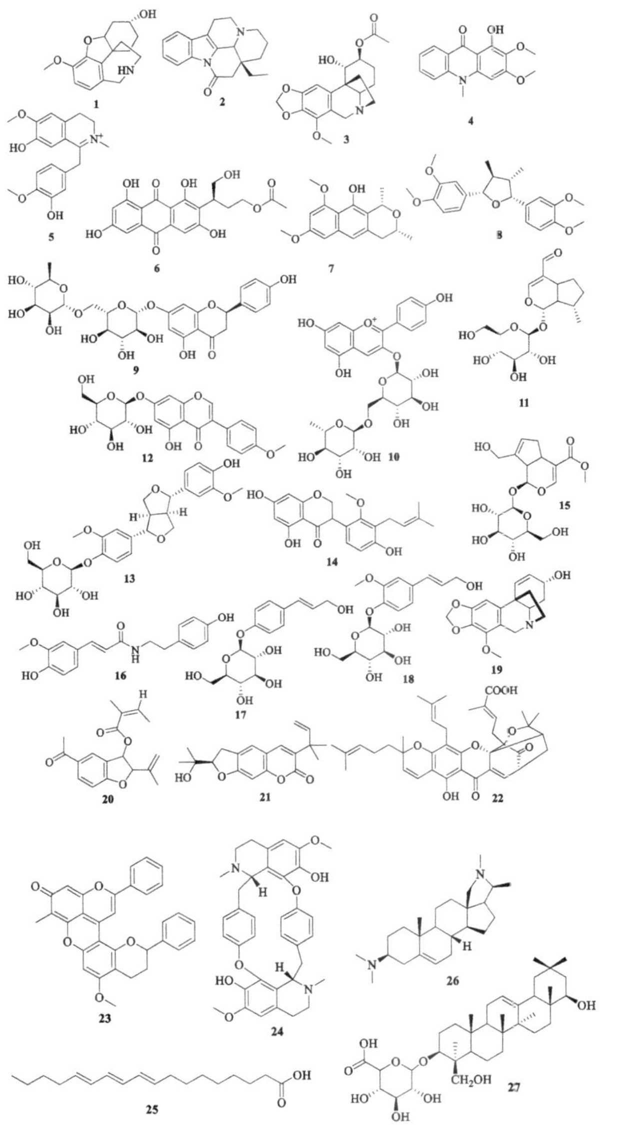

Figure 2. Structures of the dereplicated compounds from *A. alopecuroides* methanolic extract by LC-HRMS/MS.

Table 2. Docking sores and estimated absolute binding free energies (in kcal/mol) of the dereplicated structures, along with those of the reported co-crystalized inhibitors.

No.	Tentative Identification	Binding Energy (kcal/mol)	
		α-Amylase	α-Glucosidase
1	Epinorlycoramine	−7.6	−6.2
2	Eburnamonine	−8.7	−6.4
3	3-Acetylnerbowdine	−7.2	−6.6
4	Arborinine	−7.5	−6.0

Table 2. *Cont.*

No.	Tentative Identification	Binding Energy (kcal/mol)	
		α-Amylase	α-Glucosidase
5	1,2-Dehydroreticuline	−7.4	−7.4
6	Versiconol acetate	−7.5	−7.1
7	Karwinaphthol B	−7.4	−5.9
8	Veraguensin	−7.5	−6.7
9	Narirutin	−8.5	−7.9
10	Pelargonidin 3-O-rutinoside	−8.5	−8.4
11	Boschnaloside	−7.1	−5.8
12	Biochanin A-β-D-glucoside	−7.2	−6.3
13	Pinoresinol glucoside	−7.9	−6.0
14	Sophora isoflavanone A	−7.4	−9.1
15	Geniposide	−7.2	−6.5
16	N-Feruloyltyramine	−7.3	−7.6
17	4-Hydroxycinnamyl alcohol 4-D-glucoside	−7.3	−6.8
18	Coniferin	−6.9	−5.7
19	Powelline	−6.7	−6.5
20	Toxyl angelate	−7.2	−6.4
21	Heliettin	−7.2	−7.1
22	Gambogic acid	−7.2	−7.1
23	Dracorubin	−7.8	−8.3
24	Isochondrodendrine	−9.1	−7.0
25	Punicic acid	−5.6	−5.7
26	Conessine	−4.8	−5.6
27	Soyasapogenol B 3-O-D-glucuronide	−4.5	−4.7
STD	Montbretin A	−8.1	....
	Miglitol	....	−8.0

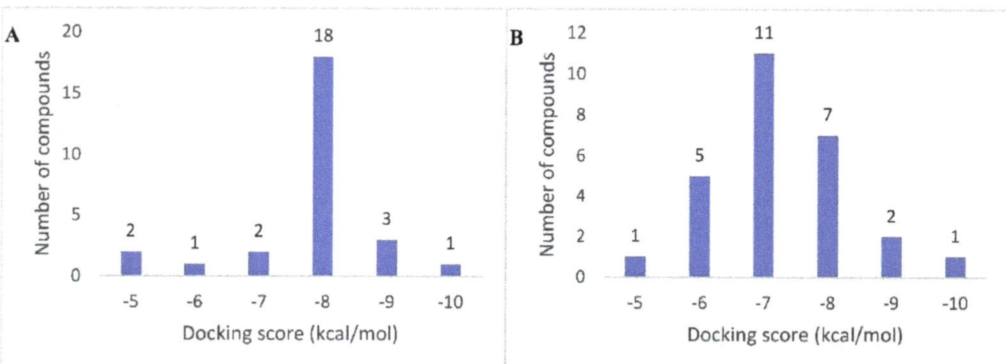

Figure 3. Docking score distribution of the dereplicated metabolites in AAE against both human α-amylase (**A**) and α-glucosidase (**B**) (PDB: 4W93 and 3L4W, respectively).

4. Discussion

This study is a continuation of our previous research dealing with the potential antidiabetic agents from halophytes. Previously, AAE was acknowledged for potent antidiabetic activity through both in vitro and in vivo investigations. The current study adds more information about the metabolic profiling of AAE and highlights the most promising metabolites expected to be responsible for the recorded antidiabetic activity.

Metabolomic profiling of the methanolic extract of *A. alopecuroides*

For tentative identification of the components of AAE, we adopted a dereplication strategy using the hyphenated technique: liquid chromatography coupled with tandem mass spectrometry (LC-HRMS/MS) (Figures 1 and 2). The dereplication process led to the recognition of 27 compounds for the first time from AAE. The detected metabolites belong to different chemical classes: alkaloids, flavonoids, lignans, iridoid glycosides, anthraquinones, benzisochromans, furanochomarine, triterpenoid saponins, acids, and benzofurans (Table 1, Figure 2). Amid the detected metabolites, alkaloids represent the most prevailing chemical class. These alkaloids belong to different types, and among them the most abundant is the isoquinoline-type alkaloids (Figure 1). Epinorlycoramine (**1**) is a galanthamine-type alkaloid previously isolated from *Narcissus leonensis* whole plant [16], while Eburnamonine (**2**), an eburnan-type alkaloid, was isolated from *Kopsia larutensis* leaves [17]. From the crinine-type alkaloids, two alkaloids were detected: 3-Acetylnerbowdine (**3**), which was reported in *Nerine bowdenii* bulbs using GC-MS analysis [18,19], and Powelline (**19**), reported in *Crinum latifolium* leaves using GC-MS analysis [40]. Arborinine (**4**), a nitrogen-containing drug similar to anthracene classified as acridone alkaloid, was previously isolated from the ethyl acetate extract of the *Glycosmis parva* plant [20]. The quaternary isoquinoline alkaloid 1,2-Dehydroreticuline (**5**), previously isolated from *Xylopia parviflora* root and bark [21,22], was also identified. N-Feruloyltyramine (**16**) is a phenolic amide alkaloid previously detected in *Lycium barbarum* fruits and *A. alopecuroides* [10,35]. Isochondrodendrine (**24**) is bis benzyli-soquinolinic alkaloids previously reported in the *Cissampelos pareira* plant [46,51]. Finally, Conessine (**26**), a pentacyclic steroidal alkaloid, has been isolated from *Holarrhena floribunda* G. Don. [48].

Another biologically important and widely prevailing chemical class is the flavonoids that were represented by five compounds from different flavonoid subclasses. Narirutin (**9**) is a flavanone common in the citrus family and reported for potent antidiabetic activity using in vitro and docking studies [27]. Pelargonidin 3-*O*-rutinoside (**10**), an anthocyanin with potent antidiabetic activity depicted through the inhibition of α-glucosidase and α-amylase enzymes, was isolated from strawberries [28]. Biochanin A-β-D-glucoside (**12**) is an isoflavone previously isolated from *Trifolium pratense* L. [31], while Sophora isoflavanone A (**14**) is a pterocarpan previously identified from *Sophora tomentosa* [33]. Finally, Dracorubin (**23**) was recognized as the major red coloring matter in the tree *Dracaena draco* resin [45].

LC-HRMS/MS results also characterized four lignans. Veraguensin (**8**) is a lignan compound previously identified in *N. turbacensis* (Kunth) Nees leaves and root bark [25,26]. Pinoresinol glucoside (**12**) was isolated from *Prunus domestica* [32] and was stated to exhibit potent antioxidant activity and powerful in vitro antihyperglycemic and hepatoprotective effects. 4-Hydroxycinnamyl alcohol 4-D-glucoside "4-*O*-β-D-glucopyranosyl-*p*-coumaric acid" (**17**) is a phenolic acid derivative that was isolated and identified in the flaxseed phenolic rich fraction [36,37]. Coniferin (**18**), is a phenolic glycoside previously isolated from the bark of *Paulownia tomentosa* [38,39].

Furthermore, the two iridoid glycosides Boschnaloside (**11**), previously reported from *Euphrasia pectinata* aerial parts [29,30], and Geniposide (**15**), isolated from *Gardenia jasminoides* Ellis fruit [34], were detected in AAE.

An additional seven compounds from variable secondary metabolites classes were also detected in AAE and dereplicated as: the anthraquinone compound Versiconol acetate (**6**), previously recognized in the cultures of *Aspergillus parasiticus* after using the insecticide dichlorvos [52,53]; the dimethyl benzisochroman compound Karwinaphthol B (**7**), isolated from *Karwinskia humboldtiana* roots [24]; the natural benzofuran Toxyl angelate (**20**), previously isolated from *Isocoma wrightii* plant [41]; the furanocoumarine compound Heliettin (**21**), previously isolated from the stem bark of *Helietta longifolia* Britt and *Helietta apiculata* [42,43]; Gambogic acid (**22**), previously isolated from *Garcinia hanburyi* plant [44]; the unsaturated fatty acid Punicic acid (**25**), previously isolated from pomegranate seed oil [47]; and finally, the triterpenoid saponin Soyasapogenol B 3-*O*-D-glucuronide (**27**), previously isolated from aerial parts of *Lathylus palustris* L. [49,50].

Molecular docking study of *A. alopecuroides* metabolites for inhibition of α-amylase and α-glucosidase enzymes

In order to highlight the probably bioactive metabolites in AAE, all the dereplicated compounds were subjected to molecular docking study against both human α-amylase and α-glucosidase enzymes. All the dereplicated compounds displayed binding energies within the range of −4.5 to −9.1 Kcal/mol with the two enzymes (Table 2). The two alkaloids Eburnamonine (**2**) and Isochondrodendrine (**24**) as well as the two flavonoids Narirutin (**9**) and Pelargonidin 3-*O*-rutinoside (**10**) achieved the best binding scores with α-amylase enzyme. These compounds showed various binding modes inside the enzyme active site. Narirutin and Pelargonidin 3-*O*-rutinoside binding poses were comparable with that of the co-crystalized inhibitor Montbretin A (Figure 4B,C,E), where they established multiple H-bonds with TYR-151, ASP-197, HIS-201, GLU-233, HiS-299, and ASP-300. In addition, Narirutin established further hydrophobic interactions with TRP-58 and TRP-59. Narirutin was previously reported to have a potent role in diabetes management and control of its complications. This effect was confirmed via in vitro and docking studies against eight target proteins including α-amylase and α-glucosidase. In this report, Narirutin displayed hydrogen bonding interactions with both enzymes [13,15,27]. Moreover, variable flavonoids were previously tested for α-amylase and α-glucosidase inhibitory activity using in vitro testing and molecular docking approaches. The ligand–enzyme complexes for these compounds were studied, and it was concluded that the interactions occur mainly through H-bonding [54,55].

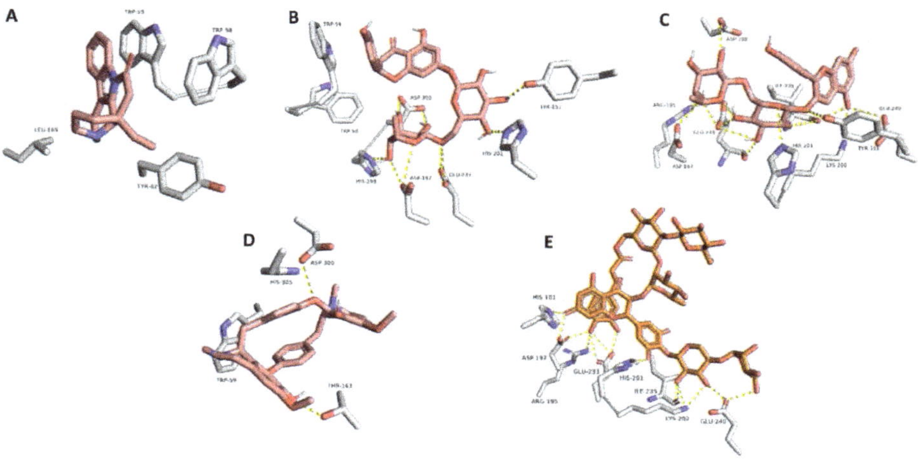

Figure 4. Docking poses of compounds with docking scores <−8.0 kcal/mol (i.e., Eburnamonine, Narirutin, Pelargonidin 3-*O*-rutinoside, and Isochondrodendrine along with the co-crystalized inhibitor Montbretin A) inside the human α-amylase (**A–E**, respectively).

On the other side, the alkaloid Eburnamonine (**2**) established four hydrophobic interactions only inside the enzyme's active site with TRP-58 TRP-59, TYR-62, and LEU-165 without any H-bonds (Figure 4A). Eburnamonine is an alkaloid that was previously isolated from several Vinica species and stated to contribute to the recorded antidiabetic effect of the total extract, via increasing hepatic utilization of glucose, suppressing the gluconeogenic enzymes, and regulation of insulin secretion, glucose, and lipid metabolism [17,56,57]. One more alkaloid, the isoquinoline alkaloid Isochondrodendrine (**24**), showed a remarkable result where it achieved the highest docking score (−9.1 kcal/mol) among all tested compounds. It established two H-bonds with THR-163 and ASP-300 together with a single hydrophobic interaction with TRP-59 (Figure 4D). Notably, this is the first report for the α-amylase enzyme inhibitory potential of Isochondrodendrine (**24**). However, other alkaloids, e.g., Topetecan and Cathine, were previously studied, and docking results concluded potent inhibitory activity against α-amylase enzyme [58].

Regarding the α-glucosidase enzyme, the three flavonoids Pelargonidin 3-O-rutinoside (**10**), Sophora isoflavanone A (**14**), and Dracorubin (**23**) achieved the best scores for binding affinity. They showed different binding interactions inside the enzyme's active site (Figure 5). Pelargonidin 3-*O*-rutinoside and Sophora isoflavanone A established interactions highly similar to that of the co-crystalized inhibitor Miglitol, where H-bonds were the predominant, e.g., with ASP-203, ASP-327, TRP-406, ASP-443, ASN-449, ARG-526, ASP-542, and HIS-600 (Figure 5A,B). On the other hand, Dracorubin's major interactions were hydrophobic (e.g., with TRP-406, PHE-450, and LYS-480) in addition to a single H-bond with GLN-603 (Figure 5C). It is worth mentioning that this is the first report on the anti-enzyme activity of both compounds Sophora isoflavanone A and Dracorubin. On the other side, the anthocyanin compound, Pelargonidin 3-*O*-rutinoside, was previously reported to be a potent novel α-glucosidase inhibitor that can improve postprandial hyperglycemia [28,59]. Herein, Pelargonidin 3-*O*-rutinoside showed a promising dual inhibitory activity against both enzymes α-amylase and α-glucosidase, expressed as binding energy (−8.5 and −8.4 kcal/mol, respectively), which was better than that of both co-crystalized inhibitors of the two corresponding enzymes (−8.1 and −8.0 kcal, mol, respectively). The current results augmented the previous finding for α-glucosidase inhibitory activity in addition to providing further proof of the α-amylase inhibitory effect. Accordingly, this study nominated Pelargonidin 3-*O*-rutinoside as a potential antidiabetic agent.

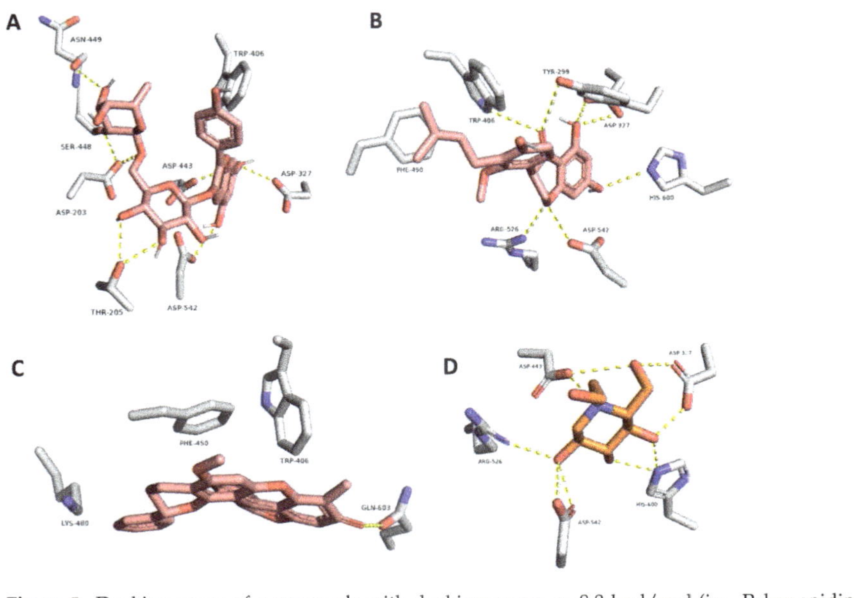

Figure 5. Docking poses of compounds with docking scores <−8.0 kcal/mol (i.e., Pelargonidin 3-*O*-rutinoside, Sophora isoflavanone A, and Dracorubin along with the co-crystalized inhibitor miglitol) inside the human α-glucosidase (**A–D**, respectively).

Bridging the metabolomic profiling of *A. alopecuroides* with its biological activity.

In a research program dedicated to investigation of the biological potential and the phytochemical content of halophytes, *A. alopecuroides* was acknowledged for its characteristic antidiabetic activity [7]. Hence, it was crucial to explore the phytoconstituents in this species that might be responsible for such characteristic activity. In order to achieve this goal, the LC-HRMS/MS technique was employed. The current results addressed the richness of AAE with a wide variety of secondary metabolite classes. Among them, alkaloids and flavonoids are the most characteristic. Afterwards, the molecular docking technique was used to assess the antidiabetic potential of all dereplicated compounds. Docking results indicated the probable potential of most of the annonated compounds (binding energies ranging from −5 to −9); however, the most characteristic results were recorded by alkaloid and flavonoid constituents. Some of these constituents, e.g., Eburnamonine, Narirutin, and Pelargonidin 3-*O*-rutinoside, were previously reported for such activity, thus giving an interpretation for the observed antidiabetic activity of the crude extract. Other metabolites, such as Isochondrodendrine (−9.1 Kcal/mol, α-amylase), Sophora isoflavanone A (−9.1 Kcal/mol, α-glucosidase), and Dracorubin (−8.3 Kcal/mol, α-amylase), were noted for the first time as potential antidiabetic compounds. This finding adds more explanation for the observed activity of the total extract. Among the remaining dereplicated compounds, Pinoresinol glucoside was previously stated as a potent antidiabetic natural entity [32]. Herein, Pinoresinol glucoside displayed good inhibitory activity against α-amylase enzyme (−7.9 Kcal/mol) and α-glucosidase (−6.0 Kcal/mol). Other compounds such as Heliettin, 1,2-Dehydroreticuline, Epinorlycoramine, N-Feruloyltyramine, and Veraguensin also displayed good activity, expressed as binding energy in the range of −7.5 to −7.9 Kcal/mol. In conclusion, AAE contains a powerful mixture of phytoconstituents that could be considered, either individually or collectively, as a probable antidiabetic agents.

Supplementary Materials: The following supporting information can be downloaded at: https://www.mdpi.com/article/10.3390/life12111852/s1. Detailed material and method of docking study. References [60–63] are cited in the supplementary materials. Figure S1: 2D diagram of the binding interactions of the twenty-seven compounds detected in *A. alopecuroides* against α-Glucosidase enzyme. Figure S2: 2D diagram of the binding interactions of the twenty-seven compounds detected in *A. alopecuroides* against α-Amylase enzyme.

Author Contributions: Conceptualization, E.A.; investigation, E.A. and M.S.A.-B.; methodology, E.A. and M.H.A.H.; software, M.S.A.-B. and H.A.M.; writing—original draft preparation, E.A.; writing—review and editing, E.A. and M.H.A.H.; supervision, E.A. All authors have read and agreed to the published version of the manuscript.

Funding: The research was funded by the Deanship of Scientific research, Qassim University, grant number (10122-pharmacy-2020-1-3-I).

Institutional Review Board Statement: Not applicable.

Informed Consent Statement: Not applicable.

Data Availability Statement: All Data are contained within the article and Supplementary Materials.

Acknowledgments: The authors gratefully acknowledge Qassim University, represented by the Deanship of Scientific research, for financial support of this research under the number (10122-pharmacy-2020-1-3-I) during the academic year 1441AH-2021 AD.

Conflicts of Interest: The authors declare no conflict of interest.

References

1. Belayneh, Y.M.; Birru, E.M. Antidiabetic activities of hydromethanolic leaf extract of *Calpurnia aurea* (Ait.) Benth. Subspecies *aurea* (Fabaceae) in mice. *Evid. Based Complement. Altern. Med.* **2018**, *2018*, 3509073. [CrossRef] [PubMed]
2. Alqahtani, A.S.; Hidayathulla, S.; Rehman, T.; ElGamal, A.A.; Al-Massarani, S.; Razmovski-Naumovski, V.; Alqahtani, M.S.; El Dib, R.A.; AlAjmi, M.F. Alpha-Amylase and Alpha-Glucosidase Enzyme Inhibition and Antioxidant Potential of 3-Oxolupenal and Katononic Acid Isolated from Nuxia oppositifolia. *Biomolecules* **2019**, *10*, 61. [CrossRef] [PubMed]

3. Kajaria, D.; Tiwari, S.; Tripathi, J.; Tripathi, Y.; Ranjana. In-vitro α amylase and glycosidase inhibitory effect of ethanolic extract of antiasthmatic drug—Shirishadi. *J. Am. Pharm. Technol. Res.* **2013**, *4*, 206. [CrossRef] [PubMed]
4. Yao, X.; Zhu, L.; Chen, Y.; Tian, J.; Wang, Y. In vivo and in vitro antioxidant activity and α-glucosidase, α-amylase inhibitory effects of flavonoids from Cichorium glandulosum seeds. *Food Chem.* **2013**, *139*, 59–66. [CrossRef]
5. Nair, S.S.; Kavrekar, V.; Mishra, A. In vitro studies on alpha amylase and alpha glucosidase inhibitory activities of selected plant extracts. *Eur. J. Exp. Biol.* **2013**, *3*, 128–132.
6. Mumtaz, M.W.; Al-Zuaidy, M.H.; Hamid, A.A.; Danish, M.; Akhtar, M.T.; Mukhtar, H. Metabolite profiling and inhibitory properties of leaf extracts of *Ficus benjamina* towards α-glucosidase and α-amylase. *Int. J. Food Prop.* **2018**, *21*, 1560–1574. [CrossRef]
7. Amin, E.; Abdel-Bakky, M.S.; Darwish, M.A.; Mohammed, H.A.; Chigurupati, S.; Qureshi, K.A.; Hassan, M.H. The Glycemic Control Potential of Some Amaranthaceae Plants, with Particular Reference to In Vivo Antidiabetic Potential of Agathophora alopecuroides. *Molecules* **2022**, *27*, 973. [CrossRef]
8. Amin, E.; Abdel-Bakky, M.S.; Mohammed, H.A.; Chigrupati, S.; Qureshi, K.A.; Hassan, M.H.A. Phytochemical Analysis and Evaluation of the Antioxidant and Antimicrobial Activities of Five Halophytes from Qassim Flora. *Pol. J. Environ. Stud.* **2022**, *31*, 3005–3012. [CrossRef]
9. Chikhi, I.; Allali, H.; Dib, M.E.A.; Medjdoub, H.; Tabti, B. Antidiabetic activity of aqueous leaf extract of *Atriplex halimus* L. (Chenopodiaceae) in streptozotocin–induced diabetic rats. *Asian Pac. J. Trop. Dis.* **2014**, *4*, 181–184. [CrossRef]
10. Othman, A.; Sayed, A.M.; Amen, Y.; Shimizu, K. Possible neuroprotective effects of amide alkaloids from *Bassia indica* and *Agathophora alopecuroides*: In vitro and in silico investigations. *RSC Adv.* **2022**, *12*, 18746–18758. [CrossRef]
11. Sashidhara, K.V.; Rosaiah, J.N. Various Dereplication Strategies Using LC-MS for Rapid Natural Product Lead Identification and Drug Discovery. *Nat. Prod. Commun.* **2007**, *2*, 193–202. [CrossRef]
12. Ferreira, L.G.; Dos Santos, R.N.; Oliva, G.; Andricopulo, A.D. Molecular Docking and Structure-Based Drug Design Strategies. *Molecules* **2015**, *20*, 13384–13421. [CrossRef] [PubMed]
13. Seeliger, D.; de Groot, B.L. Ligand docking and binding site analysis with PyMOL and Autodock/Vina. *J. Comput. Aided Mol. Des.* **2010**, *24*, 417–422. [CrossRef] [PubMed]
14. Sim, L.; Jayakanthan, K.; Mohan, S.; Nasi, R.; Johnston, B.D.; Pinto, B.M.; Rose, D.R. New Glucosidase Inhibitors from an Ayurvedic Herbal Treatment for Type 2 Diabetes: Structures and Inhibition of Human Intestinal Maltase-Glucoamylase with Compounds from *Salacia reticulata*. *Biochemistry* **2009**, *49*, 443–451. [CrossRef] [PubMed]
15. Williams, L.K.; Zhang, X.; Caner, S.; Tysoe, C.; Nguyen, N.T.; Wicki, J.; Williams, D.E.; Coleman, J.; McNeill, J.H.; Yuen, V.; et al. The amylase inhibitor montbretin A reveals a new glycosidase inhibition motif. *Nat. Chem. Biol.* **2015**, *11*, 691–696. [CrossRef] [PubMed]
16. Bastida, J.; Viladomat, F.; Bergoñon, S.; Fernandez, J.M.; Codina, C.; Rubiralta, M.; Quirion, J.-C. Alkaloids from Narcissus leonensis. *Phytochemistry* **1993**, *34*, 1656–1658. [CrossRef]
17. Kam, T.-S.; Tan, P.-S.; Chuah, C.-H. Alkaloids from leaves of Kopsia larutensis. *Phytochemistry* **1992**, *31*, 2936–2938. [CrossRef]
18. Cahlíková, L.; Vaněčková, N.; Šafratová, M.; Breiterová, K.; Blunden, G.; Hulcová, D.; Opletal, L. The Genus Nerine Herb. (Amaryllidaceae): Ethnobotany, Phytochemistry, and Biological Activity. *Molecules* **2019**, *24*, 4238. [CrossRef]
19. Cahlíková, L.; Zavadil, S.; Macáková, K.; Valterova, I.; Kulhánková, A.; Hostalkova, A.; Kuneš, J.; Opletal, L. Isolation and Cholinesterase Activity of Amaryllidaceae Alkaloids from Nerine bowdenii. *Nat. Prod. Commun.* **2011**, *6*, 1827–1830. [CrossRef]
20. Piboonprai, K.; Khumkhrong, P.; Khongkow, M.; Yata, T.; Ruangrungsi, N.; Chansriniyom, C.; Iempridee, T. Anticancer activity of arborinine from Glycosmis parva leaf extract in human cervical cancer cells. *Biochem. Biophys. Res. Commun.* **2018**, *500*, 866–872. [CrossRef]
21. Hagel, J.M.; Facchini, P.J. Benzylisoquinoline Alkaloid Metabolism: A Century of Discovery and a Brave New World. *Plant Cell Physiol.* **2013**, *54*, 647–672. [CrossRef] [PubMed]
22. Nishiyama, Y.; Moriyasu, M.; Ichimaru, M.; Iwasa, K.; Kato, A.; Mathenge, S.G.; Mutiso, P.B.C.; Juma, F.D. Quaternary isoquinoline alkaloids from Xylopia parviflora. *Phytochemistry* **2004**, *65*, 939–944. [CrossRef] [PubMed]
23. Steyn, P.S.; Vleggaar, R.; Wessels, P.L.; Cole, R.J.; Scott, D.B. Structure and carbon-13 nuclear magnetic resonance assignments of versiconal acetate, versiconol acetate, and versiconol, metabolites from cultures of Aspergillus parasiticus treated with dichlorvos. *J. Chem. Soc. Perkin Trans. 1* **1979**, 451–459. [CrossRef]
24. Mitscher, L.A.; Gollapudi, S.R.; Oburn, D.S.; Drake, S. Antimicrobial agents from higher plants: Two dimethylbenzisochromans from Karwinskia humboldtiana. *Phytochemistry* **1985**, *24*, 1681–1683. [CrossRef]
25. Li, Y.; Xie, S.; Ying, J.; Wei, W.; Gao, K. Chemical Structures of Lignans and Neolignans Isolated from Lauraceae. *Molecules* **2018**, *23*, 3164. [CrossRef]
26. Macías-Villamizar, V.; Cuca-Suárez, L.; González, F.V.; Rodríguez, S. Lignoids Isolated from *Nectandra turbacensis* (Kunth) Nees (Lauraceae). *Rec. Nat. Prod.* **2016**, *10*, 654.
27. Qurtam, A.A.; Mechchate, H.; Es-Safi, I.; Al-Zharani, M.; Nasr, F.A.; Noman, O.M.; Aleissa, M.; Imtara, H.; Aleissa, A.M.; Bouhrim, M.; et al. Citrus Flavanone Narirutin, In Vitro and In Silico Mechanistic Antidiabetic Potential. *Pharmaceutics* **2021**, *13*, 1818. [CrossRef]
28. Xu, Y.; Xie, L.; Xie, J.; Liu, Y.; Chen, W. Pelargonidin-3-*O*-rutinoside as a novel α-glucosidase inhibitor for improving postprandial hyperglycemia. *Chem. Commun.* **2018**, *55*, 39–42. [CrossRef]

29. Viljoen, A.; Mncwangi, N.; Vermaak, I. Anti-Inflammatory Iridoids of Botanical Origin. *Curr. Med. Chem.* **2012**, *19*, 2104–2127. [CrossRef]
30. Ersöz, T.; Berkman, M.; Taşdemir, D.; Çaliş, İ.; Ireland, C.M. Iridoid and phenylethanoid glycosides from *Euphrasia pectinata*. *Turk. J. Chem.* **2002**, *26*, 178–188.
31. Toebes, A.H.W.; de Boer, V.; Verkleij, J.A.C.; Lingeman, H.; Ernst, W.H.O. Extraction of Isoflavone Malonylglucosides from *Trifolium pratense* L. *J. Agric. Food Chem.* **2005**, *53*, 4660–4666. [CrossRef] [PubMed]
32. Youssef, F.S.; Ashour, M.L.; El-Beshbishy, H.A.; Ahmed Hamza, A.; Singab, A.N.B.; Wink, M. Pinoresinol-4-O-β-D-glucopyranoside: A lignan from prunes (*Prunus domestica*) attenuates oxidative stress, hyperglycaemia and hepatic toxicity in vitro and in vivo. *J. Pharm. Pharmacol.* **2020**, *72*, 1830–1839. [CrossRef] [PubMed]
33. Kinoshita, T.; Ichinose, K.; Takahashi, C.; Ho, F.-C.; Wu, J.-B.; Sankawa, U. Chemical studies on Sophora tomentosa: The isolation of a new class of isoflavonoid. *Chem. Pharm. Bull.* **1990**, *38*, 2756–2759. [CrossRef]
34. Liang, Z.; Yang, M.; Xu, X.; Xie, Z.; Huang, J.; Li, X.; Yang, D. Isolation and purification of geniposide, crocin-1, and geniposidic acid from the fruit of *Gardenia jasminoides* Ellis by high-speed counter-current chromatography. *Sep. Sci. Technol.* **2014**, *49*, 1427–1433. [CrossRef]
35. Forino, M.; Tartaglione, L.; Dell'Aversano, C.; Ciminiello, P. NMR-based identification of the phenolic profile of fruits of *Lycium barbarum* (goji berries). Isolation and structural determination of a novel N-feruloyl tyramine dimer as the most abundant antioxidant polyphenol of goji berries. *Food Chem.* **2016**, *194*, 1254–1259. [CrossRef]
36. Pei, K.; Ou, J.; Huang, J.; Ou, S. p-Coumaric acid and its conjugates: Dietary sources, pharmacokinetic properties and biological activities. *J. Sci. Food Agric.* **2016**, *96*, 2952–2962. [CrossRef]
37. Johnsson, P.; Peerlkamp, N.; Kamal-Eldin, A.; Andersson, R.E.; Andersson, R.; Lundgren, L.N.; Åman, P. Polymeric fractions containing phenol glucosides in flaxseed. *Food Chem.* **2002**, *76*, 207–212. [CrossRef]
38. Terazawa, M.; Okuyama, H.; Miyake, M. Isolation of coniferin and syringin from the cambial tissue and inner-bark of some angiospermous woods. *J. Jpn. Wood Res. Soc.* **1984**, *30*, 409–412.
39. Sticher, O.; Lahloub, M.F. Phenolic Glycosides of Paulownia tomentosa Bark. *Planta Med.* **1982**, *46*, 145–148. [CrossRef]
40. Tram, N.T.N.; Mitova, M.; Bankova, V.; Handjieva, N.; Popov, S.S. GC-MS of *Crinum latifolium* L. alkaloids. *Z. Nat. C* **2002**, *57*, 239–242. [CrossRef]
41. Zalkow, L.; Ekpo, B.; Gelbaum, L.; Harris III, R.; Keinan, E.; Novak Jr, J.; Ramming, C.; Van Derveer, D. The benzofurans of Isocoma wrightii. Structure and stereochemistry. *J. Nat. Prod.* **1979**, *42*, 203–219. [CrossRef] [PubMed]
42. De Moura, N.F.; Simionatto, E.; Porto, C.; Hoelzel, S.C.S.; Dessoy, E.C.S.; Zanatta, N.; Morel, A.F. Quinoline Alkaloids, Coumarins and Volatile Constituents of *Helietta longifoliata*. *Planta Med.* **2002**, *68*, 631–634. [CrossRef] [PubMed]
43. Ferreira, M.E.; de Arias, A.R.; Yaluff, G.; de Bilbao, N.V.; Nakayama, H.; Torres, S.; Schinini, A.; Guy, I.; Heinzen, H.; Fournet, A. Antileishmanial activity of furoquinolines and coumarins from *Helietta apiculata*. *Phytomedicine* **2010**, *17*, 375–378. [CrossRef] [PubMed]
44. Lee, P.N.H.; Ho, W.S. Antiproliferative activity of gambogic acid isolated from *Garcinia hanburyi* in Hep3B and Huh7 cancer cells. *Oncol. Rep.* **2013**, *29*, 1744–1750. [CrossRef] [PubMed]
45. Melo, M.J.; Sousa, M.; Parola, A.J.; de Melo, J.S.S.; Catarino, F.; Marçalo, J.; Pina, F. Identification of 7, 4′-Dihydroxy-5-methoxyflavylium in "Dragon's Blood": To Be or Not to Be an Anthocyanin. *Chem. Eur. J.* **2007**, *13*, 1417–1422. [CrossRef] [PubMed]
46. He, J.; Li, F.; Yan, X.; Cheng, Q.; Xue, R.; Yu, H.; Li, Z.; Wang, C. Phytochemical Constituents and Biological Activities of Plants from the *Genus cissampelos*. *Chem. Biodivers.* **2021**, *18*, e2100358. [CrossRef] [PubMed]
47. Zielińska, A.; Wójcicki, K.; Klensporf-Pawlik, D.; Marzec, M.; Lucarini, M.; Durazzo, A.; Fonseca, J.; Santini, A.; Nowak, I.; Souto, E.B. Cold-Pressed Pomegranate Seed Oil: Study of Punicic Acid Properties by Coupling of GC/FID and FTIR. *Molecules* **2022**, *27*, 5863. [CrossRef]
48. Lannang, A.M.; Anjum, S.; Tangmouo, J.G.; Krohn, K.; Choudhary, M.I. Conessine isolated from *Holarrhena floribunda*. *Acta Crystallogr. Sect. E Struct. Rep. Online* **2007**, *63*, o4398. [CrossRef]
49. Ikeda, T.; Udayama, M.; Okawa, M.; Arao, T.; Kinjo, J.; Nohara, T. Partial Hydrolysis of Soyasaponin I and the Hepatoprotective Effects of the Hydrolytic Products. Studies on the Hepatoprotective Drugs. Part IV. (Studies on the Constituents of the Leguminous Plants. Part LVII.) Study of the Structure-Hepatoprotective Relationship of Soyasapogenol B Analogs. *Chem. Pharm. Bull.* **1998**, *46*, 359–361. [CrossRef]
50. Udayama, M.; Ohkawa, M.; Yoshida, N.; Kinjo, J.; Nohara, T. Structures of Three New Oleanene Glucuronides Isolated from Lathyrus palustris var. pilosus and Hepatoprotective Activity. *Chem. Pharm. Bull.* **1998**, *46*, 1412–1415. [CrossRef]
51. Dwuma-Badu, D.; Ayim, J.S.; Mingle, C.; Tackie, A.; Slatkin, D.; Knapp, J.; Schiff, P., Jr. Alkaloids of *Cissampelos pareira*. *Phytochemistry* **1975**, *14*, 2520–2521. [CrossRef]
52. Steyn, P.S.; Vleggaar, R.; Wessels, P.L.; Scott, D.B. Biosynthesis of versiconal acetate, versiconol acetate, and versiconol, metabolites from cultures of Aspergillus parasiticus treated with dichlorvos. The role of versiconal acetate in aflatoxin biosynthesis. *J. Chem. Soc. Perkin Trans. 1* **1979**, 460–463. [CrossRef]
53. Yao, R.C.; Hsieh, D.P. Step of dichlorvos inhibition in the pathway of aflatoxin biosynthesis. *Appl. Microbiol.* **1974**, *28*, 52–57. [CrossRef]

54. Jadalla, B.M.I.S.; Moser, J.J.; Sharma, R.; Etsassala, N.G.E.R.; Egieyeh, S.A.; Badmus, J.A.; Marnewick, J.L.; Beukes, D.; Cupido, C.N.; Hussein, A.A. In Vitro Alpha-Glucosidase and Alpha-Amylase Inhibitory Activities and Antioxidant Capacity of *Helichrysum cymosum* and *Helichrysum pandurifolium* Schrank Constituents. *Separations* **2022**, *9*, 190. [CrossRef]
55. Ahmed, S.; Ali, C.; Ruma, R.A.; Mahmud, S.; Paul, G.K.; Saleh, A.; Alshahrani, M.M.; Obaidullah, A.J.; Biswas, S.K.; Rahman, M.; et al. Molecular Docking and Dynamics Simulation of Natural Compounds from Betel Leaves (*Piper betle* L.) for Investigating the Potential Inhibition of Alpha-Amylase and Alpha-Glucosidase of Type 2 Diabetes. *Molecules* **2022**, *27*, 4526. [CrossRef]
56. Chikezie, P.C.; Ojiako, O.A.; Nwufo, K.C. Overview of anti-diabetic medicinal plants: The Nigerian research experience. *J. Diabetes Metab.* **2015**, *6*, 546. [CrossRef]
57. Pereira, A.S.; Haan, H.D.; Peña-García, J.; Moreno, M.M.; Pérez-Sánchez, H.; Apostolides, Z. Exploring African Medicinal Plants for Potential Anti-Diabetic Compounds with the DIA-DB Inverse Virtual Screening Web Server. *Molecules* **2019**, *24*, 2002. [CrossRef]
58. Akshatha, J.V.; SantoshKumar, H.S.; Prakash, H.S.; Nalini, M.S. In silico docking studies of α-amylase inhibitors from the anti-diabetic plant Leucas ciliata Benth. and an endophyte, Streptomyces longisporoflavus. *3 Biotech* **2021**, *11*, 51. [CrossRef]
59. Oliveira, H.; Fernandes, A.; Brás, N.F.; Mateus, N.; de Freitas, V.; Fernandes, I. Anthocyanins as antidiabetic agents—In vitro and in silico approaches of preventive and therapeutic effects. *Molecules* **2020**, *25*, 3813. [CrossRef]
60. O'Boyle, N.M.; Banck, M.; James, C.A.; Morley, C.; Vandermeersch, T.; Hutchison, G.R. Open Babel: An open chemical toolbox. *Cheminform. J.* **2011**, *3*, 33. [CrossRef]
61. Morris, G.M.; Huey, R.; Lindstrom, W.; Sanner, M.F.; Belew, R.K.; Goodsell, D.S.; Olson, A.J. AutoDock4 and AutoDockTools4: Automated docking with selective receptor flexibility. *J. Comput. Chem.* **2009**, *30*, 2785. [CrossRef] [PubMed]
62. Eastman, P.; Friedrichs, M.S.; Chodera, J.D.; Radmer, R.J.; Bruns, C.M.; Ku, J.P.; Beauchamp, K.A.; Lane, T.J.; Wang, L.; Shukla, D.; et al. OpenMM 4: A Reusable, Extensible, Hardware Independent Library for High Performance Molecular Simulation. *J. Chem. Theory Comput.* **2013**, *9*, 461. [CrossRef] [PubMed]
63. Dallakyan, S.; Olson, A.J. *Chemical Biology*; Hempel, J.E., Williams, C.H., Hong, C.C., Eds.; Springer: New York, NY, USA, 2015; pp. 243–250. [CrossRef]

life
MDPI

Article

Comprehensive Metabolite Profiling of Cinnamon (*Cinnamomum zeylanicum*) Leaf Oil Using LC-HR/MS, GC/MS, and GC-FID: Determination of Antiglaucoma, Antioxidant, Anticholinergic, and Antidiabetic Profiles

Muzaffer Mutlu [1], Zeynebe Bingol [2], Eda Mehtap Uc [3], Ekrem Köksal [4], Ahmet C. Goren [5], Saleh H. Alwasel [6] and İlhami Gulcin [3,*]

1 Vocational School of Applied Sciences, Gelişim University, Istanbul 34315, Turkey
2 Department of Medical Services and Techniques, Tokat Vocational School of Health Services, Gaziosmanpasa University, Tokat 60250, Turkey
3 Department of Chemistry, Faculty of Science, Ataturk University, Erzurum 25240, Turkey
4 Department of Chemistry, Faculty of Science and Arts, Erzincan Binali Yildirim University, Erzincan 24100, Turkey
5 Department Chemistry, Faculty of Sciences, Gebze Technical University, Kocaeli 41400, Turkey
6 Department of Zoology, College of Science, King Saud University, Riyadh 11362, Saudi Arabia
* Correspondence: igulcin@atauni.edu.tr; Tel.: +90-4422314375

Abstract: In this study, for the first time, the antioxidant and antidiabetic properties of the essential oil from cinnamon (*Cinnamomum zeylanicum*) leaves were evaluated and investigated using various bioanalytical methods. In addition, the inhibitory effects of cinnamon oil on carbonic anhydrase II (hCA II), acetylcholinesterase (AChE), and α-amylase, which are associated with various metabolic diseases, were determined. Further, the phenolic contents of the essential oil were determined using LC-HRMS chromatography. Twenty-seven phenolic molecules were detected in cinnamon oil. Moreover, the amount and chemical profile of the essential oils present in cinnamon oil was determined using GC/MS and GC-FID analyses. (*E*)-cinnamaldehyde (72.98%), benzyl benzoate (4.01%), and *trans*-Cinnamyl acetate (3.36%) were the most common essential oils in cinnamon leaf oil. The radical scavenging activities of cinnamon oil were investigated using 1,1-diphenyl-2-picryl-hydrazil ($DPPH^{\bullet}$), 2,2′-azino-bis(3-ethylbenzthiazoline-6-sulfonic acid), and ($ABTS^{\bullet+}$) bioanalytical scavenging methods, which revealed its strong radical scavenging abilities ($DPPH^{\bullet}$, IC_{50}: 4.78 μg/mL; and $ABTS^{\bullet+}$, IC_{50}: 5.21 μg/mL). Similarly, the reducing capacities for iron (Fe^{3+}), copper (Cu^{2+}), and Fe^{3+}-2,4,6-tri(2-pyridyl)-S-triazine (TPTZ) were investigated. Cinnamon oil also exhibited highly effective inhibition against hCA II (IC_{50}: 243.24 μg/mL), AChE (IC_{50}: 16.03 μg/mL), and α-amylase (IC_{50}: 7.54μg/mL). This multidisciplinary study will be useful and pave the way for further studies for the determination of antioxidant properties and enzyme inhibition profiles of medically and industrially important plants and their oils.

Keywords: *Cinnamomum zeylanicum*; polyphenol; cinnamon oil; antioxidant activity; LC-HRMS; GCMS; enzyme inhibition

check for updates

Citation: Mutlu, M.; Bingol, Z.; Uc, E.M.; Köksal, E.; Goren, A.C.; Alwasel, S.H.; Gulcin, İ. Comprehensive Metabolite Profiling of Cinnamon (*Cinnamomum zeylanicum*) Leaf Oil Using LC-HR/MS, GC/MS, and GC-FID: Determination of Antiglaucoma, Antioxidant, Anticholinergic, and Antidiabetic Profiles. *Life* **2023**, *13*, 136. https://doi.org/10.3390/life13010136

Academic Editor: Jianfeng Xu

Received: 16 November 2022
Revised: 19 December 2022
Accepted: 28 December 2022
Published: 3 January 2023

1. Introduction

Plants and their biologically active compounds are among the leading potential resources for the development of new products for different industrial applications [1]. At present, several biologically active compounds derived from plants are being used effectively in various applications, especially in pharmacology [2]. Plants act as living chemical factories that produce several secondary metabolites and are beneficial to the environment. Medicinal plants are of crucial importance, especially in the health, cosmetic, food, and pharmaceutical industries. Turkey has a rich plant biodiversity that can be exploited for use in these areas [3,4]. The main reasons for the use of secondary metabolites in folk medicine are that they are harmless and biologically active. The notable properties of these metabolites include antidiabetic, anti-inflammatory, antioxidant, and neuroprotective effects [5,6].

Naturally occurring antioxidants derived from several plants can act as reactive oxygen species (ROS) or free-radical scavengers or metal chelators in various biological systems [7,8]. Excessive formation of ROS and free radicals in living organisms can induce various chronic and degenerative diseases, including age-related pathologies, cardiovascular disorders, cancer, immunodeficiency syndrome, arteriosclerosis, obesity, and diabetes, by acting on several biomolecules, such as carbohydrates, proteins, lipids, and nucleic acids [9,10]. The presence of antioxidants in biological systems, even at low concentrations, is important to eliminate the adverse effects of ROS and free radicals [11,12]. The biological activities of phenolic compounds obtained from plant sources, particularly their antioxidant properties, have been intensively investigated [13]. Plants exhibit extensive and excellent biological activities [14]. Owing to their biological activities, all living systems, including plants, can remain healthy. In addition, plants contain biologically active regulators and, thus, have a significant therapeutic effect on some chronic diseases [15]. Plant-derived phenolics are in great demand and are being used worldwide because they are less toxic. Therefore, there is a great incentive and demand for plants to treat different diseases. Several herbal medicines that are widely used in the treatment of Alzheimer's disease (AD) are also known [16].

The genus *Cinnamomum* belongs to the fairly large family Lauraceae, comprising more than 30 genera and approximately 2000–2500 species. *Cinnamomum* species are used as natural antioxidant sources because of their high phenolic content due to the presence of proanthocyanidins and flavonoids. Species of the genus *Cinnamomum* include *C. cassia*, *C. zeylanicum* (*C. verum*), *C. loureiroi*, and *C. burmannii*, which are the most-known ones [17]. However, *C. zeylanicum* is the most studied species. Dried aromatic peel is the main commercial product of *C. zeylanicum* [18], with various therapeutic effects, including antifungal activity. It was reported that that *C. zeylanicum* oil is safe for human consumption. The essential oil of *C. zeylanicum* has antibacterial effects that can combat drug-resistant and ulcer-causing *H. pylori*. *C. zeylanicum* has long been used as a health-promoting agent [19]. Additionally, its essential oil has a protective effect against gastroenteritis [20]. Essential oils from the bark of *C. zeylanicum* are used for their antioxidant, anti-Alzheimer's disease, anti-skin-whitening, and antidiabetic activities. The leaves and bark of *C. zeylanicum* are commercially valuable owing to their high contents of (*E*)-cinnamylacetate and eugenol (*E*)-cinnamaldehyde. In addition, (*E*)-cinnamaldehyde, the major compound found in the bark oil from *C. zeylanicum*, has various bioactivities [21]. The root of *C. zeylanicum* contains camphor as the main compound.

AD is a neurodegenerative disease that causes cognitive impairment, behavioral abnormalities, and dementia in elderly individuals [22]. Although complete cure of AD is not currently possible, its clinical course can be altered or slowed. Patients with AD generally exhibit three characteristic pathological symptoms, including dementia due to hyper phosphorylation of tau, low acetylcholine (ACh) levels, and amyloid-β aggregation, causing loss of neuronal functions [23]. Of these, the cholinergic hypothesis involving low ACh levels is a clinically viable medical strategy. Thus, strategies to inhibit cholinergic acetylcholinesterase (AChE) in brain tissue have gained importance [24,25]. AChE, a hydrolytic enzyme, converts ACh to acetate and choline (Ch), whereas butyrylcholinesterase (BChE) converts butyrylcholine (BCh) to butyrate and Ch [26]. AD is clinically affected by brain tissue damage caused by ACh deficiency [27]. In addition, many factors, such as environmental conditions and genetic predisposition, greatly affect the variability in the amount of ACh [28]. Particularly, AChE inhibitors of natural origin increase cholinergic transmission, exerting therapeutic effects on patients with AD [29–31].

The use of plant and plant-derived components also affects biological and pathological events related to diabetes complications [32]. Metformin, a plant-derived compound, is widely used to decrease blood sugar levels and indirectly reduce diabetes mellitus (DM), a global health problem provoked by hyperglycemia and characterized by the absence of or resistance to insulin [33]. As a result of this chronic metabolic disease, unusual sugar levels in blood plasma cause neuropathy, cardiovascular disorders, retinopathy, and nephropathy [34]. Hydrolysis of dietary polysaccharides into glucose leads to an increase in postprandial glucose levels [35]. Patients with diabetes have adversity in blood sugar transport to cells after digestion. In this case, the blood sugar level remains high while the cells are starved [36]. Additionally, α-amylase and α-glucosidase enzymes hydrolyze polysaccharides into monosaccharides for absorption from the small intestine. Therefore, their inhibition is of great importance in the treatment of type-2 diabetes mellitus (T2DM) [37,38]. Thus, digestive enzyme inhibitors (DEIs) obtained from natural sources can be used for the treatment of hyperglycemia and T2DM [39]. DEIs act as antidiabetic drugs and decrease blood sugar levels by reducing intestinal carbohydrate absorption. Thus, DEIs compete with oligosaccharides for binding to the active site of the enzyme. Acarbose is one of the best examples of this form of inhibition [40].

Carbonic anhydrases (CAs) catalyze the reversible hydration and dehydration of carbon dioxide (CO_2) to protons and bicarbonate (HCO_3^-) [41]. Since the approval of dorzolamide in 1995, CA isoenzyme inhibitors (CAIs) have been widely used to treat glaucoma [42]. Five years later, in 2000, a second clinical CA inhibitor, brinzolamide, was used in the treatment of glaucoma in many European countries. Glaucoma causes irreversible peripheral vision loss and, consequently, high intraocular pressure (IOP), causing blindness [43]. Glaucoma, an eye disease that damages the optic nerve and causes vision loss, is one of the leading causes of blindness worldwide. It is also estimated that the number of people with glaucoma will exceed 115 million worldwide by 2040 [44,45]. Although laser, surgical, and pharmacological treatments are available for glaucoma, the clinical application of CA II isoenzyme inhibitors in epithelial cells in the ciliary body reduces aqueous humor secretion and consequently reduces IOP [46]. CA II inhibitors, including methazolamide, acetazolamide, dichlorphenamide, and ethoxzolamide, are pressure-lowering systemic drugs for glaucoma treatment [47]. However, these inhibitors inhibit the localization of CA isoforms other than those in the eye, causing undesirable side effects, including paresthesia, fatigue, and serious concerns [48,49]. To avoid these undesirable side effects, inhibitors with topical activity and natural origin are preferred. Therefore, we present cinnamon leaf oil for the treatment of glaucoma to the attention of researchers working on this subject.

In our current study, we evaluated the chemical composition and antioxidant, anti-Alzheimer's disease, and antidiabetic effects of cinnamon leaf oil. We also identified the polyphenol and essential oil contents of cinnamon leaf oil using LC-HRMS and GS-MS/FID.

2. Materials and Methods

2.1. Chemicals

2,2′-azino-bis(3-ethylbenzthiazoline-6-sulfonic acid) (ABTS), α-tocopherol, 2,9-dimethyl-1,10-phenanthroline (neocuproine), Trolox, butylated hydroxyanisole (BHA), 1,1-diphenyl-2-picryl-hydrazyl ($DPPH^•$), and butylated hydroxytoluene (BHT) were purchased from Sigma-Aldrich Chemie GmbH (Steinheim, Germany). Ascorbic acid, fumaric acid, chlorogenic acid, caffeic acid, naringin, vanillic acid, syringic acid, rutin, rosmarinic acid, p-coumaric acid, salicylic acid, quercetin, luteolin, naringenin, chrysin, and emodin were purchased from Sigma-Aldrich. Hyperoside, luteolin 7-glycoside, orientin, (+)-trans-taxifolin, quercitrin, hispidulin, apigenin, hederagenin, and acacetin were obtained from TRC, Canada. Verbascoside and luteolin-7-rutinoside were purchased from HWI Analytik GMBH and Carbosynth Ltd., respectively. Hesperidin was purchased from J&K Co., Ltd. Isosakuranetin, dihydrokaempferol, and penduletin were purchased from Phytolab. Apigenin 7-glucoside was obtained from the EDQM CS. Myricetin was purchased from Carl Roth GmbH & Co. Nepetin and caffeic acid phenethylester were purchased from Supelco and the European Pharmacopoeia, respectively.

2.2. Preparation of Cinnamon (C. zeylanicum) Leaf Oil

Cinnamon (*C. zeylanicum*) leaves were obtained from a local market. Cinnamon leaf oil was prepared using the steam distillation method. This method is a multistage continuous distillation process using steam as the stripping gas to obtain the oils. Steam is applied directly to the plant. The vapor mixture is collected and condensed to obtain a liquid in which water and oil form two different layers. The upper part of these layers is essential oil containing hydrophobic compounds, and the lower part is a hydrolysate or hydrosol containing hydrophilic components. Polar compounds remaining in the water can be recovered by the cohobating process. Dried cinnamon leaves were placed on a grid above the steam inlet of the steam distillation unit. Steam was supplied to the unit for approximately 2 h. The mixture containing water vapor and volatilized essential oils was condensed using a cooler and collected in the collection container. The cinnamon leaf oil accumulated in the collection container formed a separate layer from that of water owing to the density difference and was extracted.

2.3. Polyphenolic Composition Using LC-HRMS Analysis

LC-HRMS experiments were carried out using a Thermo Orbitrap Q-Exactive mass spectrometer (Thermo Fisher Scientific Inc., Waltham, MA, USA), equipped with a Troyasil C18 column (150 × 3 mm i.d., 3 µm particle size) (Istanbul, Turkey). The mobile phases A and B were composed of 1% formic acid–water and 1% formic acid–methanol, respectively. The gradient programs were 0–1.00 min, 50% A and 50% B; 1.01–6.00 min, 100% B; and finally, 6.01–10 min, 50% A and 50% B [29]. The flow rate of the mobile phase was 0.35 mL/min, and the column temperature was set to 22 °C. Environmental conditions were set as 22.0 ± 5.0 °C temperature and 50 ± 15% relative humidity [50]. According to our previous experiences and reported data from the literature, an acidified methanol and water gradient was determined to be the best solvent system to obtain suitable ionization abundance and separation of compounds in HPLC. Because the electrospray ionization (ESI) source provides one of the best ionizations for small and relatively polar compounds, we chose the ESI source for the applied method. The ions between *m/z* 85 and 1500 were scanned in the high-resolution mode of the instrument [50]. Compounds were identified by comparing their retention time and HRMS data with those of standard compounds (in the range of purity 95–99%; see Section 2.1). Dihydrocapsaicin (purity 95%) was used as an internal standard for LC-HRMS measurements to reduce the repeatability problem caused by external effects, such as ionization repeatability, in mass spectrometry measurements. The mass parameters of each target compound are listed in Table 1. Further details of the LC-HRMS method, uncertainty evaluation methodology, and confirmation parameters for phenolics were provided in detail previously [50–52].

Validation of the LC-HRMS method was carried out using analytical standards of target compounds using negative or positive ions, listed in Table 1. Dihydrocapsaicin was used as the internal standard. Method validation parameters were selectivity, linearity, recovery, repeatability, intermediate precision, limit of detection (LOD), and limit of quantification (LOQ). The LODs of the method for individual compounds were determined using the following equation: LOD or LOQ = $\kappa SDa/b$, where LOQ is 3 and $\kappa = 3$ for LOD. Also, SDa represents the standard deviation of the intercept, and b represents the slope. The detailed validation procedure and uncertainty assessment methodology of the applied method was reported in our previous paper [49–52].

Table 1. Chemical composition and validation parameters of cinnamon (*C. zeylanicum*) leaf oil (mg/L oil) obtained using LC-HRMS.

Phenolic Compounds	Molecular Formula	m/z	Ionization Mode	Linear Range	Linear Regression Equation	LOD/LOQ	R^2	Recovery	Phenolics	U%
Ascorbic acid	$C_6H_8O_6$	175.0248	Negative	0.5–10	y = 0.00347x − 0.00137	0.39/1.29	0.9988	96.20	4.33	3.94
Epigallocatechin	$C_{15}H_{14}O_7$	307.0812	Positive	0.3–5	y = 0.00317x + 0.000443	0.17/0.57	0.9947	102.22	1.38	3.09
Chlorogenic acid	$C_{16}H_{18}O_9$	353.0878	Negative	0.05–10	y = 0.00817x + 0.000163	0.02/0.06	0.9994	96.68	0.60	3.58
Fumaric acid	$C_4H_4O_4$	115.0037	Negative	0.1–10	y = 0.00061x − 0.0000329	0.05/0.17	0.9991	97.13	4.39	2.88
Verbascoside	$C_{29}H_{36}O_{15}$	623.1981	Negative	0.1–10	y = 0.00758x + 0.000563	0.03/0.1	0.9995	96.19	0.05	2.93
Orientin	$C_{21}H_{20}O_{11}$	447.0933	Negative	0.1–10	y = 0.00757x + 0.000347	0.01/0.03	0.9993	96.22	0.31	3.67
Caffeic acid	$C_9H_8O_4$	179.0350	Negative	0.3–10	y = 0.0304x + 0.00366	0.08/0.27	0.9993	94.51	0.46	3.74
Luteolin-7-rutinoside	$C_{27}H_{30}O_{15}$	593.1512	Negative	0.1–10	y = 0.00879x + 0.000739	0.01/0.03	0.9988	93.05	-	3.06
Luteolin 7-glycoside	$C_{21}H_{20}O_{11}$	447.0933	Negative	0.1–7	y = 0.0162x + 0.00226	0.01/0.03	0.9961	96.31	0.30	4.14
Rutin	$C_{27}H_{30}O_{16}$	609.1461	Negative	0.05–10	y = 0.00329x − 0.00005576	0.01/0.03	0.999	96.97	0.26	3.07
Rosmarinic acid	$C_{18}H_{16}O_8$	359.0772	Negative	0.05–10	y = 0.00717x − 0.0003067	0.01/0.03	0.9992	99.85	3.25	3.77
Hyperoside	$C_{21}H_{20}O_{12}$	463.0882	Negative	0.05–10	y = 0.0072x − 0.00003096	0.01/0.03	0.9995	96.62	0.56	3.46
Apigenin 7-glycoside	$C_{21}H_{20}O_{10}$	431.0984	Negative	0.3–7	y = 0.0246x + 0.00306	0.01/0.03	0.9962	96.07	0.01	2.86
Ellagic acid	$C_{14}H_6O_8$	300.9990	Negative	0.05–10	y = 0.0085x − 0.000612	0.03/1	0.9994	101.49	0.27	3.59
Quercitrin	$C_{21}H_{20}O_{11}$	447.0933	Negative	0.05–10	y = 0.0179 + 0.0003331	0.01/0.03	0.999	97.00	0.93	4.20
Quercetin	$C_{15}H_{10}O_7$	301.0354	Negative	0.1–10	y = 0.0509x + 0.00467	0.01/0.03	0.9978	96.41	0.06	3.78
Herniarin	$C_{10}H_8O_3$	177.0546	Positive	0.1–7	y = 0.309x + 0.0266	0.01/0.03	0.9983	92.92	0.03	2.95
Salicylic acid	$C_7H_6O_3$	137.0244	Negative	0.3–10	y = 0.0361x + 0.00245	0.01/0.03	0.9982	92.88	7.82	3.89
Naringenin	$C_{15}H_{12}O_5$	271.0612	Negative	0.1–10	y = 0.0281x + 0.00182	0.01/0.03	0.9995	86.65	0.38	1.89
Luteolin	$C_{15}H_{10}O_6$	285.0405	Negative	0.1–10	y = 0.117x + 0.00848	0.01/0.03	0.9981	96.98	1.65	4.20
Apigenin	$C_{15}H_{10}O_5$	269.0456	Negative	0.3–10	y = 0.104x + 0.0199	0.01/0.03	0.9998	81.55	0.40	3.42
Hispidulin	$C_{16}H_{12}O_6$	301.0707	Positive	0.05–10	y = 0.02614x + 0.0003114	0.01/0.03	0.9993	98.36	6.61	2.87
Isosakuranetin	$C_{16}H_{14}O_5$	285.0769	Negative	0.05–10	y = 0.0235x + 0.000561	0.01/0.03	0.9992	96.56	9.98	3.41
Penduletin	$C_{18}H_{16}O_7$	343.0823	Negative	0.3–10	y = 0.0258x + 0.00253	0.01/0.03	0.9991	83.43	0.71	3.20
CAPE	$C_{17}H_{16}O_4$	283.0976	Negative	0.3–7	y = 0.255x + 0.0477	0.01/0.03	0.9964	94.42	0.13	3.13
Chrysin	$C_{15}H_{10}O_4$	253.0506	Negative	0.05–7	y = 0.0964x − 0.0002622	0.01/0.03	0.999	87.92	0.17	3.24
Quillaic acid	$C_{30}H_{46}O_5$	485.3273	Negative	0.05–10	y = 0.00781x − 0.0001318	0.01/0.03	0.9992	90.29	1.47	2.56
Caryophyllene oxide	$C_{15}H_{24}O$	221.1900	Positive	3–7	y = 0.00151x + 0.00692		0.9909	96.87	1.53	4.05

2.4. Essential Oil Isolation and GC/MS and GC-FID Analyses of Cinnamon Leaf Oil

The extract was dried over anhydrous $CaCl_2$, and the essential oil was stored at 4 °C until the GC-MS/FID measurements. The yield of oil was 1.52%. The GC-MS analysis was carried out using a Thermo Scientific Trace GC 1310 connected to a Thermo TSQ 9610 MS system on a DB-5 capillary column (60 m × 0.25 mm, 0.25 mm film thickness) with helium as the carrier gas (0.8 mL/min). GC oven temperature was maintained at 80 °C for 10 min, programmed to 280 °C at a rate of 4 °C/min, and kept constant at 280 °C for 5 min. The split ratio was adjusted to 1:20. The injector temperature was set to 250 °C. The mass spectra were recorded at 70 eV. The mass range was *m/z* 35–650. GC-FID analysis was performed using a Thermo Scientific Trace GC 1310 instrument. The FID detector temperature was 280 °C. To obtain the same elution order as GC-MS, simultaneous auto-injection was performed in duplicate on the same column under the same operational conditions. The relative percentage of the separated compounds was calculated from the FID chromatograms [53–55]. Alkanes were used as reference points in the calculation of Kovats Indices (KI). Compounds were identified by comparing their retention times and mass spectra with those obtained from authentic samples and/or the NIST and Wiley spectra, as well as literature data [54–57].

2.5. Reducing Ability Assays

The Fe^{3+} reduction potential of cinnamon leaf oil was determined using the $Fe^{3+}(CN^-)_6$ complex reduction method [58]. For this purpose, various concentrations of cinnamon leaf oil were transferred to test tubes, and 2.5 mL of phosphate buffer (pH 6.6, 0.2 M) and 2.5 mL of $[K_3Fe(CN)_6]$ solutions (1%) were added. Then, the mixture was vortexed and incubated at 50°C for 25 min. A portion of trichloroacetic acid (10%, 2.5 mL) was added. Then, 2.5 mL of upper layers of the solutions was mixed with 2.5 mL distilled water and 0.5 mL $FeCl_3$ (0.1%). The absorbance of reducing effects of cinnamon leaf oil was spectrophotometrically recorded at 700 nm.

After all the necessary experimental procedures, the absorbance of Cu^{2+} reducing ability of cinnamon leaf oil was determined according to a prior study [59]. For this purpose, 0.25 mL $CuCl_2$ solution (10 mM), 0.25 mL ethanolic neocuproine solution (7.5×10^{-3} M), and 250 μL NH_4Ac buffer solution (1.0 M) and different concentrations (10–30 μg/mL) of test tubes containing cinnamon leaf oil were transferred. Then, the total volume was increased to 2 mL with distilled water and their absorbance values were recorded at 450 nm after 30 min of incubation.

Lastly, the Fe^{3+}-TPTZ complex reducing ability reduction ability cinnamon leaf oil was performed according to a previous study [60]. For this, 2.25 mL of TPTZ solution (10 mM in 40 mM HCl) was freshly prepared and transferred to 2.5 mL of acetate buffer (0.3 M, pH 3.6) and 2.25 mL of $FeCl_3$ solution (20 mM). Then, different concentrations of cinnamon leaf oil were transferred and incubated at 37 °C for 25 min. Finally, the absorbance of reducing power of cinnamon leaf oil was spectrophotometrically measured at 593 nm. All experiments related to reducing abilities were repeated three times and the results were given as the arithmetic mean of these repetitions.

2.6. Radical Scavenging Activities

The radical scavenging potential of cinnamon leaf oil was evaluated using the DPPH radical according to the Blois method [61]. Briefly, 1 mL of $DPPH^\bullet$ solution (0.1 mM), which was prepared in ethanol and possessed a blue color, was added to the cinnamon leaf oil at different concentrations (10–30 μg/mL). Then, they were incubated at room temperature for 25 min and their absorbance values were recorded at 517 nm. ABTS radical scavenging ability of cinnamon leaf oil was realized according to Gulcin's method [62]. Firstly, aqueous solution of ABTS (7.0 mM) was oxidized by oxidants such as $K_2S_2O_8$ (2.5 mM) for generation of its radical cation ($ABTS^{\bullet+}$). The $ABTS^{\bullet+}$ solution was diluted with a phosphate buffer (0.1 M, pH 7.4) prior to use, adjusting the absorbance value of the control to 0.750 ± 0.025 at 734 nm. Then, 1 mL of $ABTS^{\bullet+}$ solution was added to 3 mL of cinnamon leaf oil at different concentrations (10–30 μg/mL). After 30 min, the remaining absorbance of $ABTS^{\bullet+}$ was measured at 734 nm [63].

The radical scavenging potential (RSC) of cinnamon leaf oil was calculated using the formula: RSC (%) = $(1 - A_c/A_s) \times 100$, where A_c and A_s are the absorbance values of the control and sample, respectively. In addition, IC_{50} was obtained from the graphs as μg/mL [64].

2.7. *Acetylcholinesterase Inhibition Assay*

The inhibitory effects of cinnamon leaf oil on AChE from horse serum and *Electrophorus electricus* were evaluated according to the methods described in our previous studies [65]. Acetylthiocholine iodide (AChI) and 5,5′-dithio-bis-(2-nitrobenzoic acid) (DTNB) were used as substrates [66]. Briefly, 1 mL of Tris/HCl buffer (1.0 M, pH 8.0), 10 μL of different concentrations of cinnamon leaf oil, and 50 μL AChE were mixed in a test tube. Then, the sample was incubated at 25 °C for 15 min and 50 μL of DTNB solution (0.5 mM,) was transferred. Then, the reaction was started by adding 50 μL of AChI solution (10 mM), and absorbance was recorded at 412 nm [67].

2.8. *α-Amylase Inhibition Assay*

The inhibitory potential of cinnamon leaf oil on α-amylase was determined using a starch substrate according to the Xiao procedure [68]. First, 1 g starch was dissolved in 50 mL NaOH solution (0.4 M) and heated at 80 °C for 20 min. After cooling, the pH was adjusted to 6.9, and the volume was adjusted to 100 mL using distilled water. Next, 35 μL of starch solution, 35 μL of phosphate buffer (pH 6.9), and 5 μL of the cinnamon leaf oil solutions were mixed. After incubation at 37 °C for 20 min, 20 μL of α-amylase solution was added and incubated again for 20 min. The reaction was completed by adding 50 μL of 0.1 M HCl, and absorbance was measured at 580 nm [69].

2.9. *hCA II Inhibition Assay*

Human erythrocytes were used as a source of CA II [70]. CA II was purified to high purity using the Sepharose-4B-tyrosine-sulfanilamide affinity column technique [71,72]. The protein content at each step of the purification assay was determined according to the Bradford method [73], and bovine serum albumin was used as the standard protein [74]. The purity of hCA II was determined using SDS-PAGE, as described in our previous work [75]. During hCA II purification and inhibition studies, esterase activity assay was measured spectrophotometrically at 348 nm [76].

2.10. *Determination of IC_{50} Value*

IC_{50} values were calculated to determine the potency of the inhibitory effects of cinnamon leaf oil. IC_{50} values were calculated from the graphs derived from the enzyme activity corresponding to increasing amounts of cinnamon leaf oil [77].

2.11. *Statistical Analysis*

Statistical analyses were performed using Student's *t*-test (GraphPad Prism 6, GraphPad, La Jolla, CA, USA, Software 7.0). The data are presented as means ± standard deviations (SD). The minimum significance level was set at $p < 0.05$.

3. Results

3.1. *Polyphenolic Composition of Cinnamon Leaf Oil*

The LC-HRMS assay was validated by determining the linearity, selectivity, precision, recovery, matrix effect, accuracy, and stability of the analytes [78–80]. In this study, 28 phenolic compounds were tentatively identified from cinnamon leaf oil. The LC-HRMS results showed that cinnamon leaf oil is rich in hispidulin (9.98 mg/L oil), herniarin (7.82 mg/L oil), and apigenin (6.61 mg/L oil). The cinnamon leaf oil was obtained by liquid-liquid extraction in order to determine the secondary metabolite profile of the species. A total of 100 mg of cinnamon leaf oil was dissolved by mobile phase B (1% formic acid in methanol) in a 4 mL volumetric flask and kept in an ultrasonic bath for 10 min. Then, 100 µL of internal standard (dihydrocapsaicin solution in methanol) was adjusted to volume with mobile phase B. The final solution was filtered through a 0.45 µm Millipore Millex-HV filter and the final solution (1 mL) was transferred into a capped auto sampler vial, from which 2 µL of sample was injected into LC for each run. The samples in the auto sampler were kept at 15 °C during the experiment [81].

GC-MS was used to perform accurate qualitative analysis of the contents of different aromatic compounds and essential oils [80]. Table 2 presents the relative information on the aromatic components of cinnamon leaf oil. In the present study, 17 volatile components were identified in cinnamon leaf oil samples. Among these, *E*-cinnamaldehyde (72.98%), benzyl benzoate (4.01%), β-caryophyllene (3.45%), and *trans*-cinnamylacetate (3.36%) were the most abundant compounds in cinnamon leaf oil (Figure 1 and Table 2).

Table 2. Chemical composition of the essential oil obtained from cinnamon (*C. zeylanicum*) leaf using GC-MS.

RT	Essential Oils	Contents (%)
937	α-Pinene	1.00
953	Camphene	0.34
986	β-Pinene	0.38
1008	Phellandrene	0.70
1026	*p*-Cymene	1.48
1097	Linalool	1.80
1193	α-Terpineol	0.48
1235	Z-Cinnamaldehyde	1.10
1281	Safrole	1.18
1284	*E*-Cinnamaldehyde	72.98
1365	Eugenol	1.48
1376	α-Copaene	0.77
1420	β-Caryophyllene	3.45
1433	*trans*-Cinnamylacetate	3.36
1458	α-Humulene	0.63
1525	Acetyleugenol	1.58
1586	(−)-Caryophyllene oxide	0.98
	Total	97.70

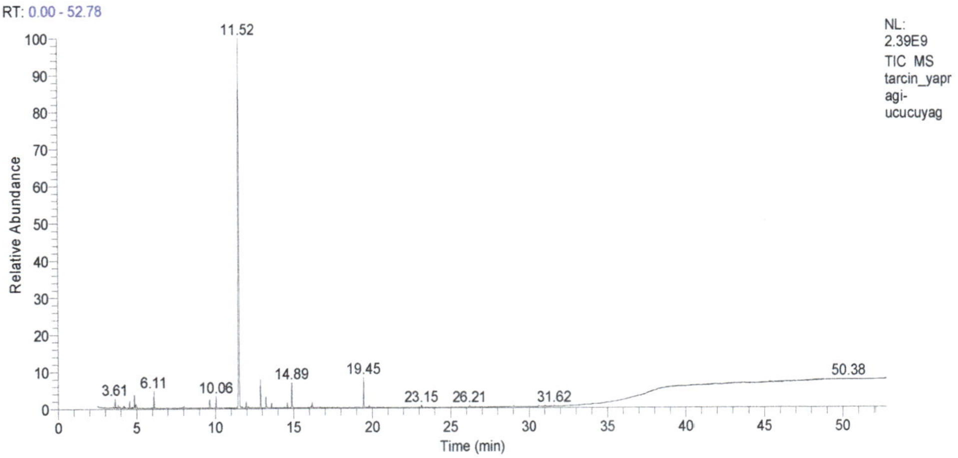

Figure 1. Compounds and their % ratios determined using GCMS analysis of the oil sample obtained from cinnamon (*C. zeylanicum*) leaf.

3.2. Reducing Ability of Cinnamon Leaf Oil

Cinnamon (*C. zeylanicum*) leaf oil demonstrated strong and effective reducing capacity in $Fe[Fe(CN^-)_6]_3$, Fe^{3+}-TPTZ, and Cu^{2+} reduction assays [82]. First, an Fe^{3+}–Fe^{2+} conversion assay was performed to measure the reducing ability of cinnamon leaf oil (Figure 2A and Table 3). Cinnamon leaf oil and standards at a concentration of 50 μg/mL (r^2 = 0.9804) exhibited Fe^{3+} reducing ability ($p < 0.01$) in the following order: ascorbic acid (2.298±0.086, r^2 = 0.9659) ≥ BHA (2.292 ± 0.012, r^2 = 0.9993) ≥ cinnamon leaf oil (2.190 ± 0.039, r^2 = 0.9741) ≥ BHT (2.136 ± 0.090, r^2 = 0.9957) > Trolox (1.514 ± 0.066, r^2 = 0.9963) > α-tocopherol (0.862 ± 0.038, r^2 = 0.9996). The increased absorbance reflects the complex formation and an increased reducing ability (Figure 1A).

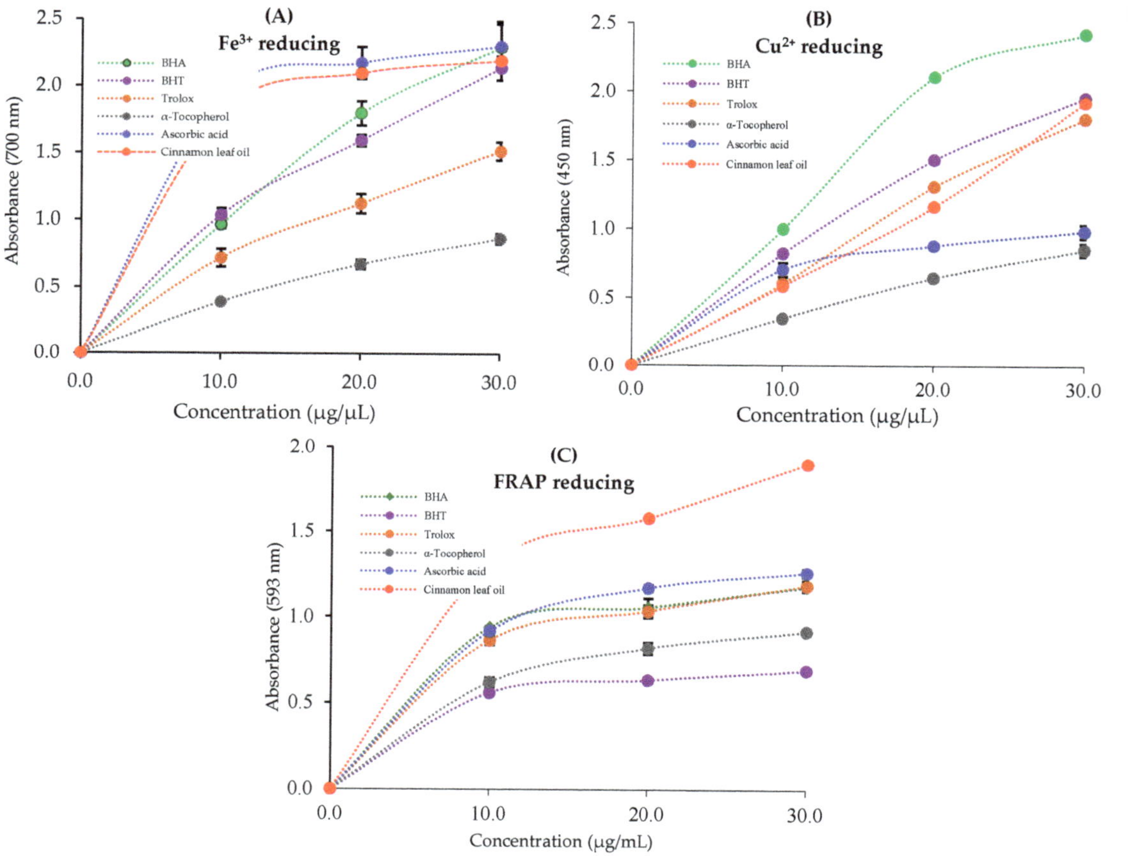

Figure 2. Fe^{3+}, Cu^{2+}, and Fe^{3+}-TPTZ reducing ability of cinnamon (*C. zeylanicum*) leaf oil and standards.

In addition to the Fe^{3+}-reducing effect, the Fe^{3+}-TPTZ and Cu^{2+}-reducing abilities of cinnamon leaf oil were investigated and are summarized in Figure 2B, C and Table 3 (r^2 = 0.9783). Cinnamon leaf oil showed high absorbance values at the tested concentrations. Cinnamon leaf oil and standards at a concentration of 30 μg/mL reduced Cu^{2+} ions in the following order (Figure 2B): BHA (2.292±0.012, r^2 = 0.9993) > BHT (1.953 ± 0.045, r^2 = 0.9998) > cinnamon leaf oil (1.918 ± 0.031, r^2 = 0.9992) > Trolox (1.800 ± 0.096, r^2 = 0.9974) > ascorbic acid (0.983 ± 0.048, r^2 = 0.9822) > α-tocopherol (0.851 ± 0.046, r^2 = 0.9994).

Table 3. Fe^{3+}, Cu^{2+}, and Fe^{3+}-TPTZ reducing ability of cinnamon oil and standards at 50 μg/mL concentration (BHA: butylated hydroxyanisole, BHT: butylated hydroxytoluene).

Antioxidants	Fe^{3+} Reducing *		Cu^{2+} Reducing *		Fe^{3+}-TPTZ Reducing *	
	λ_{700}	r^2	λ_{450}	r^2	λ_{593}	r^2
BHA	2.292 ± 0.012	0.9993	2.418 ± 0.018	0.9887	1.172 ± 0.014	0.9605
BHT	2.136 ± 0.090	0.9957	1.953 ± 0.045	0.9998	0.690 ± 0.008	0.9645
Trolox	1.514 ± 0.066	0.9963	1.800 ± 0.096	0.9974	1.180 ± 0.032	0.9732
α-Tocopherol	0.862 ± 0.038	0.9996	0.851 ± 0.046	0.9994	0.918 ± 0.011	0.9904
Ascorbic acid	2.298 ± 0.086	0.9659	0.983 ± 0.048	0.9822	1.257 ± 0.024	0.9869
Cinnamon leaf oil	2.190 ± 0.039	0.9741	1.918 ± 0.031	0.9992	1.900 ± 0.021	0.9725

* All values are averages of three parallel measurements (n = 3) and presented as mean ± SD.

Cinnamon (*C. zeylanicum*) leaf oil had an effective reducing potential in this reduction assay (Table 3 and Figure 2C). Test materials containing cinnamon leaf oil and standards declined in the following order of FRAP reducing capacity: cinnamon leaf oil (1.900 ± 0.021, r^2 = 0.9725) > ascorbic acid (1.257 ± 0.024, r^2 = 0.9869) > Trolox (1.180 ± 0.032, r^2 = 0.9732) ≥ BHA (1.172 ± 0.014, r^2 = 0.9605) > α-tocopherol (0.918 ± 0.011, r^2 = 0.9904) > BHT (0.690 ± 0.008, r^2 = 0.9645).

3.3. *Radicals Scavenging Effect of Cinnamon Leaf Oil*

The IC_{50} values of the DPPH scavenging effects of cinnamon (*C. zeylanicum*) leaf oil and standard radical scavengers were in the following order: 4.78 μg/mL for cinnamon leaf oil (r^2 = 0.9949) < 5.82 μg/mL for ascorbic acid (r^2 = 0.9668) < 6.03 μg/mL for Trolox (r^2 = 0.9925) < 6.86 μg/mL for BHA (r^2 = 0.9957) < 7.70 μg/mL for α-tocopherol (r^2 = 0.9961) < 49.50 μg/mL (r^2 = 0.9810) for BHT. The low EC_{50} values reflect high and effective $DPPH^{\bullet}$ scavenging capabilities (Table 4 and Figure 3A).

Table 4. IC_{50} values (μg/mL) of 1,1-diphenyl-2-picryl-hydrazyl free radical ($DPPH^{\bullet}$) scavenging and 2,2′-azino-bis(3-ethylbenzthiazoline-6-sulfonic acid ($ABTS^{\bullet+}$) scavenging activities of cinnamon (*C. zeylanicum*) leaf oil and standards.

Antioxidants	$DPPH^{\bullet}$ Scavenging		$ABTS^{\bullet+}$ Scavenging	
	IC_{50}	r^2	IC_{50}	r^2
BHA	6.86	0.9949	6.35	0.9746
BHT	49.50	0.9957	12.60	0.9995
Trolox	6.03	0.9925	16.50	0.9775
α-Tocopherol	7.70	0.9961	18.72	0.9347
Ascorbic acid	5.82	0.9668	11.74	0.9983
Cinnamon leaf oil	4.78	0.9344	5.21	0.9563

As shown in Figure 3B, cinnamon (*C. zeylanicum*) leaf oil exhibited effective ABTS radical scavenging ability in a concentration-dependent manner (10–20 μg/mL, $p < 0.001$). The IC_{50} value of cinnamon leaf oil in the $ABTS^{\bullet+}$ removal assay was calculated as 12.53 μg/mL (r^2 = 0.9827) (Table 4). When EC_{50} values were evaluated for standard molecules, the following sequence was observed: 6.35 μg/mL for BHA (r^2 = 0.9746) < 11.74 μg/mL, for ascorbic acid (r^2 = 0.9983) < 12.60 μg/mL for BHT (r^2 = 0.9995) < 16.50 μg/mL for Trolox (r^2 = 0.9775) < 18.72 μg/mL for α-tocopherol (r^2 = 0.9347) (Figure 3B).

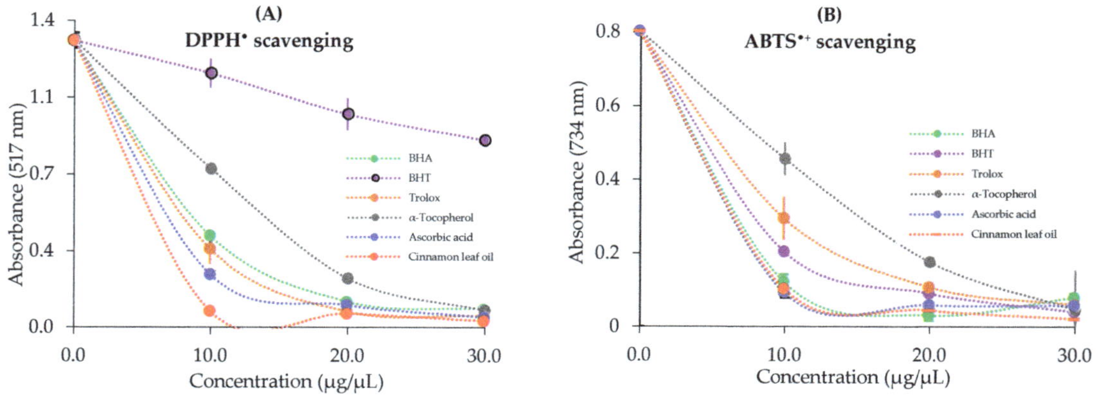

Figure 3. 1,1-diphenyl-2-picryl-hydrazyl free radical (DPPH•) scavenging and 2,2′-azino-bis(3-ethylbenzthiazoline-6-sulfonic acid (ABTS•+) scavenging effects of cinnamon (*C. zeylanicum*) leaf oil and standards.

3.4. Inhibition of Enzymes by Cinnamon Leaf Oil

The inhibition results for the enzymes used are listed in Table 5. Cinnamon (*C. zeylanicum*) leaf oil inhibited AChE with an IC_{50} value of 16.03 μM (r^2 = 0.9874). The tacrine standard inhibitor, used for comparison, inhibited AChE with an IC_{50} value of 7.58 μM (r^2 = 0.9074). As seen in Table 4, cinnamon leaf oil had a moderate inhibitory effect on α-amylase, with an IC_{50} value of 553.07 μM (r^2 = 0.9685). The IC_{50} value of cinnamon leaf oil against cytosolic and predominant hCA II isoenzyme was calculated as 25.41 ± 1.10 nM (Table 5). Acetazolamide (AZA), a clinical CA isoenzyme inhibitor used for comparison, inhibited cytosolic and predominant hCA II isoforms with an IC_{50} value of 4.41 ± 0.35 nM.

Table 5. The half maximal inhibition concentration (IC_{50}; μg/mL) values of cinnamon (*C. zeylanicum*) leaf oil against acetylcholinesterase, α-amylase, and carbonic anhydrase II enzymes.

Enzymes	Cinnamon Leaf Oil		Standard Inhibitors	
	IC_{50}	r^2	IC_{50}	r^2
α-Amylase [1]	553.07	0.9058	7.54	0.9074
Acetylcholinesterase [2]	16.03	0.9874	8.82	0.9836
Carbonic anhydrase II [3]	243.24	0.9092	9.96	0.9930

[1] Acarbose was used as a positive control for α-glycosidase and α-amylase enzymes. [2] Tacrine was used as a positive control for acetylcholinesterase and butyrylcholinesterase enzymes. [3] Acetazolamide (AZA) was used as a positive control for carbonic anhydrase II isoenzyme.

4. Discussion

Herbal extracts are commonly used in traditional and modern medicines owing to their beneficial health effects and healing properties, such as anti-aging, anti-fatigue, anti-cancer, and anti-hepatitis effects, as well as their ability to reduce blood serum lipid and glucose levels, and immunomodulation [83]. More than 25% of the drugs available worldwide are prepared from medicinal plants. Plants are rich in phenolic compounds, which are natural and active secondary plant metabolites. Recently, they have attracted great interest in pharmaceutical applications [84]. Currently, an increasing amount of scientific research is being conducted on the health benefits of naturally derived phenolics and flavonoids [36]. Consuming plants and foods rich in phenolic compounds has beneficial biological effects, including antioxidant, antimicrobial, and anti-inflammatory effects [85]. Numerous bioanalytical analyses are available for identifying and establishing a link between phenolic compounds and their applications. Due to its rich active phenolic content, the antioxidant activity of cinnamon leaf oil is expected to be high.

Owing to its high sensitivity, LC-HRMS is one of the most widely used chromatographic methods for the quantitative analysis of phenolic compounds in plant extracts [86]. Moreover, the optimization of various experimental parameters, such as the extraction solvent, column, mobile phase, and LC-HRMS conditions, is extremely important in the analysis of the phenolic content of oils obtained from plants [87]. The phenolic compounds are presumably responsible for the antioxidant ability of cinnamon leaf oil [88,89]. Phenolics extracted from plant materials have many effective biological functions, such as metal binding, ROS scavenging, reduction, and H-atom donation [90]. The most abundant phenolic compounds in plant oil were found to be isosakuranetin, hispidulin, and chlorogenic acid. Isosakuranetin is a 4′-O-methylated flavonoid, found in citrus fruits. It is consumed daily in the human diet as a constituent of a variety of plant components. Isosakuranetin has been previously described to have antimycobacterial, antioxidant, antifungal, antibacterial, anticancer, neuroprotective, and antiallergic properties [91]. Hispidulin (4′,5,7-trihydroxy-6-methoxyflavone) is an abundantly found flavonoid in medicinal plants, which has been widely used in medicine as an antioxidant [92]. Hispidulin was measured as the second dominant phenolic compound (6.61 mg/L oil) in cinnamon leaf oil. Chlorogenic acid, which is one of the abundant phenolic acids in plants, especially in coffee and tea, has been reported to have strong antioxidant activity as well as many pharmacological activities in many preclinical and clinical studies [93].

The reducing capacity of plant-derived oils contributes to their biological activities. Because of their high reducing capacity, these oils reduce oxidative stress and neutralize ROS [94]. The ferric ion (Fe^{3+}) reduction method can be used to directly determine the reducing ability of plant-derived oils or extracts. The addition of cinnamon leaf oil to solutions containing Fe^{3+} ions under experimental conditions causes the formation of blue-colored $Fe_4[Fe(CN^-)_6]_3$, which can absorb light at 700 nm [95]. The color of the test mixture, along with the formation of this chromophore complex, reveals the reducing capability of the plant extracts in different color scales from yellow to green [96]. Cinnamon leaf oil demonstrated a strong and effective reducing capacity in $Fe_4[Fe(CN^-)_6]_3$, Fe^{3+}-TPTZ, and Cu^{2+} reduction assays. The results demonstrate that cinnamon leaf oil has an effective Fe^{3+}-reducing ability and that the e-donor property neutralizes the harmful effects of ROS and free radicals. This reductive potential was higher than that of BHT, Trolox, and α-tocopherol but close to that of BHA and ascorbic acid. The CUPRAC method, another reduction test used in our study, is a low-cost, selective, stable, and fast method for plant extracts and oils, independent of chemical components and hydrophobicity [97]. The final reduction test used in this study was the FRAP reduction method. As with the prior reduction assays, high absorbance values reflect the high reduction potential of the complex. In addition, the FRAP method should be applied in acidic media to maintain the solubility of Fe^{3+} ions [98,99].

The $ABTS^{\bullet+}$ and DPPH· removal assays are the most useful and widely used radical removal assays. They are used to evaluate the radical scavenging capacity of materials in the food and pharmacology sectors. In a recent study, IC_{50} values of methanol and aqueous extracts of *M. pulegium* were calculated as 18.52 and 15.37 μg/mL, respectively [100]. The IC_{50} values for aqueous and ethanol extracts of *C. verum*, another cinnamon type, were determined to be 15.71 and 21.25 μg/mL, respectively [48]. In a prior study, IC_{50} values for methanol and aqueous extracts of *S. pilifera*, widely distributed in Anatolia, were 7.05 and 8.56 μg/mL, respectively [34]. The chromogenic ABTS radical scavenging assay can be easily applied to hydrophilic and lipophilic antioxidant compounds. The values for methanol and aqueous extracts of the *M. pulegium* plant were recorded as 7.92 and 9.37 μg/mL, respectively [100], and 3.52 and 4.76 μg/mL for methanol and aqueous extracts of *S. pilifera*, respectively [34]. In a recent study, the values were found to be 5.52 and 5.79 μg/mL for water and ethanol extracts of *C. verum*, respectively [48].

When the enzyme inhibition results were evaluated and compared with standard inhibitors, cinnamon leaf oil was found to have an effective inhibition capacity against AChE, α-amylase, and hCA II, which are associated with global metabolic diseases, such as AD and T2DM. Enzyme inhibition is an effective approach in the treatment of several diseases [101]. Based on these results, cinnamon leaf oil was a more effective and suitable inhibitor of AChE than tacrine. In a recent study by our group, methanol and water extracts of the *M. pulegium* had IC_{50} values of 40.76 and 53.31 μg/mL, respectively, for AChE [100]. In another study, water and ethanol extracts of *C. verum* had an inhibitory effect on AChE with Ki values of 221.33 μg/mL and 110.26 μg/mL, respectively [48]. Similarly, ethanolic (IC_{50}: 23.01 μg/mL) and water extracts (IC_{50}: 25.66 μg/mL) of *Rhus coriaria* have a highly effective inhibitory action on AChE [102]. The inhibition of α-amylase by natural products can effectively reduce blood glucose levels, particularly in PPGL and T2DM. Cinnamon leaf oil showed a greater inhibitory effect (IC_{50}: 22.800 mM) than acarbose, a standard inhibitor [103]. In a recent study, ethanol and dichloromethane/methanol extracts of bark and leaf of authenticated Ceylon cinnamon (*C. zeylanicum*) were studied for α-amylase, α-glucosidase, and inhibition effects. It was reported that these extracts exhibited IC_{50} values of 26.62–36.09 and 214–215 and μg/mL against AChE and α-amylase, respectively [104]. The inhibitory effects on digestive enzymes, including α-amylase, can reduce blood glucose levels. Moreover, this inhibition may have a significant therapeutic effect on regulating hyperglycemia associated with diabetes.

Plants are rich in biologically active phenolic compounds. Phenolic compounds are slightly acidic and convert into water-soluble phenolate anions by losing at least one proton (H^+) from their hydroxyl groups (-OH). Plants rich in phenolics effectively inhibit CA [105]. In addition, phenolic compounds found in plants can inhibit CA isoforms due to the coordination of the Zn^{2+} ion in the active sites to phenolic -OH, -COOH, and -OCH_3 groups in their phenolic rings. This physiologically dominant isoform is found almost everywhere in cells and is associated with many diseases such as glaucoma, edema, and epilepsy [106].

5. Conclusions

Cinnamon leaf oil was evaluated using several in vitro bioanalytical assays for its antioxidant properties and inhibitory effects on metabolic enzymes, such as AChE, α-amylase, and CA II, which are associated with diabetes, Alzheimer's disease, and glaucoma. In addition, the potential active ingredients present in cinnamon leaf oil were determined using LC-HR/MS, GC/MS, and GC-FID. The most abundant phenolic compounds in the plant oil were found to be isosakuranetin, hispidulin, and chlorogenic acid. Moreover, as a result of GC-FID analysis of cinnamon leaf oil, it was determined that 72.98% of the existing oil was E-Cinnamaldehyde. The results show that cinnamon leaf oil can be a rich and useful source of biologically important biomolecules. In addition, cinnamon oil is rich in natural phenolic compounds, such as *E*-cinnamaldehyde, benzyl benzoate, β-caryophyllene, and *trans*-cinnamylacetate. LC-HR/MS analysis showed that cinnamon leaf oil has high amounts of natural phenolic compounds, such as luteo-lin-7-glycoside, ellagic acid, and naringenin.

Author Contributions: Conceptualization, M.M., E.K., A.C.G. and İ.G.; methodology and investigation, M.M., Z.B., E.M.U., E.K. and A.C.G.; software, validation, and visualization, A.C.G. and İ.G.; resources, S.H.A. and İ.G.; data curation, writing—original draft preparation, writing—review and editing, supervision, funding, and acquisition, İ.G., S.H.A. and A.C.G. All authors have read and agreed to the published version of the manuscript.

Funding: This study received no external funding.

Institutional Review Board Statement: Not applicable.

Informed Consent Statement: Not applicable.

Data Availability Statement: Data are available in a publicly accessible repository.

Acknowledgments: S.A. would like to extend his sincere appreciation to the Researchers Support Program (RSP-2023/59) of King Saud University, Saudi Arabia.

Conflicts of Interest: The authors declare no conflict of interest.

References

1. Topal, M.; Gocer, H.; Topal, F.; Kalin, P.; Polat Kose, P.; Gulcin, I.; Cakmak, K.C.; Kucuk, M.; Durmaz, L.; Goren, A.C.; et al. Antioxidant, antiradical and anticholinergic properties of cynarin purified from the illyrian thistle (*Onopordum illyricum* L.). *J. Enzym. Inhib. Med. Chem.* **2016**, *31*, 266–275. [CrossRef] [PubMed]
2. Koksal, E.; Bursal, E.; Gulcin, I.; Korkmaz, M.; Caglayan, C.; Goren, A.C.; Alwasel, S.H. Antioxidant activity and polyphenol content of Turkish thyme (*Thymus vulgaris*) monitored by LC-MS/MS. *Int. J. Food Prop.* **2017**, *20*, 514–525. [CrossRef]
3. Gulcin, I.; Beydemir, Ş.; Sat, I.G.; Küfrevioglu, O.I. Evaluation of antioxidant activity of cornelian cherry (*Cornus mas* L.). *Acta Aliment. Hung.* **2005**, *34*, 193–202. [CrossRef]
4. Gulcin, I.; Sat, I.G.; Beydemir, Ş.; Kufrevioglu, O.I. Evaluation of the in vitro antioxidant properties of extracts of broccoli (*Brassica oleracea* L.). *Ital. J. Food Sci.* **2004**, *16*, 17–30.
5. Gulcin, I.; Topal, F.; Ozturk Sarikaya, S.B.; Bursal, E.; Goren, A.C.; Bilsel, M. Polyphenol contents and antioxidant properties of medlar (*Mespilus germanica* L.). *Rec. Nat. Prod.* **2011**, *5*, 158–175.
6. Serbetci Tohma, H.; Gulcin, I. Antioxidant and radical scavenging activity of aerial parts and roots of Turkish liquorice (*Glycyrrhiza glabra* L.). *Int. J. Food Prop.* **2010**, *13*, 657–671. [CrossRef]
7. Elmastas, M.; Türkekul, I.; Ozturk, L.; Gulcin, I.; Isıldak, O.; Aboul-Enein, H.Y. The antioxidant activity of two wild edible mushrooms (*Morchella vulgaris* and *Morchella esculanta*). *Comb. Chem. High Throughput Screen.* **2006**, *9*, 443–448. [CrossRef]
8. Gulcin, I.; Tel, A.Z.; Kirecci, E. Antioxidant, antimicrobial, antifungal and antiradical activities of *Cyclotrichium niveum* (Boiss.) Manden and Scheng. *Int. J. Food Prop.* **2008**, *11*, 450–471. [CrossRef]
9. Ak, T.; Gulcin, I. Antioxidant and radical scavenging properties of curcumin. *Chem. Biol. Interact.* **2008**, *174*, 27–37. [CrossRef]
10. Talaz, O.; Gulcin, I.; Göksu, S.; Saracoglu, N. Antioxidant activity of 5,10-dihydroindeno[1,2-b]indoles containing substituents on dihydroindeno part. *Bioorg. Med. Chem.* **2009**, *17*, 6583–6589. [CrossRef]
11. Gulcin, I. Antioxidant activity of eugenol-a structure and activity relationship study. *J. Med. Food* **2011**, *14*, 975–985. [CrossRef] [PubMed]
12. Kiziltas, H.; Goren, A.C.; Alwasel, S.; Gulcin, I. Sahlep (*Dactylorhiza osmanica*): Comprehensive metabolic profiling of Acantholimon caryophyllaceum using LC-HRMS and evaluation of antioxidant activities, enzyme inhibition properties and molecular docking studies. *S. Afr. J. Bot.* **2022**, *151*, 743–751. [CrossRef]
13. Gulcin, I.; Oktay, M.; Kirecci, E.; Kufrevioglu, O.I. Screening of antioxidant and antimicrobial activities of anise (*Pimpinella anisum* L.) seed extracts. *Food Chem.* **2003**, *83*, 371–382. [CrossRef]
14. Elmastas, M.; Celik, S.M.; Genc, N.; Aksit, H.; Erenler, R.; Gulcin, I. Antioxidant activity of an Anatolian herbal tea-*Origanum minutiflorum*: Isolation and characterization of its secondary metabolites. *Int. J. Food Prop.* **2018**, *21*, 374–384. [CrossRef]
15. Koksal, E.; Bursal, E.; Dikici, E.; Tozoglu, F.; Gulcin, I. Antioxidant activity of *Melissa officinalis* leaves. *J. Med. Plants Res.* **2011**, *5*, 217–222.
16. Gulcin, I.; Sat, I.G.; Beydemir, Ş.; Elmastas, M.; Kufrevioglu, O.I. Comparison of antioxidant activity of clove (*Eugenia caryophylata* Thunb) buds and lavender (*Lavandula stoechas* L.). *Food Chem.* **2004**, *87*, 393–400. [CrossRef]
17. Tulini, F.L.; Souza, V.B.; Echalar-Barrientos, M.A.; Thomazini, M.; Pallone, E.M.J.A.; Favaro-Trindade, C.S. Development of solid lipid microparticles loaded with a proanthocyanidin-rich cinnamon extract (*Cinnamomum zeylanicum*): Potential for increasing antioxidant content in functional foods for diabetic population. *Food Res. Int.* **2016**, *85*, 10–18. [CrossRef]
18. Hanumantha, M.; Vasudeva, R. Influence of patch geometry, post-bark-extraction-treatment on bark recovery and standardizing number of sprouts for bark harvest from coppices in *Cinnamomum zeylanicum* blume: Implications for sustainable harvesting. *Environ. Monit. Assess.* **2022**, *194*, 214. [CrossRef]
19. Ali, S.S.; Abd Elnabi, M.K.; Alkherkhisy, M.M.; Hasan, A.; Li, F.; Khalil, M.; Sun, J.; El-Zawawy, N. Exploring the potential of *Cinnamomum zeylanicum* oil against drug resistant Helicobacter pylori-producing cytotoxic genes. *J. Appl. Biomed.* **2022**, *20*, 22–36. [CrossRef]
20. Azab, S.S.; Abdel Jaleel, G.A.; Eldahshan, O.A. Anti-inflammatory and gastroprotective potential of leaf essential oil of Cinnamomum glanduliferum in ethanol-induced rat experimental gastritis. *Pharm. Biol.* **2017**, *55*, 1654–1661. [CrossRef]
21. Sihoglu Tepe, A.; Ozaslan, M. Anti-Alzheimer, anti-diabetic, skin-whitening, and antioxidant activities of the essential oil of *Cinnamomum zeylanicum*. *Ind. Crops Prod.* **2020**, *145*, 112069. [CrossRef]
22. Oztaskın, N.; Goksu, S.; Demir, Y.; Maras, A.; Gulcin, I. Synthesis of novel bromophenol including diaryl Methanes-Determination of their inhibition effects on carbonic anhydrase and acetylcholinesterase. *Molecules* **2022**, *27*, 7426. [CrossRef] [PubMed]
23. Aktas, A.; Yaklı, G.; Demir, Y.; Gulcin, I.; Aygun, M.; Gok, Y. The palladium-based complexes bearing 1,3-dibenzylbenzimidazolium with morpholine, triphenylphosphine, and pyridine derivate ligands: Synthesis, characterization, structure and enzyme inhibitions. *Heliyon* **2022**, *8*, e10625. [CrossRef] [PubMed]
24. Gocer, H.; Topal, F.; Topal, M.; Kucuk, M.; Teke, D.; Gulcin, I.; Alwasel, S.H.; Supuran, C.T. Acetylcholinesterase and carbonic anhydrase isoenzymes I and II inhibition profiles of taxifolin. *J. Enzym. Inhib. Med. Chem.* **2016**, *31*, 441–447. [CrossRef] [PubMed]
25. Bursal, E.; Taslimi, P.; Goren, A.; Gulcin, I. Assessments of anticholinergic, antidiabetic, antioxidant activities and phenolic content of *Stachys annua*. *Biocat. Agric. Biotechnol.* **2020**, *28*, 101711. [CrossRef]
26. Durmaz, L.; Kiziltas, H.; Guven, L.; Karagecili, H.; Alwasel, S.; Gulcin, I. Antioxidant, antidiabetic, anticholinergic, and antiglaucoma effects of magnofluorine. *Molecules* **2020**, *27*, 5902. [CrossRef] [PubMed]

27. Bora, R.E.; Genç Bilgicli, H.; Uc, E.M.; Alagoz, M.A.; Zengin, M.; Gulcin, I. Synthesis, characterization, evaluation of metabolic enzyme inhibitors and in silico studies of thymol based 2-amino thiol and sulfonic acid compounds. *Chem. Biol. Interact.* **2022**, *366*, 110134. [CrossRef] [PubMed]
28. Gulcin, I.; Scozzafava, A.; Supuran, C.T.; Akıncıoglu, H.; Koksal, Z.; Turkan, F.; Alwasel, S. The effect of caffeic acid phenethyl ester (CAPE) metabolic enzymes including acetylcholinesterase, butyrylcholinesterase, glutathione s-transferase, lactoperoxidase and carbonic anhydrase isoenzymes I, II, IX and XII. *J. Enzym. Inhib. Med. Chem.* **2016**, *31*, 1095–1101. [CrossRef]
29. Ozbey, F.; Taslimi, P.; Gulcin, I.; Maras, A.; Goksu, S.; Supuran, C.T. Synthesis, acetylcholinesterase, butyrilcholinesterase, carbonic anhydrase inhibitory and metal chelating properties of some novel diaryl ether. *J. Enzym. Inhib. Med. Chem.* **2016**, *31*, 79–85. [CrossRef]
30. Polat Kose, L.; Bingol, Z.; Kaya, R.; Goren, A.C.; Akincioglu, H.; Durmaz, L.; Koksal, E.; Alwasel, S.; Gulcin, I. Anticholinergic and antioxidant activities of avocado (*Folium perseae*) leaves—Phytochemical content by LC-MS/MS Analysis. *Int. J. Food Prop.* **2020**, *23*, 878–893. [CrossRef]
31. Gulcin, I.; Kaya, R.; Goren, A.C.; Akıncıoğlu, H.; Topal, M.; Bingol, Z.; Cetin Cakmak, K.; Ozturk Sarikaya, S.B.; Durmaz, L.; Alwasel, S. Anticholinergic, antidiabetic and antioxidant activities of Cinnamon (*Cinnamomum verum*) bark extracts: Polyphenol contents analysis by LC-MS/MS. *Int. J. Food Prop.* **2019**, *22*, 1511–1526. [CrossRef]
32. Aktas Anıl, D.; Polat, M.F.; Saglamtas, R.; Tarikogulları, A.H.; Alagoz, M.A.; Gulcin, I.; Algul, O.; Burmaoglu, S. Exploring enzyme inhibition profiles of novel halogenated chalcone derivatives on some metabolic enzymes: Synthesis, characterization and molecular modeling studies. *Comp. Biol. Chem.* **2022**, *100*, 107748. [CrossRef] [PubMed]
33. Erdemir, F.; Barut Celepci, D.; Aktas, A.; Gök, Y.; Kaya, R.; Taslimi, P.; Demir, Y.; Gulcin, I. Novel 2-aminopyridine liganded Pd(II) N-heterocyclic carbene complexes: Synthesis, characterization, crystal structure and bioactivity properties. *Bioorg. Chem.* **2019**, *91*, 103134. [CrossRef] [PubMed]
34. Erdemir, F.; Barut Celepci, D.; Aktas, A.; Taslimi, P.; Gok, Y.; Karabıyık, H.; Gulcin, I. 2-Hydroxyethyl substituted NHC precursors: Synthesis, characterization, crystal structure and carbonic anhydrase, α-glycosidase, butyrylcholinesterase, and acetylcholinesterase inhibitory properties. *J. Mol. Struct.* **2018**, *1155*, 797–806. [CrossRef]
35. Akter, K.; Lanza, E.A.; Martin, S.A.; Myronyuk, N.; Rua, M.; Raffa, R.B. Diabetes mellitus and Alzheimer's disease: Shared pathology and treatment? *Br. J. Clin. Pharmacol.* **2010**, *71*, 365–376. [CrossRef]
36. Gulcin, I.; Tel, A.Z.; Goren, A.C.; Taslimi, P.; Alwasel, S. Sage (*Salvia pilifera*): Determination its polyphenol contents, anticholinergic, antidiabetic and antioxidant activities. *J. Food Meas. Charact.* **2019**, *13*, 2062–2074. [CrossRef]
37. Fatah, N.H.A.; Amen, Y.; Abdel Bar, F.M.; Halim, A.F.; Saad, H.E.A. Antioxidants and α-glucosidase Inhibitors from *Lactuca serriola* L. *Rec. Nat. Prod.* **2020**, *14*, 410–415. [CrossRef]
38. Dasgin, S.; Gok, Y.; Barut Celepci, D.; Taslimi, P.; Izmirli, M.; Aktas, A.; Gulcin, I. Synthesis, characterization, crystal structure and bioactivity properties of the benzimidazole-functionalized PEPPSI type of Pd(II)NHC complexes. *J. Mol. Struct.* **2021**, *1228*, 129442. [CrossRef]
39. Takım, K.; Yigin, A.; Koyuncu, I.; Kaya, R.; Gulcin, I. Anticancer, anticholinesterase and antidiabetic activities of Tunceli garlic (*Allium tuncelianum*)—Determining its phytochemical content by LC-MS/MS analysis. *J. Food Meas. Charact.* **2021**, *15*, 3323–3335. [CrossRef]
40. Atmaca, U.; Alp, C.; Akıncıoglu, H.; Karaman, H.S.; Gulcin, I.; Celik, M. Novel hypervalent iodine catalyzed synthesis of α-sulfonoxy ketones: Biological activity and molecular docking studies. *J. Mol. Struct.* **2021**, *1239*, 130492. [CrossRef]
41. Aktas, A.; Barut Celepci, D.; Gok, Y.; Taslimi, P.; Akıncıoglu, H.; Gulcin, I. A novel Ag-N-heterocyclic carbene complex bearing the hydroxyethyl ligand: Synthesis, characterization, crystal and spectral structures and bioactivity properties. *Crystals* **2020**, *10*, 171. [CrossRef]
42. Akıncıoglu, A.; Akıncıoglu, H.; Gulcin, I.; Durdagı, S.; Supuran, C.T.; Goksu, S. Discovery of potent carbonic anhydrase and acetylcholine esterase inhibitors: Novel sulfamoylcarbamates and sulfamides derived from acetophenones. *Bioorg. Med. Chem.* **2015**, *23*, 3592–3602. [CrossRef] [PubMed]
43. Pedrood, K.; Sherefati, M.; Taslimi, P.; Mohammadi-Khanaposhtani, M.; Asgari, M.S.; Hosseini, S.; Rastegar, H.; Larijani, B.; Mahdavi, M.; Taslimi, P.; et al. Design, synthesis, characterization, enzymatic inhibition evaluations, and docking study of novel quinazolinone derivatives. *Int. J. Biol. Macromol.* **2021**, *170*, 1–12. [CrossRef] [PubMed]
44. Hashmi, S.; Khan, S.; Shafiq, Z.; Taslimi, P.; Ishaq, M.; Sadeghian, N.; Karaman, S.H.; Akhtar, N.; Islam, M.; Asari, A.; et al. Probing 4-(diethylamino)-salicylaldehyde-based thiosemicarbazones as multi-target directed ligands against cholinesterases, carbonic anhydrases and α-glycosidase enzymes. *Bioorg. Chem.* **2021**, *107*, 104554. [CrossRef] [PubMed]
45. Gulcin, I.; Trofimov, B.; Kaya, R.; Taslimi, P.; Sobenina, L.; Schmidt, E.; Petrova, O.; Malysheva, S.; Gusarova, N.; Farzaliyev, V.; et al. Synthesis of nitrogen, phosphorus, selenium and sulfur-containing heterocyclic compounds-determination of their carbonic anhydrase, acetylcholinesterase, butyrylcholinesterase and α-glucosidase inhibition properties. *Bioorg. Chem.* **2020**, *103*, 104171. [CrossRef]
46. Lolak, N.; Akocak, S.; Turkes, C.; Taslimi, P.; Isık, M.; Beydemir, S.; Gulcin, I.; Durgun, M. Synthesis, characterization, inhibition effects, and molecular docking studies as acetylcholinesterase, α-glycosidase, and carbonic anhydrase inhibitors of novel benzenesulfonamides incorporating 1,3,5-triazine structural motifs. *Bioorg. Chem.* **2020**, *100*, 103897. [CrossRef]

47. Yigit, B.; Taslimi, P.; Barut Celepci, D.; Taskin-Tok, T.; Yigit, M.; Aygun, M.; Ozdemir, I.; Gulcin, I. Novel PEPPSI-type N-heterocyclic carbene palladium(II) complexes: Synthesis, characterization, in silico studies and enzyme inhibitory properties against some metabolic enzymes. *Inorg. Chim. Acta* **2023**, *544*, 121239. [CrossRef]
48. Boztas, M.; Cetinkaya, Y.; Topal, M.; Gulcin, I.; Menzek, A.; Şahin, E.; Tanc, M.; Supuran, C.T. Synthesis and carbonic anhydrase isoenzymes I, II, IX, and XII inhibitory effects of dimethoxy-bromophenol derivatives incorporating cyclopropane moieties. *J. Med. Chem.* **2015**, *58*, 640–650. [CrossRef]
49. Akbaba, Y.; Akıncıoğlu, A.; Göçer, H.; Goksu, S.; Gulcin, I.; Supuran, C.T. Carbonic anhydrase inhibitory properties of novel sulfonamide derivatives of aminoindanes and aminotetralins. *J. Enzyme Inhib. Med. Chem.* **2014**, *29*, 35–42. [CrossRef]
50. Akbas, F.; Ozaydın, A.; Polat, E.; Onaran, I. *Lucilia sericata* larval secretions stimulating wound healing effects on rat dermal fibroblast cells. *Rec. Nat. Prod.* **2020**, *14*, 340–354. [CrossRef]
51. Kiziltas, H.; Goren, A.C.; Bingol, Z.; Alwasel, S.H.; Gulcin, I. Anticholinergic, antidiabetic and antioxidant activities of *Ferula oriantalis* L. determination of its polyphenol contents by LC-HRMS. *Rec. Nat. Prod.* **2021**, *15*, 513–528.
52. Magnusson, B.; Ornemark, U. (Eds.) *Eurachem Guide: The Fitness for Purpose of Analytical Methods—A Laboratory Guide to Method Validation and Related Topics*, 2nd ed.; LGC: Teddington, UK, 2014; ISBN 978-91-87461-59-0.
53. Patino-Bayona, W.R.; Plazas, E.; Bustos-Cortes, J.J.; Prieto-Rodríguez, J.A.; Patino-Ladino, O.J. Essential oils of three Hypericum species from Colombia: Chemical composition, insecticidal and repellent activity against *Sitophilus zeamais* Motsch. (Coleoptera: Curculionidae). *Rec. Nat. Prod.* **2021**, *15*, 111–121.
54. Suzgec-Selcuk, S.; Ozek, G.; Yur, S.; Goger, F.; Gurdal, M.B.; Toplan, G.G.; Mericli, A.H.; Baser, K.H.C. The leaf and the gall volatiles of *Salvia fruticosa* miller from Turkey: Chemical composition and biological activities. *Rec. Nat. Prod.* **2021**, *15*, 10–24. [CrossRef]
55. Altun, M.; Goren, A.C. Essential oil composition of *Satureja cuneifolia* by simultaneous distillation-extraction and thermal desorption GC-MS techniques. *J. Essent. Oil Bear. Plants* **2007**, *10*, 139–144. [CrossRef]
56. Ha, C.T.T.; Thuy, D.T.T.; Nam, V.Q.; Tam, N.K.B.; Setzer, W.N. Composition and antimicrobial activity of essential oils from leaves and twigs of *Magnolia hookeri* var. Longirostrata, D.X.LI & R. Z. Zhou and Magnolia insignis wall. in Ha Giang province of Vietnam. *Rec. Nat. Prod.* **2021**, *15*, 207–212.
57. Salinas, M.; Bec, N.; Calva, J.; Ramírez, J.; Andrade, J.M.; Vidari, G.; Larroque, C. Chemical composition and anticholinesterase activity of the essential oil from the ecuadorian plant *Salvia pichinchensis* Benth. *Rec. Nat. Prod.* **2021**, *14*, 276–2285. [CrossRef]
58. Gulcin, I.; Buyukokuroglu, M.E.; Oktay, M.; Kufrevioglu, O.I. Antioxidant and analgesic activities of turpentine of *Pinus nigra* Arn. Subsp. pallsiana (Lamb.) Holmboe. *J. Ethnopharmacol.* **2003**, *86*, 51–58. [CrossRef] [PubMed]
59. Topal, F.; Topal, M.; Gocer, H.; Kalın, P.; Kocyigit, U.M.; Gulcin, I.; Alwasel, S.H. Antioxidant activity of taxifolin: An activity-structure relationship. *J. Enzyme Inhib. Med. Chem.* **2016**, *31*, 674–683. [CrossRef]
60. Artunc, T.; Menzek, A.; Taslimi, P.; Gulcin, I.; Kazaz, C.; Sahin, E. Synthesis and antioxidant activities of phenol derivatives from 1,6-bis(dimethoxyphenyl)hexane-1,6-dione. *Bioorg. Chem.* **2020**, *100*, 103884. [CrossRef]
61. Blois, M.S. Antioxidant determinations by the use of a stable free radical. *Nature* **1958**, *26*, 1199–1200. [CrossRef]
62. Gulcin, I. Antioxidant activity of L-Adrenaline: An activity-structure insight. *Chem. Biol. Interact.* **2009**, *179*, 71–80. [CrossRef] [PubMed]
63. Gulcin, I. Antioxidant activity of food constituents: An overview. *Arch. Toxicol.* **2012**, *86*, 345–391. [CrossRef] [PubMed]
64. Gulcin, I.; Huyut, Z.; Elmastas, M.; Aboul-Enein, H.Y. Radical scavenging and antioxidant activity of tannic acid. *Arab. J. Chem.* **2010**, *3*, 43–53. [CrossRef]
65. Taslimi, P.; Gulcin, I. Antioxidant and anticholinergic properties of olivetol. *J. Food Biochem.* **2018**, *42*, e12516. [CrossRef]
66. Yigit, B.; Kaya, R.; Taslimi, P.; Isık, Y.; Karaman, M.; Yigit, M.; Ozdemir, I.; Gulcin, I. Imidazolinium chloride salts bearing wing tip groups: Synthesis, molecular docking and metabolic enzymes inhibition. *J. Mol. Struct.* **2019**, *1179*, 709–718. [CrossRef]
67. Bal, S.; Demirci, O.; Sen, B.; Taslimi, P.; Aktas, A.; Gok, Y.; Aygun, M.; Gulcin, I. PEPPSI type Pd(II)NHC complexes bearing Chloro-/fluorobenzyl group: Synthesis, characterization, crystal structures, α-glycosidase and acetylcholinesterase inhibitory properties. *Polyhedron* **2021**, *198*, 115060. [CrossRef]
68. Xiao, Z.; Storms, R.; Tsang, A. A quantitative starch–iodine method for measuring alpha-amylase and glucoamylase activities. *Anal. Biochem.* **2006**, *351*, 146–148. [CrossRef]
69. Karimov, A.; Orujova, A.; Taslimi, P.; Sadeghian, N.; Mammadov, B.; Karaman, H.S.; Farzaliyev, V.; Sujayev, A.; Tas, R.; Alwasel, S.; et al. Novel functionally substituted esters based on sodium diethyldithiocarbamate derivatives: Synthesis, characterization, biological activity and molecular docking studies. *Bioorg. Chem.* **2020**, *99*, 103762. [CrossRef]
70. Scozzafava, A.; Passaponti, M.; Supuran, C.T.; Gulcin, I. Carbonic anhydrase inhibitors: Guaiacol and catechol derivatives effectively inhibit certain human carbonic anhydrase isoenzymes (hCA I, II, IX, and XII). *J. Enzyme Inhib. Med. Chem.* **2015**, *30*, 586–591. [CrossRef]
71. Kucuk, M.; Gulcin, I. Purification and characterization of carbonic anhydrase enzyme from black sea trout (*Salmo trutta* Labrax Coruhensis) kidney and inhibition effects of some metal ions on the enzyme activity. *Environ. Toxicol. Pharmacol.* **2016**, *44*, 134–139. [CrossRef]
72. Akıncıoglu, A.; Topal, M.; Gulcin, I.; Goksu, S. Novel sulfamides and sulfonamides incorporating tetralin scaffold as carbonic anhydrase and acetylcholine esterase inhibitors. *Arch. Pharm.* **2014**, *347*, 68–76. [CrossRef] [PubMed]

73. Bradford, M. A Rapid and sensitive method for the quantification of microgram quantities of protein utilizing the principle of protein-dye binding. *Anal. Biochem.* **1976**, *72*, 248–254. [CrossRef] [PubMed]
74. Coban, T.A.; Beydemir, Ş.; Gulcin, I.; Ekinci, D. The inhibitory effect of ethanol on carbonic anhydrase isoenzymes: In vivo and in vitro studies. *J. Enzyme Inhib. Med. Chem.* **2005**, *23*, 266–270. [CrossRef] [PubMed]
75. Bicer, A.; Taslimi, P.; Yakali, G.; Gulcin, I.; Gultekin, M.S.; Turgut Cin, G. Synthesis, characterization, crystal structure of novel bis-thiomethylcyclohexanone derivatives and their inhibitory properties against some metabolic enzymes. *Bioorg. Chem.* **2019**, *82*, 393–404. [CrossRef] [PubMed]
76. Kocyigit, U.M.; Budak, Y.; Gurdere, M.B.; Erturk, F.; Yencilek, B.; Taslimi, P.; Gulcin, I.; Ceylan, M. Synthesis of chalcone-imide derivatives and investigation of their anticancer and antimicrobial activities, carbonic anhydrase and acetylcholinesterase enzymes inhibition profiles. *Arch. Physiol. Biochem.* **2018**, *124*, 61–68. [CrossRef]
77. Gulcin, I.; Alwasel, S.H. Metal ions, metal chelators and metal chelating assay as antioxidant method. *Processes* **2022**, *10*, 132. [CrossRef]
78. Hamad, H.O.; Alma, M.H.; Gulcin, I.; Yılmaz, M.A.; Karaogul, E. Evaluation of phenolic contents and bioactivity of root and nutgall extracts from Iraqian *Quercus infectoria* Olivier. *Rec. Nat. Prod.* **2017**, *11*, 205–210.
79. Han, H.; Yılmaz, H.; Gulcin, I. Antioxidant activity of flaxseed (*Linum usitatissimum* L.) and analysis of its polyphenol contents by LC-MS/MS. *Rec. Nat. Prod.* **2018**, *12*, 397–402. [CrossRef]
80. Topal, M. Secondary metabolites of ethanol extracts of *Pinus sylvestris* cones from Eastern Anatolia and their antioxidant, cholinesterase and α-glucosidase activities. *Rec. Nat. Prod.* **2019**, *14*, 129–138. [CrossRef]
81. Sarikahya, N.B.; Sumer Okkali, G.; Celenk, V.U.; Mertoglu, E.; Pekmez, M.; Arda, N.; Topcu, G.; Goren, A.C. Identification of natural compounds of *Jurinea* species by LC-HRMS and GC-FID and their bioactivities. *J. Pharm. Biomed. Anal.* **2021**, *202*, 114146. [CrossRef]
82. Gulcin, I.; Kufrevioglu, O.I.; Oktay, M.; Buyukokuroglu, M.E. Antioxidant, antimicrobial, antiulcer and analgesic activities of nettle (*Urtica dioica* L.). *J. Ethnopharmacol.* **2004**, *90*, 205–215. [CrossRef] [PubMed]
83. Gulcin, I.; Kufrevioglu, O.I.; Oktay, M. Purification and characterization of polyphenol oxidase from nettle (*Urtica dioica* L.) and inhibition effects of some chemicals on the enzyme activity. *J. Enzyme Inhib. Med. Chem.* **2005**, *20*, 297–302. [CrossRef] [PubMed]
84. Tohma, H.; Koksal, E.; Kılıc, O.; Alan, Y.; Yılmaz, M.A.; Gulcin, I.; Bursal, E.; Alwasel, S.H. RP-HPLC/MS/MS analysis of the phenolic compounds, antioxidant and antimicrobial activities of *Salvia* L. species. *Antioxidants* **2016**, *5*, 38. [CrossRef] [PubMed]
85. Sarikahya, N.B.; Goren, A.C.; Kirmizigul, S. Simultaneous determination of several flavonoids and phenolic compounds in nineteen different Cephalaria species by HPLC-MS/MS. *J. Pharm. Biomed.* **2019**, *173*, 120–125. [CrossRef] [PubMed]
86. Ozer, Z.; Carıkcı, S.; Yılmaz, H.; Kılıc, T.; Dirmenci, T.; Goren, A.C. Determination of secondary metabolites of *Origanum vulgare* subsp. hirtum and *O. vulgare* subsp. vulgare by LC-MS/MS. *J. Chem. Metrol.* **2020**, *14*, 25–34. [CrossRef]
87. Su, M.; Mi, W.; Zhang, Y.; Lv, M.; Shen, W. Determination of illegal additive-ethyl maltol in edible oil by LC-MS/MS in China. *J. Food Comp. Anal.* **2022**, *114*, 104822. [CrossRef]
88. Koksal, E.; Gulcin, I.; Ozturk Sarikaya, S.B.; Bursal, E. On the in vitro antioxidant activity of silymarin. *J. Enzyme Inhib. Med. Chem.* **2009**, *24*, 395–405. [CrossRef]
89. Gulcin, I.; Elmastas, M.; Aboul-Enein, H.Y. Antioxidant activity of clove oil-A powerful antioxidant source. *Arab. J. Chem.* **2012**, *5*, 489–499. [CrossRef]
90. Balaydin, H.T.; Gulcin, I.; Menzek, A.; Goksu, S.; Sahin, E. Synthesis and antioxidant properties of diphenylmethane derivative bromophenols including a natural product. *J. Enzyme Inhib. Med. Chem.* **2010**, *25*, 685–695. [CrossRef]
91. Vega-Villa, K.R.; Remsberg, C.M.; Podelnyk, K.L.; Davies, N.M. Stereospecific high-performance liquid chromatographic assay of isosakuranetin in rat urine. *J. Chromatogr. B* **2008**, *875*, 142–147. [CrossRef]
92. Kacper Kut, K.; Bartosz, G.; Soszynskic, M.; Sadowska-Bartosz, I. Antioxidant properties of hispidulin. *Nat. Prod. Res.* **2022**, *36*, 6401–6404.
93. Yin, X.; He, X.; Wu, L.; Yan, D.; Yan, S. Chlorogenic acid, the main antioxidant in coffee, reduces radiation-induced apoptosis and DNA damage via NF-E2-related factor 2 (Nrf2) activation in hepatocellular carcinoma. *Oxid. Med. Cell. Longev.* **2022**, *2022*, 4566949. [CrossRef] [PubMed]
94. Gulcin, I. Antioxidants and antioxidant methods-An updated overview. *Arch. Toxicol.* **2020**, *94*, 651–715. [CrossRef] [PubMed]
95. Gulcin, I.; Mshvildadze, V.; Gepdiremen, A.; Elias, R. Antioxidant activity of saponins isolated from ivy: α-Hederin, hederasaponin-C, hederacolchiside-E and hederacolchiside F. *Planta Med.* **2004**, *70*, 561–563. [CrossRef] [PubMed]
96. Gulcin, I. Antioxidant properties of resveratrol: A structure-activity insight. *Innov. Food Sci. Emerg.* **2010**, *11*, 210–218. [CrossRef]
97. Gulcin, I. Comparison of in vitro antioxidant and antiradical activities of L-tyrosine and L-Dopa. *Amino Acids* **2007**, *32*, 431–438. [CrossRef]
98. Hamide, M.; Gok, Y.; Demir, Y.; Yakali, G.; Taskin-Tok, T.; Aktas, A.; Sevincek, R.; Guzel, B.; Gulcin, I. Pentafluorobenzyl-substituted benzimidazolium salts: Synthesis, characterization, crystal structures, computational studies and inhibitory properties of some metabolic enzymes. *J. Mol. Struct.* **2022**, *1265*, 133266. [CrossRef]
99. Gulcin, I.; Mshvildadze, V.; Gepdiremen, A.; Elias, R. Antioxidant activity of a triterpenoid glycoside isolated from the berries of *Hedera colchica*: 3-O-(β-D-glucopyranosyl)-hederagenin. *Phytother. Res.* **2006**, *20*, 130–134. [CrossRef]

100. Gulcin, I.; Goren, A.C.; Taslimi, P.; Alwasel, S.H.; Kilic, O.; Bursal, E. Anticholinergic, antidiabetic and antioxidant activities of Anatolian pennyroyal (*Mentha pulegium*)-Analysis of its polyphenol contents by LC-MS/MS. *Biocat. Agric. Biotechnol.* **2020**, *23*, 101441. [CrossRef]
101. Tohma, H.; Altay, A.; Koksal, E.; Goren, A.C.; Gulcin, I. Measurement of anticancer, antidiabetic and anticholinergic properties of sumac (*Rhus coriaria*)-Analysis of its phenolic compounds by LC-MS/MS. *J. Food Meas. Charact.* **2019**, *13*, 1607–1619. [CrossRef]
102. Tao, Y.; Zhang, Y.; Cheng, Y.; Wang, Y. Rapid screening and identification of α-glucosidase inhibitors from mulberry leaves using enzyme-immobilized magnetic beads coupled with HPLC/MS and NMR. *Biomed. Chromatogr.* **2013**, *27*, 148–155. [CrossRef] [PubMed]
103. Gulcin, I.; Alici, H.A.; Cesur, M. Determination of in vitro antioxidant and radical scavenging activities of propofol. *Chem. Pharm. Bull.* **2005**, *53*, 281–285. [CrossRef] [PubMed]
104. Arachchige, S.P.G.; Abeysekera, W.P.K.M.; Ratnasooriya, W.D. Antiamylase, Anticholinesterases, antiglycation, and glycation reversing potential of bark and leaf of Ceylon cinnamon (*Cinnamomum zeylanicum* Blume) in vitro. *Evid. Based Complement. Alternat. Med.* **2017**, *2017*, 5076029. [CrossRef] [PubMed]
105. Gocer, H.; Gulcin, I. Caffeic acid phenethyl ester (CAPE): Correlation of structure and antioxidant properties. *Int. J. Food Sci. Nutr.* **2011**, *62*, 821–825. [CrossRef] [PubMed]
106. Gulcin, I. Measurement of antioxidant ability of melatonin and serotonin by the DMPD and CUPRAC methods as trolox equivalent. *J. Enzyme Inhib. Med. Chem.* **2008**, *23*, 871–876. [CrossRef]

life

MDPI

Article

GC–MS Profiling of Naturally Extracted Essential Oils: Antimicrobial and Beverage Preservative Actions

Reham F. El-Kased [1,*] and Dina M. El-Kersh [2,*]

1 Microbiology Department, Faculty of Pharmacy, The British University in Egypt, Cairo 11837, Egypt
2 Pharmacognosy Department, Faculty of Pharmacy, The British University in Egypt, Cairo 11837, Egypt
* Correspondence: reham.kased@bue.edu.eg (R.F.E.-K.); dina.elkersh@bue.edu.eg (D.M.E.-K.)

Abstract: The purpose of this study was to demonstrate the antimicrobial effects of natural essential oils (EO) and determine their preservative action. Eight natural essential oils were tested against *Staphylococcus aureus*, *Escherichia coli*, and *Candida albicans* representing gram positive, gram negative, and fungi, respectively. The plant materials were used in this study viz. *Thymus vulgaris*—thyme (TV), *Mentha virdis* (MV), *Mentha longifolia* (ML), *Rosmarinus officinalis*—rosemary (RO), *Lavandula dentata*—lavender (LD), *Origanum majorana*—oregano (OM), which belong to the Lamiaceae family. The other two plants were *Cymbopogon citratus*—lemon grass (family Poaceae) (CC), and *Eucalyptus globulus* (family Myrtaceae) (EG). Employing the disc diffusion susceptibility test, minimum inhibitory and minimum bactericidal concentrations were estimated for each oil, followed by the addition of oils to pasteurized apple juice after microbial induction. The results revealed that *thyme* oil showed the maximum zone of inhibition against all tested microbes enriched with monoterpenes class viz. eucalyptol (24.3%), thymol (17.4%), and γ-terpinene (15.2%). All other tested oils exhibited a concentration-dependent inhibition of growth and their MIC ranged from 0.1 to 100 μL/mL. The recorded minimum bactericidal concentration values were apparently double the minimum inhibitory concentration. The EO of *Mentha virdis* followed by *Mentha longifolia* showed maximum antimicrobial activity against the tested organisms in pasteurized apple juice. A gas chromatography–mass spectroscopy (GC–MS) analysis of lemon grass, thyme, and *Mentha virdis* essential oils showed their enrichment with monoterpenes class recording 97.10, 97.04, and 97.61%, respectively.

Keywords: beverage preservative; natural essential oils; antimicrobial activity

check for updates

Citation: El-Kased, R.F.; El-Kersh, D.M. GC–MS Profiling of Naturally Extracted Essential Oils: Antimicrobial and Beverage Preservative Actions. *Life* **2022**, *12*, 1587. https://doi.org/10.3390/life12101587

Academic Editors: Jianfeng Xu and Othmane Merah

Received: 7 September 2022
Accepted: 10 October 2022
Published: 12 October 2022

1. Introduction

Food poisoning is a widespread, life-threatening illness that is considered an important public health problem [1]. Its main cause is consuming food or beverages contaminated with viruses, bacteria, fungi, or their toxins. Foodborne disease incidence can be affected by climate factors such as temperature, humidity, and rainfall [2]. Eating raw vegetables, fruits, unpasteurized dairy, and raw meat or fish may increase the risk of contamination. On the other hand, some types of food might be contaminated by microbes through harvesting, improper storage, or even transport, which may lead to cross contamination upon improper cooking [3]. Mild cases of food poisoning may be improved without the need of drug treatments, others may be hospitalized. Pathogenicity of food poisoning illness may differ depending on the source of contamination and individual susceptibility. Population as immunocompromised, children, pregnant women, and the old aged, being more susceptible, show serious and life-threatening symptoms [4]. Food contaminating organisms may be bacteria, such as *Staphylococcus aureus* and *Escherichia coli*, or *Candida albicans* fungus [5]. Food contamination would be via infectious microorganisms and toxins at any stage of processing or production from the farm-to-table causing foodborne illnesses. Food poisoning symptoms appear hours or days after consuming contaminated food and frequently results in vomiting, nausea, watery or bloody diarrhea, and dehydration [6].

Gram-positive *Staphylococcus aureus (S. aureus)* bacterium normally exists as normal flora on the skin as well as mucosal membranes of healthy people. It is among the four most common bacteria causing foodborne diseases viz. *Salmonella, Clostridium perfringens,* and *Campylobacter*. *S. aureus* produces about 20 different types of enterotoxins [7]. *S. aureus* has the ability to resist some antibiotics such as methicillin resistance *S. aureus* (MRSA) [8]. *S. aureus* may lead to complicated gastrointestinal diseases with systemic progression [9].

Pathogenic *Escherichia coli* (*E. coli*) are gram-negative strains that are not affected by the acidity of stomach pH followed by the colonization in the intestine [10]. The environmental presence of *E. coli* in water and food, as natural reservoir, can widely lead to food contamination and therefore severe illnesses, including watery to bloody diarrhea with dehydration, gastro-intestinal upsets, and septicemia [11].

Candida albicans (*C. albicans*) is the most widespread fungus, causing systemic illnesses [12]. About 20 *Candida* species have been identified to cause serious infections in humans, especially in the immunocompromised [13]. *C. albicans* causes collateral damage to tissues thereby exacerbating the pathological effects of infections [14]. Consequently, the risk of devastating illnesses increases and can result in considerable mortality due to the persistence and worsening of some chronic inflammatory bowel diseases.

Chemical preservatives were first used as food preservatives but with undesirable biological effects on humans while increasing microbial resistance [15].

The use of plant extracts and spices have been of great interest as a natural alternative to food preservatives. These natural substances have been used for seasoning, flavoring, and as food preservatives by preventing or inhibiting the growth of microorganisms [16,17]. In addition, they have been utilized for their antimicrobial, anti-inflammatory, and analgesic effects [18]. Many essential oils (EO) showed broad antimicrobial activity and could be used to prevent microbial contamination, guarantee safety, quality, and prolong shelf-life depending on their chemical composition which differ according to their plant origin [19]. Essential oils are categorized as safe (GRAS) by FDA and are more accepted by consumers than conventional chemical alternatives due to their natural origin [20].

Bacteria and fungi differ in their acquiring resistance to a specific plant extract. Effectively, natural plants have gained great interest in the food industry as important, natural additives to substitute chemical products [21]. These plants are traditionally used as natural plant remedies for various purposes, including treatment of infections, pain relief, decongestant, antifungal, antiviral, antioxidant, anti-inflammatory, and antiseptic [22].

The Lamiaceae family, (formerly Labiatae) has been attracting attention for the potent antimicrobial action of its plant species and therefore preservative effect. Being enriched with a high content of effective classes with antimicrobial action places the whole family as a powerful natural preservative agents. A recent review by Ramos de Silva, et al., 2021 discussed the importance of the Lamiaceae family as a potent antimicrobial, antioxidant, and other biological activities in context to their chemical profiling [23]. Algerian Thyme "*Thymus vulgaris*" was enriched with thymol ingredient, which varied as per its region in Algeria "59.5% and 67.7%". *Origanum compactum* and *O. vulgare* were effective antimicrobial agents against *S. aureus* and *E. coli* gram-positive and negative bacteria, respectively. Both the Origanum species were chemically enriched with thymol, carvacrol, *p*-cymene, and γ-terpinene volatiles [23]. In a published study by Moumni et al., 2020, the authors worked on understanding the antimicrobial activity of some members of the Lamiaceae family cultivated in Tunisia with relation to their antimicrobial activity. The authors investigated the extracted EOs from *Rosmarinus officinalis, Thymus capitatus, Origanum majorana,* and *Salvia officinalis* by understanding their antimicrobial action on *Pseudomonas aeruginosa, Escherichia coli, Salmonella enterica, Bacillus subtilis,* and *Staphylococcus aureus*. The results proved the effectiveness of all tested EOs where the *T. capitatus* essential oil recorded the best results among other EOs. *T. capitatus* recorded MBC ranging from 0.73 to 2.94 mg/mL [24].

The aim of this study was to assess the antimicrobial as well as the preservative action of various naturally extracted essential oils by inhibiting the survival, growth, and multiplication of *S. aureus, E. coli,* and *C. albicans* present in fresh juice. The essential oils from different plant origins were extracted from members of the Lamiaceae family viz. Oregano, lavender, rosemary, lemon grass, *Mentha virdis*, *Mentha longifolia*, and thyme, in addition to Eucalyptus (Myrtaceae) and lemon grass (Poaceae) cultivated in Egypt. Further, in this study, the authors compared the activity of the extracted essential oils with limonene and eucalyptol selected as authentic volatiles being potent antimicrobial agents [25].

2. Materials and Methods

2.1. Plant Materials, Microorganisms, and a Prepared Juice Sample

2.1.1. Plant and Standard Volatiles

Eight plants were used in this study, six belong to the Lamiaceae family viz. *Thymus vulgaris*—thyme (TV), *Mentha virdis* (MV), *Mentha longifolia* (ML), *Rosmarinus officinalis*—rosemary (RO), *Lavandula dentata*—lavender (LD), *Origanum majorana*—oregano (OM). The other two plants were *Cymbopogon citratus*—lemon grass (family Poaceae) (CC), and *Eucalyptus globulus* (family Myrtaceae) (EG). All previously mentioned plants have been purchased from the Experimental Station of Medicinal and Aromatic Plants, Pharmacognosy Department, Faculty of Pharmacy, Cairo University, Giza, Egypt, in July 2018. Voucher specimens have been deposited in the Faculty of Pharmacy Herbarium, The British University in Egypt. Voucher specimen codes are as follows: Thyme, TV-H01; *Mentha virdis*, MV-H02; *Mentha longifolia*, ML-H03; rosemary, RM-H04; lavender, LO-H05; oregano, OM-H06; lemon grass, CC-H07, and eucalyptus, EG-H08. Eng. Therese Labib, a consultant of plant taxonomy at the Egyptian Ministry of Agriculture, authenticated all the plants used in this current study. Limonene and eucalyptol standards were purchased from Sigma-Aldrich Chemie GmbH, Germany.

2.1.2. Preparation of Volatile Oils

The dried grinded plants, each (500 g), were subjected to water distillation (750 mL) for 3–6 h using a Clevenger-type apparatus until plant exhaustion. For each plant, the extraction was carried out in triplets. The EOs yield and their chemical composition variation "in a quantitative not qualitative way" would vary, and so replicates of experimental procedures in EOs extraction and their GC–MS analysis would be encouraged to guarantee the reproducibility and data accuracy [26]. The yield of the volatile oils (expressed as volume mL/500 g weight) per each plant was 6.0 ± 0.10, 5.5 ± 0.21, 4.0 ± 0.12, 4.8 ± 0.11, 5.8 ± 0.14, 5.6 ± 0.30, 6.4 ± 1.20, and 5.0 ± 24 mL for the plant material (TV), (MV), (ML), (RO), (LD), (OM), (CC) and (EG), respectively. The obtained oils were placed in desiccator after collecting and kept at −4 °C in sealed vials the in dark for further analyses. The physical properties of EOs were assessed as per the Egyptian pharmacopoeia 1984 [27].

2.1.3. Microorganisms

The microorganisms studied in this research represent frequent organisms involved in infections related to healthcare and food poisoning. Thus, clinically-pure strains of *Escherichia coli* representing gram-negative bacteria were isolated on Sorbitol MacConkey agar (Difco^{TM}), *Staphylococcus aureus* representing gram-positive bacteria isolated on Mannitol salt agar (MSA) (Difco^{TM}), and *Candida albicans* fungus isolated on Sabouraud Dextrose agar (Difco^{TM}), all acquired from The British University in Egypt.

2.1.4. Juice Sample

Fresh apple juice (pH 6.5–7) was used to assess the preservative potential of the essential oils of the plants. Whole apple fruits were machine-squeezed and then underwent multiple filtration steps through filter paper until obtaining clear juice. Clear apple juice portions of 100 mL were placed in 250 mL bottles. This was followed by pasteurization of the juice by autoclaving at 70 °C for one minute followed by rapid cooling to 7 °C.

2.2. Methods of Analysis

2.2.1. The Disk Diffusion Susceptibility Test

The disk diffusion susceptibility test was performed using Mueller-Hinton Agar (MHA) to assess the antibacterial and antifungal effects of the essential oils. Three concentrations of the essential oils were tested: 100% oil, 50% oil (diluted with ethanol with ratio 1:1), and 25% oil (diluted with ethanol with ratio 1:4). Four 6 mm sterile filter paper discs saturated with 20 μL of each of the dilutions were applied on plates on which each of the test organisms were streaked at a concentration equivalent to 0.5 MacFarland standard, and a disc impregnated with 20 μL solvent alone was used as a blank. A standard antibiotic disk was applied as a positive control. The antimicrobial activity was assessed by measuring the diameters of zones of inhibition after a period of incubation (18–24 h for bacteria and 48–72 h for *C. albicans*). Measuring the diameter of zones of inhibition in millimeters was performed using a Vernier caliper (together with the diameter of the disc). Three readings average were recorded. Zones of inhibition equivalent to or more than 7 mm reflected the antimicrobial activity of essential oils against the test organisms. An activity index (AI) of the tested essential oils was calculated, where the inhibition zone diameter of the tested essential oil was divided by that of the standard antimicrobial agent, and where an activity index greater than 0.5 was regarded as significant antimicrobial activity [28].

2.2.2. Minimum Inhibitory Concentration (MIC) Determination

The MIC of essential oils was estimated by the broth dilution method for the selected test organisms which resulted in diameters of inhibition zones of more than 7 mm. Bacterial and fungal test strains dilutions were prepared using Mueller Hinton Broth (Difco^{TM} Detroit, MI, USA) and Sabouraud Dextrose Broth (Difco^{TM}, Detroit, MI, USA), respectively. Five 2-fold serial dilutions of essential oils were prepared with the highest and lowest concentrations of 2000 $\mu L \cdot mL^{-1}$ and 125 $\mu L \cdot mL^{-1}$ for bacteria and fungi. The final volume of the prepared concentrations was adjusted to the number of the test organisms. Standard antibiotics (Clotrimazole against fungi and Ofloxacin against gram-positive and gram-negative bacteria) were prepared using the same procedure. To each of the dilution and control tubes containing broth only, a standard inoculum (1.5×10^8 $CFU \cdot mL^{-1}$) was added to reach final highest and lowest concentrations of 1000 and 62.5 $\mu g \cdot mL^{-1}$ for bacteria and 10,000 and 625 $\mu g \cdot mL^{-1}$ for fungi. After the incubation time (18–24 h for bacteria, 5–10 days for fungi), the test tube with the least concentration of essential oils with no visible growth was regarded as the MIC against the test microbe. An average of three readings was recorded as MIC. In this study, a MIC of less than 100 $\mu L \cdot mL^{-1}$ was considered as good antimicrobial activity, MICs of 100–500 $\mu L \cdot mL^{-1}$ with moderate activity, MICs of 500–1000 $\mu L \cdot mL^{-1}$ with weak activity, and MICs greater than 1000 $\mu L \cdot mL^{-1}$ with no activity [28].

2.2.3. Minimum Bactericidal Concentration (MBC) and Minimum Fungicidal Concentration (MFC) Determination

MBC and MFC determination were performed by taking 0.1 mL from MIC tubes showing no observable growth was inoculated onto Mueller Hinton agar (Difco^{TM}, Detroit, MI, USA) and Sabouraud Dextrose agar (Difco^{TM}, Detroit, MI, USA) for fungi by the spread plate method. At the end of incubation time (18–24 h for bacteria and 5–10 days for fungi), the lowest concentration of essential oils with no observable growth on subculture was regarded as its MBC and MFC against the test microbe. The ratios of MFC:MIC or MBC:MIC were estimated to determine the antifungal or antibacterial activity of essential oils against the test microbes, respectively. The compound is bactericidal or fungicidal when the ratio is between 1:2 to 2:1, and it is bacteriostatic or fungistatic if the ratio is greater than 2:1 [28].

2.2.4. Induction of Microorganisms in Juice

The concentrations of the 3 different microbial strains under investigation were adjusted to 1.5×10^8 $CFU.mL^{-1}$ (0.5 McFarland) and each were grown in its relevant media.

One mL of each microbial suspension was added to 4 mL sterilized apple juice to form a total volume of 5 mL in the sterile falcon tube. This step was immediately followed by the addition of the essential oils under investigation at concentrations according to their minimum inhibitory concentration (MIC) obtained values and stored at room temperature (Supplementary Table S1). Two control tubes were prepared, one containing juice and microbial strain only and the other contained juice only.

Rigorous antiseptic measures were applied throughout the sample preparation and inoculation to prevent any possible microbial contamination.

2.2.5. Bacterial Count/Viable Count

The original sample was diluted so that a range of 30 to 300 colonies of the test bacterium are grown. A number of dilutions were cultured to be certain that a suitable number of colonies will be grown (500 μL, 200 μL, and 100 μL). Serial dilutions of the sample in sterile water were performed (1:10, 1:100, 1:1000 etc.). This was followed by cultivation on a nutrient agar dish then sealed and incubated. The media used include nutrient agar for the *S. aureus* count or MacConkey agar to count *E. coli* gram-negative bacteria or Sabouraud Dextrose agar for *C. albicans*. One set of dishes were incubated at 22 °C for 24 h and a second set at 37 °C for 24 h. Colonies are counted by eye at the end of the incubation time.

2.2.6. GC–MS analysis of Essential Oils

An analysis of each essential oil was carried out separately via gas chromatography-mass spectrometry (GC-MS) where the procedure was adopted from a previous work [29]. A system operating Shimadzu GCMS-QP2010 (Tokyo, Japan) was used with the following conditions: The column (RTX-5 MS) was used with specifications that were (30 m × 0.25 mm i.d. × 0.25 μm film thickness) (Restek Corporation, Bellefonte, Pennsylvania, USA). The starting temperature of the column was 45 °C for 2 min and then increased to 300 °C at a rate of 5 °C/min and kept steady for 5 min. The temperature of the injector was 250 °C. The flow rate of helium (carrier gas) was (1.41 mL/min). The following conditions were applied when recording the mass spectra: (equipment current) filament emission current, 60 mA; ionization voltage, 70 eV; ion source, 200 °C. Automatic injection of the essential oil was at (1 μL, 1% v/v) with a splitting ratio (1:15). The identification of volatile metabolites was performed upon comparing the mass spectra as well as the retention index with those of the National Institute of Standards and Technology's (NIST) chemistry webbook library. In addition, literature data was used to identify *n*-alkanes series by comparing their mass spectra and retention indices.

3. Results

3.1. The Disk Diffusion Susceptibility Test

The disk diffusion method was used to test the antibacterial and antifungal activities of the essential oils against the selected microorganisms: *E. coli* as gram-negative bacteria, *S. aureus* as gram-positive bacteria, and *C. albicans* representing fungi. The oils were tested in 3 concentrations: 100%; 50%, and 25%. The antibacterial and antifungal activities were estimated according to the American Society for Microbiology, where zones of diameter < 12.00 mm were considered resistant, zones of diameter ranging 13.00–14.00 mm were considered with intermediate activity, and zones of diameter more than 15.00 mm were considered susceptible. The tested oil concentrations showed variable activities according to the results shown in Table 1 and Figure 1. The results showed that thyme (100% and 50% concentrations) and limonene (all concentrations) showed the maximum activity against *E. coli*, whereas thyme (100% concentration) and lemon grass (all concentrations) showed maximum activity against *S. aureus*. Against *C. albicans*, the oils showed decreasing activity according to the following order: thyme (100% and 50% concentration), limonene (100% and 50% concentration), *Mentha virdis* (100% concentration) and lemon grass (100%

concentration), while 25% limonene and 50% lemon grass were the same in showing the least activity.

Table 1. Measurements of zones of inhibition showing the activity of different oils concentrations against the tested microorganisms.

Oil/ Microorganism		*E. coli*	*S. aureus*	*C. albicans*
	Oil Concentration (%)	**Zones of Inhibition (mm)**		
Mentha virdis	100	7 ± 0.42	10 ± 0.60	29.5 ± 1.50
	50	6 ± 0.21	12 ± 0.80	7 ± 0.58
	25	6 ± 0.23	8 ± 0.70	11.5 ± 0.95
Lavender	100	5 ± 0.12	-	-
	50	6.5 ± 0.22	-	-
	25	5.5 ± 0.13	-	-
Rosemary	100	6 ± 0.22	13 ± 1.20	-
	50	6 ± 0.40	8 ± 0.90	-
	25	2 ± 0.0	9 ± 0.80	-
Eucalyptus	100	6 ± 0.20	12 ± 0.99	12.5 ± 1.00
	50	10 ± 0.3	-	9 ± 0.95
	25	13 ± 0.35	-	8 ± 0.55
Thyme	100	27 ± 1.5	22 ± 1.20	60 ± 1.50
	50	17 ± 1.0	11 ± 0.99	45 ± 1.00
	25	3 ± 0.003	-	-
Lemon grass	100	12 ± 0.90	40 ± 1.20	25 ± 0.80
	50	13 ± 0.90	20 ± 1.20	16 ± 0.95
	25	9 ± 0.60	18 ± 1.10	10 ± 0.55
M. longifolia	100	6 ± 0.30	4 ± 0.00	9 ± 0.00
	50	6.5 ± 0.29	-	8 ± 0.85
	25	7.5 ± 0.30	10 ± 0.70	8 ± 0.95
Oregano	100	7 ± 0.30	-	-
	50	6 ± 0.28	8 ± 0.31	-
	25	8 ± 0.31	-	-
Limonene	100	36 ± 2.82	-	36 ± 1.20
	50	34 ± 1.22	10 ± 0.68	34 ± 1.30
	25	16 ± 1.10	-	16 ± 0.95

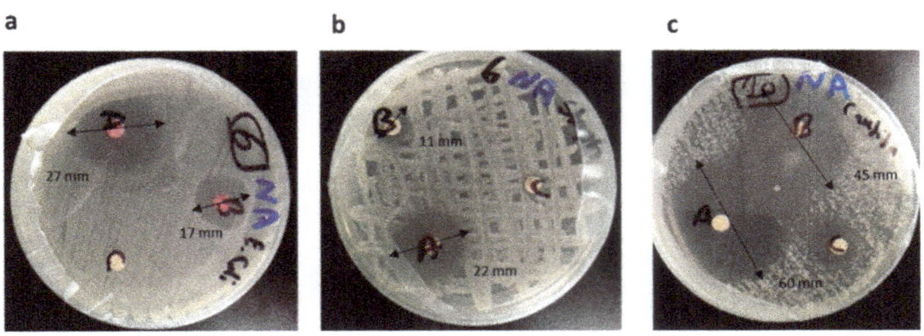

Figure 1. The disc diffusion susceptibility test of thyme against (**a**) *E coli.* (**b**) *S. aureus* and (**c**) *C. albicans.* A: 100% oil concentration, B: 50% oil concentration, and C: ethanol as negative control.

3.2. Minimum Inhibitory Concentration (MIC) Determination

The MIC of oils was determined against the selected strains (*E. coli, S. aureus,* and *C. albicans*) where the oils exhibited concentration-dependent inhibition of growth and the MIC of oils ranged from 0.1 to 100 μL/mL as shown in Table 2. The table shows that the least MIC (0.1 μL/mL) against *E. coli* was shown by *lemon grass* whereas the highest MIC

(25 µL/mL) was shown by *oregano* and *lavender*, whereas against *S. aureus*, the least MIC (0.2 µL/mL) and the highest MIC (100 µL/mL) were shown by *lemon grass* and *limonene*, respectively. *Lemon grass* and *thyme* showed the least MIC against *C. albicans* (0.8 µL/mL) and the highest MIC (12.5 µL/mL) was shown with *Mentha longifolia*.

Table 2. Minimum inhibitory concentration (MIC), minimum bactericidal concentration (MBC), minimum fungicidal concentration (MFC), and MIC:MBC ratio of oils against *E. coli*, *S. aureus*, and *C. albicans*. ND: not defined; NC: not calculated.

Oils		***E. coli***			***S. aureus***			***C. albicans***	
	MIC (µL/mL)	**MBC (µL/mL)**	**MBC: MIC**	**MIC (µL/mL)**	**MBC (µL/mL)**	**MBC: MIC**	**MIC (µL/mL)**	**MBC (µL/mL)**	**MBC: MIC**
Mentha virdis	1.6 ± 0.02	ND	NC	12.5 ± 0.85	ND	NC	3.2 ± 0.85	ND	NC
Lavender	25 ± 1.20	ND	NC	NC	NC	NC	NC	NC	NC
Rosemary	4.0 ± 0.03	ND	NC	NC	NC	NC	NC	NC	NC
Oregano	25 ± 0.95	ND	NC	6.3 ± 0.55	ND	NC	NC	NC	NC
Eucalyptus	6.3 ± 0.80	ND	NC	25 ± 1.10	ND	NC	6.3 ± 0.95	ND	NC
Thyme	1.6 ± 0.40	ND	NC	1.6 ± 0.25	3.2	2:1	0.8 ± 0.05	1.6	2:1
Lemon grass	0.1 ± 0.00	0.2	2:1	0.2 ± 0.00	0.4	2:1	0.8 ± 0.06	1.6	2:1
M. longifolia	0.2 ± 0.00	0.4	2:1	50 ± 1.95	ND	NC	12.5 ± 1.25	ND	NC
Limonene	1.6 ± 0.45	ND	NC	100 ± 2.55	NC	NC	1.6 ± 0.85	ND	NC

Minimum bactericidal concentration (MBC) and minimum fungicidal concentration (MFC):

The results in Table 2 shows that the least MBC against *E. coli* was with *lemon grass*, *Mentha longifolia* and *rosemary* at concentrations 0.2 µL/mL, 0.4 µL/mL, and 0.8 µL/mL, respectively, whereas *Mentha virdis*, *thyme* and *limonene* showed the highest MBC (3.2 µL/mL). Whereas against *S. aureus*, the least MBC was 0.4 µL/mL and 3.2 µL/mL with thyme and lemon grass, respectively, whereas MBC was highest with Eucalyptus (50 µL/mL). Regarding *C. albicans*, the least MBC (1.6 µL/mL) was determined with *thyme* and *lemon grass*, whereas *oregano*, *rosemary* and *lavender* appeared to have no fungicidal activity at all. Table 2 also shows that the recorded MBC values were apparently double the MICs.

3.3. Bacterial Count/Viable Count

Oils were added to apple juice samples according to their calculated MIC. Oils added to apple juice samples containing *E. coli* showed a decline in the number of bacteria in decreasing order: *Mentha virdis*, *Mentha longifolia* and *limonene*. Regarding the apple juices containing *S. aureus* and *C. albicans*, *Mentha virdis*, *Mentha longifolia*, and *Limonene* showed an increased number of bacteria at day 1, which declined by days 5 and 7 to less than 30 CFU (Supplementary Table S2). Other oils did not show any remarkable activity where the bacterial load increased to more than 300 CFU by time. In this study, pasteurized apple juice was expected to have a good initial microbiological quality. All essential oils concentrations added to the juice were selected according to the calculated MIC for each oil. All experiments were assessed at room temperature where the sensitivities of *E. coli*, *S. aureus*, and *C. albicans* to essential oils were in the following decreasing order: *Mentha virdis*, *M. longifolia*. and limonene.

3.4. Physical and Chemical (GC-MS) Analyses of Essential Oils

The physical properties "specific gravity, relative density, refractive index" in addition to oil appearance, color, and odor of EOs of *Cymbopogon citratus* "lemon grass", *Thymus vulgaris* "thyme" and *Mentha virdis* "mentha" volatiles have been investigated and summarized in Table 3.

Table 3. The physical properties of essential oils (average of 3 independent experiments ± SD) of *Cymbopogon citratus* "lemon grass", *Thymus vulgaris* "thyme" and *Mentha virdis* "mentha" volatiles.

Physical Property	EOs Yield (mL/500 g Plant)	Specific Gravity (25 °C)	Relative Density (g/cm^3)	Refractive Index (20 °C)	Appearance	Color	Odor
Lemon grass oil	6.4 ± 1.2	0.9012 ± 0.03	0.893 ± 0.01	1.4862 ± 0.16	Clear oil	Pale yellow	citrus
Thyme oil	6.0 ± 0.10	0.9180 ± 0.04	0.865 ± 0.06	1.4823 ± 0.15	Clear oil	Pale yellow	thyme
Mentha oil	5.5 ± 0.21	0.9160 ± 0.01	0.910 ± 0.08	1.4638 ± 0.11	Clear oil	Pale yellow	mentha

Upon an GC–MS chemical analysis of volatile oils prepared from *Cymbopogon citratus* (Poaceae), *Thymus vulgaris* (Lamiaceae), and *Mentha virdis* (Lamiaceae), a total of 24, 51, and 43 compounds have been identified in *Cymbopogon citratus, Thymus vulgaris,* and *Mentha virdis* volatile oils, respectively. It was observed that in all prepared essential oils, the major class was monoterpenes followed by sesquiterpenes. In both *Cymbopogon citratus* and *Mentha virdis* oils, the monoterpenes percentiles were 97.41% and 97.61%, respectively, followed by *Thymus vulgaris* 93.04%. The total sesquiterpenes in the three analyzed oils were 4.93, 2.4, and 0.89% for *Thymus vulgaris, Mentha virdis* and *Cymbopogon citratus* herbs, respectively. In *Cymbopogon citratus* oil, the monoterpenes class exemplified by dominance of geranial 36.35%, followed by neral 35.00%, representing monoterpene aldehydes, then β-myrcene monoterpene hydrocarbon 11.7% as presented in Figure 2a and Table 4. These results complied with previous studies representing almost the same percentiles where the identification of those monoterpenes as majors in the volatile oil of *Cymbopogon citratus* herb cultivated in Cameroon [30]. In a previous article, it was proved the potential efficacy of *Cymbopogon citratus* oil and its main citral "Geranial" being a potent antimicrobial agent against polymicrobial biofilm forming bacteria *viz.* Staphylococcus aureus and Candida species. The underlying mechanism of action was refereed to its citral volatile ingredient through reducing the biofilm mass and cell viability by interfering of nucleic acids, proteins, and carbohydrates of the biomass that lead to deformity of the biomass matrix in addition to disruptions to the biomass adhesive characters. It worth mentioning that the enrichment of the Egyptian *Cymbopogon citratus* oil with the citral volatile ingredient of this current study (36.35%) than that mentioned in Gao, et al., 2020 (29.364%) [31]. In *Thymus vulgaris* volatile oil, the identified major monoterpenes were eucalyptol as monoterpenoid oxide 24.30%, thymol monoterpenoid phenol 17.40%, and γ-terpinene monoterpene hydrocarbon 15.20% shown in Figure 2b and Table 4 where similar percentiles of the major monoterpene class except eucalyptol have been recorded in the previous literature [32]. The enrichment of *Thymus vulgaris* volatile oil with eucalyptol 24.30% counted for its potent antimicrobial activity against a myriad of pathogens as previously mentioned [33]. In a published article by Sienkiewicz et al., 2011, the authors tested the antimicrobial activity of Thymus vulgaris oil against a myriad of clinically multidrug resistant strains of *Staphylococcus, Enterococcus, Escherichia,* and *Pseudomonas* genus. The results revealed the strong antimicrobial activity of the tested thyme oil against multidrug resistant microbes previously mentioned. The authors would relate the bioactivity of thyme oil to its main volatile ingredients *p*-cymene and thymol recording 29.10 and 38.10%, respectively [34]. The GC–MS analysis of the Egyptian Thymus vulgaris oil in this current study revealed its enrichment with eucalyptol (1,8 cineole) by (24.3%), in addition to thymol (17.4%). Eucalyptol itself possessed potent antimicrobial activity as reported in previous works [34,35]. The volatile oil of *Mentha virdis* enriched with mainly carvone "monoterpene ketone" 42.50% followed by eucalyptol "monoterpene oxide" 17.40% then finally dihydrocarveol "monoterpene alcohol" 13.00% as in Figure 2c and Table 4 where these data matched with the previous literature by Mkaddem et al., 2022 [36] upon which an analysis of the oil obtained from *Mentha virdis* collected from Tunisia. In a previous article by Mkaddem et al., 2022, *Mentha virdis* oil was rich with carvone (50.47%), eucalyptol (9.14%), and limonene (4.87%) which encountered the potent antimicrobial activity of the oil against *Listeria monocytogenes* and *Klebsiella pneumoniae* bacteria [36]. It is worth mentioning in this previous study that despite the enrichment of *Mentha virdis* oil by carvone up to 50.47%, still the oil obtained from the

Egyptian *Mentha virdis* was richer with other oxygenated monoterpenes "eucalyptol 17.40% and dihydrocarveol 13.00%" than the Tunisian one. The major volatile structures of each oil are illustrated in Supplementary Figure S1.

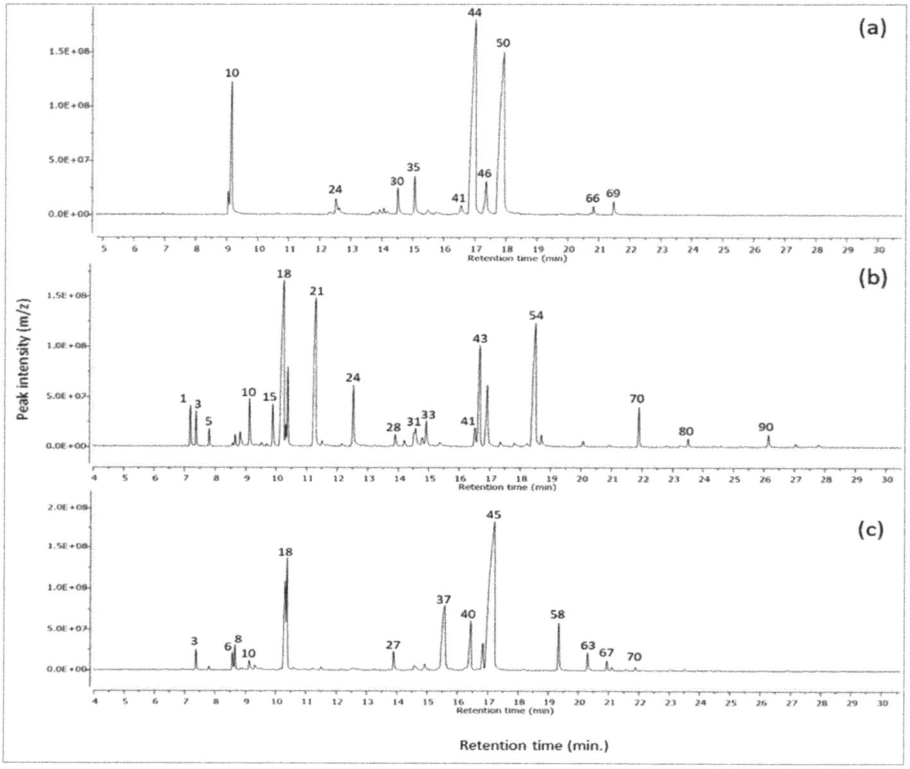

Figure 2. GC–MS chromatogram of volatile oil of (**a**): *Cymbopogon citratus* "lemon grass", (**b**): *Thymus vulgaris* "thyme" and (**c**): *Mentha virdis* "mentha" herbs.

Table 4. GC–MS analysis of essential oils (average of 3 independent runs ± SD) of *Cymbopogon citratus* "lemon grass", *Thymus vulgaris* "thyme" and *Mentha virdis* "mentha" volatiles. Volatiles are listed regarding their elution on GC column (RTX-5). **RT**: retention time (min), **KI obs.**: Kovat's index observed practically (RTX-5 GC column) related to C_8-C_{28} n-alkanes series. KI. Lit.: Kovat's index reported from the previous literature. MS: The volatile components identification was based on mass spectral data. KI: The volatile components identification was based on comparing KI published in mass spectral library of National Institute of Standards and Technology (NIST). Bold values are the major constituents in the volatile oil.

Peak No.	Rt. (Min.)	Name of Compound	KI (Obs.)	KI (Lit.)	Area (%) Lemon Grass Oil	Thyme Oil	Mentha Oil	Chemical Class	Identification
1	7.17	β-Thujene	910	902	_	2.09 ± 0.720	_	Monoterpene	MS, KI
2	7.19	α-thujene	910	905	_	_	0.04 ± 0.001	Monoterpenes	MS, KI
3	7.37	α-pinene	917	917	_	1.68 ± 0.0320	1.2 ± 0.030	Monoterpenes	MS, KI
4	7.72	2,4(10)-Thujadiene	930	946	_	0.04 ± 0.0050	_	Monoterpene	MS, KI

Table 4. *Cont.*

Peak No.	Rt. (Min.)	Name of Compound	KI (Obs.)	KI (Lit.)	Area (%) Lemon Grass Oil	Area (%) Thyme Oil	Area (%) Mentha Oil	Chemical Class	Identification
5	7.81	Camphene	933	929	–	0.86 ± 0.020	0.21 ± 0.004	Monoterpenoids	MS, KI
6	8.57	Sabinene	961	964	–	–	1.07 ± 0.560	Monoterpenes	MS, KI
7	8.58	4(10)-Thujene	961	969	–	0.19 ± 0.001	–	Monoterpene	MS, KI
8	8.65	β-pinene	964	965	–	0.62 ± 0.050	1.54 ± 0.580	Monoterpenes	MS, KI
9	8.82	Vinyl amyl carbinol	970	969	–	1.23 ± 0.063	–	Alkenyl alcohol	MS, KI
10	9.16	β-Myrcene	982	985	**11.69 ± 1.300**	3.36 ± 0.730	0.94 ± 0.020	Monoterpenes	MS, KI
11	9.33	3-octanol	988	988	–	0.15 ± 0.003	–	Aliphatic alcohol	MS, KI
12	9.49	pseudolimonene	994	996	–	–	0.3 ± 0.002	Monoterpenes	MS, KI
13	9.51	a-Phellandrene	995	994	–	0.22 ± 0.002	–	Monoterpene	MS, KI
14	9.69	3-Carene	1001	1004	–	0.14 ± 0.003	–	Monoterpene	MS, KI
15	9.9	4-Carene	1008	1011	–	2.53 ± 0.430	–	Monoterpene	MS, KI
16	9.9	2-carene	1008	1010	–	–	0.07 ± 0.001	Monoterpenes	MS, KI
17	10.17	O-cymene	1017	1021	–	–	0.12 ± 0.001	Monoterpenes	MS, KI
18	10.39	Eucalyptol	1024	1020	0.05 ± 0.010	**24.30 ± 3.530**	**17.40 ± 3.220**	Monoterpene oxide	MS, KI
19	10.59	trans-β-ocimene	1030	1042	–	–	0.19 ± 0.030	Monoterpenes	MS, KI
20	10.93	β-ocimene	1041	1046	0.26 ± 0.020	0.04	0.07 ± 0.005	Monoterpenes	MS, KI
21	11.23	γ-terpinene	1051	1055	–	**15.2 ± 1.720**	0.17 ± 0.020	Monoterpenes	MS, KI
22	11.49	trans-sabinene hydrate	1059	1051	–	1.2 ± 0.040	0.33 ± 0.060	Monoterpenes	MS, KI
23	11.67	Linalool oxide	1065	1071	–	0.28 ± 0.002	–	Monoterpene	MS, KI
24	12.52	β-linalool	1092	1098	1.58 ± 0.340	4.79 ± 0.050	0.64 ± 0.030	Monoterpenes	MS, KI
25	12.63	Cis-verbenol	1096	1095	0.90 ± 0.030	–	–	Monoterpenes	MS, KI
26	13.74	Pinen-3-ol	1131	1131	–	–	0.12 ± 0.006	Monoterpene ketone	MS, KI
27	13.89	Camphor isomer	1136	1141	–	–	1.50 ± 0.400	Monoterpenoid ketone	MS, KI
28	13.9	Camphor	1137	1131	0.40 ± 0.010	0.97 ± 0.050	–	Monoterpene ketone	MS, KI
29	14.19	Isomenthone	1146	1148	–	0.60 ± 0.030	–	Monoterpenoid	MS, KI
30	14.52	t-verbenol	1156	1148	2.27 ± 0.010	–	–	Monoterpenes	MS, KI
31	14.59	Isoborneol	1056	1159	–	2.5 ± 0.060	0.86 ± 0.300	Monoterpene alcohol	MS, KI
32	14.79	Menthol	1165	1164	–	0.88 ± 0.040	–	Monoterpenoid	MS, KI
33	14.92	Terpinen-4-ol	1170	1177	–	2.19 ± 0.060	0.77 ± 0.030	Monoterpene alcohol	MS, KI
34	15.1	Levomenthol	1175	1172	–	0.03 ± 0.005	–	Monoterpenoid	MS, KI
35	15.13	Isoneral	1174	1174	2.91 ± 0.02	–	–		MS, KI
36	15.39	α-Terpineol	1185	1189	–	0.56 ± 0.005	–	Monoterpene alcohol	MS, KI
37	15.58	Dihydrocarveol	1191	1202	–	–	**12.9 ± 2.600**	Monoterpene alcohol	MS, KI
38	15.61	Trans-Dihydrocarvone	1192	1195	–	0.14 ± 0.003	0.35 ± 0.030	Monoterpenoid	MS, KI
39	15.764	Decanal	1197	1207	0.42 ± 0.020	–	–	Monoterpenes	MS, KI
40	16.45	trans-carveol	1220	1219	–	–	5.99 ± 1.600	Monoterpene alcohol	MS, KI
41	16.518	Citronellol	1223	1228	1.23 ± 0.040	1.62 ± 0.040	–	Monoterpenoid	MS, KI
42	16.84	carveol	1234	1246	–	–	2.65 ± 0.860	Monoterpene alcohol	MS, KI
43	16.921	Isothymol methyl ether	1237	1244	–	6.14 ± 0.720	–	Aromatic monoterpenoid	MS, KI
44	17.02	Neral (β-citral)	1240	1240	**34.99 ± 3.530**	–	–	Monoterpene aldehyde	MS, KI
45	17.22	carvone	1247	1248	–	–	**42.5 ± 6.800**	Monoterpene ketone	MS, KI
46	17.35	cis-Geraniol	1252	1254	4.24 ± 0.520	0.39 ± 0.002	–	Monoterpene alcohol	MS, KI
47	17.52	Piperitone	1258	1254	–	0.08 ± 0.003	–	Monoterpene ketone	MS, KI
48	17.7	Carvenone	1263	1258	–	0.01 ± 0.002	–	Methane monoterpenoid	MS, KI
49	17.81	Citronellyl formate	1268	1273	–	0.4 ± 0.030	–	fatty alcohol ester	MS, KI
50	17.94	Geranial	1272	1277	**36.35 ± 4.230**	–	–	Monoterpene aldehyde	MS, KI
51	18.16	Bornyl acetate	1280	1284	–	–	0.1 ± 0.003	Monoterpenes	MS, KI
52	18.38	2-Undecanone	1288	1292	0.08 ± 0.010	–	–	Organic ketone	MS, KI
53	18.47	Carvacrol	1299	1300	–	1.54 ± 0.021	–	Monoterpenoid phenol	MS, KI
54	18.5	Thymol	1292	1292	–	**17.4 ± 3.410**	–	Monoterpenoid phenol	MS, KI
55	18.76	Dihydrocarvenyl acetate	1301	1304	–	–	0.03 ± 0.001	Monoterpenes	MS, KI
56	18.77	Undecanal	1301	1303	0.02 ± 0.001	–	–	Organic aldehyde	MS, KI
57	18.92	Isopulegyl acetate	1307	1335	–	–	0.03 ± 0.004	Monoterpenes	MS, KI
58	19.36	Dihydrocarvyl acetate	1322	1344	–	–	3.97 ± 0.053	Monoterpenes	MS, KI
59	19.43	Geranic acid	1355	1347	0.06 ± 0.002	–	–	Poly unsaturated fatty acid	MS, KI
60	20.02	Citronellol acetate	1345	1355	0.02 ± 0.002	–	–	Sesquiterpenes	MS, KI
61	20.071	Thymyl acetate	1347	1349	–	0.45 ± 0.001	–	Monoterpene	MS, KI
62	20.31	Geranic acid isomer	1355	1355	0.25 ± 0.030	–	–	Poly unsaturated fatty acid	MS, KI
63	20.31	Carvyl acetate	1355	1346	–	–	1.51 ± 0.090	Monoterpenes	MS, KI
64	20.67	Isobornyl propionate	1368	1388	–	0.06 ± 0.003	–	Sesquiterpene	MS, KI
65	20.67	α-copaene	1368	1372	–	–	0.02 ± 0.004	Sesquiterpene	MS, KI
66	20.83	Geranyl acetate	1374	1368	0.77 ± 0.021	–	–	Sesquiterpenes	MS, KI
67	20.93	β-Bourbonene	1377	1385	–	0.15 ± 0.002	0.76 ± 0.050	Sesquiterpene	MS, KI
68	21.1	β-Elemene	1383	1384	–	–	0.29 ± 0.040	Sesquiterpene	MS, KI
69	21.48	Methyl eugenol	1396	1395	1.29 ± 0.310	–	–	Phenyl propanoid	MS, KI
70	21.89	Caryophyllene	1411	1418	0.02 ± 0.004	2.46 ± 0.540	0.46 ± 0.030	Sesquiterpenes	MS, KI

Table 4. *Cont.*

Peak No.	Rt. (Min.)	Name of Compound	KI (Obs.)	KI (Lit.)	Area (%)			Chemical Class	Identification
					Lemon Grass Oil	Thyme Oil	Mentha Oil		
71	22.14	α-copaene	1421	1380	_	0.04 ± 0.002	_	Hydrocarbon	MS, KI
72	22.26	t-α-Bergamotene	1426	1436	0.03 ± 0.001	_	_	Bicyclic monoterpenes	MS, KI
73	22.425	Citronellyl propionate	1433	1444	_	0.01 ± 0.001	_	fatty alcohol ester	MS, KI
74	22.54	Germacrene D	1437	1477	_	0.01 ± 0.002	0.13 ± 0.007	Sesquiterpene	MS, KI
75	22.79	cis-a-Bisabolene	1488	1518	0.02 ± 0.004	_	_	Sesquiterpenes	MS, KI
76	22.79	Humulene	1447	1442	_	0.12 ± 0.006	0.05 ± 0.000	Sesquiterpene	MS, KI
77	23.04	(+)-epi-Bicyclosesquiphellandrene	1457	1452	_	_	0.12 ± 0.020	Sesquiterpene	MS, KI
78	23.241	Geranyl isovalerate	1464	1582	_	0.14 ± 0.003	_	fatty alcohol ester	MS, KI
79	23.34	(Z,Z)-α-Farnesene	1468	1506	_	_	0.01 ± 0.001	Sesquiterpene	MS, KI
80	23.38	γ-Muurolene	1470	1477	_	0.66 ± 0.007	_	Sesquiterpene	MS, KI
81	23.42	Ylangene	1471	1470	_	_	0.02 ± 0.003	Sesquiterpene	MS, KI
82	23.52	Germacrene D isomer	1482	1475	_	_	0.16 ± 0.001	Sesquiterpene	MS, KI
83	23.73	α-Guaiene	1483	1490	_	_	0.03 ± 0.000	Sesquiterpene	MS, KI
84	24.35	γ-Muurolene	1508	1491	_	_	0.1 ± 0.000	Sesquiterpene	MS, KI
85	24.36	γ- Cadinene	1508	1513	_	0.08 ± 0.001	_	Sesquiterpene	MS, KI
86	24.58	Delta- Cadinene	1517	1523	_	0.11 ± 0.000	_	Sesquiterpene	MS, KI
87	24.59	Cis-Calamenene	1517	1531	_	_	0.16 ± 0.000	Sesquiterpene	MS, KI
88	25.93	α-Bourbonene	1569	1531	_	0.02 ± 0.005	_	Sesquiterpene	MS, KI
89	26.01	Spathulenol	1573	1576	_	0.02 ± 0.001	_	Sesquiterpene	MS, KI
90	26.15	Caryophyllene oxide	1578	1577	0.06 ± 0.002	0.97 ± 0.130	0.08 ± 0.001	Sesquiterpene	MS, KI
91	26.9	Eudesmol	1608	1602	_	0.02 ± 0.001	_	Sesquiterpene	MS, KI
92	27.04	Cadinol	1614	1635	_	0.24 ± 0.060	_	Sesquiterpene	MS, KI
Total monoterpenes (%)					97.10	93.04	97.61		
Total sesquiterpenes (%)					0.89	4.92	2.39		
Miscellaneous (%)					1.70	1.97	_		
Total percentage (Approximated %)					100	100	100		

(_): indicate absence of volatile component.

4. Discussion

E. coli, *S. aureus*, and *C. albicans* are common pathogens causing serious systemic infections in humans. Since the continuous development of antimicrobial resistance, natural products and essential oils have been studied as alternatives for the treatment of infections acquired in healthcare [16,37,38].

Essential oils are important sources of new antimicrobial agents particularly against bacterial pathogens. In vitro studies in this research demonstrated that the tested essential oils inhibited microbial growth with variable effectiveness [31].

The tested oils exhibited concentration dependent inhibition of growth and the MIC of oils ranged from 0.1 to 100 μL/mL. MBC/MFC was assessed to demonstrate the least concentration of oils resulting in microbial viability reduction of 99.90% of the initial count. The MBC readings were double the MICs, which indicates that the bactericidal activities of the oils occur at concentrations higher than its growth inhibitory concentrations.

It has been observed the enrichment of the analyzed volatiles of lemon grass, thyme and mentha with oxygenated monoterpenes mainly geranial and neral with a total percent of 74.34%, eucalyptol and thymol with total amount of 41.70% whereas carvone, eucalyptol, and dihydrocarveol with a total content of 72.80%, respectively. Volatile oils enriched with oxygenated monoterpenes encountered the oil as being more potent as antimicrobial rather than monoterpene hydrocarbons [39]. Allenspach et al., 2020 in a recent article for absolute quantification of terpenes in conifer species, the authors implemented a validated simple method for quantification of different terpenes viz. α and β-pinenes, camphene, 3-carene, limonene, bornyl acetate, β-caryophyllene, and borneol. Antibacterial activity of conifer essential oil proved its efficacy on both *E-coli* and *S. aureus* as gram-negative and positive bacteria, respectively [40]. In this current study, the GC–MS analysis proved the presence of such terpenes although in low percentile as listed in Table 3, still may be related to the antibacterial activity of the tested oils in this study. It is worth mentioning that previous studies worked on the mechanism of actions being antibacterial for each component in the volatile oil. In a previous work by Oz, et al., 2015, it was mentioned that the presence of an aromatic ring and a polar functional group of a volatile constituent "thymol" could lead to rupture of the bacterial cell membrane leading to release of the vital cell constituents [41]. A previous published review by Wińska et al., 2019 mentioned the effectiveness of essential oils as antimicrobial agents, specifically thyme and mentha in context to their volatile chemical profiling and major phytochemical ingredients. The antimicrobial activity of thyme would be referred to its enrichment with thymol (36–55%) and *p*-cymene (15–28%) whereas mentha antimicrobial activity referred to its higher percentile of menthol (30–55%) and menthone (14–32%) [42]. Citral "neral and geranial isomers", have been approved by the U.S Food and Drug administration as being safe, so its use as natural preservative and flavoring agent due to its antibacterial activity against gram-negative and positive bacteria as *Escherichia coli* and *Staphylococcus aureus*, respectively [43,44]. Eucalyptol "1,8 cineole" antibacterial mechanism was summarized in a previous review where its effect mainly was due to changes in both size and shape of gram- positive and gram negative bacterial cell which ends with apoptosis [45]. α-pinene was detected in both thyme and mentha oils ca. 1.68 and 1.2%, respectively. In a recent published review by Allenspach and Steuer, 2021, the authors summarized the different biological activities of α pinene, of which they mentioned its positive antimicrobial activity on both gram-positive and negative bacteria *viz.* methicillin-resistant Staphylococcus aureus (MRSA), *Escherichia coli* (*E. coli*), and antifungal activity against Candida species [46]. Concerning carvone antibacterial mechanism of action, in a previous study, carvone enriched in the oil of *Mentha spicata* causes instability of phospholipid bilayer structure as well as interacts between the bacterial membrane enzymes and proteins [47].

Finally, the volumes of oils added ranged from 0.8 μL to 400 μL in 5 mL juice (with a range from 0.016% to 8%) (Supplementary Table S1). Concerning the taste of drinks after the addition of the EOs; EOs improve the flavor, odor, and color when added to foods. Many individual EOs are approved food flavorings and impart a certain flavor to foods, as well as delaying food spoilage without changing the organoleptic properties of the food. However, certain strategies could be implemented to decrease the organoleptic effects, if found, by optimizing the food/beverage formulation or by combining the essential oils and/or their active constituents with other means of sterilization such as pH or heat treatment (when applicable) [48].

5. Conclusions

Chemical preservatives have been utilized as food preservatives, but they turned to have undesirable biological effects on humans and increase in microbial resistance. This study demonstrates that natural volatile oils extracted from plants exhibited a concentration-dependent inhibition of microbial growth and offer potential antimicrobial activity against common food spoilage bacteria and fungi. For future research, the authors would encourage a larger scale comparative study of natural essential oils to chemical ones commonly used in the market as well as preparing a commercial naturally preserved effective product to replace chemical preservatives.

Supplementary Materials: The following supporting information can be downloaded at: https://www.mdpi.com/article/10.3390/life12101587/s1, Table S1: Oils were added with concentrations as shown in the table based on their MIC results. Table S2: Mean readings of colonies count on days 1, 5 and 7. Figure S1: Major identified volatiles in herbs of *Cymbopogon citratus*, *Thymus vulgaris* and *Mentha virdis*.

Author Contributions: Conceptualization, R.F.E.-K. and D.M.E.-K.; methodology, R.F.E.-K. and D.M.E.-K.; formal analysis, investigation, and resources, R.F.E.-K. and D.M.E.-K.; writing—original draft preparation, writing—review and editing, R.F.E.-K. and D.M.E.-K. All authors have read and agreed to the published version of the manuscript.

Funding: This research received no external funding.

Institutional Review Board Statement: Not applicable.

Informed Consent Statement: Not applicable.

Data Availability Statement: Data are available with the authors' reham.kased@bue.edu.eg and dina.elkersh@bue.edu.eg.

Acknowledgments: The authors would like to thank Ahmed Rafaat, Dana Abdel Rahman, Farah El Araishy, Hoda Fahmy, Radwa Tarek, Rawan Mobasher, Sara Abdel Hay, Eman Ahmed, Amar Shouman, Nour Abou El-Ala, Farah Ahmed, Mahmoud Magdy and Nadin Sakr for collecting the samples.

Conflicts of Interest: The authors declare no conflict of interest.

References

1. Yang, S.C.; Lin, C.H.; Aljuffali, I.A.; Fang, J.Y. Current pathogenic Escherichia coli foodborne outbreak cases and therapy development. *Arch. Microbiol.* **2017**, *199*, 811–825. [CrossRef] [PubMed]
2. Park, M.S.; Park, K.H.; Bahk, G.J. Interrelationships between Multiple Climatic Factors and Incidence of Foodborne Diseases. *Int. J. Environ. Res. Public Health* **2018**, *15*, 2482. [CrossRef] [PubMed]
3. Valero, A.; Rodríguez, M.-Y.; Posada-Izquierdo, G.D.; Pérez-Rodríguez, F.; Carrasco, E.; García-Gimeno, R.M. Risk factors influencing microbial contamination in food service centers. In *Significance, Prevention and Control of Food Related Diseases*; IntechOpen: London, UK, 2016; pp. 27–58.
4. World Health Organization. *The Burden of Foodborne Diseases in the WHO European Region*; WHO Regional Office for Europe: Copenhagen, Denmark, 2017.
5. Liu, Q.; Meng, X.; Li, Y.; Zhao, C.-N.; Tang, G.-Y.; Li, H.-B. Antibacterial and antifungal activities of spices. *Int. J. Mol. Sci.* **2017**, *18*, 1283. [CrossRef]
6. Bintsis, T. Foodborne pathogens. *AIMS Microbiol.* **2017**, *3*, 529. [CrossRef] [PubMed]
7. Hennekinne, J.A.; De Buyser, M.L.; Dragacci, S. Staphylococcus aureus and its food poisoning toxins: Characterization and outbreak investigation. *Fems. Microbiol. Rev.* **2012**, *36*, 815–836. [CrossRef] [PubMed]
8. Leong, H.N.; Kurup, A.; Tan, M.Y.; Kwa, A.L.H.; Liau, K.H.; Wilcox, M.H. Management of complicated skin and soft tissue infections with a special focus on the role of newer antibiotics. *Infect. Drug Resist.* **2018**, *11*, 1959–1973. [CrossRef] [PubMed]
9. de Barros, J.C.; da Conceicao, M.L.; Neto, N.J.G.; da Costa, A.C.V.; Siqueira, J.P.; Basilio, I.D.; de Souza, E.L. Interference of *Origanum vulgare* L. essential oil on the growth and some physiological characteristics of Staphylococcus aureus strains isolated from foods. *Lwt-Food Sci. Technol.* **2009**, *42*, 1139–1143. [CrossRef]
10. Jubelin, G.; Desvaux, M.; Schuller, S.; Etienne-Mesmin, L.; Muniesa, M.; Blanquet-Diot, S. Modulation of Enterohaemorrhagic Escherichia coli Survival and Virulence in the Human Gastrointestinal Tract. *Microorganisms* **2018**, *6*, 115. [CrossRef] [PubMed]
11. Kolenda, R.; Burdukiewicz, M.; Schierack, P. A systematic review and meta-analysis of the epidemiology of pathogenic Escherichia coli of calves and the role of calves as reservoirs for human pathogenic *E. coli*. *Front. Cell Infect. Mi* **2015**, *5*, 23. [CrossRef] [PubMed]
12. Chong, P.P.; Chin, V.K.; Wong, W.F.; Madhavan, P.; Yong, V.C.; Looi, C.Y. Transcriptomic and Genomic Approaches for Unravelling Candida albicans Biofilm Formation and Drug Resistance-An Update. *Genes* **2018**, *9*, 540. [CrossRef] [PubMed]
13. Mayer, F.L.; Wilson, D.; Hube, B. Candida albicans pathogenicity mechanisms. *Virulence* **2013**, *4*, 119–128. [CrossRef] [PubMed]
14. Poulain, D. Candida albicans, plasticity and pathogenesis. *Crit. Rev. Microbiol.* **2015**, *41*, 208–217. [CrossRef]
15. Bondi, M.; Lauková, A.; de Niederhausern, S.; Messi, P.; Papadopoulou, C. Natural preservatives to improve food quality and safety. *Hindawi J. Food Qual.* **2017**, *2017*, 1–3. [CrossRef]
16. El-Kased, R.F. Natural antibacterial remedy for respiratory tract infections. *Asian Pac. J. Trop Bio.* **2016**, *6*, 270–274. [CrossRef]
17. Yuste, J.; Fung, D.Y.C. Evaluation of Salmonella typhimurium, Yersinia enterocolitica and Staphylococcus aureus counts in apple juice with cinnamon, by conventional media and thin agar layer method. *Food Microbiol.* **2003**, *20*, 365–370. [CrossRef]
18. Trinetta, V.; Morgan, M.T.; Coupland, J.N.; Yucel, U. Essential Oils Against Pathogen and Spoilage Microorganisms of Fruit Juices: Use of Versatile Antimicrobial Delivery Systems. *J. Food Sci.* **2017**, *82*, 471–476. [CrossRef] [PubMed]
19. Mota, V.D.; Turrini, R.N.T.; Poveda, V.D. Antimicrobial activity of Eucalyptus globulus oil, xylitol and papain: A pilot study. *Rev. Esc. Enferm. Usp.* **2015**, *49*, 215–219. [CrossRef] [PubMed]
20. Adelakun, O.E.; Oyelade, O.J.; Olanipekun, B.F. Use of essential oils in food preservation. In *Essential Oils in Food Preservation, Flavor and Safety*; Elsevier: Amsterdam, The Netherlands, 2016; pp. 71–84.
21. Beya, M.M.; Netzel, M.E.; Sultanbawa, Y.; Smyth, H.; Hoffman, L.C. Plant-based phenolic molecules as natural preservatives in comminuted meats: A review. *Antioxidants* **2021**, *10*, 263. [CrossRef] [PubMed]
22. Mintah, S.O.; Asafo-Agyei, T.; Archer, M.-A.; Junior, P.A.-A.; Boamah, D.; Kumadoh, D.; Appiah, A.; Ocloo, A.; Boakye, Y.D.; Agyare, C. Medicinal plants for treatment of prevalent diseases. In *Pharmacognosy-Medicinal Plants*; IntechOpen: Rijeka, Croatia, 2019.
23. Ramos da Silva, L.R.; Ferreira, O.O.; Cruz, J.N.; de Jesus Pereira Franco, C.; Oliveira dos Anjos, T.; Cascaes, M.M.; Almeida da Costa, W.; Helena de Aguiar Andrade, E.; Santana de Oliveira, M. Lamiaceae Essential Oils, Phytochemical Profile, Antioxidant, and Biological Activities. *Evid.-Based Complementary Altern. Med.* **2021**, *2021*, 6748052. [CrossRef] [PubMed]

24. Moumni, S.; Elaissi, A.; Trabelsi, A.; Merghni, A.; Chraief, I.; Jelassi, B.; Chemli, R.; Ferchichi, S. Correlation between chemical composition and antibacterial activity of some Lamiaceae species essential oils from Tunisia. *BMC Complementary Med. Ther.* **2020**, *20*, 1–15. [CrossRef] [PubMed]
25. Miguel, M.G.; Gago, C.; Antunes, M.D.; Lagoas, S.; Faleiro, M.L.; Megias, C.; Cortes-Giraldo, I.; Vioque, J.; Figueiredo, A.C. Antibacterial, Antioxidant, and Antiproliferative Activities of Corymbia citriodora and the Essential Oils of Eight Eucalyptus Species. *Medicines* **2018**, *5*, 61. [CrossRef] [PubMed]
26. Ribeiro, B.S.; Ferreira, M.d.F.; Moreira, J.L.; Santos, L. Simultaneous Distillation–Extraction of Essential Oils from *Rosmarinus officinalis* L. *Cosmetics* **2021**, *8*, 117. [CrossRef]
27. Pharmacopoeia, E. *Egyptian Pharmacopoeia, General Organization for Governmental*; Printing Office, Ministry of Health: Cairo, Egypt, 1984; pp. 31–33.
28. Padla, E.P.; Solis, L.T.; Ragasa, C.Y. Antibacterial and antifungal properties of ent-kaurenoic acid from Smallanthus sonchifolius. *Chin. J. Nat. Med.* **2012**, *10*, 408–414. [CrossRef]
29. Al-Sayed, E.; Gad, H.A.; El-Kersh, D.M. Characterization of Four Piper Essential Oils (GC/MS and ATR-IR) Coupled to Chemometrics and Their anti-Helicobacter pylori Activity. *Acs Omega* **2021**, *6*, 25652–25663. [CrossRef] [PubMed]
30. Fokom, R.; Adamou, S.; Essono, D.; Ngwasiri, D.P.; Eke, P.; Mofor, C.T.; Tchoumbougnang, F.; Fekam, B.F.; Zollo, P.H.A.; Nwaga, D.; et al. Growth, essential oil content, chemical composition and antioxidant properties of lemongrass as affected by harvest period and arbuscular mycorrhizal fungi in field conditions. *Ind. Crop. Prod.* **2019**, *138*, 111477. [CrossRef]
31. Gao, S.; Liu, G.; Li, J.; Chen, J.; Li, L.; Li, Z.; Zhang, X.; Zhang, S.; Thorne, R.F.; Zhang, S. Antimicrobial activity of lemongrass essential oil (Cymbopogon flexuosus) and its active component citral against dual-species biofilms of Staphylococcus aureus and Candida species. *Front. Cell Infect. Mi* **2020**, *10*, 603858. [CrossRef] [PubMed]
32. Hudaib, M.; Speroni, E.; Di Pietra, A.M.; Cavrini, V. GC/MS evaluation of thyme (Thymus vulgaris L.) oil composition and variations during the vegetative cycle. *J. Pharm. Biomed.* **2002**, *29*, 691–700. [CrossRef]
33. Safaei-Ghomi, J.; Ahd, A.A. Antimicrobial and antifungal properties of the essential oil and methanol extracts of Eucalyptus largiflorens and Eucalyptus intertexta. *Pharm. Mag.* **2010**, *6*, 172–175. [CrossRef]
34. Sienkiewicz, M.; Łysakowska, M.; Denys, P.; Kowalczyk, E. The antimicrobial activity of thyme essential oil against multidrug resistant clinical bacterial strains. *Microb. Drug Resist.* **2012**, *18*, 137–148. [CrossRef]
35. Shala, A.Y.; Gururani, M.A. Phytochemical Properties and Diverse Beneficial Roles of Eucalyptus globulus Labill.: A Review. *Horticulturae* **2021**, *7*, 450. [CrossRef]
36. Mkaddem, M.; Bouajila, J.; Ennajar, M.; Lebrihi, A.; Mathieu, F.; Romdhane, M. Chemical Composition and Antimicrobial and Antioxidant Activities of Mentha (*longifolia* L. and *viridis*) Essential Oils. *J. Food Sci.* **2009**, *74*, M358–M363. [CrossRef] [PubMed]
37. El-Kased, R.F.; Amer, R.I.; Attia, D.; Elmazar, M.M. Honey-based hydrogel: In vitro and comparative In vivo evaluation for burn wound healing. *Sci. Rep.* **2017**, *7*, 1–11.
38. Emad, A.M.; Rasheed, D.M.; El-Kased, R.F.; El-Kersh, D.M. Antioxidant, Antimicrobial Activities and Characterization of Polyphenol-Enriched Extract of Egyptian Celery (Apium graveolens L., Apiaceae) Aerial Parts via UPLC/ESI/TOF-MS. *Molecules* **2022**, *27*, 698. [CrossRef] [PubMed]
39. Guimaraes, A.C.; Meireles, L.M.; Lemos, M.F.; Guimaraes, M.C.C.; Endringer, D.C.; Fronza, M.; Scherer, R. Antibacterial Activity of Terpenes and Terpenoids Present in Essential Oils. *Molecules* **2019**, *24*, 2471. [CrossRef]
40. Allenspach, M.D.; Valder, C.; Steuer, C. Absolute quantification of terpenes in conifer-derived essential oils and their antibacterial activity. *J. Anal. Sci. Technol.* **2020**, *11*, 12. [CrossRef]
41. Oz, M.; Lozon, Y.; Sultan, A.; Yang, K.H.S.; Galadari, S. Effects of monoterpenes on ion channels of excitable cells. *Pharm. Ther.* **2015**, *152*, 83–97. [CrossRef]
42. Wińska, K.; Mączka, W.; Łyczko, J.; Grabarczyk, M.; Czubaszek, A.; Szumny, A. Essential oils as antimicrobial agents—Myth or real alternative? *Molecules* **2019**, *24*, 2130. [CrossRef]
43. Fisher, K.; Phillips, C. Potential antimicrobial uses of essential oils in food: Is citrus the answer? *Trends Food Sci. Technol.* **2008**, *19*, 156–164. [CrossRef]
44. Somolinos, M.; Garcia, D.; Manas, P.; Condon, S.; Pagan, R. Effect of environmental factors and cell physiological state on Pulsed Electric Fields resistance and repair capacity of various strains of Escherichia coli. *Int. J. Food Microbiol.* **2008**, *124*, 260–267. [CrossRef]
45. Mączka, W.; Duda-Madej, A.; Górny, A.; Grabarczyk, M.; Wińska, K. Can eucalyptol replace antibiotics? *Molecules* **2021**, *26*, 4933. [CrossRef]
46. Allenspach, M.; Steuer, C. α-Pinene: A never-ending story. *Phytochemistry* **2021**, *190*, 112857. [CrossRef] [PubMed]
47. Shahbazi, Y. Chemical Composition and In Vitro Antibacterial Activity of Mentha spicata Essential Oil against Common Food-Borne Pathogenic Bacteria. *J. Pathog.* **2015**, *2015*, 916305. [CrossRef] [PubMed]
48. Mariod, A.A. Effect of essential oils on organoleptic (smell, taste, and texture) properties of food. In *Essential Oils in Food Preservation, Flavor and Safety*; Elsevier Inc.: Amsterdam, The Netherlands, 2016; pp. 131–137.

life

Article

Phenolic and Antioxidant Compound Accumulation of *Quercus robur* Bark Diverges Based on Tree Genotype, Phenology and Extraction Method

Vaida Sirgedaitė-Šėžienė, Ieva Čėsnienė, Gabija Leleikaitė, Virgilijus Baliuckas and Dorotėja Vaitiekūnaitė *

Institute of Forestry, Lithuanian Research Centre for Agriculture and Forestry, Liepų Str. 1, LT-53101 Girionys, Lithuania
* Correspondence: doroteja.vaitiekunaite@lammc.lt

Abstract: Oak bark is a rich niche for beneficial bioactive compounds. It is known that the amount of the compounds found in plant tissues can depend on species, genotype, growth site, etc., but it is unclear whether oak phenology, i.e., late or early bud burst, can also influence the amount of phenols and antioxidants that can be extracted. We tested two *Quercus robur* populations expressing different phenology and five half-sib families in each population to see how phenology, genotype, as well as extrahent differences (75% methanol or water) can determine the total phenol, total flavonoid content, as well as antioxidant activity. Significant statistical differences were found between half-sib families of the same population, between populations representing different oak phenology and different extrahents used. We determined that the extraction of flavonoids was more favorable when using water. So was antioxidant activity using one of the indicators, when significant differences between extrahents were observed. Furthermore, in families where there was a significant difference, phenols showed better results when using methanol. Overall, late bud burst families exhibited higher levels in all parameters tested. Thus, we recommend that for further bioactive compound extraction, all these factors be noted.

Keywords: ABTS; DPPH; early bud burst; English oak; half-sib; late bud burst; pedunculate oak; total flavonoid content; total phenol content; natural products

check for updates

Citation: Sirgedaitė-Šėžienė, V.; Čėsnienė, I.; Leleikaitė, G.; Baliuckas, V.; Vaitiekūnaitė, D. Phenolic and Antioxidant Compound Accumulation of *Quercus robur* Bark Diverges Based on Tree Genotype, Phenology and Extraction Method. *Life* **2023**, *13*, 710. https://doi.org/10.3390/life13030710

Academic Editor: Jianfeng Xu

Received: 13 February 2023
Revised: 28 February 2023
Accepted: 4 March 2023
Published: 6 March 2023

1. Introduction

In recent years, there has been a lot of interest in functional compounds derived from plants and also notably from byproducts of plant-associated industries, as a way to reduce waste and provide additional valorization [1,2]. For example, wine-making and leather tanning enterprises have used oaks (*Quercus*) for centuries. It is known that specific compounds from trees in the oak genus facilitate taste in aged alcoholic drinks and are responsible for altering the protein structure in the hides to induce durability [3,4]. However, research shows that oaks contain many useful compounds that may be utilized in multiple ways in varied other fields, i.e., as medicine, supplements, and additives. Oak flour has been used for centuries and recently has been reintroduced as a bioactive ingredient in human foodstuffs [5]. Furthermore, Gamboa-Gomez et al. demonstrated that oak leaf infusions can be used as additives that could potentially be anti-hyperglycemic and have antioxidative effects in mice [6]. Oak leaf extracts were also shown to reduce lipid oxidation, increase antioxidant capacity and reduce bacterial growth in meat [7,8]. Moreover, oak leaf extracts were shown to modify rumen fermentation, thus alleviating the oxidative imbalance ruminant animals face [9]. As an additive to common carp fish food, oak leaf extracts were shown to stimulate antioxidant and immune system of the carp and reduce stress [10].

Thus, as can be seen, oak tissues contain many beneficial compounds; however, more and more research is geared toward compounds that could be located in oak bark specifically, as it has been singled out as a potential byproduct for valorization [11–14]. Mirski et al. showed that ground oak bark can be used as a filler in plywood adhesives [11]. Furthermore, oak bark-based animal feed additives were reported to positively affect chicken immunity [15]. Moreover, oak bark-derived tannins were suggested for use in medicinal topical creams meant for allergy treatment [16]. Oak bark extracts were also shown to work in treating periodontal disease [17]. Additionally, multiple studies reported that oak bark derivatives were demonstrated to have antibacterial properties [13,18–20]. Similarly, as with oak leaves, oak bark derivatives were reported to positively affect digestion in the rumen [21]. Oak bark compounds were also reported to have been successfully used as an effective additive in yogurt production [22].

Oaks are rich in phenolic compounds [1,20], that can correlate positively with antioxidative capacity [23]. Several studies report on the composition of oak bark extracts. Elansary et al. found high levels of antioxidant phenols such as ellagic, gallic, protocatechuic, vanillic and caffeic acids and catechin derivatives in oak bark. Perhaps most notably, high levels of ellagic acid were reported [24]. Recently ellagic acid has been linked with multiple health benefits and as such is a prized bioactive compound [25]. Complimentary to Elansary et al., Ucar and Ucar also report on catechin and ellagic acid [24,26]. Ucar and Ucar also observed sitosterol and quercitol as notable oak bark derivatives [26]. While at the moment research into the direct effects these compounds may have on human health are scarce, preliminary data suggest that catechin, sitosterol, and quercitol may reduce blood pressure [27], have antidiabetic properties [28], and have cholesterol-lowering properties [29], respectively. Furthermore, a 2015 study refers to the anti-quorum sensing and antimicrobial capabilities of oak bark extracts and identified at least two compounds that were responsible—1,2,3-benzenetriol and 4-propyl-1,3-benzenediol [30]. Since this field is not well researched, it is possible to find even more of these beneficial bioactive compounds in oak bark.

It has been observed that the amount of varied compounds within plant tissues are potentially determined by a multitude of factors, such as genotype [31,32], species [4,33], leaf age and seasonal variations [34,35], growth site [35], stress [36], and extraction methodology [13,14,20,37].

Oaks can have phenotype differences, i.e., they may be late or early in terms of bud burst/flushing. This is a natural mechanism that may help protect early bud burst oak trees from herbivore or pathogen attacks and also protect late bud burst oaks from frost damage as well as certain herbivore attacks [38–40].

Since phenolic compounds and antioxidants are involved in protection from both biotic and abiotic stressors [36,41], it is worth investigating whether either early or late bud burst oak phenotypes may produce larger amounts of them in their tissues and thus the bark of oaks with this phenotype would potentially be more beneficial for extracting bioactive compounds of phenolic or antioxidant origin.

Based on all these data, in order to optimize the process of extracting bioactive compounds from oak bark, it is important to accurately determine all the possible factors that may impact polyphenol or antioxidant production. We hypothesized that *Quercus robur* (pedunculate oak) bark phenolics and antioxidant activity levels were determined not only by genotype (different half-sib families) but also by tree phenology (early and late bud flushing/burst time). We also looked into the variation that could be introduced in this process due to the use of different extrahents, i.e., methanol and distilled water. This may be important, as some extracted compounds may be used for foodstuffs and thus methanol may not be appropriate due to health or other concerns.

We found significant differences in phenolics and antioxidant extraction efficiency from oak bark between half-sib families of the same population, between populations representing different oak phenology, and different extrahents used.

2. Materials and Methods

2.1. Test Subjects and Tree Phenology Evaluation

Ten half-sib families from two *Q. robur* populations—Josvainiai and Dūkštos—(five from each) were studied when trees were 24 years old. All trees were growing in the same growth site. The stage of budburst was recorded and a value from 0 to 6 was assigned, with the larger figure corresponding to a more advanced stage of bud or leaf development (0 means bud stage and 6—fully developed leaf) (Table 1). Growth cessation was estimated by recording autumn leaf coloring stage. Again, values from 1 to 5 were assigned, with 5 corresponding to the stage when all leaves were lost.

Table 1. The budburst stages of half-sib families of pedunculate oak (arranged from earliest (Josvainiai, Jox) to latest (Dūkštos, D)).

Josvainiai Oak Population	Budburst Stage	Standart Deviation	Dūkštos Oak Population	Budburst Stage	Standart Deviation
Jox8	5.4	0.5	D31	3.2	1.3
Jox7	5.0	0.6	D22	2.8	1.4
Jox1	4.8	0.5	D72	2.8	1.4
Jox6	4.0	0.9	D61	2.2	1.6
Jox3	3.6	1.0	Dx1	2.1	1.3

2.2. Sample Collection

The raw material needed for research was collected using a cordless drill in July of 2020. The top layers of bark were collected. Five trees from each half-sib family represent five biological replicates. In total, samples studied were: 2 populations × 5 half-sib families × 5 biological replicates. The collected plant material was dried at 40 °C for 24 h before use in testing. A detailed description is shown Table 2.

Table 2. Used variables for the evaluation of biological compounds and antioxidant activity changes in oak bark: different populations, different half-sib families, and sample extraction method.

Population		Half-Sib Families (Genotype)		Extrahent	
		Josvainiai (Jox):	Dūkštos (D):		
Early burst phenology: Josvainiai (Jox)	Late burst phenology: Dūkštos (D)	1. Jox8 2. Jox7 3. Jox1 4. Jox6 5. Jox3	1. D31 2. D22 3. D72 4. D61 5. Dx1	MeOH (75% in water)	dH_2O

2.3. Extract Preparation

Extracts were prepared from 0.5 g of air-dried bark samples homogenized using an A11 basic analytical mill (Laboratory Equipment, Staufen, Germany), which was shaken with 10 mL of either 75% methanol (MeOH) or 10 mL of distilled water (dH_2O) for 24 h at room temperature using a Kuhner Shaker X electronic shaker (Adolf Kühner AG, Birsfelden, Switzerland). The obtained extracts were filtered through Whatman no. 1 filter paper, with a retention of 5–8 μm.

2.4. Quantification of Total Phenol Content

Total phenol content (TPC) was determined using Folin-Ciocalteu reagent according to Slinkard and Singleton's method [42]. The reaction mixture used in this study is detailed in Table 3. The absorbance was measured using Synergy HT Multi-Mode Microplate Reader (BioTek Instruments, Inc., Bad Friedrichshall, Germany) at 760 nm against the reagent blank (MeOH for methanol extracts and dH_2O for aqueous extracts). The phenol content was expressed as chlorogenic acid per gram of weight of bark (mg CAE/g). The standard calibration curve equation used for MeOH samples: y = 5.5358x − 0.0423 (R^2 = 0.9975); for dH_2O samples: y = 5.5x − 0.0451 (R^2 = 0.999).

Table 3. Reaction mixture used to evaluate total phenolic content (TPC) and total flavonoid content (TFC).

Total Phenolic Content (TPC)	Total Flavonoid Content (TFC)
Reaction mixture: 100 µL sample + 2500 µL dH_2O + 100 µL *Folin–Ciocalteu* reagent (2 N) (wait 6 min) + 5000 µL Na_2CO_3 (25%, *w/v*). The mixture was left for 30 min at room temperature.	Reaction mixture: 1000 µL sample + 300 µL $NaNO_2$ (5%, *w/v*) (wait 5 min) + 500 µL $AlCl_3$ (2%, *w/v*) (wait 6 min) + 500 µL NaOH (1M).

2.5. Quantification of Total Flavonoid Content

The total flavonoid content (TFC) in the extracts was determined according to a method described in Lučinskaitė et al. [43]. The reaction mixture used in this study is detailed in Table 3. The absorbance of the mixture was recorded at 470 nm on the Synergy HT Multi-Mode Microplate Reader. The same blanks (MeOH and dH_2O) as those used for TPC were used here as well. The flavonoid content was expressed in milligrams of catechin per gram of weight of bark (mg CE/g). The standard calibration curve equation for MeOH samples: y = 11.616x + 0 0634 (R^2 = 0.9983); for dH_2O samples: y = 10.201x + 0.0527 (R^2 = 0.9983).

2.6. Quantification of Antioxidant Activity

2.6.1. DPPH (2,2-Diphenyl-1-picryl-hydrazyl-hydrate)

Total free radical scavenging capacity of the extracts from different *Q. robur* samples were estimated according to Ragaee et al.'s [44] method. The reaction mixture used in this study is detailed in Table 4. Absorbance was measured at 515 nm using Genesys 6 spectrophotometer (Thermo Spectronic, Waltham, MA, USA) against an equal amount of DPPH and 75% methanol as a blank (or water and DPPH in the case of dH_2O extracts). The standard calibration curve equation was y = 0.2074x − 0.004 (R^2 = 0.9907). The radical scavenging activity was calculated as antioxidant Trolox equivalents per gram of sample and calculated to Equation (1):

$$TE = (c \times V)/m \tag{1}$$

where c =Trolox concentration (mM/mL); V = the extract volume (mL); m = the sample amount (g).

Table 4. Reaction mixtures used to evaluate antioxidant activity with two different methods (DPPH and ABTS).

Radical Scavenging Activity	
DPPH Method	**ABTS Method**
Reaction mixture: 100 µL samples + 400 µL MeOH (75%) + 1000 µL DPPH solution (0.1 mM). Mixture was incubated at room temperature in the dark for 16 min. DPPH solution preparation: 11.8 mg was dissolved in 300 mL MeOH (100%).	Reaction mixture: 50 µL samples + 2000 µL ABTS solution. Mixture was incubated at room temperature in the dark for 10 min. ABTS solution preparation: 56 mg of ABTS (>99%, Fluka, Buchs, Germany) was dissolved in 50 mL of dH_2O. ABTS radical cation was prepared by reacting ABTS stock solution with 200 µL of $K_2S_2O_8$ (70 nM). The mixture was held in the dark at room temperature for 16 h before it was used.

2.6.2. ABTS (2,2′-Azino-bis(3-ethylbenzothiazoline-6-sulfonic acid)

Free radical scavenging activity in plant extracts was determined by ABTS radical cation decolorization assay [43]. The reaction mixture used in this study is detailed in Table 4. After 16 h, the mixture was diluted with dH_2O until it reached 0.700 ± 0.2 absorbance (734 nm). dH_2O was used as a blank. Absorbance was measured at 734 nm using Genesys 6 spectrophotometer (Thermo Spectronic) (ABTS and 75% methanol as a blank or water and ABTS in the case of dH_2O extracts). Trolox was used as the standard. Twenty-five milligrams of Trolox (97%, Sigma-Aldrich, St. Louis, MO, USA) was dissolved in 75% MeOH (LaboChema, Vilnius, Lithuania). The stock standard of Trolox was 1 mg/mL, and 1, 2, 3, 4, or 5 mL of stock standard was used, diluted with 10 mL of 80% (v/v) ethanol, to determine the effect of varying the Trolox concentration. The standard calibration curve equation was y = 0.2734x + 0.0304 (R^2 = 0.9842). The radical scavenging activity was calculated as antioxidant Trolox equivalent per gram of sample and calculated to Equation (1).

2.7. *Statistical Analysis*

Group means and standard errors were calculated using Microsoft Excel. Statistical data analysis was performed using the SPSS program (IBM, version 28.0.1.1.). The Kruskal–Wallis H test was used for analysis as a non-parametric alternative to one-way ANOVA. During this test, differences are determined by comparing the mean ranks of groups. A post hoc Dunn's test was performed to indicate differences between individual pairs [32,45,46].

The importance or random and fixed effects on variance were analyzed using SAS software (SAS Institute Inc. 2002–2012, version 9.4). SAS UNIVARIATE procedure was used to check if residuals follow normal distribution. Data on TFC in dH_2O extracts showed significant deviation from the normal distribution, thus logarithmic transformation was applied to get appropriate normality. Test for homogeneity of trait variance was done with GLM procedure Levene's Test. Tukey's studentized range (HSD) test in GLM procedure was used to carry out multiple comparisons between traits.

The variance components were calculated using the SAS MIXED procedure (REML method):

$$Y_{ijkl} = \mu + R_i + P_j + F_k + E_{ijkl} \tag{2}$$

where μ is the grand mean, R_i is the fixed effect of replicate i, P_j is the random effect of population j, F_k is the random effect of family k, and E_{ijkl} is the residual error. Standard errors of the estimates of variance components were calculated by Taylor expansions and the asymptotic covariance matrix of the estimates was obtained from MIXED procedure [47].

3. Results

Based on bud burst phenology evaluation it was concluded that Josvainiai (Jox) population is of early bud burst phenology and Dūkštos (D) population is of the late bud burst phenology.

3.1. *Total Phenol (TPC) and Total Flavonoid (TFC) Content*

It was shown that secondary metabolite (TPC and TFC) content in the oak bark varied significantly between populations, half-sib families, and when using different extrahents (Figure 1).

Our results showed that TPC in both populations (Josvainiai and Dūkštos) was similar (from 13.71 mg/g to 18.71 mg/g). The highest amount of TPC was observed in Josvainiai population, when the extrahent was MeOH—18.71 mg/g (Jox1 family). Bark samples from Dūkštos population family D61 had similar amount of TPC—18.93 mg/g. In Jox (Josvainiai) population TPC variation between half-sib families was not significant, irrespective of extrahent used. Furthermore, in Jox (Josvainiai) population TFC varied significantly between families, irrespective of extrahent used. In Dūkštos (D) population, significant differences between families were noted in TPC extracted with water and TFC extracted with methanol.

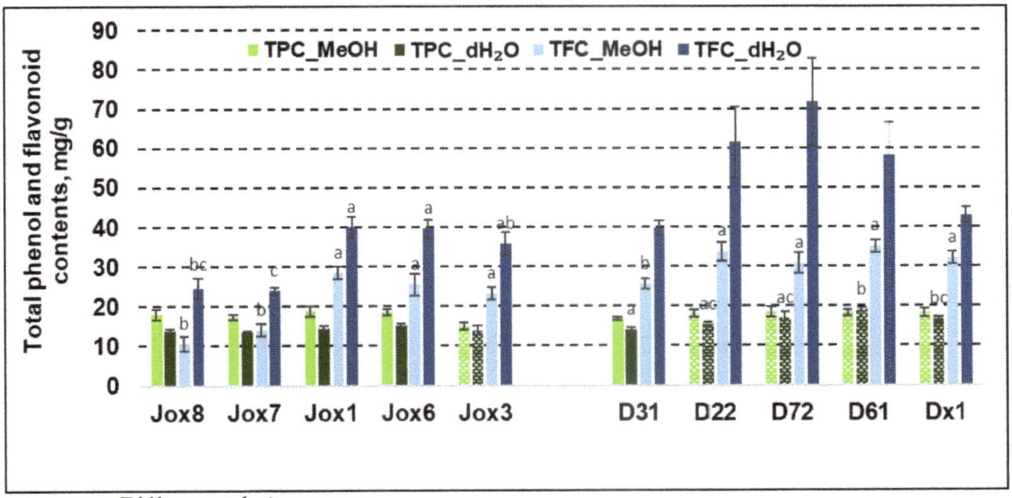

Differences between populations	Extrahents	Significance, p value
TPC	dH2O	****
	MeOH	NS
TFC	dH2O	****
	MeOH	****

Figure 1. Means (mM/g fresh weight) ± SE of total phenol (TPC) and total flavonoid (TFC) content in 10 *Q. robur* families. Significance was calculated using the Kruskal–Wallis H test for ranks and post hoc Dunn's test for pairs ($p < 0.05$). Different letters next to the same colors indicate significant differences between families, solid colors indicate significant differences between extrahents and differences among the studied populations are noted in the table (p values: ****—≤0.0001, NS—not significant). Different populations are indicated by group designations Jox and D.

It was noted that late budburst population (Dūkštos, D) had higher amount of TFC, compared to Josvaniai (Jox) population, irrespective of extrahents used. The highest amount of TFC in the oak bark samples was found in Dūkštos population, where dH_2O was used for extraction—71.61 mg/g—family D72. Meanwhile, the highest amount of TFC in Josvainiai (Jox) population, the same as TPC, was observed in the Jox1 family—39.95 mg/g. In addition, our study showed that TPC variation within individuals from the same half-sib family using different extrahents was significant. For example, it was determined that Josvainiai (Jox) population exhibited significant TPC variation in four out of five tested families. When looking at significant differences between extrahents, TPC extraction using methanol was more effective. It was noted that TPC accumulation in Dūkštos (D) population was significantly different between extrahents only in one family (D31). TFC variation using different extrahents was significantly different in both populations and in all families. In all cases, better TFC values were achieved using water as an extrahent rather than methanol. The highest difference between extrahents in Josvainiai (Jox) population was observed in family Jox6. Meanwhile, the highest difference between extrahents in Dūkštos (D) population was noted in family D72.

Statistical analysis showed that differences between two populations (as they represent oak phenology) also were significant in most cases. As shown in Figure 1, TFC differences between populations were more noticeable, compared to TPC variation. When using water as an extrahent, the amount of phenols and flavonoids in both populations were statistically different. Meanwhile, when using methanol, the difference was only noted in TFC.

Analysis of random and fixed effects on variance showed that genotype (half-sib family) had a significant effect on TFC extraction by methanol and by water, while tree phenology had no significant effect. This can only partly be explained by within-group variation as can be seen from replicate significance levels. At the same time, TPC was unaffected by either oak tree phenology or genotype (Table 5).

Table 5. Variance components for random effects as percent from the total variation and significance of the fixed effect. Level of significance is denoted by *: $0.05 > p > 0.01$, **: $0.01 > p > 0.001$, ***: $p < 0.001$, ns—not significant.

Trait	Variance Components of Random Effects, %		Significance of Fixed Effect
	Phenology (Population)	half-Sib Family (Genotype)	Replicate (Individual)
DPPH_MeOH	20.0 ± 39.1	34.5 ± 18.8 *	ns
ABTS_MeOH	42.4 ± 69.6	32.3 ± 17.0 *	ns
TPC_MeOH	0.0	3.5 ± 4.7	***
TFC_MeOH	38.1 ± 61.1	22.6 ± 12.6 *	ns
DPPH_dH$_2$O	0.0	18.4 ± 11.3 *	**
ABTS_dH$_2$O	0.0	41.7 ± 21.5	*
TPC_dH$_2$O	17.2 ± 28.5	9.8 ± 7.3	*
TFC_dH$_2$O	37.9 ± 59.2	16.7 ± 9.9 *	ns

3.2. Radical Scavenging Activity

As with phenolics, our results show that oak population, sample extraction method and half-sib family had significant impact on free radical scavenging activity (Figure 2).

Antioxidant activity in oak bark was evaluated by DPPH and ABTS radical scavenging assays. Radical scavenging activity varied between 235.28 mM/g and 481.33 mM/g (DPPH method) and from 172.03 mM/g to 1376.22 mM/g (ABTS method) in Jox (Josvainiai) population. In late budburst population (Dūkštos) antioxidant activity varied between 337.50 mM/g and 481.33 mM/g (DPPH analysis) and from 652.23 mM/g to 1490.17 mM/g (ABTS analysis), showing significant variation in both populations between assays and within the results from either assay. Analysis showed that radical scavenging activity as measured by the ABTS method had more obvious differences between populations, compared to the DPPH assay.

Significant differences between half-sib families in the same population were observed in both assays and when using both extrahents. DPPH assay resulted in significant differences between extrahents in five families. In three families, better results were achieved when using methanol, in the other two—when using water. Family Jox1 exhibited highest antioxidant capacity in Josvainiai population, while family D61 was the best in Dūkštos population. On the other hand, antioxidant activity when using ABTS assay was significantly different in six families. In five of them, higher values were achieved when using water. In three families, Jox3, Jox6, and D31, radical scavenging activity as measured by ABTS assay, were especially high.

Results showed that differences between two populations (as they represent oak phenology) also were significant in most cases, similarly as with TPC and TFC. Except in the case of radical scavenging activity as shown in Figure 2, differences between populations in antioxidant activity measured by DPPH assay were not as big and in the results of ABTS assay. When using water as an extrahent, the free radical scavenging activity as measured by DPPH assay was not significantly different between populations, but it was different when using methanol as an extrahent. ABTS assay resulted in significant differences between populations regardless of extrahent used.

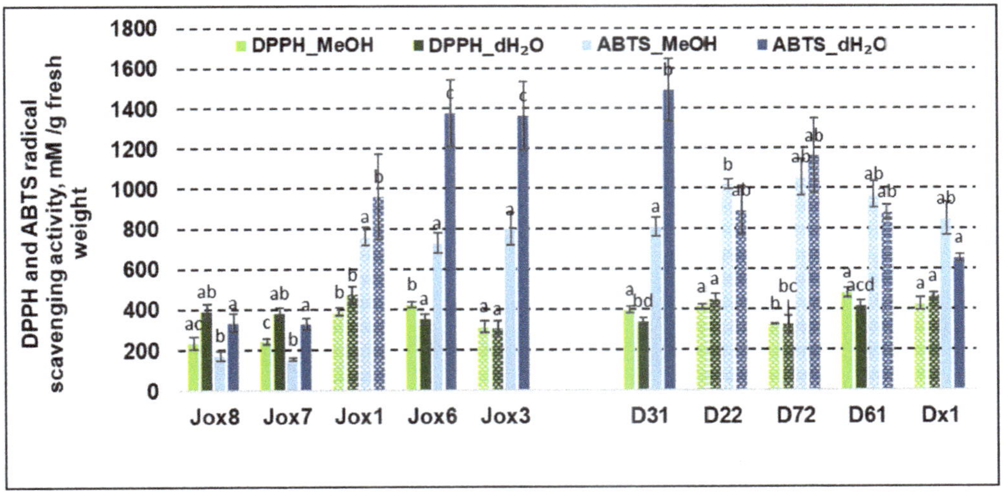

Differences between populations	Extrahents	Significance, p value
DPPH	dH_2O	NS
	MeOH	****
ABTS	dH_2O	**
	MeOH	****

Figure 2. Means (mM/g fresh weight) ± SE of DPPH and ABTS scavenging activity in 10 *Q. robur* families. Significance was calculated using the Kruskal–Wallis H test for ranks and post hoc Dunn's test for pairs ($p < 0.05$). Different letters next to the same colors indicate significant differences between families, solid colors indicate significant differences between extrahents, and differences among the studied population are noted in the table (p values: **—≤0.01, ****—≤0.0001, NS—not significant). Different populations are indicated by group designations Jox and D.

Furthermore, analysis of random and fixed effects on variance showed that genotype (half-sib family) had a significant effect on DPPH extraction by both means. Tree phenology had no significant effect on this. ABTS extraction by methanol was also affected by genotype, but not phenology. In addition, neither genotype nor tree phenology affected ABTS extraction by water (Table 5). All in all, ~96% of all variance in methanol extracts was determined by tree genotype (TPC, TFC, ABTS, DPPH). Methanol extracts were also less affected by replicate variation. Moreover, only around 35% of variance was determined by genotype in water extraction for all 4 tests. Water extraction was also more affected by replicate variation.

4. Discussion

As noted previously, multiple factors can determine the amounts of bioactive compounds a plant produces at one time. It has been shown that different species of oak synthesize different amounts of tested compounds, such as *Q. alba*, *Q. robur*, and *Q. petraea* as noted by Cabrita et al. [48] and Jordao et al. [4]. Phenolic acids, aldehydes and furanic derivatives were tested by Cabrita et al. Among them was the previously discussed ellagic acid [48]. Fernández de Simón et al. observed that the same wine aged in barrels of *Q. robur* and *Q. petraea* barrels (same species trees from different origin sites) for 21 months exhibited statistically significant differences in terms of ellagic acid and trans-resveratrol [49]. It is noteworthy that ellagitannins, precursors of ellagic acid, were found to be largely responsible for the antioxidant activity of oak wood [50]. Additionally, Prida and Puech showed that *Q. robur* grown in France and Eastern Europe diverged on the basis of their biochemistry, of which several chemicals were predominantly responsible, i.e., eugenol, 2-phenylethanol, vanillin, vescalagin, cis/trans-whiskey lactone ratio, and roburin B. In the same study, the authors also noted the vast differences between oak species [51]. Moreover, Kovalikova et al. also reported on this issue. She noted that *Q. robur* from different locations have different amounts of soluble phenols and flavonoids [35]. Another study on 30 *Q. robur* full sibs showed that phenols, flavonoids, tannins, and lignins were not affected by tree genotype all that much (up to 10% $\pm$ SE of total variation) [52]. Our own work with other tree species has shown that even when grown in carefully monitored and unified in vitro conditions two different genotypes of *Populus* spp. can produce different amounts of phenols, flavonoids, and photosynthesis pigments [31]. Furthermore, we observed the same tendencies in ash, spruce, and pine trees as well as blueberries and lingonberries [32,43,47,53,54]. All these examples align perfectly with the findings of this study, whereby we observed statistically significant differences in the amounts of TPC, TFC and antioxidant scavenging capacity diverging in different half-sib families grown in the same site and belonging to the same population.

In this work we also looked into the use of two extrahents—methanol and water. Results show that all four tests exhibited statistically significant differences in the amount of phenols, flavonoids, and antioxidants extracted. Similar results were noted in a review by Ignat et al. Their analysis of multiple works showed that methanol extraction is oftentimes more effective, but is less commonly used in food industries due to methanol toxicity [55]. Lavado et al. tested how different extracts of cork oak behaved and observed that water:ethanol extracts exhibited higher antioxidant activity than using just water or just ethanol. These mixed extrahents allowed for greater amounts of phenols, flavonoids, and condensed tannins, but not tocopherols. In the case of the latter, pure ethanol extrahents worked best [7]. Similarly, Šukele et al. presents results on using different extrahents on oak bark. They report acetone and 30% ethanol having the best outcome in terms of using these extracts for growth control of *Streptococcus agalactiae*, *Streptococcus uberis*, *Serratia liquefaciens*, and *Staphylococcus aureus* [18]. In a 2018 article Valencia-Aviles et al. report that hot water extracts were more effective that using 90% ethanol for both phenols and antioxidants [37]. Furthermore, Arina and Harisun observed that even using the same extrahent, different extraction temperatures still significantly affected the outcome [56]. All in all, these results collectively show that it is important to see which extraction methodology and which extrahent is more appropriate to use in any given case.

The most unique aspect of our investigation was the determination of whether *Q. robur* phenology can be a determining factor in the amount of bioactive compounds a tree produces. This is different from the effect tree genotype has, at tree phenotype can be determined both by genotype and by environmental conditions [57,58]. Previously, this characteristic has been shown to impact enzymatic activities in oak symbiotic fungus *Lactarius quietus*. Specifically, enzymes that contribute to the degradation and mobilization of carbon-rich components of the dead plant [59]. More importantly, it was demonstrated by Barber and Fahey that *Quercus alba* expressed differences in leaf antioxidant capacity of phenolics depending on oak phenology, i.e., early bud burst oaks had lower oxidative capacity in the first weeks of leaf growth as compared to the late bud burst oaks [60]. In our study, we observed significant differences in TPC (just water), TFC (both extrahents), DPPH (just methanol), and ABTS (both extrahents) parameters between both populations. Furthermore, it could be said that overall Josvainiai population (early bud burst) did produce less phenolics and expressed lower antioxidant capacity than Dūkštos population (late bud burst). This is a directly comparable result to that of Barber and Fahey, but it is worth noting they worked with a different oak species, and as was noted before, different oak species diverge in bioactive compound production.

5. Conclusions

The amount of total phenols, total flavonoids, and the antioxidant scavenging activity as expressed by ABTS and DPPH assays are different between the bark of *Quercus robur* genotypes (half-sib families), oaks with different phonologies (early or late bud burst populations), and different extrahents (75% methanol and distilled water). Overall, late bud burst population exhibited higher values in all parameters measured. Thus, in order to optimize extraction of desired bioactive compounds of phenolic origin from *Q. robur* bark, it is pertinent to take these factors into account. We would also like to emphasize that oak bark has a huge potential to be used as a natural product in supplement, additive, and other industries.

Author Contributions: Conceptualization, V.S.-Š. and V.B.; methodology, V.S.-Š., V.B. and D.V.; software, I.Č. and D.V.; validation, I.Č., D.V. and V.S.-Š.; formal analysis, I.Č., G.L. and V.S.-Š.; investigation, I.Č., G.L., V.B. and V.S.-Š.; resources, V.B. and V.S.-Š.; data curation, I.Č. and D.V.; writing—original draft preparation, D.V. and I.Č.; writing—review and editing, D.V. and I.Č.; visualization, D.V. and I.Č.; supervision, V.S.-Š.; project administration, V.S.-Š. All authors have read and agreed to the published version of the manuscript.

Funding: This research received no external funding.

Institutional Review Board Statement: Not applicable.

Informed Consent Statement: Not applicable.

Data Availability Statement: All data will be made available upon request.

Conflicts of Interest: The authors declare no conflict of interest.

References

1. Ianni, F.; Segoloni, E.; Blasi, F.; Di Maria, F. Low-Molecular-Weight Phenols Recovery by Eco-Friendly Extraction from *Quercus* spp. Wastes: An Analytical and Biomass-Sustainability Evaluation. *Processes* **2020**, *8*, 387. [CrossRef]
2. Dong, L.; Qin, C.; Li, Y.; Wu, Z.; Liu, L. Oat Phenolic Compounds Regulate Metabolic Syndrome in High Fat Diet-Fed Mice via Gut Microbiota. *Food Biosci.* **2022**, *50*, 101946. [CrossRef]
3. Ayalew, A.A.; Wodag, A.F. Extraction and Chromatographic Analysis of Ethiopian Oak Bark Plant for Leather Tanning Applications. *Chem. Afr.* **2023**. [CrossRef]
4. Jordão, A.M.; Correia, A.C.; DelCampo, R.; González SanJosé, M.L. Antioxidant Capacity, Scavenger Activity, and Ellagitannins Content from Commercial Oak Pieces Used in Winemaking. *Eur. Food Res. Technol.* **2012**, *235*, 817–825. [CrossRef]
5. Parsaei, M.; Goli, M.; Abbasi, H. Oak Flour as a Replacement of Wheat and Corn Flour to Improve Biscuit Antioxidant Activity. *Food Sci. Nutr.* **2018**, *6*, 253–258. [CrossRef]
6. Gamboa-Gómez, C.I.; Simental-Mendía, L.E.; González-Laredo, R.F.; Alcantar-Orozco, E.J.; Monserrat-Juarez, V.H.; Ramírez-España, J.C.; Gallegos-Infante, J.A.; Moreno-Jiménez, M.R.; Rocha-Guzmán, N.E. In Vitro and in Vivo Assessment of Anti-Hyperglycemic and Antioxidant Effects of Oak Leaves (*Quercus convallata* and *Quercus arizonica*) Infusions and Fermented Beverages. *Food Res. Int.* **2017**, *102*, 690–699. [CrossRef] [PubMed]
7. Lavado, G.; Ladero, L.; Cava, R. Cork Oak (*Quercus suber* L.) Leaf Extracts Potential Use as Natural Antioxidants in Cooked Meat. *Ind. Crops Prod.* **2021**, *160*, 113086. [CrossRef]
8. Soriano, A.; Alañón, M.E.; Alarcón, M.; García-Ruíz, A.; Díaz-Maroto, M.C.; Pérez-Coello, M.S. Oak Wood Extracts as Natural Antioxidants to Increase Shelf Life of Raw Pork Patties in Modified Atmosphere Packaging. *Food Res. Int.* **2018**, *111*, 524–533. [CrossRef]
9. Formato, M.; Vastolo, A.; Piccolella, S.; Calabr, S.; Cutrignelli, M.I.; Zidorn, C.; Pacifico, S. Antioxidants in Animal Nutrition: UHPLC-ESI-Q q TOF Analysis and Effects on In Vitro Rumen Fermentation of Oak Leaf Extracts. *Antioxidants* **2022**, *11*, 2366. [CrossRef]
10. Paray, B.A.; Hoseini, S.M.; Hoseinifar, S.H.; Van Doan, H. Effects of Dietary Oak (*Quercus castaneifolia*) Leaf Extract on Growth, Antioxidant, and Immune Characteristics and Responses to Crowding Stress in Common Carp (*Cyprinus carpio*). *Aquaculture* **2020**, *524*, 735276. [CrossRef]
11. Mirski, R.; Kawalerczyk, J.; Dziurka, D.; Siuda, J.; Wieruszewski, M. The Application of Oak Bark Powder as a Filler for Melamine-Urea-Formaldehyde Adhesive in Plywood Manufacturing. *Forests* **2020**, *11*, 1249. [CrossRef]
12. Hamad, A.M.A.; Ates, S.; Olgun, Ç.; Gür, M. Chemical Composition and Antioxidant Properties of Some Industrial Tree Bark Extracts. *BioResources* **2019**, *14*, 5657–5671. [CrossRef]
13. Nisca, A.; Ștefănescu, R.; Moldovan, C.; Mocan, A.; Mare, A.D.; Ciurea, C.N.; Man, A.; Muntean, D.L.; Tanase, C. Optimization of Microwave Assisted Extraction Conditions to Improve Phenolic Content and In Vitro Antioxidant and Anti-Microbial Activity in *Quercus Cerris* Bark Extracts. *Plants* **2022**, *11*, 240. [CrossRef] [PubMed]
14. Ștefănescu, R.; Ciurea, C.N.; Mare, A.D.; Man, A.; Nisca, A.; Nicolescu, A.; Mocan, A.; Babotă, M.; Coman, N.A.; Tanase, C. *Quercus Robur* Older Bark—A Source of Polyphenolic Extracts with Biological Activities. *Appl. Sci.* **2022**, *12*, 11738. [CrossRef]

15. Fisinin, V.I.; Ushakov, A.S.; Duskaev, G.K.; Kazachkova, N.M.; Nurzhanov, B.S.; Rakhmatullin, S.G.; Levakhin, G.I. Mixtures of Biologically Active Substances of Oak Bark Extracts Change Immunological and Productive Indicators of Broilers. *Sel'skokhozyaistvennaya Biol.* **2018**, *53*, 385–392. [CrossRef]
16. Lorenz, P.; Heinrich, M.; Garcia-Käufer, M.; Grunewald, F.; Messerschmidt, S.; Herrick, A.; Gruber, K.; Beckmann, C.; Knoedler, M.; Huber, R.; et al. Constituents from Oak Bark (*Quercus robur* L.) Inhibit Degranulation and Allergic Mediator Release from Basophils and Mast Cells in Vitro. *J. Ethnopharmacol.* **2016**, *194*, 642–650. [CrossRef]
17. Tsubanova, N.; Zhurenko, D.; Khokhlenkova, N.; Artiukh, T. Screening Study for Finding the Optimal Combination Gel Composition for the Treatment of Periodontal Disease, Which Contains Extracts of *Aloe vera* and Oak Bark. *Asian J. Pharm.* **2017**, *11*, 353–357.
18. Šukele, R.; Skadiņš, I.; Koka, R.; Bandere, D. Antibacterial Effects of Oak Bark (*Quercus robur*) and Heather Herb (*Calluna vulgaris* L.) Extracts against the Causative Bacteria of Bovine Mastitis. *Vet. World* **2022**, *15*, 2315–2322. [CrossRef]
19. Vasilchenko, A.S.; Poshvina, D.V.; Sidorov, R.Y.; Iashnikov, A.V.; Rogozhin, E.A.; Vasilchenko, A.V. Oak Bark (*Quercus* sp. *Cortex*) Protects Plants through the Inhibition of Quorum Sensing Mediated Virulence of *Pectobacterium carotovorum*. *World J. Microbiol. Biotechnol.* **2022**, *38*, 184. [CrossRef]
20. Tanase, C.; Nicolescu, A.; Nisca, A.; Ştefănescu, R.; Babotă, M.; Mare, A.D.; Ciurea, C.N.; Man, A. Biological Activity of Bark Extracts from Northern Red Oak (*Quercus Rubra* L.): An Antioxidant, Antimicrobial and Enzymatic Inhibitory Evaluation. *Plants* **2022**, *11*, 2357. [CrossRef]
21. Atlanderova, K.N.; Makaeva, A.M.; Sizova, E.A.; Duskaev, G.K. Stimulation of Ruminal Digestion of Young Cattle with Oak Bark Extract (*Quercus cortex*). In *IOP Conference Series: Earth and Environmental Science*; IOP Publishing: Bristol, UK, 2019; Volume 341. [CrossRef]
22. Valencia-Avilés, E.; García-Pérez, M.E.; Garnica-Romo, M.G.; de Dios Figueroa-Cárdenas, J.; Paciulli, M.; Martínez-Flores, H.E. Chemical Composition, Physicochemical Evaluation and Sensory Analysis of Yogurt Added with Extract of Polyphenolic Compounds from *Quercus Crassifolia* Oak Bark. *Funct. Foods Health Dis.* **2022**, *12*, 502–517. [CrossRef]
23. Zhang, Y.; Li, Y.; Ren, X.; Zhang, X.; Wu, Z.; Liu, L. The Positive Correlation of Antioxidant Activity and Prebiotic Effect about Oat Phenolic Compounds. *Food Chem.* **2023**, *402*, 134231. [CrossRef] [PubMed]
24. Elansary, H.O.; Szopa, A.; Kubica, P.; Ekiert, H.; Mattar, M.A.; Al-Yafrasi, M.A.; El-Ansary, D.O.; Zin El-Abedin, T.K.; Yessoufou, K. Polyphenol Profile and Pharmaceutical Potential of *Quercus* Spp. Bark Extracts. *Plants* **2019**, *8*, 486. [CrossRef] [PubMed]
25. Ríos, J.L.; Giner, R.M.; Marín, M.; Recio, M.C. A Pharmacological Update of Ellagic Acid. *Planta Med.* **2018**, *84*, 1068–1093. [CrossRef] [PubMed]
26. Ucar, M.B.; Ucar, G. Characterization of Methanol Extracts from *Quercus Hartwissiana* Wood and Bark. *Chem. Nat. Compd.* **2011**, *47*, 697–703. [CrossRef]
27. Ellinger, S.; Reusch, A.; Stehle, P.; Helfrich, H.P. Epicatechin Ingested via Cocoa Products Reduces Blood Pressure in Humans: A Nonlinear Regression Model with a Bayesian Approach. *Am. J. Clin. Nutr.* **2012**, *95*, 1365–1377. [CrossRef]
28. Worawalai, W.; Sompornpisut, P.; Wacharasindhu, S.; Phuwapraisirisan, P. Quercitol: From a Taxonomic Marker of the Genus *Quercus* to a Versatile Chiral Building Block of Antidiabetic Agents. *J. Agric. Food Chem.* **2018**, *66*, 5741–5745. [CrossRef]
29. Rudkowska, I.; AbuMweis, S.S.; Jones, P.J.H.; Nicolle, C. Cholesterol-Lowering Efficacy of Plant Sterols in Low-Fat Yogurt Consumed as a Snack or with a Meal. *J. Am. Coll. Nutr.* **2008**, *27*, 588–595. [CrossRef]
30. Deryabin, D.G.; Tolmacheva, A.A. Antibacterial and Anti-Quorum Sensing Molecular Composition Derived from *Quercus cortex* (Oak Bark) Extract. *Molecules* **2015**, *20*, 17093–17108. [CrossRef]
31. Striganavičiūtė, G.; Žiauka, J.; Sirgedaitė-Šėžienė, V.; Vaitiekūnaitė, D. Impact of Plant-Associated Bacteria on the In Vitro Growth and Pathogenic Resistance against *Phellinus Tremulae* of Different Aspen (*Populus*) Genotypes. *Microorganisms* **2021**, *9*, 1901. [CrossRef]
32. Beniušytė, E.; Čėsnienė, I.; Sirgedaitė-Šėžienė, V.; Vaitiekūnaitė, D. Genotype-Dependent Jasmonic Acid Effect on *Pinus Sylvestris* L. Growth and Induced Systemic Resistance Indicators. *Plants* **2023**, *12*, 255. [CrossRef] [PubMed]
33. Garcia, R.; Soares, B.; Dias, C.B.; Freitas, A.M.C.; Cabrita, M.J. Phenolic and Furanic Compounds of Portuguese Chestnut and French, American and Portuguese Oak Wood Chips. *Eur. Food Res. Technol.* **2012**, *235*, 457–467. [CrossRef]
34. Louis, J.; Meyer, S.; Maunoury-Danger, F.; Fresneau, C.; Meudec, E.; Cerovic, Z.G. Seasonal Changes in Optically Assessed Epidermal Phenolic Compounds and Chlorophyll Contents in Leaves of Sessile Oak (*Quercus petraea*): Towards Signatures of Phenological Stage. *Funct. Plant Biol.* **2009**, *36*, 732–741. [CrossRef] [PubMed]
35. Kovalikova, Z.; Lnenicka, J.; Andrys, R. The Influence of Locality on Phenolic Profile and Antioxidant Capacity of Bud Extracts. *Foods* **2021**, *10*, 1608. [CrossRef]
36. Sirgedaitė-šėžienė, V.; Laužikė, K.; Uselis, N.; Samuolienė, G. Metabolic Response of *Malus Domestica* Borkh Cv. Rubin Apple to Canopy Training Treatments in Intensive Orchards. *Horticulturae* **2022**, *8*, 300. [CrossRef]
37. Valencia-Avilés, E.; García-Pérez, M.E.; Garnica-Romo, M.G.; Figueroa-Cárdenas, J.d.D.; Meléndez-Herrera, E.; Salgado-Garciglia, R.; Martínez-Flores, H.E. Antioxidant Properties of Polyphenolic Extracts from *Quercus laurina*, *Quercus crassifolia*, and *Quercus scytophylla* Bark. *Antioxidants* **2018**, *7*, 81. [CrossRef]
38. Dantec, C.F.; Ducasse, H.; Capdevielle, X.; Fabreguettes, O.; Delzon, S.; Desprez-Loustau, M.L. Escape of Spring Frost and Disease through Phenological Variations in Oak Populations along Elevation Gradients. *J. Ecol.* **2015**, *103*, 1044–1056. [CrossRef]

39. Crawley, M.J.; Akhteruzzaman, M. Individual Variation in the Phenology of Oak Trees and Its Consequences for Herbivorous Insects. *Funct. Ecol.* **1988**, *2*, 409. [CrossRef]
40. Wesołowski, T.; Rowiński, P. Late Leaf Development in Pedunculate Oak (*Quercus robur*): An Antiherbivore Defence? *Scand. J. For. Res.* **2008**, *23*, 386–394. [CrossRef]
41. Eyles, A.; Bonello, P.; Ganley, R.; Mohammed, C. Induced Resistance to Pests and Pathogens in Trees. *New Phytol.* **2010**, *185*, 893–908. [CrossRef]
42. Singleton, V.L.; Orthhofer, R.; Lamuela-Raventos, R.M. Analysis of Total Phenols and Other Oxidation Substrates and Antioxidants by Means of Folin-Ciocalteu Reagent. *Methods Enzymol.* **1999**, *299*, 152–178. [CrossRef]
43. Lučinskaitė, I.; Laužikė, K.; Žiauka, J.; Baliuckas, V.; Čėsna, V.; Sirgedaitė-Šėžienė, V. Assessment of Biologically Active Compounds, Organic Acids and Antioxidant Activity in Needle Extracts of Different Norway Spruce (*Picea abies* (L.) H. Karst) Half-sib Families. *Wood Sci. Technol.* **2021**, *55*, 1221–1235. [CrossRef]
44. Ragaee, S.; Abdel-Aal, E.S.M.; Noaman, M. Antioxidant Activity and Nutrient Composition of Selected Cereals for Food Use. *Food Chem.* **2006**, *98*, 32–38. [CrossRef]
45. Becker, R.; Ulrich, K.; Behrendt, U.; Kube, M.; Ulrich, A. Analyzing Ash Leaf-Colonizing Fungal Communities for Their Biological Control of *Hymenoscyphus fraxineus*. *Front. Microbiol.* **2020**, *11*, 590944. [CrossRef]
46. Jové, P.; Olivella, M.; Cano, L. Study of the Variability in Chemical Composition of Bark Layers of *Quercus suber* L. from Different Production Areas. *BioResources* **2011**, *6*, 1806–1815.
47. Šėžienė, V.; Lučinskaitė, I.; Mildažienė, V.; Ivankov, A.; Koga, K.; Shiratani, M.; Laužikė, K.; Baliuckas, V. Changes in Content of Bioactive Compounds and Antioxidant Activity Induced in Needles of Different Half-Sib Families of Norway Spruce (*Picea abies* (L.) H. Karst) by Seed Treatment with Cold Plasma. *Antioxidants* **2022**, *11*, 1558. [CrossRef] [PubMed]
48. Cabrita, M.J.; Barrocas Dias, C.; Costa Freitas, A. Phenolic Acids, Phenolic Aldehydes and Furanic Derivatives in Oak Chips: American vs. French Oaks. *S. Afr. J. Enol. Vitic.* **2011**, *32*, 204–210. [CrossRef]
49. Fernández de Simón, B.; Hernández, T.; Cadahía, E.; Dueñas, M.; Estrella, I. Phenolic Compounds in a Spanish Red Wine Aged in Barrels Made of Spanish, French and American Oak Wood. *Eur. Food Res. Technol.* **2003**, *216*, 150–156. [CrossRef]
50. Alañón, M.E.; Castro-Vázquez, L.; Díaz-Maroto, M.C.; Gordon, M.H.; Pérez-Coello, M.S. A Study of the Antioxidant Capacity of Oak Wood Used in Wine Ageing and the Correlation with Polyphenol Composition. *Food Chem.* **2011**, *128*, 997–1002. [CrossRef]
51. Prida, A.; Puech, J. Influence of Geographical Origin and Botanical Species on the Content of Extractives in American, French, and East European Oak Woods. *J. Agric. Food Chem.* **2006**, *54*, 8115–8126. [CrossRef]
52. Damestoy, T.; Brachi, B.; Moreira, X.; Jactel, H.; Plomion, C.; Castagneyrol, B. Oak Genotype and Phenolic Compounds Differently Affect the Performance of Two Insect Herbivores with Contrasting Diet Breadth. *Tree Physiol.* **2019**, *39*, 615–627. [CrossRef] [PubMed]
53. Striganavičiūtė, G.; Žiauka, J.; Sirgedaitė-Šėžienė, V.; Vaitiekūnaitė, D. Priming of Resistance-Related Phenolics: A Study of Plant-Associated Bacteria and *Hymenoscyphus fraxineus*. *Microorganisms* **2021**, *9*, 2504. [CrossRef] [PubMed]
54. Dróżdż, P.; Šėžienė, V.; Pyrzynska, K. Phytochemical Properties and Antioxidant Activities of Extracts from Wild Blueberries and Lingonberries. *Plant Foods Hum. Nutr.* **2017**, *72*, 360–364. [CrossRef] [PubMed]
55. Ignat, I.; Volf, I.; Popa, V.I. A Critical Review of Methods for Characterisation of Polyphenolic Compounds in Fruits and Vegetables. *Food Chem.* **2011**, *126*, 1821–1835. [CrossRef]
56. Iylia Arina, M.Z.; Harisun, Y. Effect of Extraction Temperatures on Tannin Content and Antioxidant Activity of *Quercus infectoria* (Manjakani). *Biocatal. Agric. Biotechnol.* **2019**, *19*, 101104. [CrossRef]
57. Ekholm, A.; Tack, A.J.M.; Pulkkinen, P.; Roslin, T. Host Plant Phenology, Insect Outbreaks and Herbivore Communities—The Importance of Timing. *J. Anim. Ecol.* **2020**, *89*, 829–841. [CrossRef]
58. Kleinschmit, J. Intraspecific Variation of Growth and Adaptive Traits in European Oak Species. *Ann. Des Sci. For.* **1993**, *50*, 166s–185s. [CrossRef]
59. Courty, P.E.; Bréda, N.; Garbaye, J. Relation between Oak Tree Phenology and the Secretion of Organic Matter Degrading Enzymes by *Lactarius Quietus* Ectomycorrhizas before and during Bud Break. *Soil Biol. Biochem.* **2007**, *39*, 1655–1663. [CrossRef]
60. Barber, N.A.; Fahey, R.T. Consequences of Phenology Variation and Oxidative Defenses in *Quercus*. *Chemoecology* **2015**, *25*, 261–270. [CrossRef]

life

MDPI

Communication

Caucasian Blueberry: Comparative Study of Phenolic Compounds and Neuroprotective and Antioxidant Potential of *Vaccinium myrtillus* and *Vaccinium arctostaphylos* Leaves

Arnold A. Shamilov [1], Daniil N. Olennikov [2,*], Dmitryi I. Pozdnyakov [3], Valentina N. Bubenchikova [4], Ekaterina R. Garsiya [1] and Mikhail V. Larskii [5]

[1] Department of Pharmacognosy, Botany and Technology of Phytopreparations, Pyatigorsk Medical-Pharmaceutical Institute, Branch of Volgograd State Medical University, Ministry of Health of Russian Federation, 11 Kalinina Avenue, 357500 Pyatigorsk, Russia
[2] Laboratory of Biomedical Research, Institute of General and Experimental Biology, Siberian Division, Russian Academy of Science, 6 Sakhyanovoy Street, 670047 Ulan-Ude, Russia
[3] Department of Pharmacology and Clinical Pharmacology, Pyatigorsk Medical-Pharmaceutical Institute, Branch of Volgograd State Medical University, Ministry of Health of Russian Federation, 11 Kalinina Avenue, 357500 Pyatigorsk, Russia
[4] Department of Pharmacognosy and Botany, Kursk State Medical University, Ministry of Health of Russian Federation, 3 Karl Marks Street, 305000 Kursk, Russia
[5] Department of Pharmaceutical Chemistry, Pyatigorsk Medical-Pharmaceutical Institute, Branch of Volgograd State Medical University, Ministry of Health of Russian Federation, 11 Kalinina Avenue, 357500 Pyatigorsk, Russia
* Correspondence: olennikovdn@mail.ru; Tel.: +7-902-160-06-27

Citation: Shamilov, A.A.; Olennikov, D.N.; Pozdnyakov, D.I.; Bubenchikova, V.N.; Garsiya, E.R.; Larskii, M.V. Caucasian Blueberry: Comparative Study of Phenolic Compounds and Neuroprotective and Antioxidant Potential of *Vaccinium myrtillus* and *Vaccinium arctostaphylos* Leaves. *Life* **2022**, *12*, 2079. https://doi.org/10.3390/life12122079

Academic Editor: Jianfeng Xu

Received: 12 November 2022
Accepted: 8 December 2022
Published: 11 December 2022

Abstract: (1) Background: Two Caucasian blueberries *Vaccinium myrtillus* L. and *Vaccinium arctostaphylos* L. are famous berry bushes growing in the Caucasus region and used to treat neurological diseases, but the chemistry and bioactivity of leaf extracts are still poorly studied. (2) Methods: Phenolic compounds of *V. myrtillus* and *V. arctostaphylos* leaf extracts were profiled and quantified by HPLC–PDA–ESI–tQ–MS. The neurotropic potential of *Vaccinium* extracts was studied using the model of middle cerebral artery permanent occlusion to determine cerebral blood flow, the area of the brain tissue necrosis, and antioxidant enzyme activity (including superoxide dismutase, succinate dehydrogenase, and cytochrome C oxidase), as well as the concentration of TBARS. (3) Results: Hydroxycinnamates and flavonoids were identified in the leaves of *V. myrtillus* and *V. arctostaphylos*, and the dominant metabolite of both extracts was 5-*O*-caffeoylquinic acid in the amount of 105–226 mg/g. The studied extracts enhanced the cerebral hemodynamics and decreased the frequency of necrotic and lipooxidative processes in the brain tissue, accompanied by an increase in the activity of antioxidant enzymes. The positive effect of *V. arctostaphylos* was stronger and exceeded the effectiveness of *Ginkgo biloba* standardized extract. (4) Conclusion: The leaf extracts of Caucasian blueberries *V. myrtillus* and *V. arctostaphylos* as a new source of hydroxycinnamates demonstrated a protective effect of the brain ischemia pathology and can be used as therapeutic agents to treat neurological diseases.

Keywords: blueberry; *Vaccinium myrtillus*; *Vaccinium arctostaphylos*; phenolic compounds; antioxidant activity; neuroprotective activity

1. Introduction

Currently, the *Vaccinium* L. genus includes over 170 species [1], and there are 14 native species in the Russian Federation, including European blueberry (*V. myrtillus* L.), which grows throughout Europe outside of Central Asia, and Caucasian blueberry (*V. arctostaphylos* L.), which is an endemic plant of Transcaucasia and North Caucasus [2,3]. The fruits of *V. myrtillus* are included in pharmacopoeias of different countries as an astringent, hypolipidemic, and vision-improving remedy [4]. It is also known that an aqueous decoction of the fruit has antiprotozoal, antitumor, and cytotoxic activity [5–7]. Literature data indicate that *V. myrtillus* is a rich source of phenolic acids [8,9], flavonoids, and phenolic glycosides, as well as triterpenoids, carotenoids, organic acids, carbohydrates, and higher fatty acids [10,11].

In contrast to the European blueberry, *V. arctostaphylos* is not a pharmacopoeial plant due to insufficient knowledge of its chemical composition and pharmacological activity. Botanically, the Caucasian blueberry is a shrub or small tree up to 2–3 m high with rounded branches and large oblong leaves. Flowers in the raceme and the corolla are large whitish red. Fruits are large, globular, and black with a smooth surface. The leaves of *V. arctostaphylos* are up to 10 cm length compared to the leaves of *V. myrtillus*, which are up to 3 cm (Figure 1). To date, the chemical composition of fruits was only slightly studied [12,13], and there is a need for new studies on this promising source of bioactive metabolites. Additionally, some phenolics have been identified in the leaves [14], and hypoglycemic and hypotensive activities have been demonstrated for the leaf extracts [15,16].

(**a**) (**b**) (**c**) (**d**)

Figure 1. *Vaccinium* spp. in their natural habitat: *Vaccinium arctostaphylos* L. (Caucasian blueberry, bank of Belaya river, Khamyshki village, Maykop District, Republic of Adygea) (**a**); *Vaccinium myrtillus* L. (bank of the Cherek River, Kabardino–Balkarian Republic) (**b**). Dried leaves of *V. arctostaphylos* (**c**) and *V. myrtillus* (**d**).

Phenolics have a significant effect on the human organism and are primarily used as effective remedies for preventing a wide range of diseases from atherosclerosis to ischemic stroke. Phenolic compounds and total phenolics-based phytocompositions are increasingly used in modern neurology practice, for example, for the treatment and prevention of cerebral ischemia and neurodegenerative diseases [17,18]. Previously, it has been shown that the addition of polyphenols to the diet contributes to the development of a number of therapeutic benefits in patients with Parkinson's disease, apparently by reducing oxidative damage of the *substantia nigra* neurons [19]. In addition, a number of studies show that the administration of polyphenolic substances significantly reduces damage to neurons during ischemia [20]. Moreover, polyphenolic substances are often considered as compounds that eliminate the hyperproduction of ROS and oxidative stress [21,22], and their neuroprotective effects have been clinically confirmed [21]. The results of a prospective cohort study by Gao et al. demonstrated a significant reduction in the risk of stroke when large amounts of dietary polyphenols are consumed, particularly citrus fruits and grapes [23]. Of note, some gender differences were identified in this work: the pharmacological efficacy of polyphenols was higher in women than in men [23].

Disorders of cerebral circulation represent a considerable medical and social problem. According to the World Health Organization, stroke ranks second among the causes of mortality and first among the causes of primary disability [24]. At the same time, approximately 80% of all stroke cases are due to ischemic brain damage [25]. One of the therapeutic options for the adjuvant treatment of cerebral ischemia is neuroprotection, in which a certain share is provided by herbal medicines. For example, some traditional Chinese medicine products have a pronounced neuroprotective effect, and the polyvalence of their action is important, affecting many pathogenetic mechanisms of cellular damage in ischemia: from oxidative stress to metabolic homeostasis and excitotoxicity [26].

The genus *Vaccinium* L., along with classical plant sources of antioxidants, can serve as a basis for the search for new biologically active compounds with neuroprotective activity; for example, the neuroprotective properties of anthocyanins of blueberries have been established [27]. In leaves, the main phenolic groups are proantocyanidines, flavonols, and hydroxycinnamic acids with potential activity in therapy of Alzheimer's and Parkinson's disease [28].

As part of the ongoing study of *Vaccinium* species [29,30], in this study, the phenolic compounds from *V. myrtillus* and *V. arctostaphylos* leaves were profiled and quantified by HPLC–PDA–ESI–tQ–MS assay, and the antioxidant and neuroprotection potential of total blueberry leave extracts was estimated in brain ischemia in vivo experiments.

2. Materials and Methods

2.1. Plant Material and Chemicals

Samples of *Vaccinium myrtillus* L. leaves were collected on the bank of the Cherek River, Kabardino–Balkarian Republic (22 September 2019, 43°57′63.24″ N, 42°59′23.81″ E, 1905 m a.s.l.; Russia; flowering stage; 5 locations, 10 samples; voucher No PALE 005073-01–PALE 005073-10). The species were authenticated by Mikhail Goncharov, Ph.D. Biology (Saint Petersburg State Chemical and Pharmaceutical University, Saint Petersburg, Russia). Samples of the *V. arctostaphylos* L. leaves were collected from an alpine meadow in the Republic of Adygea, Khamyshki village, Maykop District (26 September 2020, 44°02′341″ N, 40°10′035″ E, 698 m a.s.l.; Russia; flowering stage; 5 locations, 10 samples; voucher No PALE 005082-01–PALE 005082). The species were authenticated by Dmitryi Shilnikov, Ph.D. Biology (Ecological and Botanical Station "Pyatigorsk", Botanical Institute, Russian Academy of Sciences, Pyatigorsk, Russia). Plant material was dried at 40 °C (ventilated heat oven; 8–10 days) and stored at 3–4 °C before analysis. Selected chemicals were purchased from Sigma–Aldrich (St. Louis, MO, USA): acetonitrile for HPLC (Cat. No. 34851, ≥99.9%); formic acid (Cat. No. F0507, ≥95%); 4-*O*-caffeoylquinic acid (CAS 905-99-7, Sigma-Aldrich No. 65969, ≥98%); 5-*O*-caffeoylquinic acid (CAS 906-33-2, Sigma-Aldrich No. 94419, ≥98%); caffeic acid (CAS 331-39-5, Sigma-Aldrich No. C0625, ≥98%); 4,5-di-*O*-caffeoylquinic acid (CAS 14535-61-3, Sigma-Aldrich No. PHL80427, ≥95%); rutin (CAS 207671-50-9, Sigma-Aldrich No. R5143, ≥94%); hyperoside (CAS 482-36-0, Sigma-Aldrich No. 83388, ≥97%); isoquercitrin (CAS 482-35-9, Sigma-Aldrich No. 16,654, ≥98%); guaiaverin (CAS 22255-13-6, Sigma-Aldrich No. 94821, ≥95%); avicularin (CAS 572-30-5, Sigma-Aldrich No. PHL80361, ≥95%); quercitrin (CAS 849061-97-8, Sigma-Aldrich No. 337951, ≥95%); quercetin-3-*O*-(6″-*O*-acetyl)-glucoside (CAS 54542-51-7, No.1099, ≥85%, Extrasynthese, Lyon, France).

2.2. Plant Extract Preparation

Plant extract for HPLC analysis was prepared using 200 mg sample of milled material (particle size 0.125 μm) treated by 2 mL of 70% ethanol and sonication (ultrasonic bath, 30 min, 50 °C, ultrasound power 100 W, frequency 35 kHz; ×3). The ethanolic extract was centrifuged at 6000 × *g* (10 min), filtered through 0.22-μm syringe filters into a measuring flask (10 mL) and the final volume was filled up to 10 mL. The final extract was stored at 2 °C before analysis. Plant extract for the pharmacological study was prepared using the same procedure from 150 g-portion of material and drying in a vacuum oven. The total yields of *V. myrtillus* and *V. arctostaphylos* extracts were 49.3% and 40.1% of plant material weight, respectively.

2.3. *High-Performance Liquid Chromatography with Photodiode Array Detection and Electrospray Ionization Triple Quadrupole Mass Spectrometric Detection (HPLC–PDA–ESI–QQQ–MS)*

High-performance liquid chromatography with photodiode array detection and electrospray ionization triple quadrupole mass spectrometric detection (HPLC–PDA–ESI–QQQ–MS) with LC-20 Prominence liquid chromatograph coupled with an SPD-M30A photodiode array detector, LCMS 8050 triple-quadrupole mass spectrometer and GLC Mastro C_{18} column (2.1 × 150 mm, 3 μm; all Shimadzu, Columbia, MD, USA) were used for the profiling and quantification of the plant extracts, as described previously [31]. The similarity of retention time, ultraviolet and mass spectral data was used for the identification of metabolites after comparison with the reference standards and literature data. The metrics of calibration curves as correlation coefficient (r^2), standard deviation (S_{YX}), limits of detection (LOD), limits of quantification (LOQ), and linear range were found for calibration solution (1, 10, 25, 50, 100 μg/mL) of 11 reference standards (4-*O*-caffeoylquinic acid, 5-*O*-caffeoylquinic acid, caffeic acid, 4,5-di-*O*-caffeoylquinic acid, rutin, hyperoside, isoquercitrin, guaiaverin, avicularin, quercitrin, quercetin-3-*O*-(6″-*O*-acetyl)-glucoside) using the known method [32] (Table 1).

Table 1. Regression equations, correlation coefficients (r^2), standard deviation (S_{YX}), limits of detection (LOD), limits of quantification (LOQ), and linear ranges for 11 reference standards.

Compound	Regression Equation [a]		r^2	S_{YX}	LOD/LOQ (μg/mL)	Linear Range (μg/mL)
	a	b × 10^6				
4-*O*-Caffeoylquinic acid	0.162	−0.011	0.9973	1.83×10^{-2}	0.37/1.12	2–100
5-*O*-Caffeoylquinic acid	0.150	−0.010	0.9982	1.67×10^{-2}	0.36/1.11	2–100
Caffeic acid	0.189	−0.017	0.9988	1.26×10^{-2}	0.22/0.67	1–100
Quercetin 3-*O*-rutinoside (rutin)	0.085	−0.062	0.9873	3.89×10^{-2}	1.51/4.57	5–100
Quercetin 3-*O*-galactoside (hyperoside)	0.090	−0.052	0.9891	3.52×10^{-2}	1.29/3.91	4–100
Quercetin 3-*O*-glucoside (isoquercitrin)	0.098	−0.053	0.9889	3.37×10^{-2}	1.14/3.43	4–100
Quercetin 3-*O*-arabinopyranoside (guaiaverin)	0.121	−0.031	0.9953	2.59×10^{-2}	0.70/2.14	3–100
Quercetin 3-*O*-arabinofuranoside (avicularin)	0.114	−0.026	0.9927	2.80×10^{-2}	0.81/2.46	3–100
Quercetin 3-*O*-rhamnoside (quercitrin)	0.109	−0.043	0.9912	3.16×10^{-2}	0.96/2.90	3–100
Quercetin 3-*O*-(6″-acetyl)-glucoside	0.095	−0.050	0.9880	3.28×10^{-2}	1.14/3.45	4–100
4,5-Di-*O*-caffeoylquinic acid	0.163	−0.016	0.9975	1.93×10^{-2}	0.39/1.18	2–100

[a] Regression equation: $y = a \cdot x + b$.

2.4. *Neuroprotective Activity*

2.4.1. Animals

The study was used 100 male Wistar rats, 180–200 g weight. The animals were purchased from the laboratory animal nursery "Rappolovo" (Russia, Leningrad region) and during the experiment were kept in polypropylene boxes by 5 rats in controlled conditions of detention: air temperature 20 ± 20 °C, humidity—55–65%, daily cycle 12 h a day/12 h a night. The treatment of laboratory animals and their maintenance met the requirements of Directive 2010/63/EU of the European Parliament and of the Council of the European Union of 22 September 2010 for the protection of animals used for scientific purposes. The local Ethics committee (Protocol of meeting No. 10 of 20 August 2021) approved the research concept.

2.4.2. Brain Ischemia Model

Permanent focal cerebral ischemia was reproduced by right sided thermocoagulation of the middle cerebral artery (MCA). In this research, the concept irreversible technique was used. Rats were anesthetized by chloral hydrate (350 mg/kg, intraperitoneal). After, the area from the right side of the eye was depilated, skin was cut, and soft tissues were pushed apart, exposing the process of the zygomatic bone, which was deleted. Next, a trepanation hole was made with a drill and the MCA was burned with a thermocoagulator under the place of its intersection with the olfactory tract. If possible, the anatomic structure of soft tissues was restored. The suture was processed with a 10% povidone-iodine solution [33].

2.4.3. Study Design

Experimental animals were divided into 5 equal groups of 20 individuals (the deviation of the body weight of rats in one experimental group did not exceed 10%). The first group was sham operated (SO), to whose animals all sequential surgical manipulations were applied with the exception of arterial coagulation; negative control (NC)—animals with focal cerebral ischemia, but deprived of pharmacological support (this group received purified water throughout the study); EGB761 is a group of rats that were treated with a reference—standardized extract of *Ginkgo biloba* (EGB761; obtained from Hunan Warrant Pharmaceuticals, Changsha, China) at a dose of 35 mg/kg [34]; VM is a group of animals that received extract from *V. myrtillus*; VA is a group of animals that received extract from *V. arctostaphylos*. The administered dose for the studied extracts was selected on the basis of previous studies and amounted to 50 mg/kg. The referent and the studied extracts were administered for 3 days after ischemia, orally, once a day. On the 4th day of the experiment, a change in the level of local cerebral blood flow was evaluated in rats, after which the animals were decapitated and the brain was extracted, in the tissue of which the change in the necrosis zone was evaluated (in 10 rats from the group). The activity of superoxide dismutase, cytochrome-c oxidase and succinate dehydrogenase was determined in the remaining 10 individuals from the group. Changes in the concentration of 2-thiobarbituric acid active substances (TBARS) in brain tissue were also evaluated. All tests were performed once for each animal. A total of 20 tests were performed for one experimental group.

2.4.4. Cerebral Blood Flow Evaluation

The average systolic velocity of cerebral blood flow was recorded in the of the MCA area, for which a trepanation hole was made in this animal's brain region. Rats were anesthetized by chloral hydrate. The average systolic velocity was recorded using an ultrasound Doppler device, a sensor USOP-010-01 with a work frequency of 25 MHz and a MM-D-K-Minimax Doppler v.1.7. (Saint Petersburg, Russia) software for Windows, using mathematical processing of the fast Fourier transform signal to transform the color spectrum of the blood flow distribution [35].

2.4.5. Biomaterial Sampling and Preparation

Animal brains were used as biomaterial for analysis. The brain was homogenized in a mechanical homogenizer Potter type in a cold system of buffer solutions that consist of 1 mM EGTA, 215 mM mannitol, 75 mM sucrose, 0.1% bovine serum albumin solution with a 20 mM HEPES. The pH of buffer solution was 7.2. To obtain a primary homogenate, the ratio of tissue mass/volume of buffer solution 1:7 was used. Primary brain homogenate was separated into two parts. The first of them was used to determine the concentration of TBARS. The second part of the homogenate was used to determine superoxide dismutase, cytochrome c oxidase and succinate dehydrogenase activity. For that, homogenate was centrifuged 2 min at 1100× *g* acceleration. The supernatant was moved to Eppendorf-type test tubes and a percoll solution 10% concentration was layered, after which mixture was re-centrifuged at 18,000× *g* for 10 min. The secondary supernatant was deleted, the precipitate was resuspended in a buffer solution and re-centrifuged at 10,000× *g* for 5 min

2.4.6. Evaluation of Necrosis Zone

The size of the necrosis area of the brain was determined by the triphenyl tetrazolium method (n = 10 from each group). The brain was extracted from skull, the cerebellum was cut off. After, the hemispheres were separated. Both hemispheres were weighed. Next, the hemispheres were separately homogenized in PBS and placed in test-tubes with 10 mL of

1% triphenyltetrazolium chloride solution in a PBS with pH 7.4. The test-tubes were placed in a water bath for 20 min at a temperature of 37 ° C. Next, the probes were centrifuged in the mode of 3000 RCF/10 min and the resulting supernatant was removed. A total of 3 mL of phosphate buffer and 3 mL of cooled chloroform were added to the precipitate. Shaken for 2 min. Chloroform extract of formazan was obtained for 15 min at 4 ° C, shaking the mixture every 5 min for 30 s. The optical density (492 nm) against pure chloroform was centrifuged and measured. The necrosis zone area was expressed as a percentage of the total mass of the hemispheres and was calculated by formula:

$$x = 100 - \frac{\varepsilon_1 M_1 + \varepsilon_2 M_2}{\varepsilon_1 (M_1 + M_2)} 100\%,$$

where x is the size of the necrosis zone as a percentage of the total mass of the brain; ε_1 is the optical density of the sample with an intact hemisphere; ε_2 is the optical density of the sample with a damaged hemisphere; M_1 is the mass of the intact hemisphere; M_2 is the mass of the damaged hemisphere.

2.5. *Antioxidant Potential*

2.5.1. Thiobarbituric Acid Reactive Substances (TBARS) Evaluation

The concentration of TBARS was evaluated in the brain homogenate by spectrophotometric method. The principle of the method is based on the spectrophotometric (at 532 nm) determination of colored products of the condensation reaction between unsaturated aldehydes and 2-thiobarbituric acid. In this case, the absorbance of the solution is correlated to the concentration of malondialdehyde. The amount of TBARS was calculated by the malondialdehyde molar extinction coefficient (1.56×10^5 l $\times$ mol^{-1} $\times$ cm^{-1}) value. The obtained results were expressed in nmol/mg protein. Protein content was determined by the Bradford method. Absorbance was recorded on Infinite F50 reader (Tecan Austria GmbH, Grödig, Austria) [36].

2.5.2. Superoxide Dismutase (SOD) Activity

The activity of SOD was estimated by the xanthine-xanthine oxidase method. Principal of a method is based on the determination of products of reaction of the dismutation of the superoxide radical that a was formed during the complex oxidation and reduction of xanthine and 2-(4-iodophenyl)-3-(4-nitrophenol)-5-phenyltetrazolium chloride, respectively. The reaction mixture consisted of 0.05 mM of xanthine; 0.025 mM of 2-(4-iodophenyl)-3-(4-nitrophenol)-5-phenyltetrazolium chloride; 0.94 mM of EDTA, 80 U/L of xanthine oxidase, 40 mM of CAPS (N-cyclohexyl-3-aminopropanesulfonic acid) buffer. The optical density of the mixture was estimated at 505 nm. SOD activity was expressed in units/protein mg. Protein content was determined by the Bradford method. Absorbance was recorded on Infinite F50 reader (Tecan Austria GmbH, Grödig, Austria) [37].

2.5.3. Succinate Dehydrogenase (SDH) Activity

The SDH activity was evaluated by spectrophotometric method which based on the reaction of the reduction of dichlorophenolindophenol that was catalyzed by succinate. To inhibit a mitochondrial complex I activity rotenone was added to the analyzed mixture. Absorbance of the probes was estimated at 600 nm [23]. During the analysis, standard kits from Abcam (ab110217/MS643) were used. Absorbance was recorded on Infinite F50 reader (Tecan Austria GmbH, Grödig, Austria) [38].

2.5.4. Cytochrome-C-Oxidase (COX) Activity

The COX activity was evaluated in the reaction of cytochrome C (II) oxidation when KCN was added in an incubation mixture. Absorbance was estimated at 500 nm (24). Standard kits from Abcam (ab110217/MS643) were used in the analysis. Absorbance was recorded on Infinite F50 reader (Tecan Austria GmbH, Grödig, Austria) [39].

2.6. Statistical Analysis

Statistical analysis of chemical data was performed by one-way analysis of variance, and the significance of the mean difference was determined by Duncan's multiple range test. Differences at $p < 0.05$ were considered statistically significant. The results are presented as mean values ± standard deviations (S.D.). The linear regression analysis and generation of calibration graphs were conducted using Advanced Grapher 2.2 (Alentum Software Inc., Ramat-Gan, Israel). The statistical analysis of the results of neuroprotective activity evaluation was processed by using the STATISTICA 6.0 software (Dell Technologies Inc., Round Rock, TX, USA). The obtained data were expressed as M ± SEM (mean ± standard error of the mean). The normality was assessed by the Shapiro–Wilk test, the uniformity of variance was estimated by the Leven's test. The ANOVA with post hoc Newman-Keusle test was used in post-processing analysis. A critical level of significance was $p < 0.05$.

3. Results

3.1. Phenolic Compounds of *V. myrtillus* and *V. arctostaphylos* Leaves

Nine phenolic compounds, including the four phenolic acids of 4-*O*-caffeoylquinic, 5-*O*-caffeoylquinic, caffeic, and 4,5-di-*O*-caffeoylquinic acids as well as five flavonoid quercetin glycosides (quercetin-3-*O*-galactoside, quercetin-3-*O*-glucoside, quercetin-3-*O*-arabinopyranoside, quercetin-3-*O*-arabinofuranoside, quercetin-3-*O*-rhamnoside), were identified in *V. myrtillus* leaves (Figure 2, Table 2).

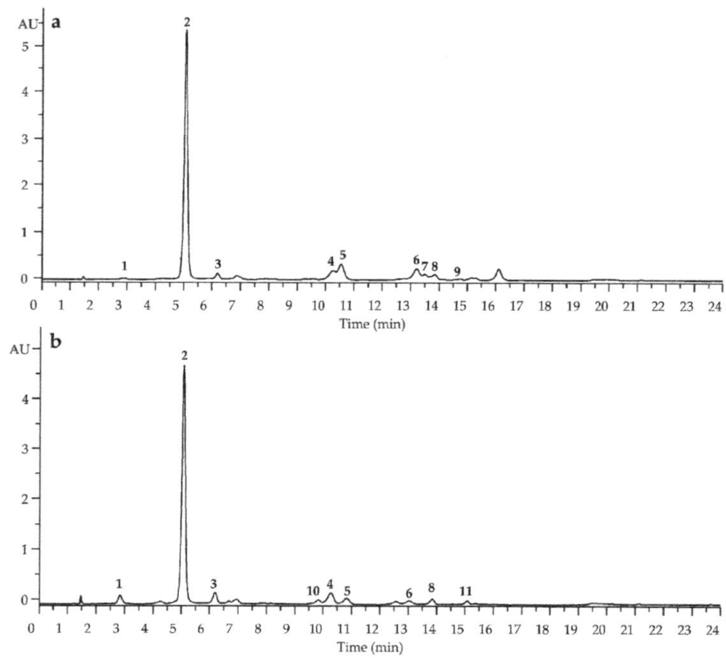

Figure 2. *Cont.*

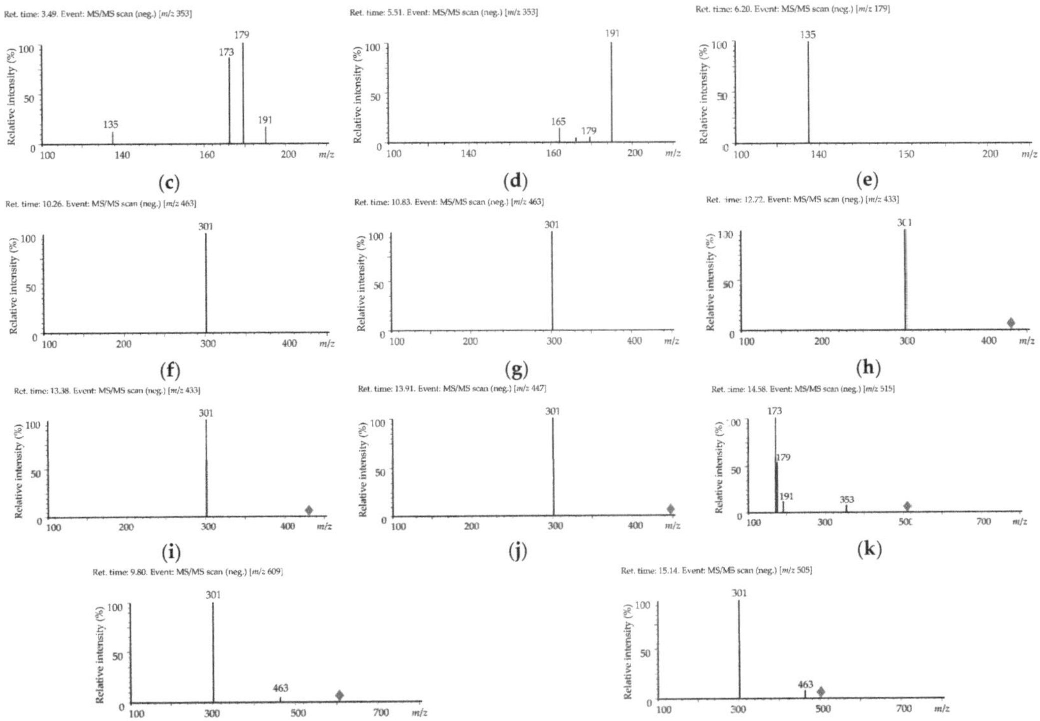

Figure 2. High-performance liquid chromatography with photodiode array detection (HPLC-PDA) chromatograms (330 nm) of leaf extracts of *V. myrtillus* (**a**) and *V. arctostaphylos* (**b**) and mass spectra of compounds **1** (**c**), **2** (**d**), **3** (**e**), **4** (**f**), **5** (**g**), **6** (**h**), **7** (**i**), **8** (**j**), **9** (**k**), **10** (**l**), and **11** (**m**). Compounds are numbered as listed in Table 2. Blue square indicates location of $[M-H]^-$ ion.

Table 2. Retention times (t), ultraviolet (UV), and mass spectral (ESI-MS, MS/MS) data of compounds **1–11** were found for leaf extracts of *V. myrtillus* and *V. arctostaphylos*.

No	t, min	Compound *	UV, λ_{max}, nm	ESI-MS, $[M-H]^-$, *m/z*	MS/MS, *m/z*
1	3.49	4-*O*-Caffeoylquinic acid	322	353	[353]: 191, 179, 173, 135
2	5.51	5-*O*-Caffeoylquinic acid	322	353	[353]: 191, 179, 165
3	6.20	Caffeic acid	320	179	[179]: 135
4	10.26	Quercetin 3-*O*-galactoside (hyperoside)	255, 267, 355	463	[463]: 301
5	10.83	Quercetin 3-*O*-glucoside (isoquercitrin)	255, 267, 356	463	[463]: 301
6	12.72	Quercetin 3-*O*-arabinopyranoside (guaiaverin)	255, 267, 354	433	[433]: 301
7	13.38	Quercetin 3-*O*-arabinofuranoside (avicularin)	255, 267, 354	433	[433]: 301
8	13.91	Quercetin 3-*O*-rhamnoside (quercitrin)	255, 267, 352	447	[447]: 301
9	14.58	3,5-Di-*O*-caffeoylquinic acid	322	515	[515]: 353, 191, 179, 173
10	9.80	Quercetin 3-*O*-rutinoside (rutin)	255, 267, 355	609	[609]: 463, 301
11	15.14	Quercetin 3-*O*-(6″-acetyl)-glucoside	256, 268, 351	505	[505]: 463, 301

* Identification was performed by comparison of obtained retention times, UV, MS, and MS/MS with reference standards.

Ten compounds of phenolic origin were found in *V. arctostaphylos* leave extracts, among them three phenolic acids (4-*O*-caffeoylquinic, 5-*O*-caffeoylquinic, and caffeic acids) and six quercetin glycosides [quercetin-3-*O*-rutinoside, quercetin-3-*O*-galactoside, quercetin-3-*O*-glucoside, quercetin-3-*O*-arabinopyranoside, quercetin-3-*O*-rhamnoside and quercetin-3-*O*-(6″-acetyl)-glucoside] (Figure 3).

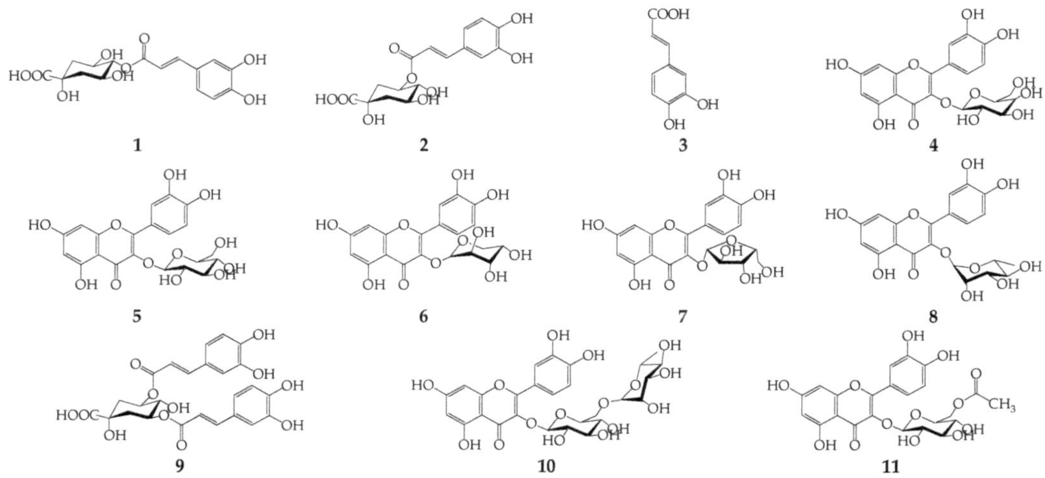

Figure 3. Compounds **1–11** found in leaf extracts of *V. myrtillus* and *V. arctostaphylos*.

Caffeoylquinic acids and caffeic acid, the known metabolites of *V. myrtillus* leaves [40], were also found in *V. arctostaphylos* leaves of Georgian origin [41]. In a previous study, the basic quercetin glycosides in *V. myrtillus* leaves from Finland were rutin, hyperoside, guaiaverin, avicularin, and quercitrin [42], and the only known flavonoid in *V. arctostaphylos* leaves is quercetin [43]. Quercetin 3-*O*-(6″-acetyl)-glucoside was found in *V. myrtillus* leaves for the first time as well as quercetin glycosides, which were newly discovered in *V. arctostaphylos* leaves.

Results of the quantification of compounds **1–11** in *V. myrtillus* and *V. arctostaphylos* leaf extracts demonstrated high contents of 5-*O*-caffeoylquinic acid (226.85 mg/g in *V. myrtillus*, 105.32 mg/g in *V. arctostaphylos*), quercetin 3-*O*-glucoside (12.02 mg/g in *V. myrtillus*), and quercetin 3-*O*-galactoside (12.02 mg/g in *V. arctostaphylos*) (Table 3). Earlier, chlorogenic acid was detected as the main phenolic compound in the *V. myrtillus* leaves (approximately 52–84% of the total amount of phenols) [44,45], and quercetin and kaempferol derivatives were the predominant phenolic groups (39.2%) [46]. The water–alcoholic extract of *V. arctostaphylos* leaves from Iran contained chlorogenic acid and flavonoids at 97.2 mg/g and 2.2–22.4 mg/g, respectively [15,47].

Table 3. Contents of compounds **1–11** in leaf extracts of *V. myrtillus* and *V. arctostaphylos* in mg/g of dry weight ± S.D.

Compound	*V. myrtillus*	*V. arctostaphylos*
4-*O*-Caffeoylquinic acid	<0.01	8.01 ± 0.14
5-*O*-Caffeoylquinic acid	226.85 ± 5.21	105.32 ± 2.41
Caffeic acid	<0.01	4.29 ± 0.08
Quercetin 3-*O*-rutinoside (rutin)	<0.01	1.46 ± 0.03
Quercetin 3-*O*-galactoside (hyperoside)	4.69 ± 0.09	3.99 ± 0.06
Quercetin 3-*O*-glucoside (isoquercitrin)	12.02 ± 0.24	2.38 ± 0.05
Quercetin 3-*O*-arabinopyranoside (guaiaverin)	1.34 ± 0.02	1.29 ± 0.02
Quercetin 3-*O*-arabinofuranoside (avicularin)	<0.01	<0.01
Quercetin 3-*O*-rhamnoside (quercitrin)	2.77 ± 0.05	1.23 ± 0.02
Quercetin 3-*O*-(6″-acetyl)-glucoside	<0.01	0.70 ± 0.01
4,5-Di-*O*-caffeoylquinic acid	<0.01	<0.01

3.2. *Neuroprotective Activity of V. myrtillus and V. arctostaphylos Leaf Extracts*

The study of neuroprotective activity of extracts from *V. myrtillus* and *V. arctostaphylos* showed that their administration to animals in the post-ischemic period contributed to an increase in the level of cerebral blood flow (Figure 4) compared to that of untreated rats by 99.3% ($p < 0.05$) and 115.0% ($p < 0.05$), respectively, while the use of the EGB761 reference increased this indicator by 79.3% ($p < 0.05$). Of note, in animals that received extraction from *V. arctostaphylos*, the level of cerebral blood flow was higher in the rats injected with the reference by 19.9% ($p < 0.05$).

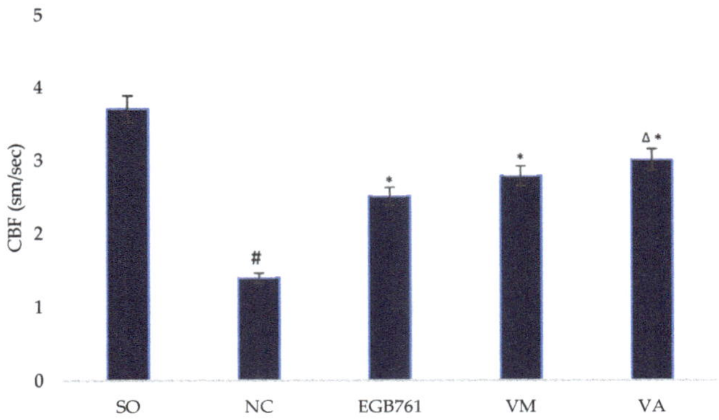

Figure 4. The effect of *V. myrtillus* and *V. arctostaphylos* leaf extracts on the cerebral hemodynamics in rats with cerebral ischemia. Experimental groups: SO—sham-operated animals; NC—negative control animals; EGB761—EGB761 reference remedy group; VM—*V. myrtillus* group; VA—*V. arctostaphylos* group. Hashtag (#) indicates significant difference ($p < 0.05$) vs. the SO group; asterisk (*) indicates significant difference ($p < 0.05$) vs. the NC group; delta (Δ) indicates significant difference ($p < 0.05$) vs. the EGB761 group.

Additionally, the use of the studied extracts led to a decrease in the brain necrosis zone (Figure 5), while against the background of the administration of the extract from *V. myrtillus*, this indicator decreased by 28.2% ($p < 0.05$), and for *V. arctostaphylos*, it decreased by 42.2% ($p < 0.05$). At the same time, the use of the reference contributed to a decrease in the brain necrosis zone by 34.3% ($p < 0.05$), which was higher than the same indicator for the group of animals that were injected with extraction from *V. arctostaphylos* by 12.1% ($p < 0.05$).

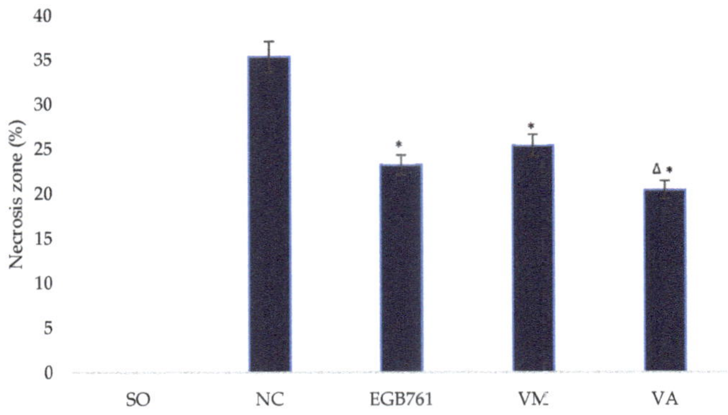

Figure 5. The effect of *V. myrtillus* and *V. arctostaphylos* leaf extracts on the necrosis zone in rats with cerebral ischemia. Experimental groups: SO—sham-operated animals; NC—negative control animals; EGB761—EGB761 reference remedy group; VM—*V. myrtillus* group; VA—*V. arctostaphylos* group. Asterisk (*) indicates significant difference ($p < 0.05$) vs. the NC group; delta (Δ) indicates significant difference ($p < 0.05$) vs. the EGB761 group.

3.3. Antioixdant Potential of V. myrtillus and V. arctostaphylos Leaf Extracts

The study of changes in the antioxidant balance of brain tissue with the administration of the studied extracts showed that the use of extracts from *V. myrtillus* and *V. arctostaphylos* contributed to an increase in the activity of SOD by 4.1% ($p < 0.05$) and 76.4% ($p < 0.05$), respectively, and reduced the content of TBARS by 55.1% ($p < 0.05$) and 64.3% ($p < 0.05$), respectively (Table 4).

Table 4. Effect of *V. myrtillus* and *V. arctostaphylos* leaf extracts on activity of mitochondrial enzymes and TBARS in rats with cerebral ischemia.

Experimental Group	SOD, U/Protein mg	COX, U/Protein mg	SDH, U/Protein mg	TBARS, μM/Protein mg
Sham-operated animals	300.5 ± 7.1(29)	4.3 ± 0.0(47)	2.7 ± 0.062	2.4 ± 0.124
Negative control group	125.6 ± 9.5(41) #	2.1 ± 0.0(14) #	1.1 ± 0.097 #	9.8 ± 0.663 #
EGB761 reference group	190.5 ± 11.2(84) *	2.5 ± 0.1(17) *	1.6 ± 0.018 *	5.2 ± 0.541 *
V. myrtillus group	181.0 ± 10.9(37) *	2.6 ± 0.0(37) *	1.5 ± 0.09 *	4.4 ± 0.364 *
V. arctostaphylos group	221.5 ± 10.2(32) *	3.0 ± 0.0(58) * Δα	2.1 ± 0.071 * Δα	3.5 ± 0.193 *

Hashtag (#) indicates significant difference ($p < 0.05$) vs. sham-operated animals' group; asterisk (*) indicates significant difference ($p < 0.05$) vs. negative control group; delta (Δ) indicates significant difference ($p < 0.05$) vs. the EGB761 reference group; alpha (α) indicates significant difference ($p < 0.05$) vs. the *V. myrtillus* group.

At the same time, the use of EGB761 contributed to an increase in SOD activity by 51.7% ($p < 0.05$) and a decrease in TBARS by 46.9% ($p < 0.05$), which was statistically significantly lower than similar indicators of animals receiving extraction from *V. arctostaphylos*. Analysis of changes in the activity of mitochondrial enzymes showed that when extracts from *V. myrtillus* L. and *V. arctostaphylos* were used, as well as the reference, the activity of COX increased (relative to the NC group of rats) by 23.8% ($p < 0.05$), 42.9% ($p < 0.05$), and 19.0% ($p < 0.05$), respectively, whereas SDH activity increased by 36.4% ($p < 0.05$), 45.5% ($p < 0.05$), and 90.9% ($p < 0.05$). Of note, in rats injected with the extract from *V. arctostaphylos*, the activities of both COX and SDH were higher than in animals treated with the reference and extract from *V. myrtillus* ($p < 0.05$).

4. Discussion

A detailed study of plant metabolites opens possibilities for selection of precise and accurate methods to include in pharmacopeial practice. The results obtained in this study show that the major and marker compound for both *Vaccinium* species was 5-*O*-caffeoylquinic acid (chlorogenic acid). In predicting the pharmacological activity, phenolic acids are mostly represented by chlorogenic acid isomers (4-*O*-caffeoylquinic and 5-*O*-caffeoylquinic acids). Moreover, the main aglycone of flavonoids is quercetin, and its glycosides are presented by monosides (quercetin-3-*O*-galactoside, quercetin-3-*O*-glucoside, quercetin-3-*O*-arabinopyranoside, quercetin-3-*O*-arabinofuranoside, quercetin-3-*O*-rhamnoside, and quercetin-3-*O*-(6″-acetyl)-glucoside) and by only one bioside (quercetin-3-*O*-rutinoside). Chlorogenic acids and quercetin glycosides are known neuroprotective agents. Thus, chlorogenic acid shows a protective effect on the nitrosative stress induced by sodium nitroprusside [48]. In vivo, chlorogenic acid decreased the infarct area in the hypoxic–ischemic brains of rats [49]. Additionally, chlorogenic acid may be a potential agent to treat Parkinson's disease in the apoptosis-mediated neuronal senescence of dopaminergic neurons [50]. The neuroprotective effects of quercetin may be used in the treatment of Alzheimer's disease via its antioxidant properties and providing an inhibitory effect of fibril formation of amyloid-β aggregates [51]. These effects are useful in the therapy of pediatric neurological diseases such as central nervous system tumors, autism, and hyperactivity disorder [52]. Quercetin and rutin show neuroprotective effects in the cerebral ischemia model in rats by its 4-oxo group and 2,3-double bond in the C ring [53]. Thus, the potential neuroprotective effects of *Vaccinium* extracts may be related to the significant amount of these phenolics in the *Vaccinium* leaves.

Increasingly, natural remedies are being used as adjuvant neuroprotective therapy for cerebrovascular disorders [54]. The neuroprotective effects of traditional Chinese medicine (for example, ginseng preparations, as well as *Scutellaria baicalensis*, *Withania somnifera*, *Phyllanthus emblica*, and *Viscum album* [55]) are widely known. It has been established that the use of phytomedicines has a pleiotropic effect on brain tissue, while eliminating almost all main pathogenetic pathways of the "ischemic cascade": neuroinflammation, apoptosis, energy deficiency, glutamate-calcium excitotoxicity, hyperlactatemia, and acidosis [55]. However, as a rule, remedies based on phytocompositions are known as effective antioxidants [55,56].

The concept of phytotherapeutic intervention in ischemic stroke is currently being actively studied; however, there is no doubt that this approach is only auxiliary and cannot fully replace the "gold standard" of treatment of ischemic brain damage—effective thrombolysis [57]. The rapidly developing direction of phytotherapeutic neuroprotection makes it necessary to search for new drugs that meet the basic postulates of this area of medicine and pharmacology [58]. In this work, possible neuroprotective properties of extracts obtained from two species of representatives of the heather family, *V. myrtillus* and *V. arctostaphylos*, were studied. Blueberry-based products are primarily known as effective biologically active additives with pronounced antioxidant activity and tropicity in relation to blood vessels and the cardiovascular system as a whole, which has made it possible to use these products in the correction of vascular complications of diabetes mellitus, atherosclerosis, and glaucoma [58]. At the same time, a similar therapeutic profile of extracts from blueberries suggests the presence of neuroprotective activity. During the study, it was found that the administration of extracts obtained from *V. myrtillus* and *V. arctostaphylos* to animals with focal ischemia contributed to the improvement of cerebral hemodynamics in the post-ischemic period, which may be associated with the restoration of the function of collateral blood supply to the ischemic zone. A decrease in the necrotic focus was also shown in rats receiving the analyzed extracts compared with untreated animals. Because one of the significant mechanisms of ischemic brain damage is the development of oxidative stress, we studied the effect of extracts from blueberries on changes in the parameters characterizing the development of this pathological process—the activity of superoxide dismutase and the final concentration of peroxidation products represented by TBARS [36]. As a result, the introduction of extracts from *V. myrtillus* and *V. arctostaphylos* increased the activity of superoxide dismutase and reduced the intensity of free radical lipid oxidation reactions. An important aspect of brain ischemia is a violation of the energy state of cells with a sharp drop in ATP synthesis and the induction of irreversible processes in the form of necroptosis or apoptosis [59]. In this regard, the effect of extracts from *V. myrtillus* and *V. arctostaphylos* on the activity of two integral enzymatic indicators of the metabolic status of the cell, cytochrome c oxidase and succinate dehydrogenase, was evaluated. The administration of the analyzed extracts led to an increase in the activity of these enzymes, and the effect of the administration of the extract obtained from *V. arctostaphylos* exceeded that of the reference, the classical neuroprotective herbal remedy *Ginkgo biloba* extract (EGB761). Thus, based on the overall results of the set of pharmacological tests of extracts from *V. myrtillus* and *V. arctostaphylos*, it can be assumed that these objects have neuroprotective activity, which can be realized through antioxidant and metabolic action. At the same time, a more promising therapeutic agent is the extraction from *V. arctostaphylos*, the use of which led to a statistically significantly better pharmacological response than the introduction of the reference and extract obtained from *V. myrtillus*.

Author Contributions: Conceptualization, A.A.S. and D.N.O.; methodology, A.A.S.; software, A.A.S.; validation, D.N.O., D.I.P. and E.R.G.; formal analysis, V.N.B., E.R.G. and M.V.L.; investigation, A.A.S., D.N.O., D.I.P., E.R.G., M.V.L.; resources, D.N.O., E.R.G. and M.V.L.; data curation, D.I.P., V.N.B. and M.V.L.; writing—original draft preparation, A.A.S., E.R.G.; writing—review and editing, D.N.O.; visualization, A.A.S., E.R.G.; supervision, D.N.O.; project administration, A.A.S.; funding acquisition, D.N.O. All authors have read and agreed to the published version of the manuscript.

Funding: This research was funded by the Ministry of Education and Science of Russia, grant numbers FSRG-2020-0019, 121030100227-7.

Institutional Review Board Statement: The study was conducted according to the guidelines of the Declaration of Helsinki, and approved by the Russian Health Ministry (protocol code 708H, 23 August 2010) and the Ethics Committee of Institute of General and Experimental Biology (protocol code 10, 20 August 2021).

Informed Consent Statement: Not applicable.

Data Availability Statement: Data are contained within the article.

Acknowledgments: The authors acknowledge the Buryat Research Resource Center for the technical support in chromatographic and mass-spectrometric research.

Conflicts of Interest: The authors declare no conflict of interest. The funders had no role in the design of the study, in the collection, analyses, or interpretation of data, in the writing of the manuscript, or in the decision to publish the results.

References

1. WHO: The World Flora Online. Available online: http://www.worldfloraonline.org/ (accessed on 12 November 2022).
2. Takhtadjan, A. *Flowering Plants*, 2nd ed.; Springer Science & Business Media: Berlin/Heidelberg, Germany, 2009; pp. 188–193, ISBN 978-1-4020-9608-2.
3. Czerepanov, S.K. *Vascular Plants of Russia and Neighboring Countries*; Word and Family-95: Saint-Petersburg, Russia, 1995; p. 424.
4. Kasotea, D.M.; Duncanb, G.J.; Neacsua, M.; Russella, W.R. Rapid method for quantification of anthocyanidins and anthocyanins in human biological samples. *Food Chem.* **2019**, *290*, 56–63. [CrossRef]
5. Chan, S.W.; Tomlinson, B. Effects of bilberry supplementation on metabolic and cardiovascular disease risk. *Molecules* **2020**, *25*, 1653. [CrossRef]
6. Satoh, Y.; Ishihara, K. Investigation of the antimicrobial activity of bilberry (*Vaccinium myrtillus* L.) extract against periodontopathic bacteria. *J. Oral Biosci.* **2020**, *62*, 169–174. [CrossRef]
7. Sezer, E.D.; Oktay, L.M.; Karadadaş, E.; Memmedov, H.; Gunel, N.S.; Sözmen, E. Assessing anticancer potential of blueberry flavonoids, quercetin, kaempferol, and gentisic acid, through oxidative stress and apoptosis parameters on HCT-116 cells. *J. Med. Food.* **2019**, *22*, 1118–1126. [CrossRef]
8. Vaneková, Z.; Vanek, M.; Škvarenina, J.; Nagy, M. The influence of local habitat and microclimate on the levels of secondary metabolites in slovak bilberry (*Vaccinium myrtillus* L.) Fruits. *Plants* **2020**, *9*, 436. [CrossRef]
9. Zheng, W.; Wang, S.Y. Oxygen radical absorbing capacity of phenolics in blueberries, cranberries, chokeberries, and lingonberries. *J. Agric. Food Chem.* **2003**, *51*, 502–509. [CrossRef]
10. Viljanen, K.; Kylli, P.; Kivikari, R.; Heinonen, M. Inhibition of protein and lipid oxidation in liposomes by berry phenolics. *J. Agric. Food Chem.* **2004**, *52*, 7419–7424. [CrossRef]
11. Puupponen-Pimiä, R.; Nohynek, L.; Hartmann-Schmidlin, S.; Kähkönen, M.; Heinonen, M.; Määttä-Riihinen, K.; Oksman-Caldentey, K.M. Berry phenolics selectively inhibit the growth of intestinal pathogens. *J. Appl. Microbiol.* **2005**, *98*, 991–1000. [CrossRef]
12. Ayaz, F.A.; Ayaz, S.H.; Gruz, J.; Novak, O.; Strnad, M. Separation, characterization, and quantitation of phenolic acids in a little-known blueberry (*Vaccinium arctostaphylos* L.) fruit by HPLC-MS. *J. Agric. Food Chem.* **2005**, *53*, 8116–8122. [CrossRef]
13. Lätti, A.K.; Kainulainen, P.S.; Hayirlioglu-Ayaz, S.; Ayaz, F.A.; Riihinen, K.R. Characterization of anthocyanins in Caucasian blueberries (*Vaccinium arctostaphylos* L.) native to Turkey. *J. Agric. Food Chem.* **2009**, *57*, 5244–5249. [CrossRef]
14. Chkhikvishvili, I.D.; Kharebava, G.I. Chicoric and chlorogenic acids in plant species from Georgia. *Appl. Biochem. Microbiol.* **2001**, *37*, 188–191. [CrossRef]
15. Mohtashami, R.; Huseini, H.F.; Nabati, F.; Hajiaghaee, R.; Kianbakht, S. Effects of standardized hydro-alcoholic extract of *Vaccinium arctostaphylos* leaf on hypertension and biochemical parameters in hypertensive hyperlipidemic type 2 diabetic patients: A randomized, double-blind and placebo-controlled clinical trial. *Avicenna J. Phytomed.* **2019**, *9*, 44–53.
16. Nickavar, B.; Amin, G. Enzyme assay guided isolation of an α-amylase inhibitor flavonoid from *Vaccinium arctostaphylos* leaves. *Iran J. Pharm. Res.* **2011**, *10*, 849–853.
17. Yang, L.; Wang, Z.M.; Wang, Y.; Li, R.S.; Wang, F.; Wang, K. Phenolic constituents with neuroprotective activities from *Hypericum wightianum*. *Phytochemistry* **2019**, *165*, 112049. [CrossRef]
18. Xie, Y.; Yang, W.; Tang, F.; Chen, X.; Ren, L. Antibacterial activities of flavonoids: Structure-activity relationship and mechanism. *Curr. Med. Chem.* **2015**, *22*, 132–149. [CrossRef]
19. Aryal, S.; Skinner, T.; Bridges, B.; Weber, J.T. The pathology of Parkinson's disease and potential benefit of dietary polyphenols. *Molecules* **2020**, *25*, 4382. [CrossRef]

20. Kempuraj, D.; Thangavel, R.; Kempuraj, D.D.; Ahmed, M.E.; Selvakumar, G.P.; Raikwar, S.P.; Zaheer, S.A.; Iyer, S.S.; Govindarajan, R.; Chandrasekaran, P.N.; et al. Neuroprotective effects of flavone luteolin in neuroinflammation and neurotrauma. *Biofactors* **2021**, *47*, 190–197. [CrossRef]
21. Zhou, Y.; Zhang, S.; Fan, X. Role of polyphenols as antioxidant supplementation in ischemic stroke. *Oxid. Med. Cell Longev.* **2021**, *2021*, 5471347. [CrossRef]
22. Xie, Y.; Wang, H.; He, Z. Recent advances in polyphenols improving vascular endothelial dysfunction induced by endogenous toxicity. *J. Appl. Toxicol.* **2021**, *41*, 701–712. [CrossRef]
23. Gao, Q.; Dong, J.Y.; Cui, R.; Muraki, I.; Yamagishi, K.; Sawada, N.; Iso, H.; Tsugane, S.; Japan Public Health Center-based Prospective study group. consumption of flavonoid-rich fruits, flavonoids from fruits and stroke risk: A prospective cohort study. *Br. J. Nutr.* **2021**, *126*, 1717–1724. [CrossRef]
24. Kalt, W.; Cassidy, A.; Howard, L.R.; Krikorian, R.; Stull, A.J.; Tremblay, F.; Zamora-Ros, R. Recent research on the health benefits of blueberries and their anthocyanins. *Adv. Nutr.* **2020**, *11*, 224–236. [CrossRef]
25. Rabinstein, A.A. Update on treatment of acute ischemic stroke. *Continuum* **2020**, *26*, 268–286. [CrossRef]
26. Ting, H.C.; Chang, C.Y.; Lu, K.Y.; Chuang, H.M.; Tsai, S.F.; Huang, M.H.; Liu, C.A.; Lin, S.Z.; Harn, H.J. Targeting cellular stress mechanisms and metabolic homeostasis by chinese herbal drugs for neuroprotection. *Molecules* **2018**, *23*, 259. [CrossRef]
27. Riihinen, K.; Jaakola, L.; Kärenlampi, S.; Hohtola, A. Organ-specific distribution of phenolic compounds in bilberry (*Vaccinium myrtillus*) and 'northblue' blueberry (*Vaccinium corymbosum* × *V. angustifolium*). *Food Chem.* **2008**, *110*, 156–160. [CrossRef]
28. Rahman, M.M.; Ichiyanagi, T.; Komiyama, T.; Sato, S.; Konishi, T. Effects of anthocyanins on psychological stress-induced oxidative stress and neurotransmitter status. *J. Agric. Food Chem.* **2008**, *56*, 7545–7550. [CrossRef]
29. Olennikov, D.N.; Shamilov, A.A. New compounds from *Vaccinium vitis-idaea*. *Chem. Nat. Compd.* **2022**, *58*, 240–244. [CrossRef]
30. Olennikov, D.N.; Shamilov, A.A. Catechin-O-rhamnosides from *Vaccinium vitis-idaea* stems. *Chem. Nat. Compd.* **2022**, *58*, 269–273. [CrossRef]
31. Shamilov, A.A.; Olennikov, D.N.; Pozdnyakov, D.I.; Bubenchikova, V.N.; Garsiya, E.R. Investigation of phenolic compounds at the leaves and shoots *Arctostaphylos* spp. and their antioxidant and antityrosinase activities. *Nat. Prod. Res.* **2022**, *36*, 6312–6317. [CrossRef]
32. Kashchenko, N.I.; Jafarova, G.S.; Isaev, J.I.; Olennikov, D.N.; Chirikova, N.K. Caucasian dragonheads: Phenolic compounds, polysaccharides, and bioactivity *of Dracocephalum austriacum* and *Dracocephalum botryoides*. *Plants* **2022**, *11*, 2126. [CrossRef]
33. Pozdnyakov, D.I. 4-Hydroxy-3,5-di-tret-butyl cinnamic acid restores the activity of the hippocampal mitochondria in rats under permanent focal cerebral ischemia. *Iran. J. Basic. Med. Sci.* **2021**, *24*, 1590–1601. [CrossRef]
34. Wang, N.; Chen, X.; Geng, D.; Huang, H.; Zhou, H. *Ginkgo biloba* leaf extract improves the cognitive abilities of rats with D-galactose induced dementia. *J. Biomed. Res.* **2013**, *27*, 29–36. [CrossRef] [PubMed]
35. Voronkov, A.V.; Dyakova, I.N.; Pozdnyakov, D.I. The influence of natural compounds of polyphenolic structure on the vasodilating function of the vascular endothelium of the rat brain in conditions of its focal ischemia. *Eksp. Klin. Farmakol.* **2016**, *79*, 7–9. [PubMed]
36. Janero, D.R. Malondialdehyde and thiobarbituric acid-reactivity as diagnostic indices of lipid peroxidation and peroxidative tissue injury. *Free Rad. Biol. Med.* **1990**, *9*, 515–540. [CrossRef] [PubMed]
37. Woolliams, J.A.; Wiener, G.; Anderson, P.H.; McMurray, C.H. Variation in the activities of glutathione peroxidase and superoxide dismutase and in the concentration of copper in the blood in various breed crosses of sheep. *Res. Vet. Sci.* **1983**, *34*, 253–256. [CrossRef] [PubMed]
38. Wang, H.; Huwaimel, B.; Verma, K.; Miller, J.; Germain, T.M.; Kinarivala, N.; Pappas, D.; Brookes, P.S.; Trippier, P.C. Synthesis and antineoplastic evaluation of mitochondrial complex II (succinate dehydrogenase) inhibitors derived from atpenin A5. *ChemMedChem* **2017**, *12*, 1033–1044. [CrossRef]
39. Li, Y.; D'Aurelio, M.; Deng, J.H.; Park, J.S.; Manfredi, G.; Hu, P.; Lu, J.; Bai, Y. An assembled complex IV maintains the stability and activity of complex I in mammalian mitochondria. *J. Biol. Chem.* **2007**, *282*, 17557–17562. [CrossRef]
40. Ștefănescu, B.E.; Szabo, K.; Mocan, A.; Crişan, G. Phenolic compounds from five Ericaceae species leaves and their related bioavailability and health benefits. *Molecules* **2019**, *24*, 2046. [CrossRef]
41. Mzhavanadze, V.V.; Targamadze, I.L.; Dranik, L.I. Polyphenols of the leaves of *Vaccinium arctostaphylos*. *Chem. Nat. Compd.* **1971**, *7*, 536–537. [CrossRef]
42. Liu, P.; Lindstedt, A.; Markkinen, N.; Sinkkonen, J.; Suomela, J.-P.; Yang, B. Characterization of metabolite profiles of leaves of bilberry (*Vaccinium myrtillus* L.) and lingonberry (*Vaccinium vitis-idaea* L.). *J. Agric. Food Chem.* **2014**, *62*, 12015–12026. [CrossRef]
43. Mzhavanadze, V.V.; Targamadze, I.L.; Dranik, L.I. Phenolic compounds of the leaves of *Vaccinium arctostaphylos*. *Chem. Nat. Compd.* **1972**, *8*, 125–126. [CrossRef]
44. Tadić, V.M.; Nešić, I.; Martinović, M.; Rój, E.; Brašanac-Vukanović, S.; Maksimović, S.; Žugić, A. Old plant, new possibilities: Wild bilberry (*Vaccinium myrtillus* L., Ericaceae) in topical skin preparation. *Antioxidants* **2021**, *10*, 465. [CrossRef] [PubMed]
45. Ieri, F.; Martini, S.; Innocwnti, M.; Mulinacci, N. Phenolic distribution in liquid preparations of *Vaccinium myrtillus* L. and *Vaccinium vitis idaea* L. *Phytochem. Anal.* **2013**, *24*, 467–475. [CrossRef] [PubMed]
46. Hasaloo, T.; Sepehrifar, R.; Hajimehdipoor, H. Levels of phenolic compounds and their effects on antioxidant capacity of wild *Vaccinium arctostaphylos* L. (Qare-Qat) collected from different regions of Iran. *Turk. J. Biol.* **2011**, *35*, 13. [CrossRef]

47. Oszmianski, J.; Wojdyło, A.; Gorzelany, J.; Kapusta, I. Identification and characterization of low molecular weight polyphenols in berry leaf extracts by HPLC-DAD and LC-ESI/MS). *J. Agric. Food Chem.* **2011**, *59*, 12830–12835. [CrossRef]
48. Taram, F.; Winter, A.N.; Linseman, D.A. Neuroprotection comparison of chlorogenic acid and its metabolites against mechanistically distinct cell death-inducing agents in cultured cerebellar granule neurons. *Brain Res.* **2016**, *1648*, 69–80. [CrossRef]
49. Zheng, Y.; Li, L.; Chen, B.; Fang, Y.; Lin, W.; Zhang, T.; Feng, X.; Tao, X.; Wu, Y.; Fu, X.; et al. Chlorogenic acid exerts neuroprotective effect against hypoxia-ischemia brain injury in neonatal rats by activating Sirt1 to regulate the Nrf2-NF-κB signaling pathway. *Cell Commun. Signal.* **2022**, *20*, 84. [CrossRef]
50. Singh, S.S.; Rai, S.N.; Birla, H.; Zahra, W.; Rathore, A.S.; Dilnashin, H.; Singh, R.; Singh, S.P. Neuroprotective Effect of chlorogenic acid on mitochondrial dysfunction-mediated apoptotic death of DA neurons in a Parkinsonian mouse model. *Oxid. Med. Cell Longev.* **2020**, *2020*, 6571484. [CrossRef]
51. Khan, H.; Ullah, H.; Aschner, M.; Cheang, W.S.; Akkol, E.K. Neuroprotective effects of quercetin in Alzheimer's disease. *Biomolecules* **2019**, *10*, 59. [CrossRef]
52. Alvarez-Arellano, L.; Salazar-García, M.; Corona, J.C. Neuroprotective effects of quercetin in pediatric neurological diseases. *Molecules* **2020**, *25*, 5597. [CrossRef]
53. Pu, F.; Mishima, K.; Irie, K.; Motohashi, K.; Tanaka, Y.; Orito, K.; Egawa, T.; Kitamura, Y.; Egashira, N.; Iwasaki, K.; et al. Neuroprotective effects of quercetin and rutin on spatial memory impairment in an 8-arm radial maze task and neuronal death induced by repeated cerebral ischemia in rats. *J. Pharmacol. Sci.* **2007**, *104*, 329–334. [CrossRef]
54. Gaire, B.P. Herbal medicine in ischemic stroke: Challenges and prospective. *Chin. J. Integr. Med.* **2018**, *24*, 243–246. [CrossRef] [PubMed]
55. Mishra, A.; Mishra, P.S.; Bandopadhyay, R.; Khurana, N.; Angelopoulou, E.; Paudel, Y.N.; Piperi, C. Neuroprotective potential of chrysin: Mechanistic insights and therapeutic potential for neurological disorders. *Molecules* **2021**, *26*, 6456. [CrossRef] [PubMed]
56. Wang, J.; Zhang, J.; Li, S.; Huang, C.; Xie, Y.; Cao, Y. Anthocyanins decrease the internalization of TiO2 nanoparticles into 3D Caco-2 spheroids. *Food Chem.* **2020**, *331*, 127360. [CrossRef] [PubMed]
57. Wu, S.; Wu, B.; Liu, M.; Chen, Z.; Wang, W.; Anderson, C.S.; Sandercock, P.; Wang, Y.; Huang, Y.; Cui, L.; et al. China stroke study collaboration. Stroke in China: Advances and challenges in epidemiology, prevention, and management. *Lancet Neurol.* **2019**, *18*, 394–405. [CrossRef]
58. Silva, S.; Costa, E.M.; Veiga, M.; Morais, R.M.; Calhau, C.; Pintado, M. Health promoting properties of blueberries: A review. *Crit. Rev. Food Sci. Nutr.* **2020**, *60*, 181–200. [CrossRef]
59. He, Z.; Ning, N.; Zhou, Q.; Khoshnam, S.E.; Farzaneh, M. Mitochondria as a therapeutic target for ischemic stroke. *Free Radic. Biol. Med.* **2020**, *146*, 45–58. [CrossRef]

life MDPI

Article

Variation in Phenolic Profile, Antioxidant, and Anti-Inflammatory Activities of *Salvadora oleoides* Decene. and *Salvadora persica* L. Fruits and Aerial Part Extracts

Arifa Khanam [1], Ashfaq Ahmad [2], Neelam Iftikhar [1], Qasim Ali [3], Tabinda Fatima [4], Farhan Khashim Alswailmi [2], Abdullah Ijaz Hussain [1,*], Sulaiman Mohammed Abdullah Alnasser [5,*] and Jamshaid Akhtar [6]

1 Department of Chemistry, Government College University Faisalabad, Faisalabad 38000, Pakistan
2 Department of Pharmacy practice, College of Pharmacy, University of Hafr Al Batin, Hafr Al Batin 31991, Saudi Arabia
3 Department of Botany, Government College University Faisalabad, Faisalabad 38000, Pakistan
4 Department of Pharmaceutical Chemistry, College of Pharmacy, University of Hafr Al Batin, Hafr Al Batin 31991, Saudi Arabia
5 Department of Pharmacology and Toxicology, Unaizah College of Pharmacy, Qassim University, Buraidah 51452, Saudi Arabia
6 Department of Internal Medicine, Allama Lqbal Medical College, Lahore 54700, Pakistan
* Correspondence: abdullahijaz@gcuf.edu.pk (A.I.H.); sm.alnasser@qu.edu.sa (S.M.A.A.)

Abstract: (1) Background: The objective of this study was to investigate the potential of *Salvadora oleoides* (*S. oleoides*) and *Salvadora persica* (*S. persica*) polyphenols as antioxidant and anti-inflammatory agents. (2) Methods: Aerial parts and fruits of *S. oleoides* and *S. persica* were collected from the periphery of District Bhakkar, Punjab, Pakistan. Methanol extracts were prepared using the Soxhlet extraction technique. Extract yield varied from 8.15 to 19.6 g/100 g dry plant material. RP-HPLC revealed the detection of thirteen phenolic aids and five flavonoids. Gallic acid, hydroxy benzoic acid, chlorogenic acid, and cinamic acid were the major phenolic acids, whereas catechin, rutin, and myricetin were the flavonoids detected. (3) Results: Maximum total phenolic contents (TPCs) (22.2 mg/g of dry plant material) and total flavonoid contents (TFCs) (6.17 mg/g of dry plant material) were found in the fruit extract of *S. persica*, and the minimum TPC (11.9 mg/g) and TFC (1.72 mg/g) were found in the aerial part of *S. oleoides*. The fruit extract of *S. persica* showed the highest DPPH radical scavenging activity. In vivo anti-inflammatory activity of all the extracts was performed on albumin-induced rat paw edema that was comparable with the standard indomethacin; *S. persica* fruit extract showed remarkable anti-inflammatory activity. Analgesic activity of aerial part and fruit extracts of *S. oleoides* and *S. persica* was investigated using a mouse model, and the results showed that maximum possible analgesia of fruit extracts of *S. persica* was 53.44%, which is better than the PC group (52.98%). (4) Conclusions: The variations in the antioxidant, anti-inflammatory, and analgesic activities of methanolic extracts of *S. oleoides* and *S. persica* were found to be significant, and they have therapeutic potential as antioxidant, analgesic, and anti-inflammatory agents.

Keywords: phenolic acid; flavonoids; DPPH radical scavenging activity; anti-inflammatory activity; maximum possible analgesia

check for updates

Citation: Khanam, A.; Ahmad, A.; Iftikhar, N.; Ali, Q.; Fatima, T.; Alswailmi, F.K.; Hussain, A.I.; Alnasser, S.M.A.; Akhtar, J. Variation in Phenolic Profile, Antioxidant, and Anti-Inflammatory Activities of *Salvadora oleoides* Decene. and *Salvadora persica* L. Fruits and Aerial Part Extracts. *Life* **2022**, *12*, 1446. https://doi.org/10.3390/life12091446

Academic Editor: Jianfeng Xu

Received: 30 August 2022
Accepted: 15 September 2022
Published: 18 September 2022

1. Introduction

Reactive oxygen species produced during different metabolic activities cause oxidative damages that are responsible for many diseases including, Parkinson's and Alzheimer's diseases, tumor formation, heart diseases, nervous disorders, pulmonary disorders, rheumatic premature aging, and inflammatory diseases [1,2]. Inflammation has multiple actions in the process of growth and differentiation, as well as in lymphoid and non-lymphoid cells, and regulates their production during injury and infections [3]. Chronic inflammation can cause the destruction of tissues and cells [4]. Antioxidants, especially natural compounds, can combat against oxidative stress and inflammation that provide relief against degenerative diseases [5–7].

The Asia-Pacific region is rich in plant genetic resources, including less known food plants and underutilized species that serve as a means of survival during times of famine, shocks, drought, and risks. They also contribute immensely to family food security, which can supplement nutritional requirements due to their better nutritional value [8]. Furthermore, these medicinal plants are potential sources of natural products that are generally recognized as safe (GRAS) [7]. The standardized extracts of these plants/isolated compounds, especially the polyphenols, provide unlimited opportunities for new drug discoveries [5,6,9]. Huge proportions of the global population are now more dependent on natural and alternative traditional medicines as primary health care [9,10].

The family Salvadoraceae has some underutilized species that are little known, despite their wide distribution in arid areas around the world [11]. In Pakistan, the genera Salvadora has two underutilized species, i.e., *Salvadora persica (S. persica)* and *Salvadora oleoides (S. oleoides)* [12]. *S. persica* (arak, jhak, pilu) is a small tree or shrub known by the name miswak [13]. Pharmacological studies indicated that the *S. persica* plant possesses alexiteric, analgesic, anti-inflammatory, anti-microbial, anti-plaque, astringent, diuretic aphrodisiac anti-pyretic, and bitter stomachic activities [14]. In folk literature, it is used by various populations worldwide, especially by Muslims, and it cited in the Holy Quran [15,16]. *S. oleoides* (Jall, Mitha Jall, Peelu, Pilu, Khabbar) is an oil-yielding medicinal and multi-purpose tree, which is reported to possess anti hypoglycemic, hypolipidemic, analgesic, anti-inflammatory and antimicrobial activities [17]. *S. oleoides* plant extract contains terpenoids, phenolic compounds, alkaloid, glycosides, and flavonoids, which are frequently used against many microbial activities [18].

Currently, due to the many side effects of synthetic medicines, the public is now again shifting towards natural products due to their GRAS status and negligible side effects. Hence, there is contemporary need to explore more underutilized species having strong traditional use against specific diseases [19]. To the best of our knowledge, no previous reports have been presented in the literature on the comparative evaluation of antioxidant and anti-inflammatory activities of aerial parts and fruits of *S. oleoides* and *S. persica*. Therefore, this study was planned to investigate the antioxidant and anti-inflammatory activities of polyphenol-rich fractions of *S. oleoides* and *S. persica* aerial parts and fruits. Moreover, simultaneous quantifications of phenolic acids and flavonoids were performed using reverse-phase high-performance liquid chromatography (Rp-HPLC). The in vivo anti-inflammatory and analgesic activities were assessed using albumin-induced inflammation in rat paw and the measurement of maximum possible analgesia in mice models.

2. Materials and Methods

2.1. Collection and Identification of Plant Materials

Fruits and aerial parts (10 kG) of *S. oleoides* Decene. and *S. persica* L. were collected from the periphery of District Bhakkar during summer in May and June, 2018. The samples were further authenticated by the Taxonomist, Department of Botany, Government College University Faisalabad, Pakistan. The samples were transferred to the Natural Product and Synthetic Chemistry Lab, Government College University, for further investigations.

2.2. Reagents and Chemicals

All the chemicals and reagents, including Folin–Ciocalteu phenolic reagent, gallic acid, 2, 2-diphenyl-1 picrylhydrazyl (DPPH), catechin, gallic acid, phenolic acid and flavonoid standards, ascorbic acid, and linoleic acid (75%), were purchased from Sigma-Aldrich Co., (St Louis, MO, USA). All the solvents, butylated hydroxyl toluene (BHT), hydrochloric acid, sodium carbonate, sodium hydroxide, aluminum chloride, and sodium nitrite, were purchased from Merck (Darmstadt, Germany). All chemicals used, including the solvents, were of analytical grade and used without further purification.

2.3. Preparation of Plant Extracts

Plant materials were dried in the shade and then powdered (80 mesh) using an electric grinder (BL 999SP, LG, Frankfurt, Germany). Extractions were carried out using the 500 mL Soxhlet extractor with absolute methanol, as reported previously [20]. Briefly, 50 g of powdered material was taken in a thimble and extracted with 300 mL of absolute methanol for 8 h on Soxhlet extraction apparatus. The extracts were then concentrated using a rotary evaporator (Eyela, SB-651, Rikakikai Co., Ltd., Tokyo, Japan) under reduced pressure and stored in a refrigerator at 4 °C for further studies. The percentage yield of each extract was determined using the formula given below.

$$\text{Yield}\left(\frac{\text{g}}{100\ \text{g}}\right) = \frac{Weight\ of\ dry\ extract}{Weight\ of\ dry\ plant\ material} \times 100$$

2.4. Qualitative and Quantitative Analysis of Phenolic Acids and Flavonoids Simultaneously

Standard stock solutions of all the phenolic acid and flavonoids available were prepared fresh by dissolving the compounds in methanol (10 mg/mL). Working solutions (0.2–1.0 mg/mL) were prepared and standard curves were constructed by plotting concentrations against peak areas. The extracts were prepared as reported previously and filtered through a 0.45 µm non-pyrogenic filter (Minisart, Satorius Stedim Biotech GmbH, Goettingen, Germany) prior to injection [7].

The HPLC analysis was performed with the Perkin Elmer system (Perkin Elmer, Japan) equipped with gradient model binary pump systems, and a UV/Visible detector. The injection mode was manual, and the degasser (DGU-20A5) system was intact. The column oven was installed and equipped with hypersil GOLD C18 column (250 × 4.6 mm internal diameter, 5 mm particle size) (Thermo Fischer Scientific Inc., Waltham, MA, USA) supported with a guard column and a non-linear gradient consisting of solvent A (acetonitrile: methanol, 70:30) and solvent B (water with 0.5% glacial acetic acid). The quantification was based on an external standard method, whereas the analytes were identified by matching the retention times and spiking the samples with the standard.

2.5. In Vitro Antioxidant Analysis

2.5.1. Total Phenolic and Total Flavonoid Contents

Total phenolic contents (TPCs) and total flavonoid contents (TFCs) of the prepared extracts were measured according to the Folin–Ciocalteu phenol reagent and aluminum chloride colorimetric assays, respectively, as reported by Hussain et al. [7]; several different dilutions (10–80 ppm) of gallic acid were prepared to create a calibration curve (Y = 0.0265X − 0.1834). Similarly, different dilutions of catechin (10–160 ppm) were prepared and the calibration curve was derived (Y = 0.0063X − 0.023). The TPC and TFC of the extracts were calculated from respective curves and reported as mg per gram of dry matter, expressed as gallic acid equivalent (GAE) and catechin equivalent (CE), respectively.

2.5.2. DPPH Radical Scavenging Assay

DPPH (2, 2-Diphenyl-1-picrylhydrazyl) radical scavenging activity was assayed by the method reported previously [7]. Briefly, 10 µg/mL concentrations of extracts and BHT were mixed with 2 mL of 90 µM DPPH. The solution was incubated at room temperature for half an hour. The absorbance was read at 517 nm, and scavenging in terms of the percentage was calculated as follows:

$$\text{Scavenging}(\%) = \frac{Absorbance\ of\ DPPH\ solution - Absorbance\ of\ sapmle\ solution}{Absorbance\ of\ DPPH\ solution} \times 100$$

2.6. *In Vivo Evaluation of Anti-Inflammatory and Analgesic Activities*

For in vivo experiments, all procedures were conducted in conformity with the international guidelines on the ethical use of animals, and the study was approved by Institutional Ethical Committee for Animal Care at Government College University Faisalabad (Study No 19680/IRB No 680).

2.6.1. Animals

Adult male Wistar rats (weighing approximately 130–160 g) and mice weighing approximately 30–40 g were collected from the animal house of the Department of Physiology, Government College University, Faisalabad. Before the start of experiment, the animals were adapted for a week at the standard conditions (26 °C ± 2 °C temperature; 40–60% ambient humidity). Rats were housed in elevated wire cages with free access to food and water. The intake of food was measured on a daily basis for each animal.

2.6.2. Anti-Inflammatory Activity

To examine the anti-inflammatory activity of *S. oleoides* and *S. persica* aerial parts and fruit extracts in rats, thirty-six albino male Westar rats were randomly divided into the following six groups, all having six rats in each group (n = 6).

NC GROUP: Normal control group, which received no treatment.
PC GROUP: Positive control group, which were supplemented with a dose of 10 mg/kg body weight (bw) of indomethacin served as a standard drug.
G1 GROUP: Treatment group-1, which were supplemented with 250 mg/kg bw extract of *S. oleoides* aerial parts.
G2 GROUP: Treatment group-2, which were supplemented with 250 mg/kg bw extract of *S. persica* aerial parts.
G3 GROUP: Treatment group-3, which were supplemented with 250 mg/kg bw extract of *S. oleoides* fruits.
G4 GROUP: Treatment group-4, which were supplemented with 250 mg/kg bw extract of *S. persica* fruits.

All the rat groups were provided water ad libitum and normal feed (approx. 20 g/rat/day). For measuring the anti-inflammatory activity, extract doses of *S. oleoides* and *S. persica* were administrated orally through gavage feeding tube (16–18 cm) for 7 days, while positive controls were orally administered indomethacin 10 mg/kg bw for 7 days before the induction of inflammation [21]. After 7 days of treatment, inflammation was induced in the right paw by injecting the 0.1 mL egg albumin in each treatment and PC group. The inflammation was measured for 4 h by screw gauge. The percentage Edema Inhibition was calculated using the NC group as a standard.

2.6.3. Analgesic Activity

To evaluate the analgesic activity, the mice were randomly divided into the following six groups, all having five rats in each group (n = 5).

NC GROUP: Normal Control, which received no treatment.
PC GROUP: Positive Control, which were supplemented with a dose of 10 mg/kg of Diclofenac sodium served as standard.
G1 GROUP: Treatment group-1, which were supplemented with 250 mg/kg bw extract of *S. oleoides* aerial parts
G2 GROUP: Treatment group-2, which were supplemented with 250 mg/kg bw extract of *S. persica* aerial parts
G3 GROUP: Treatment group-3, which were supplemented with 250 mg/kg bw extract of *S. oleoides* fruits
G4 GROUP: Treatment group-4, which were supplemented with 250 mg/kg bw extract of *S. persica* fruits.

Analgesic activity was performed as described by Hossain et al. [22]. Mice were fasted for 12 h with adequate clean water. Mice were placed on hot plates and the temperature was maintained at 55 ± 1 °C. Latency period, or the pain reaction time determined with a stopwatch, was recorded, which represented the time taken for the mice to react to the pain stimulus. The time in seconds was measured with a stopwatch by observing discomfort reactions, such as licking paws or jumping. The first reading was taken just before the administration of the drug and later at 0, 60, 120, 180 and 240 min before and after the drug administration. The cutoff time was fixed for 12 s to prevent any type of injury. This served as the control pain reaction time. The maximum possible analgesia (MPA) was calculated as follows:

$$\text{MPA } (\%) = \frac{Reaction\ time\ for\ treatment - Reaction\ time\ for\ saline}{12 - Reaction\ time\ for\ saline} \times 100$$

2.7. Statistical Analysis

All the experiments were performed in three replicates and data are reported as the mean ± standard deviation (SD). Statistical analysis was performed by means of the statistical package, STATISTICA (Stat Sift Inc., Tulsa, OK, USA). Data from different tests were analyzed using one-way analysis of variance (ANOVA), followed by Bonferroni/Dunnett (all mean) post hoc tests; the differences between the means were considered statistically significant at probability value $p \leq 0.05$.

3. Results

3.1. Yield of Extracts

The extract yields (g/100 g) from *S. oleoides* and *S. persica* aerial parts and fruits are given in Table 1. Extract yields varied from 8.15 to 19.60 g/100 g of dry plant material. The maximum extract yield (19.60 g/100 g) was obtained from the aerial part of *S. persica*, whereas the minimum extract yield (8.15 g/100 g) was found from *S. oleoides* aerial parts. The results showed a significant ($p \leq 0.05$) difference in the yields among different extracts. Saleem et al. [23] reported a 12% extract yield from the aerial parts of *S. oleoides*. Variation in the extract yield might be due to differences in extractable components and months.

Table 1. Yield, TPC, TFC, and DPPH radical scavenging capacity of *S. oleoides* and *S. persica* aerial parts and fruit extracts.

Assays	*S. oleoides*		*S. persica*		BHT
	Aerial Parts	Fruits	Aerial Parts	Fruits	
Yield (g/100 g of dry weight)	8.15 ± 0.48 [a]	14.90 ± 0.74 [b]	19.60 ± 0.98 [c]	16.20 ± 0.81 [b]	—
TPC (mg/g, measures as gallic acid equivalent)	11.90 ± 0.50 [a]	19.20 ± 0.96 [c]	16.20 ± 0.81 [b]	22.20 ± 1.11 [d]	—
TFC (mg/g, measured as catechin equivalent)	1.72 ± 0.09 [a]	5.54 ± 0.28 [c]	2.44 ± 0.12 [b]	6.17 ± 0.31 [c]	—
DPPH scavenging (%) by 10 μg/mL extract solution	46.90 ± 2.34 [a]	52.70 ± 2.63 [b]	51.60 ± 2.58 [ab]	54.30 ± 2.72 [b]	60.80 ± 3.04 [e]

The values are the mean ± SD of three independent experiments. Different alphabet letters in superscript show significant ($p \leq 0.05$) differences among different extracts.

3.2. HPLC Results

Thirteen phenolic acids (gallic, hydroxy benzoic, chlorogenic, caffeic, syringic, vanillic, *p*-coumeric, salicylic, sinapic, ferulic, ellagic, cinamic, and benzoic acids) and five flavonoids (catechin, rutin, myricetin, quercetin, and kaempferol) were identified and quantified using Rp-HPLC from different extracts of *S. oleoides* and *S. persica*, and the data are presented in Table 2. Chlorogenic acid was the major phenolic acid from the aerial parts and fruit extracts of *S. oleoides* and *S. persica*, followed by gallic acid and hydroxyl benzoic acid. *S. oleoides* and *S. persica* fruit extracts exhibited the maximum concentrations of chlorogenic acids, i.e., 1786.0 and 1473.0 mg/100 g of extract, respectively. Gallic acid was abundantly found in the *S. oleoides* aerial part extract (1265.0 mg/100 g), followed by the *S. persica* fruit extracts (942.4 mg/100 g). *S. persica* fruit extracts also contained 942.4 mg/100 g hydroxyl benzoic acid, whereas *S. oleoides* fruit extract contained 727.0 mg/100 g hydroxyl benzoic acid. Cinamic acid was found abundantly in the aerial part extracts of *S. oleoides* (390.6 mg/100 g) and *S. persica* (471.6 mg/100 g) only (Figure 1). Catechin and myricetin were the major flavonoids detected in all extracts of *S. oleoides* and *S. persica*. *Salvadora persica* aerial part extracts were rich in myricetin, rutin, and catechin, and the contents of these were 410.5, 254.3, and 578.5 mg/100 g, respectively. *S. persica* fruit extracts contained 331.8, 275.1, and 117.2 mg/100 g myricetin, rutin, and catechin, respectively. Catechin was abundant in the aerial part extracts of *S. oleoides* (131.5 mg/100 g) and *S. persica* (578.5 mg/100 g) as compared with the fruit extracts.

Table 2. Composition (mg/100 g) of phenolic acids and flavonoids from *S. oleoides* and *S. persica* aerial parts and fruit extracts, obtained using Rp-HPLC.

Compounds	*S. oleoides*		*S. persica*	
	Aerial Parts	**Fruits**	**Aerial Parts**	**Fruits**
Gallic acid	254.4 ± 11.1 [a]	1265.0 ± 36.0 [d]	580.4 ± 20.3 [b]	942.4 ± 24.8 [c]
Hydroxy benzoic acid	254.8 ± 12.3 [a]	727.0 ± 21.2 [c]	388.9 ± 9.7 [b]	732.9 ± 16.5 [c]
Chlorogenic acid	526.4 ± 17.3 [a]	1786.0 ± 44.0 [c]	540.7 ± 21.9 [a]	1473.0 ± 48.0 [b]
Caffeic acid	16.4 ± 0.8 [a]	14.6 ± 0.8 [a]	165.9 ± 5.4 [c]	54.0 ± 1.3 [b]
Syringic acid	23.0 ± 1.2 [b]	71.6 ± 3.0 [c]	—	15.4 ± 0.8 [a]
Vanillic acid	183.1 ± 4.1 [b]	176.8 ± 5.3 [b]	—	86.5 ± 4.1 [a]
p-Coumeric acid	136.4 ± 5.8 [ab]	131.9 ± 5.2 [a]	257.4 ± 10.3 [c]	145.7 ± 5.9 [b]
Salicylic acid	239.3 ± 10.1 [c]	151.8 ± 5.9 [b]	167.5 ± 6.5 [b]	119.8 ± 5.2 [a]
Sinapic acid	69.3 ± 3.1 [c]	13.3 ± 0.8 [a]	88.3 ± 3.2 [d]	45.9 ± 2.5 [b]
Ferulic acid	22.0 ± 1.4 [a]	27.9 ± 0.9 [b]	21.6 ± 1.0 [a]	52.2 ± 2.3 [c]
Ellagic acid	22.0 ± 1.4 [b]	26.5 ± 1.3 [c]	52.5 ± 2.0 [d]	17.5 ± 0.8 [a]
Cinamic acid	390.6 ± 12.1 [b]	—	471.6 ± 14.3 [c]	33.0 ± 1.4 [a]
Benzoic acid	20.9 ± 1.2 [b]	—	24.5 ± 1.1 [c]	9.5 ± 0.6 [a]
Catechin	131.5 ± 5.90 [c]	72.0 ± 2.7 [a]	578.5 ± 25.2 [d]	117.2 ± 4.9 [b]
Rutin	302.7 ± 10.23 [d]	59.0 ± 2.2 [a]	254.3 ± 10.0 [b]	275.1 ± 9.17 [c]
Myricetin	332.0 ± 12.50 [b]	28.0 ± 1.0 [a]	410.5 ± 17.8 [c]	331.8 ± 17.6 [b]
Quercetin	24.16 ± 0.89 [b]	—	21.7 ± 0.9 [a]	21.0 ± 0.9 [a]
Kaempferol	—	—	26.2 ± 1.3 [a]	24.7 ± 1.2 [a]

The values are the mean ± SD of three independent experiments. Different alphabet letters in superscript show significant ($p \leq 0.05$) differences.

Statistical analysis showed that the concentrations of various phenolic acids and the flavonoids were significantly ($p \leq 0.05$) different among different extracts.

The phenolic and flavonoid contents were positively associated with the antioxidant activity. Both gallic acid and chlorogenic acid were the potential natural antioxidant compounds. Very few reports are available on the phenolic profile of the investigated extracts. Noumi et al. [24] also reported the separation and identification of caffeic acid, rutin trihydrate, trans-cinnamic, and gallic acids in the stems of *S. persica* using RP-HPLC.

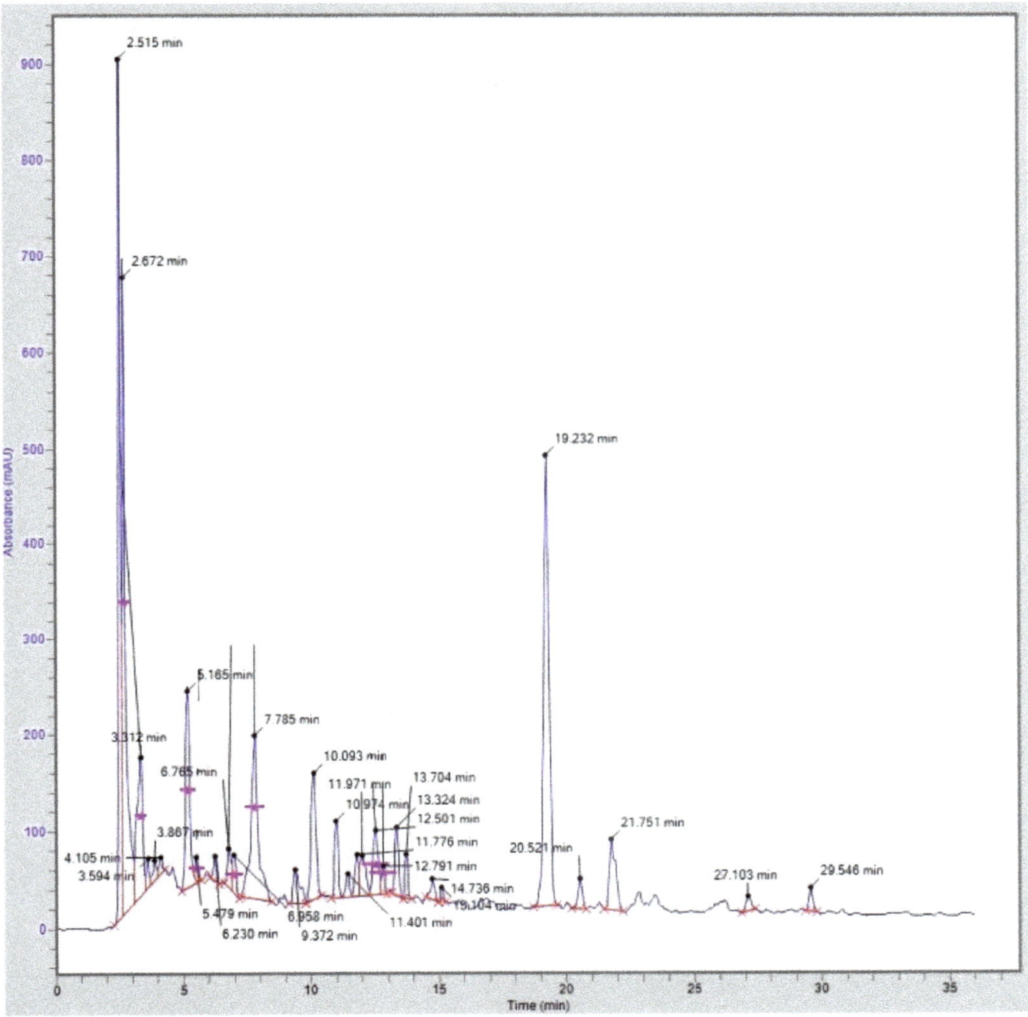

(a)

Figure 1. *Cont.*

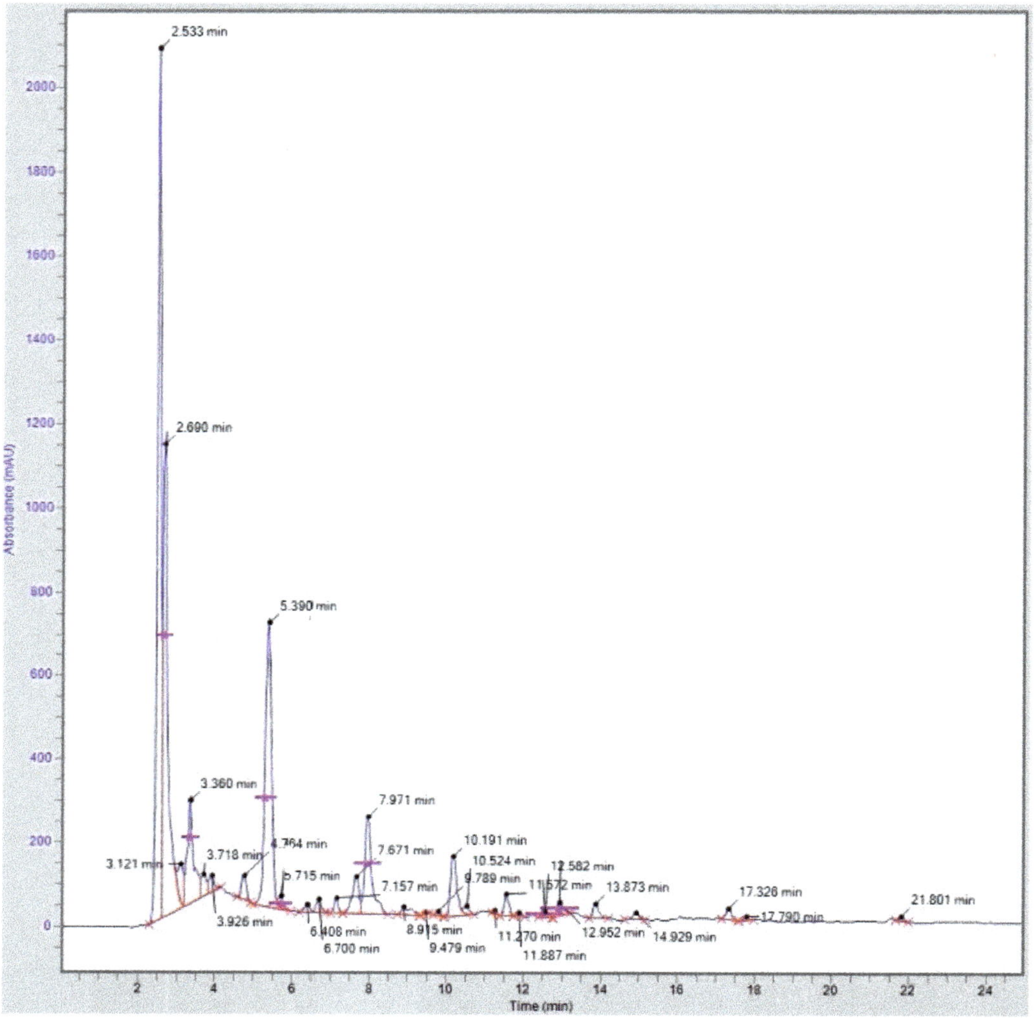

(b)

Figure 1. *Cont.*

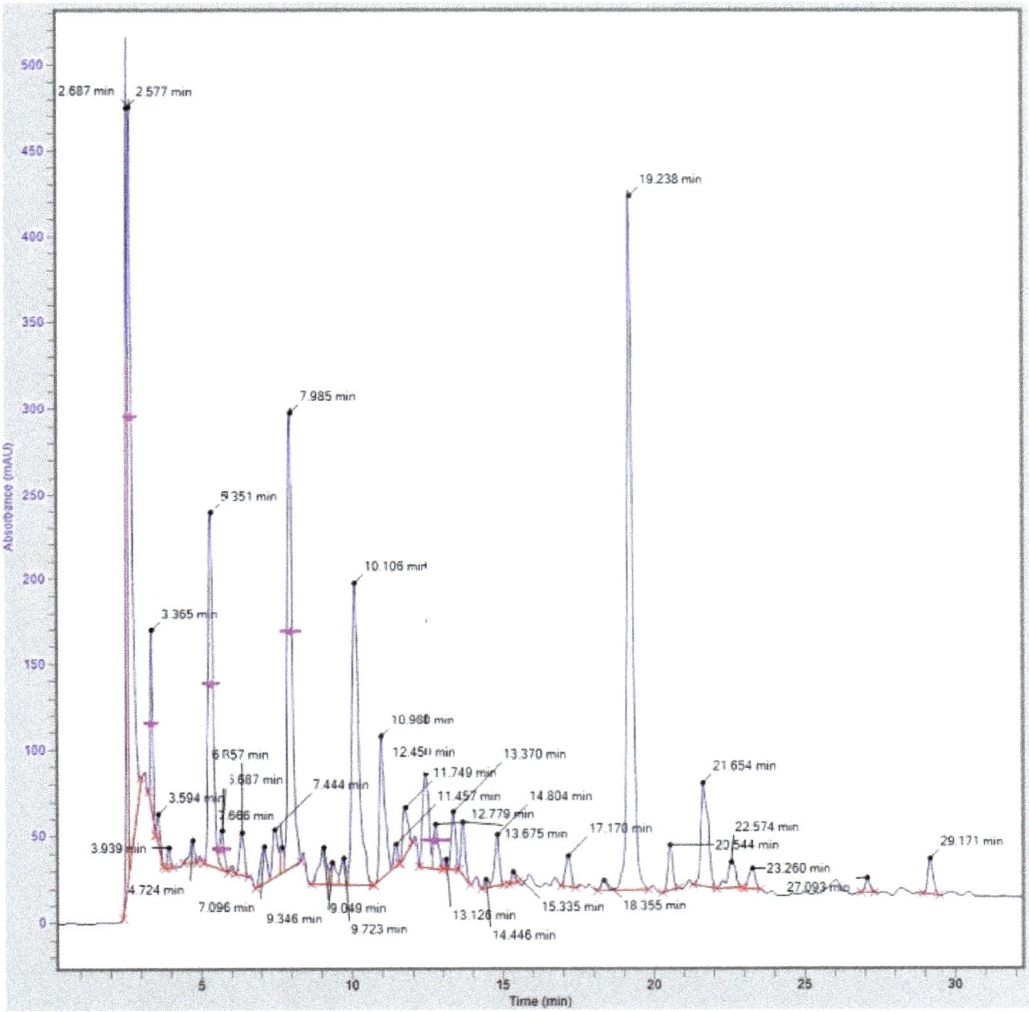

(c)

Figure 1. *Cont.*

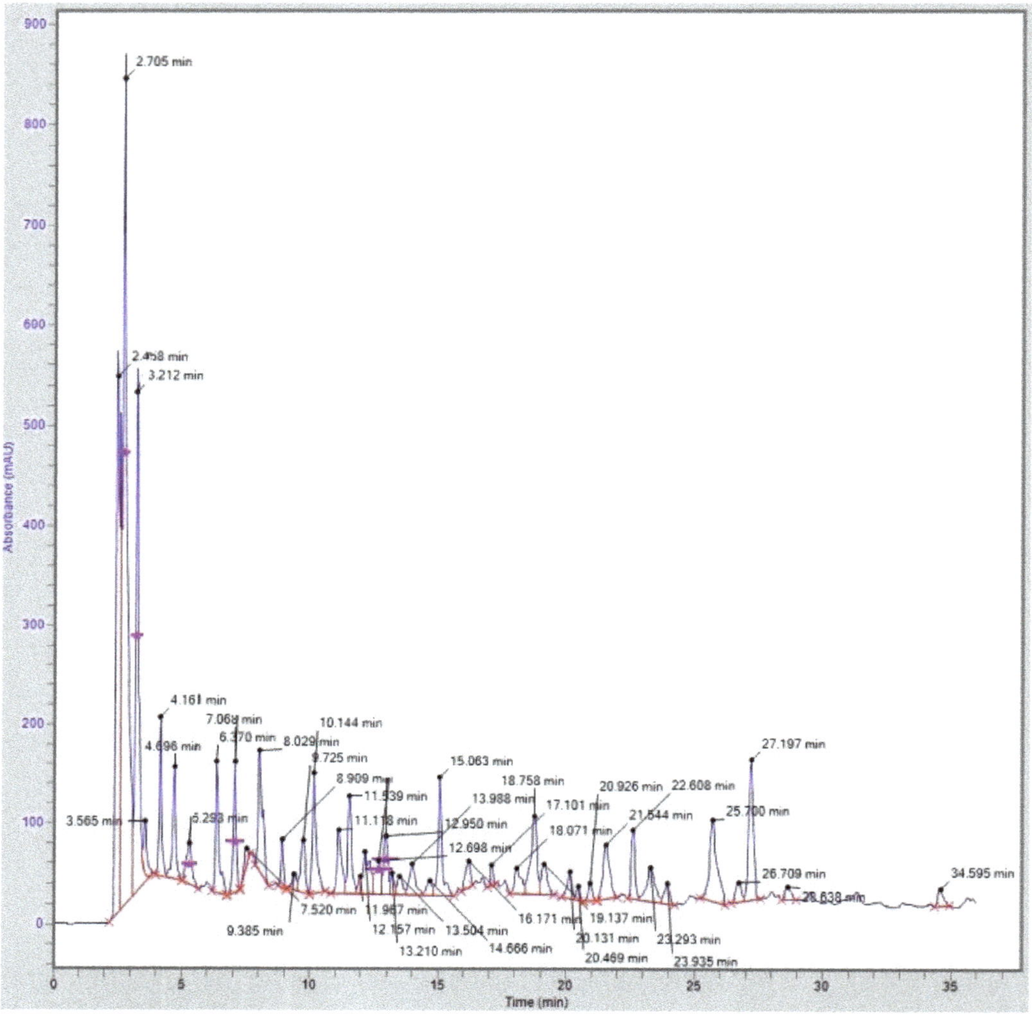

(d)

Figure 1. Typical HPLC chromatograms showing the separation of phenolic acid and flavonoids from (**a**) *S. oleoides* aerial parts; (**b**) *S. oleoides* fruit; (**c**) *S. persica* aerial parts; and (**d**) *S. persica* fruit extract.

3.3. *Evaluation of In Vitro Antioxidant Activity*

Estimation of TPC and TFC

TPC and TFC results, expressed as mg/g of dry plant material, are presented in Table 1. TPC was in the range of 11.9 to 22.2 mg TPC/g of dry plant material, measured as the GAE. The maximum TPC (22.2 mg/g of dry plant material, as the GAE) was found in fruit extracts of *S. persica*, and the minimum TPC (11.9 mg/g, as the GAE) was found in the aerial part of *S. oleoides*, followed by the aerial part extracts of *S. persica* (16.2 mg/g, as the GAE) and the fruit extracts of *S. oleoides* (19.2 mg/g, as the GAE). The TFCs in aerial parts and fruit extracts were in the range of 1.63 to 6.17 mg TFC/g of dry plant material, measured as the CE. The maximum amount of TFC (6.17 mg TFC/g of dry plant material as the CE) was exhibited by fruit extracts of *S. persica*, and the minimum TFC (1.72 mg TFC/g of dry plant material as the CE) was exhibited by aerial part extracts of *S. oleoides*.

The flavonoid contents in fruit extracts of *S. oleoides* were found to be 5.54 mg TFC/g of dry plant material and 2.44 mg TFC/g of dry plant materials in aerial part extracts of *S. persica*. Significant differences ($p \leq 0.05$) were observed in the TPC and TFC of the different extracts of *S. oleoides* and *S. persica*.

Phenolic compounds have good relationship and largely contribute to the antioxidant activity. The polar solvent, i.e., methanol, which was used has the capacity to extract more phenolics as compared with other solvents. Saleem et al. [23] reported total phenolic (0.4 mg QE/g) and total flavonoid contents (0.21 mg QE/g) of *S. oleoides* aerial parts in methanol extract which were lower than our results. Kumari et al. [25] reported the total estimated amount of phenolics in the fruit methanol extract of *S. persica* to be 120.38 mg/100 g DW, and flavonoids were estimated to be 77.59 mg/100 g DW; however, there have been no reported results on the TPC and TFC of *S. oleoides* fruit. Kaneria et al. [19] reported that *S. persica* showed higher total phenol contents as compared with *S. oleoides*. The total phenolic contents of *S. oleoides* in methanol extracts of leaves were 253.10 mg/g, whereas the total flavonoid contents were 43.65 mg/g. The total phenolic content of *S. persica* in methanol extracts of leaves was 252.770 mg/g, whereas the total flavonoid contents were estimated to be 57.94 mg/g [19], which is higher than our findings. Variation in our TFC and TPC results compared with the findings of the majority of previous studies might have been due to differences in the agro-climatic, geographical, and seasonal conditions.

3.4. DPPH Free Radical Scavenging Assay

The free radical scavenging activity of various extracts (10 μg/mL) was measured by the DPPH radical scavenging assay, and the results are presented in Table 1. The fruit extract of *S. persica* showed maximum radical scavenging activity (54.3%), whereas the aerial parts of *S. oleoides* showed minimum radical scavenging activity (46.9%) when compared with the synthetic antioxidant BHT (60.8%). Statistical analysis showed the significant ($p \leq 0.05$) differences in the radical scavenging potential of *S. persica* fruit extracts from other extracts.

The DPPH free radical scavenging capacity increases when extract concentration increases due to increases in the concentration of phenolic compounds [5]. Saleem et al. and Noumi et al. [23,24] reported the strongest scavenging DPPH assay results (51.66%) of methanolic extracts of *S. oleoides* aerial parts, which were comparable to our results. Kumari et al. [25] reported the IC_{50} for DPPH of the *S. persica* fruit (IC_{50} 307.06 μg crude methanol extract). Souri et al. [26] investigated the antioxidant activity and free radical scavenging activity on DPPH of 13 medicinal plants traditionally used in Iran. They reported that methanolic extracts of *S. persica* exhibit free radical DPPH scavenging activity with an IC_{50} value of 37.19, which is contrary to our results. The variation in DPPH radical scavenging activity might be due to different plant species, geographic regions, as well as different months for sample collection.

3.5. In Vivo Study on Animal Model

3.5.1. Anti-Inflammatory Activity

The anti-inflammatory activities of aerial part and fruit extracts of *S. persica* and *S. oleoides* were evaluated, and the results are presented in Table 3. At zero hour, mean inflammation (mm ± SD) was non-significant ($p > 0.05$) in all groups before the start of treatment.

The inflammation was significantly reduced ($p \leq 0.05$) in all treated groups, G1, G2, G3, and G4, and the positive control (PC) as compared with the negative control (NC) group one hour after the oral administration of methanolic extracts. However, reductions in inflammation were non-significant ($p > 0.05$) among all treated groups one hour after the oral administration of extracts. The inflammation was significantly reduced ($p \leq 0.05$) in all treated groups, G1, G2, G3, and G4, and the positive control group (PC) as compared with the negative control (NC) groups 2 h after the oral administration of methanol extract. However, the anti-inflammatory response was significantly reduced in G4 as compared with NC and G1 2 h after the oral administration. Inflammation was also significantly reduced in G1, G2, and G3 as compared with NC, but was non-significant between groups. The inflammation was significantly reduced ($p < 0.05$) in all treated groups, G1, G2, G3, and G4, and PC, as compared with the NC group 3 h after the oral administration of extracts. Inflammation was significantly reduced in G4 as compared with G1, G2, and G3, but was comparable to PC. Inflammation was also significantly reduced in G1, G2, and G3 as compared with NC, but less than G4 and PC. The inflammation was significantly reduced ($p < 0.05$) in all treated groups, G1, G2, G3, and G4, and PC, as compared with NC four hours after the oral administration of methanol extract. Inflammation was significantly reduced ($p < 0.05$) in G4 as compared with NC, G2, and G3. It could be concluded that the fruits of *S. persica* are more potent than the other extracts used in the study.

Table 3. Rat paw diameters and edema inhibition percentages (%) of different treatment groups at different time intervals.

Groups	Paw Diameter (mm) (Edema Inhibition %)				
	Initial Value	1 h	2 h	3 h	4 h
NC	4.14 ± 0.06 ^{a}	8.24 ± 0.21 ^{c}	6.18 ± 0.29 ^{c}	5.68 ± 0.26 ^{e}	5.21 ± 0.27 ^{e}
PC	4.17 ± 0.07 ^{a} (0.92%)	6.11 ± 0.24 ^{a} (25.73%)	5.21 ± 0.28 ^{b} (15.60%)	4.24 ± 0.10 ^{b} (25.35%)	4.06 ± 0.03 ^{b} (22.07%)
G1	4.14 ± 0.07 ^{a} (0.54%)	6.68 ± 0.19 ^{ab} (19.05%)	5.54 ± 0.13 ^{ab} (10.36%)	5.23 ± 0.04 ^{d} (7.92%)	4.15 ± 0.14 ^{bc} (20.15%)
G2	4.17 ± 0.01 ^{a} (0.82%)	6.31 ± 0.05 ^{ab} (23.42%)	5.17 ± 0.24 ^{b} (16.34%)	4.70 ± 0.14 ^{c} (17.25%)	4.55 ± 0.16 ^{d} (12.67%)
G3	4.27 ± 0.07 ^{a} (0.38%)	7.08 ± 0.38 ^{b} (14.08%)	5.24 ± 0.07 ^{b} (15.21%)	4.80 ± 0.15 ^{c} (15.49%)	4.35 ± 0.27 ^{cd} (16.51%)
G4	4.12 ± 0.04 ^{a} (0.48%)	6.44 ± 0.27 ^{a} (21.84%)	4.61 ± 0.22 ^{a} (25.40%)	4.16 ± 0.08 ^{a} (26.76%)	3.69 ± 0.15 ^{a} (29.17%)

The values are reported as the mean ± SD. Different alphabet letters in superscript show significant ($p < 0.05$) differences among different rat groups. NC, negative control (0.1 mL albumin); PC, positive control (Diclofenac sodium 10 mg/kg body weight) and treated groups: G1 (aerial part extract of *S. oleoides*, 250 mg/kg body weight), G2 (aerial part extract of *S. persica* 250 mg/kg body weight), G3 (Fruit extract of *S. oleoides* 250 mg/kg body weight), and G4 (fruit extract of *S. persica* 250 mg/kg body weight).

Polyphenols are secondary metabolites that have antioxidant capacities in addition to antiallergenic, anticancer, anti-inflammatory, anti-thrombotic, and anti-mutagenic properties [27]. The results obtained by BenSaad et al. [28] clearly indicate that ellagic acid, gallic acid, and punicalagin A & B isolated from *P. granatum* inhibited the production of NO, PGE2, and IL-6 in LPS-induced RAW267.4 macrophages. In the present study, gallic acid may have been the compounds responsible for the remarkable anti-inflammatory effect showed by *S. presica* fruit. Catechins can also inhibit the infiltration and proliferation of immune-related cells and regulate inflammation and oxidative reactions by interaction with multiple inflammation-related and oxidative-stress-related pathways [29]. Catechin was the major compound found in *S. persica* fruit and detected in all extracts by Rp-HPLC, which is why it could also be the reason for the anti-inflammatory effect shown by plant extracts. Ibrahim et al. [30] investigated the anti-inflammatory effect of aqueous alcoholic crude extract and the ethyl acetate extract of miswak sticks (*S. persica*) in carrageenan-induced rat paw edema. The inhibition percentage of inflammation was 17% for crude extract and 27% for ethyl acetate extract. Natubhai et al. [21] studied the anti-inflammatory effect of *S. oleoides* leaf extract. The alcoholic extracts of *S. oleoides* at a dose of 200 and 400 mg/kg, reduced the paw edema induced by carrageenan by 46.70 and 81.70%, respectively, whereas the water extract reduced the paw edema by 51.16 and 83.30, respectively. No reports are available in the literature on the anti-inflammatory activity of aerial parts and fruits of *S. persica* and *S. oleoides*. Baba et al. [31] investigated the anti-inflammatory activity of ethanol, acetone, and water extract of leaves, stem bark, and fruit peels of *S. persica*. The anti-inflammatory activity was compared with the standard drug (Diclofenac) at a dose of 10 mg/kg. Diclofenac showed values of 87.71 and 82.85 at a dose of 300 mg/kg, respectively.

3.5.2. Analgesic Activity

The analgesic activity of aerial part and fruit extracts of *S. oleoides* and *S. persica* was investigated using a mouse model, and the results are presented in Table 4. Mice treated with normal saline (control) did not show any significant difference in the reaction time on the experiment throughout the 240 min observation. The longest reaction time for the treated groups with the hot plate method was 180 min. The analgesic effects of Diclofenac sodium and different plant extracts could be seen from the maximum possible analgesia (MPA) graph in Figure 2. The MPA remained elevated during the observation period, reaching its peck at 180 min. The MPA of fruit extracts of *S. persica* was 53.44% that is better than PC group (52.98%). Aerial part extracts of both *S. oleoides* and *S. persica* showed significantly ($p \leq 0.05$) less MPA. Overall, the fruit extract of *S. persica* produced an excellent analgesic activity at 180 min.

Table 4. Analgesic activity of the test drug in hot plate test.

Groups	Reaction Time (Minutes)				
	Start Time	60	120	180	240
NC	3.41 ± 0.21^{a}	3.44 ± 0.32^{a}	3.46 ± 0.32^{a}	3.45 ± 0.32^{a}	3.45 ± 0.33^{a}
PC	3.43 ± 0.29^{a}	5.53 ± 0.43^{b}	6.89 ± 0.62^{c}	6.92 ± 0.65^{bc}	5.42 ± 0.49^{b}
G1	3.45 ± 0.37^{a}	5.61 ± 0.54^{b}	5.52 ± 0.47^{b}	5.85 ± 0.53^{b}	5.43 ± 0.47^{b}
G2	3.46 ± 0.31^{a}	6.21 ± 0.59^{b}	6.11 ± 0.57^{bc}	6.44 ± 0.58^{bc}	5.98 ± 0.54^{b}
G3	3.47 ± 0.34^{a}	6.33 ± 0.58^{b}	6.23 ± 0.58^{bc}	6.56 ± 0.59^{bc}	6.12 ± 0.56^{b}
G4	3.51 ± 0.32^{a}	6.45 ± 0.58^{b}	6.62 ± 0.59^{b}	6.95 ± 0.64^{c}	6.31 ± 0.59^{b}

The values are presented as the mean ± SD. Different alphabet letters in superscript show significant ($p \leq 0.05$) differences among normal and treatment groups. NC, negative control (0.1 mL albumin); PC, positive control (Diclofenac sodium 10 mg/kg body weight); and treated groups: G1 (aerial part extract of *S. oleoides*, 250 mg/kg body weight), G2 (aerial part extract of *S. persica* 250 mg/kg body weight), G3 (Fruit extract of *S. oleoides* 250 mg/kg body weight), and G4 (fruit extract of *S. persica* 250 mg/kg body weight).

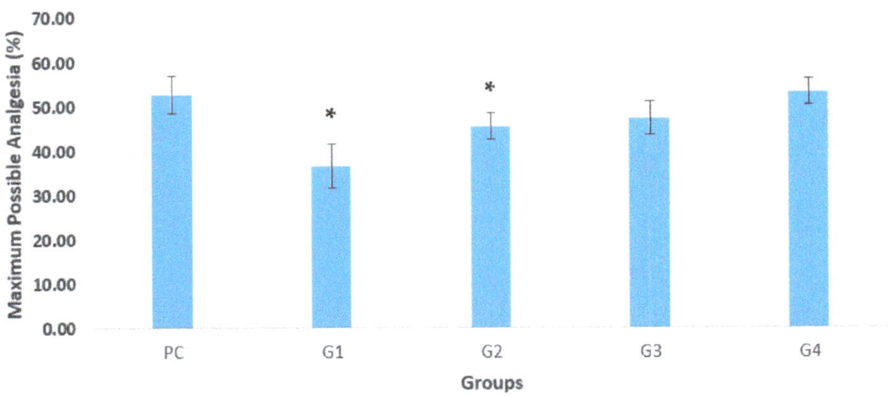

Figure 2. Maximum possible analgesia (MPA) (%) representing the effect of the drugs administered into mice, evaluated by the hot plate method. NC, normal control group; PC; positive control (Diclofenac sodium 10 mg/kg body weight); and treated groups: G1 (aerial part extract of *S. oleoides* 250 mg/kg body weight), G2 (aerial part extract of *S. persica* 250 mg/kg body weight), G3 (Fruit of *S. oleoides* 250 mg/kg body weight), and G4 (fruit of *S. persica*, 250 mg/kg body weight). * indicates significant ($p \leq 0.05$) differences from the PC group.

Hooda and Pal [32] reported the analgesic activity of hydroalcoholic extract of *S. persica* root extract on a mouse and rat model. The analgesic activity of albino mice and albino rats was evaluated using Eddy's hot plate method. In Eddy's hot plate method, the highest analgesic activity was observed at an oral concentration of 400 mg/kg for 90 min. Our results were correlated with those of Hoor et al. [33], who suggested that extracts of

S. persica possess analgesic activity. The crude extract was used in three doses of 300 mg/kg, 500 mg/kg, and 700 mg/kg of tested animals against standard aspirin, and found to be maximum at 120–150 min.

4. Conclusions

The results obtained in the present study showed the comparative assessment of TPC and TFC, and antioxidant, anti-inflammatory, and analgesic activities of *S. persica* and *S. oleoides* aerial parts and fruit extracts. *S. persica* fruit extract showed the maximum antioxidant activities in terms of TPC, TFC, and DPPH radical scavenging. Based on the results, it is concluded that fruit extracts of *S. persica* have a major role in the reduction in inflammation and exhibit better analgesic activity as well. This was due to the high antioxidant results of *S. persica* fruit extract as compared with extracts of other plant parts that were investigated, and also due to the high presence of gallic acid and catechin, as shown by the quantification of phenolic acids and flavonoids, performed using Rp-HPLC. The present study also offers the scientific evidence to use fruits of *S. persica* in treating inflammation and provides a framework for the use of *S. persica* fruits as natural antioxidants after further clinical trials. There is still a dearth of evidence for exploring its importance. However, further investigations are ongoing to determine the exact phytoconstituents that are responsible for the biological activities of methanol extracts of Salvadora species.

Author Contributions: Conceptualization, A.I.H., Q.A. and S.M.A.A.; methodology: A.K.; software, T.F.; validation, Q.A., F.K.A. and S.M.A.A.; formal analysis, A.K., N.I. and T.F.; investigation, A.K. and N.I.; resources, A.I.H.; data curation, Q.A. and A.A.; writing—original draft preparation, A.K. and A.A.; writing—review and editing, Q.A., S.M.A.A., F.K.A., J.A. and T.F.; visualization, A.I.H.; supervision, A.I.H.; project administration: A.I.H. and S.M.A.A.; funding acquisition, S.M.A.A. All authors have read and agreed to the published version of the manuscript.

Funding: This research received no external funding.

Institutional Review Board Statement: For in vivo experiments, all procedures were conducted in conformity with the guidelines prescribed by the international guidelines on the ethical use of animals and the study was approved by Institutional Ethical Committee for Animal Care at Government College University Faisalabad (Study No 19680/IRB No 680).

Informed Consent Statement: Not applicable.

Data Availability Statement: Not applicable.

Acknowledgments: The authors acknowledge the Deanship of Scientific Research, Qassim University, for funding the article process charging (APC) for publication of this project, and the team of Central Hi-Tech Lab, Government College University Faisalabad, Pakistan, for the characterization of extracts, and the Department of Physiology for assistance in in vivo analyses.

Conflicts of Interest: The authors declare no conflict of interest.

References

1. Lobo, V.; Patil, A.; Phatak, A.; Chandra, N. Free radicals, antioxidants and functional foods. Impact on human health: A review. *Pharm. Rev.* **2010**, *4*, 118–126. [CrossRef]
2. Ibrahim, I.I.; Moussa, A.A.; Chen, Z.; Zhang, J.; Cao, W.G.; Yu, C. Bioactive phenolic components and antioxidant activities of water-based extracts and flavonoid-rich fractions from *Salvadora persica* L. leaves. *Nat. Prod. Res.* **2021**, *36*, 2591–2594. [CrossRef] [PubMed]
3. Sarlis, N.J.; Chowdrey, H.S.; Stephanou, A.; Lightman, S.L. Chronic activation of the hypothalamo-pituitary-adrenal axis and loss of circadian rhythm during adjuvant-induced arthritis in the rat. *Endocrinology* **1992**, *130*, 1775–1779.
4. Dalgleish, A.G.; O'Byrne, K.J. Chronic immune activation and inflammation in the pathogenesis of AIDS and cancer. *Adv. Cancer Res.* **2002**, *84*, 231–276.
5. Sultana, B.; Anwar, F.; Przybylski, R. Antioxidant activity of phenolic components present in barks of *Azadirachta indica, Terminalia arjuna, Acacia nilotica,* and *Eugenia jambolana* Lam. trees. *Food Chem.* **2007**, *104*, 1106–1114. [CrossRef]
6. Hussain, A.I.; Anwar, F.; Shahid, M.; Ashraf, M.; Przybylski, R. Chemical composition, and antioxidant and antimicrobial activities of essential oil of spearmint (*Mentha spicata* L.) from Pakistan. *J. Essent. Oil Res.* **2010**, *22*, 78–84. [CrossRef]

7. Hussain, A.I.; Rathore, H.A.; Sattar, M.Z.; Chatha, S.A.; ud din Ahmad, F.; Ahmad, A.; Johns, E.J. Phenolic profile and antioxidant activity of various extracts from *Citrullus colocynthis* L from the Pakistani flora. *Ind. Crop. Prod.* **2013**, *45*, 416–422. [CrossRef]
8. Salvi, J.; Katewa, S.S. Underutilized wild edible plants as a potential source of alternative nutrition: A review. *Int. J. Bot. Stud.* **2016**, *1*, 32–36.
9. Cos, P.; Vlietinck, A.J.; Vanden Berghe, D.; Maes, L. Anti-infective potential of natural products: How to develop a stronger in vitro 'proof-of-concept'. *J. Ethnopharmacol.* **2006**, *106*, 290–302. [CrossRef]
10. Nehete, M.N.; Nipanikar, S.; Kanjilal, A.S.; Kanjilal, S.; Tatke, P.A. Comparative efficacy of two polyherbal creams with framycetin sulfate on diabetic wound model in rats. *J. Ayurveda Integr. Med.* **2016**, *7*, 83–87. [CrossRef]
11. Mabberley, D.J. *Mabberley's Plant-Book: A Portable Dictionary of Plants, Their Classification and Uses*; Cambridge University Press: Cambridge, UK, 2017.
12. Korejo, F.; Ali, S.A.; Shafique, H.A.; Sultana, V.; Ara, J.; Teshamul-Haque, S. Antifungal and antibacterial activity of endophytic Penicillium species isolated from *Salvadora* species. *Pak. J. Bot.* **2014**, *46*, 2313–2318.
13. Almas, K. The effect of *Salvadora persica* extract (miswak) and chlorhexidine gluconate on human dentin: A SEM study. *J. Contemp. Dent. Pract.* **2002**, *3*, 27–35. [CrossRef]
14. Mekhemar, M.; Geib, M.; Kumar, M.; Hassan, Y.; Dörfer, C. *Salvadora persica*: Nature's gift for periodontal health. *Antioxidants* **2021**, *10*, 712. [CrossRef]
15. Farag, M.; El Gamal, A.; Basudan, O. *Salvadora persica L*: Toothbrush Tree with Health Benefits and Industrial Applications–an updated evidence-based review. *Saudi Pharma. J.* **2021**, *29*, 751–763. [CrossRef]
16. Khafagi, I.; Zakaria, A.; Dewedar, A.; El-Zahdany, K. A voyage in the world of plants as mentioned in the Holy Quran. *Int J. Bot.* **2006**, *2*, 242–251. [CrossRef]
17. Samejo, M.Q.; Memon, S.; Bhanger, M.I.; Khan, K.M. Chemical constituents of essential oil of *Salvadora oleoides*. *J. Pharm. Res.* **2012**, *5*, 2366–2367.
18. Singh, M.; Goel, S.; Ajay, Y.J.; Shivakrishna, E. *Salvadora oleoides* (Meethi-JAL): A plant of ecological and medicinal importance. *Pharma Innov. Int. J.* **2021**, *10*, 1201–1204.
19. Kaneria, M.; Rakholiya, K.; Sonagara, J.; Chanda, S. Comparative Assessment of Antioxidant Activity and Phytochemical Analysis of Facultative Halophyte *Salvadora oleoides* Decne. and *Salvadora persica* L. *Am. J. Biochem. Mol. Biol.* **2017**, *7*, 102–110. [CrossRef]
20. Tatke, P.; Nehete, M.; Gabhe, S. Antioxidant, Antimicrobial and Wound Healing Activity of Salvadora persica Twig Extracts. *J. Complement. Med. Altern. Health* **2018**, *7*, 555–720. [CrossRef]
21. Natubhai, P.M.; Pandya, S.S.; Rabari, H.A. Anti-inflammatory activity of leaf extracts of *Salvadora oleoides* Decne. *Int. J. Pharm. Bio. Sci.* **2013**, *4*, 985–993.
22. Hossain, M.M.; Biva, I.J.; Jahangir, R.; Vhuiyan, M.M.I. Central nervous system depressant and analgesic activity of Aphanamixis polystachya (Wall.) parker leaf extract in mice. *Afr. J. Pharm. Pharm.* **2009**, *3*, 282–286.
23. Saleem, H.; Ahmad, I.; Zengin, G.; Mahomoodally, F.M.; Khan, K.-U.K.; Ahsan, H.M.; Ahemad, N. Comparative secondary metabolites profiling and biological activities of aerial, stem and root parts of *Salvadora oleoides* decne (Salvadoraceae). *Nat. Prod. Res.* **2020**, *34*, 3373–3377. [CrossRef]
24. Noumi, E.; Hajlaoui, H.; Trabelsi, N.; Ksouri, R.; Bakhrouf, A.; Snoussi, M. Antioxidant activities and RP-HPLC identification of polyphenols in the acetone 80 extract of *Salvadora persica*. *Afr. J. Pharm. Pharmcol.* **2011**, *5*, 966–971.
25. Kumari, A.; Parida, A.K.; Rangani, J.; Panda, A. Antioxidant activities, metabolic profiling, proximate analysis, mineral nutrient composition of *Salvadora persica* fruit unravel a potential functional food and a natural source of pharmaceuticals. *Front. Pharmcol.* **2017**, *8*, 61. [CrossRef]
26. Souri, E.; Amin, G.; Farsam, H.; Barazandeh, M. Screening of antioxidant activity and phenolic content of 24 medicinal plant extracts. *DARU J. Pharm. Sci.* **2008**, *16*, 83–87.
27. Trabelsi, N.; Megdiche, W.; Ksouri, R.; Falleh, H.; Oueslati, S.; Soumaya, B.; Abdelly, C. Solvent effects on phenolic contents and biological activities of the halophyte *Limoniastrum monopetalum* leaves. *LWT-Food Sci. Technol.* **2010**, *43*, 632–639. [CrossRef]
28. BenSaad, L.A.; Kim, K.H.; Quah, C.C.; Kim, W.R.; Shahimi, M. Anti-inflammatory potential of ellagic acid, gallic acid and punicalagin A and B isolated from *Punica granatum*. *BMC Complement. Altern. Med.* **2017**, *17*, 47. [CrossRef]
29. Fan, F.Y.; Sang, X.; Jiang, M. Catechins and their therapeutic benefits to inflammatory bowel disease. *Molecules* **2017**, *22*, 484. [CrossRef]
30. Ibrahim, A.Y.; El-Gengaihi, S.E.; Motawea, H.M.; SLEEM, A.A. Anti-inflammatory activity of *Salvadora persica* L. against carrageenan induced paw oedema in rat relevant to inflammatory cytokines. *Not. Sci. Biol.* **2011**, *3*, 22–28. [CrossRef]
31. Baba, F.K.; Ranganayakulu, G.S.; Subramanyam, P.; Muralidhara, R.D. Anti-Cancer and Anti-Inflammatory Activity of *Salvadora Persica* L. *Indo. Am. J. Pharm. Res.* **2018**, *8*, 1179–1188.
32. Hooda, M.S.; Pal, R. Analgesic activity of crude Hydro-alcoholic extract of *Salvadora persica* root. *World J. Pharm. Res.* **2017**, *6*, 895–902. [CrossRef]
33. Hoor, T.; Ahmed, M.; Shaikh, J.M.; Rehman, A.B. Analgesic activity of *Salvadora persica* in mice. *Med. Channel* **2011**, *17*, 22–24.

life

MDPI

Article

Recovery of Aconitic Acid from Sweet Sorghum Plant Extract Using a Solvent Mixture, and Its Potential Use as a Nematicide

K. Thomas Klasson [1,*], Yunci Qi [1], Gillian O. Bruni [1], Tristan T. Watson [2], Bretlyn T. Pancio [3] and Evan Terrell [1]

1 Southern Regional Research Center, Agricultural Research Service, United States Department of Agriculture, New Orleans, LA 70124, USA
2 Department of Plant Pathology and Crop Physiology, Louisiana State University, Baton Rouge, LA 70803, USA
3 Oak Ridge Institute for Science and Education Research Program at USDA, Oak Ridge, TN 37831, USA
* Correspondence: thomas.klasson@usda.gov

Abstract: *Trans*-aconitic acid (TAA) is naturally present in sweet sorghum juice and syrup, and it has been promoted as a potential biocontrol agent for nematodes. Therefore, we developed a process for the extraction of aconitic acid from sweet sorghum syrup. The process economics were evaluated, and the extract was tested for its capability to suppress the motility of the nematodes *Caenorhabditis elegans* and *Meloidogyne incognita*. Aconitic acid could be efficiently extracted from sweet sorghum syrup using acetone:butanol:ethanol mixtures, and it could be recovered from this solvent with a sodium carbonate solution, with an overall extraction and recovery efficiency of 86%. The estimated production cost was USD 16.64/kg of extract and this was highly dependent on the solvent cost, as the solvent was not recycled but was resold for recovery at a fraction of the cost. The extract was effective in reducing the motility of the parasitic *M. incognita* and causing over 78% mortality of the nematode when 2 mg/mL of TAA extract was added. However, this positive result could not conclusively be linked solely to TAA. An unidentified component (or components) in the acetone:butanol:ethanol sweet sorghum extract appears to be an effective nematode inhibitor, and it may merit further investigation. The impact of aconitic acid on *C. elegans* appeared to be entirely controlled by pH.

Keywords: acetone:butanol:ethanol; solvent extraction; process design and economics; *Caenorhabditis elegans*; *Meloidogyne incognita*

Citation: Klasson, K.T.; Qi, Y.; Bruni, G.O.; Watson, T.T.; Pancio, B.T.; Terrell, E. Recovery of Aconitic Acid from Sweet Sorghum Plant Extract Using a Solvent Mixture, and Its Potential Use as a Nematicide. *Life* **2023**, *13*, 724. https://doi.org/10.3390/life13030724

Academic Editor: Jianfeng Xu

Received: 20 January 2023
Revised: 18 February 2023
Accepted: 1 March 2023
Published: 8 March 2023

1. Introduction

Aconitic acid (prop-1-ene-1,2,3-tricarboxylic acid) is the most abundant C6 organic acid that accumulates in sugar crops such as sugarcane and sweet sorghum. Aconitic acid functions as a chemical precursor and is intermediate to several value-added chemicals and products; it was listed as a top 30 value-added chemical by the U.S. Department of Energy [1]. There is considerable interest in using aconitic acid as a chemical precursor or reactant with its three reactive carboxylic acid groups. It is especially attractive, since it can be inexpensively sourced from renewable agricultural byproducts such as sugarcane molasses and sweet sorghum syrup to produce bio-based products and chemicals with various properties that we recently reviewed [2]. For instance, aconitic acid has been used in the production of polyesters, hyper-branched polyesters, polymers, and as a chemical crosslinker, plasticizer, and grafting agent [3–7].

In addition to its numerous chemical applications, aconitic acid has several biological roles and functions that were recently summarized by Bruni and Klasson [2], and it has many bioactive functions in specific plants and microorganisms, including its impact on nematodes [8–10]. Clarke and Shepherd [8] reported that in a study of 444 inorganic and organic compounds, *trans*-aconitic acid was one of 45 compounds that stimulated the eggs of the potato cyst nematode *Heterodera rostochiensis* (now named *Globodera rostochiensis*) to hatch at 3 mM (0.500 mg/mL) but it was not among the strongest stimulants. It was even less effective in stimulating *H. schachtii* to eggs to hatch [8]. Contrary to these findings of being a stimulant, Du et al. [9] found that aconitic acid at 0.5 mg/mL killed approximately 83% of root knot nematode *Meloidogyne incognita*, and it was 92% effective against this nematode at 1.0 mg/mL. In separate research, when studying the impact of *Canavalia ensiformis* (jack bean) seed extract on *M. incognita*, Rocha et al. [10] concluded that *trans*-aconitic acid was one of the organic acid constituents in the seed extract that acted as a nematicide and that paralyzed 98% of *M. incognita* at 0.5 mg/mL. *Cis*-aconitic acid was less effective (65%), but it still showed bioactivity.

One potentially significant natural source of aconitic acid is from sugar crops, such as sugarcane and sweet sorghum. Aconitic acid accumulates in these plants through the citrate branch of the tricarboxylic acid (TCA) cycle. In sugarcane, aconitic acid can account for roughly 0.1 to 0.5% of stalk juice [11], while in cane molasses, its concentration can be as high as 1 to 5% of dissolved solids [12]. Sweet sorghum syrup has been reported to similarly contain about 1% aconitic acid [13], with juice concentrations varying between roughly 0.2 and 0.6% for sweet sorghum, depending on the cultivar [14]. In one study on the recovery of polymerization grade aconitic acid from cane molasses, Kanitkar et al. [12] reported yields of 34 to 69% when using ethyl acetate as an extractive solvent, with aconitic acid purities of up to 99.9%. These authors also reported yields and purities of 62% and 99.9%, respectively, for the recovery of aconitic acid from fermented molasses with ethanol as a co-product. Other solvents reported in the literature for aconitic acid extraction from molasses include methyl-isobutyl ketone, methyl ethyl ketone, tributyl phosphate, amines, xylene, hexane, chloroform, and alcohols [11,12,15].

When comparing three processes for aconitic acid recovery from molasses (methanol precipitation; solvent extraction; ion exchange), Regna and Bruins [16] reported that solvent extraction appears to be the most economical, with a (producer price index adjusted, [17]) cost per kg of aconitic acid in the range of USD 6.09–USD 8.61, compared to USD 7.20–USD 9.63 for methanol precipitation and USD 7.35–USD 9.94 for ion exchange. While methyl-ethyl-ketone was the primary organic solvent used in their work, the authors reported that other organic solvents (specifically butanol) could be easily substituted, with alcohol solvents having the advantage of being derivable as a product from molasses fermentation. Aconitic acid precipitate recovery from sweet sorghum juice has been reported [18,19] during clarification in juice processing, with the addition of lime and calcium chloride, where the primary goal is the improvement of sucrose crystallization. However, because of general similarities between cane molasses and sorghum syrup, the recovery of aconitic acid (where present) from these low-cost feedstocks using organic solvents should be comparably feasible.

As butanol is one of the solvents that resulted in significant aconitic extraction [11], this research focused on butanol mixtures. Butanol can be produced commercially via fermentation, but it is co-produced with acetone, ethanol, and minor quantities of organic acids [20–22]. The separate recoveries of acetone, butanol, and ethanol from the fermentation broth have been studied extensively [23–27]. In a typical acetone:butanol:ethanol (ABE) fermentation [28], the molar ratio of the products is 0.39:1:0.12 (A:B:E), but it can have other compositions as well, depending on many factors (e.g., head-space H_2 and CO, electron sinks, etc.) [29–32]. The purpose of this study was to investigate the applicability of using mixtures of acetone:butanol:ethanol for the extraction of aconitic acid from sweet sorghum syrup. This could be seen as an intermediate step to using the sweet sorghum sugars for other applications; i.e., first separating aconitic acid and then using the sugars. Additionally, it can be seen as an intermediate step for recovering and purifying acetone, butanol, and ethanol from a fermentation; i.e., the complete separation of acetone, butanol, and ethanol may not be needed for aconitic acid extraction, and it could be delayed after it has been used for aconitic acid extraction. Another goal of the study was to evaluate the capacity of the aconitic acid to inhibit the motility of free-living nematodes and parasitic nematodes. *Caenorhabditis elegans* was chosen as it does not require a permit, and because the organism is easy to store, maintain, and culture [33]; and *Meloidogyne incognita* was chosen as others reported its susceptivity to aconitic acid [9,10].

2. Materials and Methods

Commercially available sweet sorghum syrup (Village Valley Sweet Sorghum Syrup, Delta BioRenewables LLC (Memphis, TN, USA) and a non-commercial batch of sweet sorghum syrup from Delta BioRenewables [13] were used to determine the optimal aconitic acid extraction conditions, and to create a crude extract for nematode testing. The non-commercial batch was chosen for the final crude extract preparation of aconitic acid, as it contained higher levels of initial aconitic acid than the commercial product.

2.1. Aconitic Acid Extraction

Solvent ratios of acetone:butanol:ethanol used during the extraction optimization are shown in Table 1. They were chosen to represent a range of acetone:butanol ratios, representative of the products from an ABE fermentation, which can be produced at different ratios [28–32].

Table 1. Composition of acetone:butanol:ethanol solutions used in the extraction of aconitic acid from diluted sweet sorghum syrup.

Weight Fraction			Molar Fraction		
Acetone	Butanol	Ethanol	Acetone	Butanol	Ethanol
0.00	0.93	0.07	0.00	0.89	0.11
0.08	0.84	0.07	0.10	0.79	0.11
0.19	0.74	0.07	0.22	0.68	0.11
0.31	0.62	0.07	0.35	0.55	0.11

The ratios of organic solvent (ABE mixtures) to diluted sweet sorghum syrup in the extractions were 1, 2, 3, 3.5, and 4 (g organics/g syrup). At least four experiments were performed for each extraction condition. If the standard errors of the extraction efficiency were deemed as large, additional extraction experiments were performed. Statistical differences between the results were determined using Tukey's Honest Significant Difference (HSD) procedure [34].

Before extraction, sweet sorghum syrup was diluted to 50 °Brix (~50% dissolved solids), had its pH adjusted to pH 2.0 using 0.2–4 M H_2SO_4 (~0.03–0.04 mL/g syrup), and was analyzed for sucrose, glucose, fructose, and aconitic acid. The diluted syrup and organics mixture were combined in 15 mL or 50 mL centrifuge tubes (~2/3 full) and incubated for 30 min at 50 °C, after which time the mixture was briefly removed from the incubator and shaken vigorously every 10 min for 60 min. After extraction, the tubes were centrifuged at 3000 rpm (2100× *g*) for 5 min. The organic layer was removed, and the aqueous phase was analyzed for sugars and aconitic acid after a 10-fold dilution. The weights of each phase before and after extraction were recorded and used in the calculations.

For the final extractions to generate a crude extract of aconitic acid, all of the same procedures were followed as previously described, using an acetone:butanol:ethanol (ABE) mixture of 0.19:0.74:0.07 (wt:wt:wt), and a ratio of 2.5 (g organics/g diluted syrup). The back-extraction was performed using 0.4 M Na_2CO_3 and a ratio of 6 (g organics/g salt solution). To remove some of the solvents contained in the extract, the extract was placed under a vacuum on a rotary evaporator (Model R-200, Buchi Corp., New Castle, DE, USA) at 60 °C for 30 min, which resulted in an 11% mass loss.

Sucrose, glucose, fructose, acetone, butanol, ethanol, and aconitic acid were analyzed using high performance liquid chromatography using the methods and equipment previously described [13]. Typical chromatograms from sample analysis have been provided in the Supplementary Materials.

2.2. Nematicidal Assay for Caenorhabditis Elegans

Initial nematocidal assay experiments were carried out with *C. elegans*, a free-living model nematode [35,36] as it does not require a permit and the organism is easy to store, maintain, and culture [33]. *Escherichia coli* strain OP50-1, obtained from the Caenorhabditis Genetics Center (CGC, University of Minnesota, Minneapolis, MN), was cultured in Luria-Bertani (LB) medium (Becton Dickinson, Sparks, MD, USA). *C. elegans* strain N2, obtained from CGC, was routinely propagated on Nematode Growth Medium (NGM) agar [37] seeded with *E. coli*. Age-synchronized worms were obtained as previously described [37]. Briefly, gravid *C. elegans* were collected from NGM agar plates with water and treated with sodium hypochlorite to obtain the eggs, which were resuspended in S buffer [38] and allowed to hatch into L1-stage larvae by incubating overnight at 20 °C. The L1-stage larvae were collected in a pellet by centrifuging at 1200× *g* for 3 min and re-suspending the pellet in fresh S buffer with *E. coli* for 24 h at 20 °C to obtain approximately early L3-stage worms. The number of worms per mL of S buffer was estimated by spotting 10 droplets of 10 µL on a plate and averaging the number of worms per drop.

The nematocidal assay protocol was adapted from those described by Du et al. [9] and Watson [39]. After counting the number of worms per mL, the appropriate volume of worm suspension was centrifuged and resuspended with fresh S buffer to obtain approximately 1000 worms per mL. For experiments with aconitic acid solution in water, S buffer was used, while the remaining experiments used 0.1 M Na_2CO_3 (pH adjusted to the same as TAA solution or extract). In a sterile 96-well plate, 50 µL volumes of worm suspension were aliquoted (approximately 50 worms per well), followed by the addition of 50 µL of aconitic acid solution. *Trans*-aconitic acid (>98% purity) was obtained from Sigma-Aldrich (St. Louis, MO, USA). After the plates were incubated at 20 °C for 48 and 72 h, the number of motile and immotile worms were counted under a microscope. As controls, worms were treated with 4 mg/mL fluopyram (Velum Prime, Bayer Crop Science, Whippany, NJ, USA). Assays were performed in triplicate, and the statistical differences between the results were determined using Tukey's HSD procedure.

2.3. Nematode Assay for Meloidogyne Incognita

Motility assays for *M. incognita* (originally isolated from sweet potato) were performed as previously described [39]. Briefly, plate wells (a 6-well microplate) were filled with 4 mL of aqueous solutions of each treatment solution. Approximately 150 J2-stage nematodes were suspended in 10 µL of water and introduced into each well. Nematode motility and mortality were recorded at 48 and 72 h post-inoculation under a stereomicroscope. The entire experiment was performed twice, with each treatment performed in triplicate (i.e., n = 6). Fluopyram (4 mg/mL) was included as a positive control for the inhibition of motility. Significant differences were evaluated using Turkey's HSD procedure.

2.4. Process Design

The process design, sizing of equipment, and cost estimation were performed using SuperPro Designer, Version 11.2 (Intelligen, Scotch Plains, NJ, USA). The annual production of the sweet sorghum syrup, containing 65% (wt/wt) total dissolved solids (mostly sugars) and 1.2% aconitic acid was assumed to be 8,000,000 kg. This is representative of a small-sized but commercial production of syrup [40]. While this amount is generated during a short harvesting season (approximately 4 months), it was considered to be stored and available at a flow rate of 1010 kg/h for 7920 h. The syrup was heated, diluted to 50 °Brix (~50% dissolved solids), and pH adjusted to pH 2.5 before extraction with an acetone:butanol:ethanol (ABE) mixture (19%:74%:7%); a mixture found to be optimal in the preliminary studies. The organic and aqueous phases were separated, after which the aconitic acid in the organic phase was cooled and back-extracted with sodium carbonate (0.4 M). This was followed by flash evaporation to partially remove additional solvents for ABE recovery. The spent organic phase (containing most of the ABE) was processed for the recovery of ABE for extraction reuse (not modeled here). The ABE recovery was recently reported elsewhere [23–27]. The aqueous phase, containing approximately 50% sugars, can be used for an ABE fermentation (not modeled here). Any remaining unused solvent streams are passed on to solvent separation and recovery. The value of the sugar and ABE stream was valued at USD 0.46/kg of sugar [41] and 10% of the price of ABE. The prices of acetone, butanol, acetone, sodium carbonate, and sulfuric acid were obtained from Intratec [42]. All prices were adjusted to 2022 prices using the producer price index [17]. Other costs such as materials, labor, facilities, etc., were obtained directly from the SuperPro Designer software.

3. Results and Discussion

3.1. Aconitic Acid Extraction

The commercial syrup contained 5.1 g/kg, and the non-commercial syrup contained 12.0 g/kg of aconitic acid. After dilution to 50 °Brix and pH adjustment, the commercial and non-commercial syrups contained 3.3 and 9.3 g/kg of aconitic acid, respectively. The lower level of aconitic acid in the commercial syrup was likely due to additional filtration used with commercial syrup production. It has been reported that sweet sorghum syrup contains both soluble and insoluble aconitic acid [18]. These concentrations are typical, as others have reported values of 0.26–4.8 g/L in the sweet sorghum juice [14,43,44], and 10–11 g/kg of aconitic acid in the sweet sorghum syrup [13].

The extraction efficiency of the solvents to extract aconitic acid from the syrups is shown in Figure 1. As noted, the maximum extraction efficiency was obtained using an approximately 2–3.5 organics:syrup phase weight ratio. A slightly higher extraction efficiency was obtained when extracting aconitic acid from the commercial syrup than when extracting aconitic acid from the non-commercial syrup. Overall, 92–96% aconitic acid was extracted with the 2–3.5 organics:syrup ratio; and generally, the extraction efficiency increased as the acetone:butanol ratio in the solvent phase increased (Figure 1). The results compared well with Gil Zapata [11], who tested solvents for the extraction of aconitic acid from fermented and distilled sugarcane juice and molasses. He reported an extraction efficiency of 90% when pure butanol was used at pH 2.0 with an organic:aqueous ratio of 3.5. Others [15] have obtained a high extraction efficiency using amine- or phosphorous-based solvent systems, in combination with chloroform, xylene, hexane, etc. The best solvent systems extracted over 98% of the aconitic acid from sugarcane molasses at pH 1.5–1.6 [15].

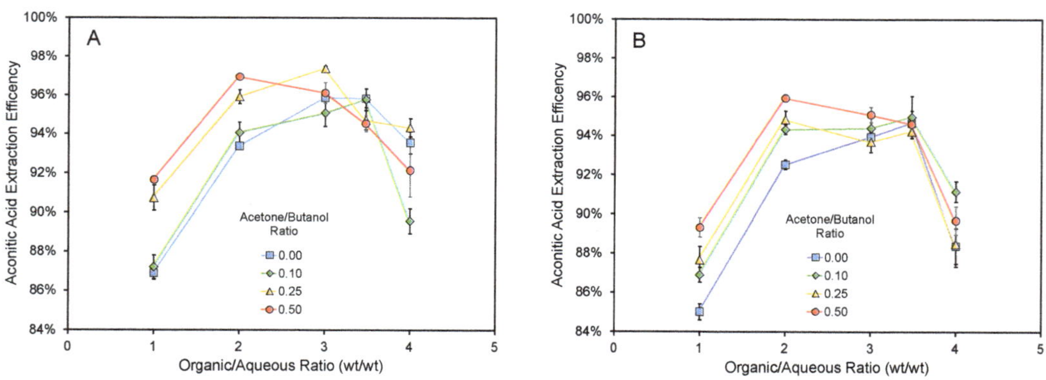

Figure 1. Aconitic acid extraction efficiency using different ABE mixtures and different solvent:syrup ratios for commercial syrup (**A**) and non-commercial syrup (**B**). The efficiency was calculated by the difference in the amount of aconitic acid present in the syrup before and after solvent extraction, accounting for weight changes to the phases after extraction. Error bars represent standard errors.

The amount of solvent lost to the spent syrup was also determined (Figure 2). Lower organics:syrup ratios led to a greater loss of solvent. At the optimal organics:syrup ratio for aconitic acid extraction (2–3.5), approximately 2–4%, 1%, and 2–6% of acetone, butanol, and ethanol were lost in the spent syrup, respectively (Figure 2). The acetone:butanol ratio did not impact on the solvent loss at the ranges studied.

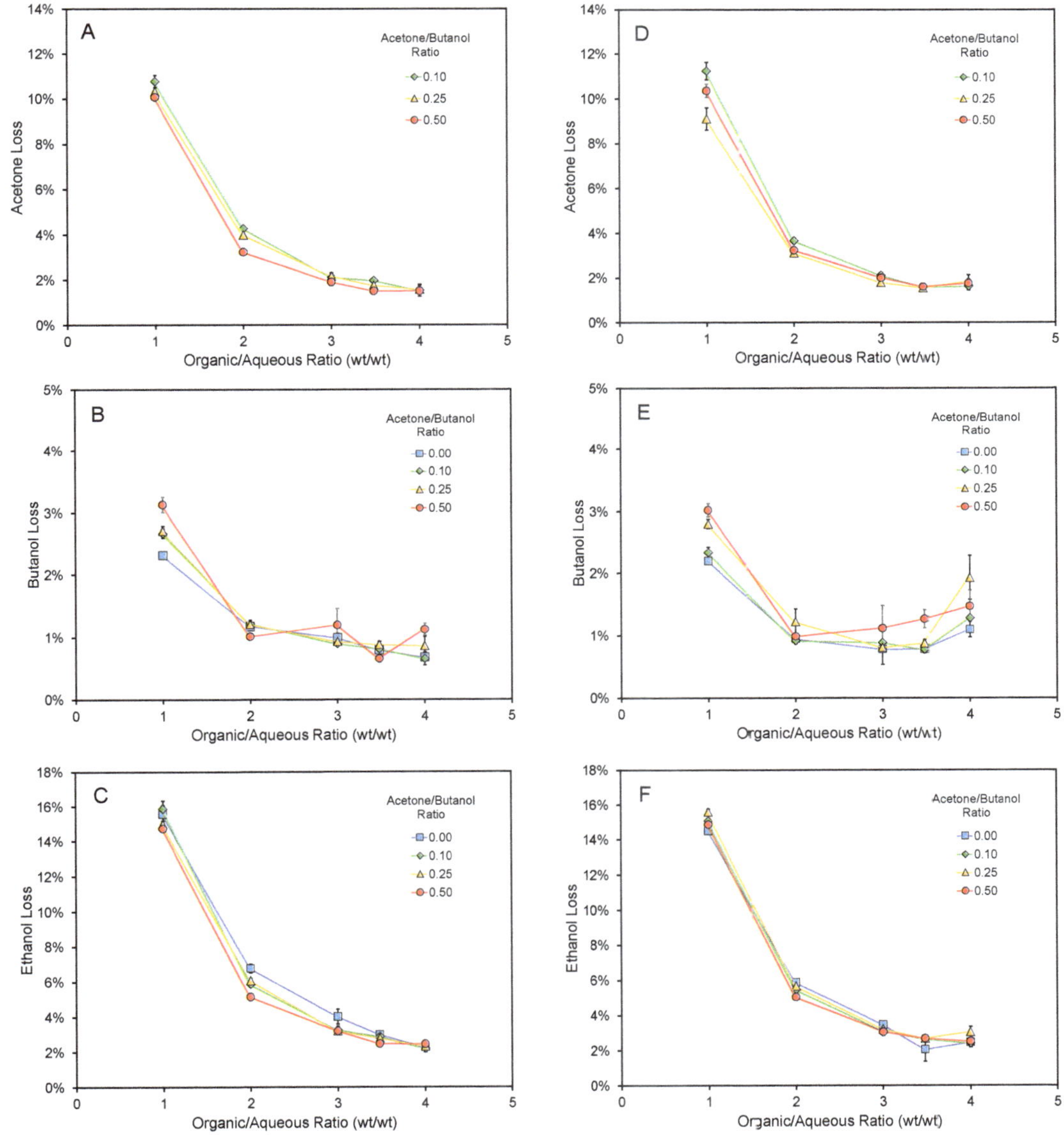

Figure 2. Solvent loss to the syrup phase during extraction with different ABE mixtures and different solvent:syrup ratios. Acetone, butanol, and ethanol losses for commercial syrup are shown in (**A**), (**B**), and (**C**), respectively. Acetone, butanol, and ethanol losses for non-commercial syrup are shown in (**D**), (**E**), and (**F**), respectively. The loss was calculated using the difference of the solvent species present in the syrup before and after solvent extraction, accounting for weight changes to the phases after extraction. Error bars represent standard errors.

To produce the extract used for nematode testing, larger quantities of the non-commercial sweet sorghum syrup, an acetone:butanol:ethanol mixture of 0.19:0.74:0.07 (wt:wt:wt), and an organic:syrup ratio of 2.5 (wt:wt) were used. This resulted in an extraction of 96% of the aconitic acid from the syrup. The back-extraction of the aconitic acid from the solvent phase was performed using 0.4 M Na_2CO_3 and a ratio of 6 (g organics/g salt solution). After solvent evaporation from the receiving salt solution and adjusting the pH to 6.5 with diluted sulfuric acid, the extract contained 45 g/L of aconitic acid. Overall, this represents 84% recovery of aconitic acid from the diluted syrup. The efficiency of Na_2CO_3 for the back-extraction of aconitic acid was also noted by Blinco et al. [15], who used 0.1–0.2 M Na_2CO_3 to back-extract 91–95% of aconitic acid from a solvent phase containing tributyl phosphate and the industrial solvent Shellsol that had previously been used to extract the aconitic acid from molasses.

3.2. Nematicidal Assays with *C. elegans*

The crude TAA extract was tested as a potential nematicide based on results previously reported by Du et al. [9]. The same results were obtained with the TAA extract and the TAA standard in Na_2CO_3 buffer at all TAA concentrations studied, resulting in approximately 22% immotile worms (Figure 3A, bar grouping 1 and 2). Less nematocidal activity was observed in assays with solutions of TAA in water, except in the case of 2 mg/mL TAA, where approximately 36% immotile worms were observed at 48 h (Figure 3A, bar grouping 3), which increased to 97% at 72 h (Figure 3A, bar grouping 4). Upon further investigations, it was determined that the extract preparation included neutralization with Na_2CO_3 and had a significant buffering capacity. While the buffering capacity was also present in the nematode growth medium, the pH was low (e.g., pH 2.7 with 2 mg/mL of TAA) in the final assay conditions when TAA in water was tested. The pH of the final assay solution was approximately pH 6 when the TAA extract was tested. Thus, it was speculated that pH may have a strong impact on the effectiveness of TAA against the nematodes.

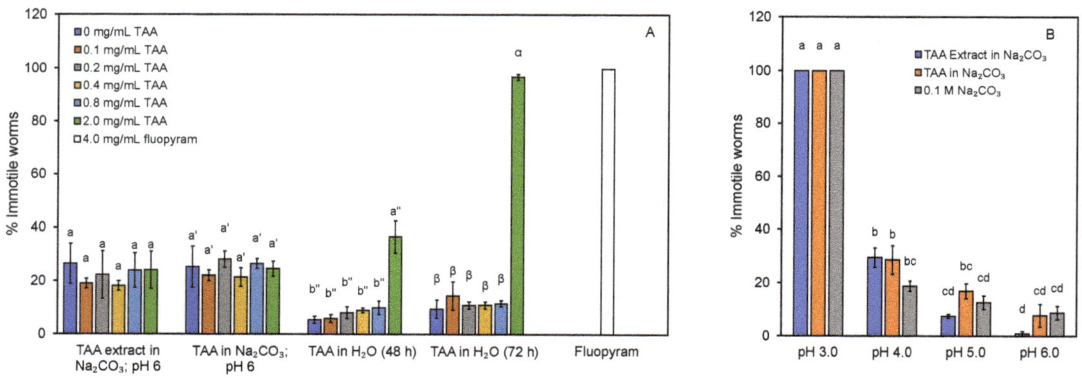

Figure 3. Nematicidal assays with TAA, TAA extract, and buffer against *C. elegans*. (**A**) The crude TAA extract, TAA in 0.1 M Na_2CO_3, and TAA solutions in water without pH adjustment. Fluopyram was used as control treatment. (**B**) The crude TAA extract, TAA in 0.1 M Na_2CO_3, and Na_2CO_3 control, adjusted to pH 3 to pH 6. Unless specified, the assays were carried out for 48 h. Percentage immotile includes dead worms. Significant differences between different treatments at the same condition in 3A and between any condition in 3B are represented by different letters above standard error bars.

The inhibitory activity of TAA against other organisms, e.g., the yeast *Saccharomyces cerevisiae*, has been reported to be pH-dependent [45]. This inhibition by undissociated organic acids is caused by an increase in transport into the cell, and a decreased internal cell pH when the organic acid dissociates and releases its protons [46]. Thus, the pH 2.7 of a 2 mg/mL solution of TAA in the assay could negatively impact on nematode viability/motility, causing the 36% and 97% immotile worms at 48 and 72 h, respectively (Figure 3A, bar groupings 3 and 4). Assays with TAA extract and TAA standard in 0.1 M Na_2CO_3 (0.1–2 mg/mL of TAA) adjusted to pH 6 showed the same nematocidal activities

(Figure 3A, bar groupings 1 and 2) as in the buffer control (0 mg/mL TAA). Subsequent tests with buffered solutions of 2 mg/mL TAA (or buffer only) adjusted to pH 3.0 to pH 6.0 suggest that nematocidal activity of TAA may be entirely attributed to its acidity during assays (Figure 3B). This is supported by Khanna et al. [47], who reported an increased mortality of *C. elegans* of below pH 3.1–3.4, depending on the environment (salinity). It is important to note that a previous report did not study or list the pH in their nematode studies with TAA [9]. Note that the majority of aconitate is fully protonated below pH 2.71 (=pK_{a1}) [45].

3.3. Nematicidal Assays with *M. incognita*

While the crude TAA extract did not appear to be an effective nematicide against *C. elegans*, a recent study suggested that TAA may affect free-living nematodes differently than parasitic nematodes such as *Meloidogyne incognita* [10]. As such, assays were performed to test the inhibitory activity of the crude TAA extract and a TAA standard, both diluted to 2 mg/mL and adjusted to pH 2.7 and pH 6, against the root knot nematode *M. incognita* (Figure 4). Nematode motility was inhibited, and mortality was high with TAA in water at pH 2.7. However, we found similar results with a Na_2CO_3 solution at pH 2.7 (without TAA). We found that the nematode was unaffected at 48 h and was modestly affected at 72 h by TAA in Na_2CO_3 at pH 6.0, but was unaffected by Na_2CO_3 at pH 6. This suggests that the impact on *M incognita* by TAA is pH-dependent. This finding contradicts a previous report that found 92.1% mortality of *M. incognita* with TAA (1 mg/mL, pH unknown) [9]. Interestingly, our study showed a significant inhibition from the TAA extract at both pH 2.7 and pH 6.0, which suggests that there are unidentified compounds in the extract that inhibit *M. incognita*. Further studies may explore this, and it is important to note that residual acetone, butanol, and ethanol (0.0, 0.5, and 0.4 mg/mL at 2 mg/mL TAA) in the crude extract may also affect nematodes. The results obtained at pH 2.7 may be less relevant, as this soil organism may not experience pH levels below pH 4 [48]. In conclusion, *M. incognita* motility and mortality were affected by the TAA extract, while *C. elegans* was not.

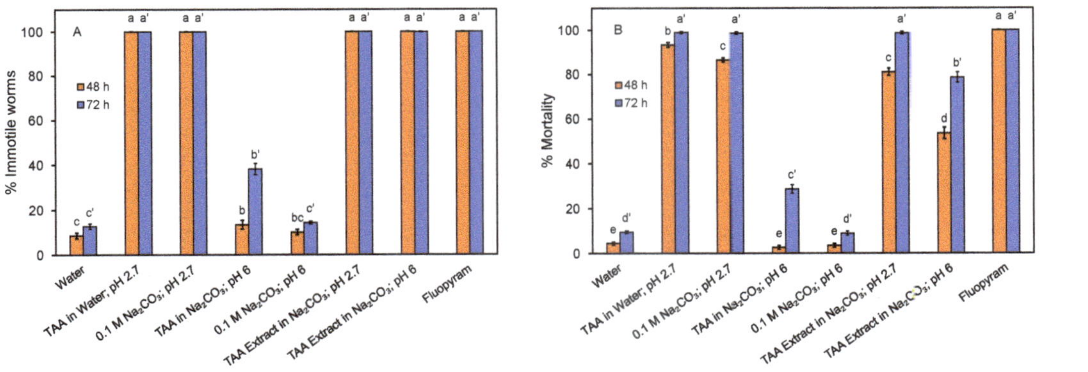

Figure 4. Nematicidal assays, mobility (**A**), and mortality (**B**), with crude TAA extract and standards versus *M. incognita*. All TAA solutions contained 2 mg/mL. Fluopyram (4 mg/mL) was used as a control. Percent immotile includes dead worms. Significant differences between treatment conditions (within each graph) are represented by different letters above standard error bars.

3.4. Process Economics of Aconitic Acid Extraction from Sweet Sorghum Syrup

SuperPro Designer was used for design and process economics. The flow sheet of the process is shown in Figure 5. Various variations to the basic flow sheet were explored, primarily for heat recovery and heat management, but very little impact was noted on the final economic analysis.

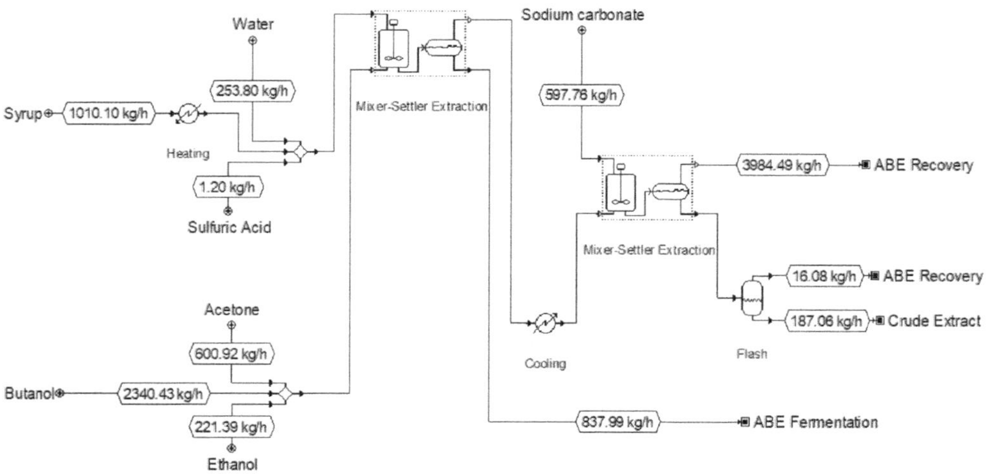

Figure 5. Process flow diagram of aconitic acid extraction and recovery from sweet sorghum syrup. The SuperPro Designer file is provided in the Supplementary Materials.

The estimated equipment cost to produce 1,481,483 kg extract per year was USD 93,000, and the total capital investment was USD 3,211,000. The significant difference between these values is due to the initial purchase of raw materials (mainly ABE). A breakdown of the capital and operating cost is shown in Table 2. The estimated net cost of production would be USD 16.64/kg of extract. It is difficult to compare this cost compared the estimated production cost developed by Regna and Bruins [16] over 50 years ago. The estimated production cost is derived mainly from solvent (ABE) use, which was not recycled in the process and was only valued at 10% of its purchase cost. It is envisioned that the extract production would be part of an integrated biorefinery producing ABE from sweet sorghum syrup. In addition, the annual production volume is low for this highly specialized product, which does not have the economic benefit of a larger production scale.

Table 2. Capital and operating cost for an extraction process to produce an aconitic acid extract.

Cost Item	Cost
Total Capital Investment	USD 3,211,000
Annual Operating Cost (AOC)	USD 29,059,000
Raw Materials	USD 26,475,000 (91% of AOC)
Labor	USD 2,108,000 (7% of AOC)
Facilities/Laboratory	USD 425,000 (1% of AOC)
Utilities	USD 51,200 (0.2% of AOC)
Credits (sugars and ABE)	USD 4,406,000 (15% of AOC)
Net Production Cost	USD 16.64/kg of extract

4. Conclusions

Aconitic acid can be efficiently extracted from sweet sorghum syrup using acetone: butanol:ethanol mixtures. An aconitic acid extract can then be recovered from this solvent with a Na_2CO_3 solution, with an overall extraction and recovery efficiency of 86%. The estimated production cost was USD 16.64/kg of extract, and this was highly dependent on the solvent cost, as the solvent was not recycled but was resold for recovery at a fraction of the cost.

The extract, containing *trans*-aconitic acid (TAA), had little impact on the motility of the model nematode *C. elegans* when compared to chemically pure TAA or the pH buffered control. In conclusion, it was determined that the low pH effect of unbuffered TAA in the

C. elegans nematocidal assay was responsible for increased motility reduction. The extract was effective in reducing the motility of the parasitic *M. incognita* and causing over 78% mortality of the nematode. However, this positive result could not be conclusively linked to TAA. Thus, in contrast to prior reports, we found that aconitic acid was not an effective inhibitor of the nematodes *C. elegans* and *M. incognita*. Finally, an unidentified component (or components) present in the acetone:butanol:ethanol sweet sorghum extract appears to be an effective inhibitor of *M. incognita*, and may merit further investigation.

Supplementary Materials: The following supporting information can be downloaded at https://www.mdpi.com/article/10.3390/life13030724/s1, SuperPro Designer Process File: MDPI-Life-Klasson-2023.spf; Figure S1: Example of HPLC chromatogram with labeled peaks.

Author Contributions: Conceptualization, K.T.K.; methodology, K.T.K., Y.Q., G.O.B. and T.T.W.; software, K.T.K. and B.T.P.; formal analysis, K.T.K., Y.Q., G.O.B., B.T.P. and T.T.W.; investigation, K.T.K., Y.Q., G.O.B., T.T.W. and B.T.P.; writing—original draft preparation, K.T.K., Y.Q., G.O.B. and E.T.; writing—review and editing, K.T.K., Y.Q., G.O.B., T.T.W., B.T.P. and E.T.; visualization, K.T.K.; supervision, K.T.K. and G.O.B. All authors have read and agreed to the published version of the manuscript.

Funding: This work was supported by the U.S. Department of Agriculture (USDA), Agricultural Research Service (ARS), under project number 6054-41000-114-000-D.

Institutional Review Board Statement: Not applicable.

Informed Consent Statement: Not applicable.

Data Availability Statement: The data presented in the figures are available from the corresponding author upon reasonable request.

Acknowledgments: Mention of trade names or commercial products in this publication is solely for the purpose of providing specific information, and does not imply recommendation or endorsement by USDA. USDA is an equal opportunity provider and employer.

Conflicts of Interest: The authors declare no conflict of interest. The authors are employed by the funding organizations (USDA and LSU). However, the funders had no role in the design of the study; in the collection, analyses, or interpretation of data or in the writing of the manuscript; but approved the decision to publish the results.

References

1. Werpy, T.; Petersen, G. *Top Value Added Chemicals from Biomass, Volume I: Results from Screening for Potential Candidates from Sugars and Synthesis Gas*; Pacific Northwst National Laboratory and the National Renewable Energy Laboratory: Washington, DC, USA, August 2004.
2. Bruni, G.O.; Klasson, K.T. Aconitic acid recovery from renewable feedstock and review of chemical and biological applications. *Foods* **2022**, *11*, 573. [CrossRef]
3. Cao, H.; Zheng, Y.; Zhou, J.; Wang, W.; Pandit, A. A novel hyperbranched polyester made from aconitic acid (B3) and di(ethylene glycol) (A2). *Polym. Int.* **2011**, *60*, 630–634. [CrossRef]
4. Gilfillan, W.N.; Doherty, W.O.S. Starch composites with aconitic acid. *Carbohydr. Polym.* **2016**, *141*, 60–67. [CrossRef] [PubMed]
5. Tzereme, A.; Christodoulou, E.; Kyzas, G.Z.; Kostoglou, M.; Bikiaris, D.N.; Lambropoulou, D.A. Chitosan grafted adsorbents for diclofenac pharmaceutical compound removal from single-component aqueous solutions and mixtures. *Polymers* **2019**, *11*, 497. [CrossRef] [PubMed]
6. Hu, L.; Bui, V.T.; Huang, L.; Singh, R.P.; Lin, H. Facilely cross-linking polybenzimidazole with polycarboxylic acids to improve H_2/CO_2 separation performance. *ACS Appl. Mater. Interfaces* **2021**, *13*, 12521–12530. [CrossRef]
7. Mai, A.Q.; Davies, J.; Nguyen, D.; Carranza, A.; Vincent, M.; Pojman, J.A. Microparticles and latexes prepared via suspension polymerization of a biobased vegetable oil and renewable carboxylic acid. *J. Appl. Polym. Sci.* **2021**, *138*, 50180. [CrossRef]
8. Clarke, A.J.; Shepherd, A.M. Hatching agents for the potato cyst-nematode, *Heterodera rostochiensis* Woll. *Ann. Appl. Biol.* **1968**, *61*, 139–149. [CrossRef]
9. Du, C.; Cao, S.; Shi, X.; Nie, X.; Zheng, J.; Deng, Y.; Ruan, L.; Peng, D.; Sun, M. Genetic and biochemical characterization of a gene operon for *trans*-aconitic acid, a novel nematicide from *Bacillus thuringiensis*. *J. Biol. Chem.* **2017**, *292*, 3517–3530. [CrossRef]
10. Rocha, T.L.; Soll, C.B.; Boughton, B.A.; Silva, T.S.; Oldach, K.; Firmino, A.A.P.; Callahan, D.L.; Sheedy, J.; Silveira, E.R.; Carneiro, R.M.D.G.; et al. Prospection and identification of nematotoxic compounds from *Canavalia ensiformis* seeds effective in the control of the root knot nematode *Meloidogyne incognita*. *Biotechnol. Res. Innov.* **2017**, *1*, 87–100. [CrossRef]

11. Gil Zapata, N.J. Aconitic Acid from Sugarcane: Production and Industrial Application. Doctoral Dissertation, Louisiana State University, Baton Rouge, LA, USA, December 2007.
12. Kanitkar, A.; Aita, G.; Madsen, L. The recovery of polymerization grade aconitic acid from sugarcane molasses. *J. Chem. Technol. Biotechnol.* **2013**, *88*, 2188–2192. [CrossRef]
13. Klasson, K.T.; Qureshi, N.; Powell, R.; Heckemeyer, M.; Eggleston, G. Fermentation of sweet sorghum syrup to butanol in the presence of natural nutrients and inhibitors. *Sugar Tech* **2018**, *20*, 224–234. [CrossRef]
14. Uchimiya, M.; Knoll, J.E.; Anderson, W.F.; Harris-Shultz, K.R. Chemical analysis of fermentable sugars and secondary products in 23 sweet sorghum cultivars. *J. Agric. Food Chem.* **2017**, *65*, 7629–7637. [CrossRef] [PubMed]
15. Blinco, J.; Griffin, G.J.; Hermans, J. Extraction of aconitic acid from cane molasses by solvent extraction. In Proceedings of the Chemeca 2000: Opportunities and Challenges for the Resource and Processing Industries, Perth Western Australia, 9–12 July 2000; pp. 644–650.
16. Regna, E.A.; Bruins, P.F. Recovery of aconitic acid from molasses. *Ind. Eng. Chem.* **1956**, *48*, 1268–1277. [CrossRef]
17. Producer Price Index by Commodity: Chemicals and Allied Products: Industrial Chemicals. Available online: https://fred.stlouisfed.org/series/WPS061 (accessed on 12 December 2022).
18. Ventre, E.K.; Ambler, J.A.; Henry, H.C.; Byall, S.; Paine, H.S. Extraction of aconitic acid from sorgo. *Ind. Eng. Chem.* **1946**, *38*, 201–204. [CrossRef]
19. Eggleston, G.; Cole, M.; Andrzejewski, B. New commercially viable processing technologies for the production of sugar feedstocks from sweet sorghum (*Sorghum bicolor* L. Moench) for manufacture of biofuels and bioproducts. *Sugar Tech* **2013**, *15*, 232–249. [CrossRef]
20. Woods, D.R. The genetic engineering of microbial solvent production. *Trends Biotechnol.* **1995**, *13*, 259–264. [CrossRef]
21. Green, E.M. Fermentative production of butanol-the industrial perspective. *Curr. Opin. Biotechnol.* **2011**, *22*, 337–343. [CrossRef]
22. Gi Moon, H.; Jang, Y.S.; Cho, C.; Lee, J.; Binkley, R.; Lee, S.Y. One hundred years of clostridial butanol fermentation. *FEMS Microbiol. Lett.* **2016**, *363*, fnw001. [CrossRef]
23. Friedl, A. Downstream process options for the ABE fermentation. *FEMS Microbiol. Lett.* **2016**, *363*, fnw073. [CrossRef]
24. Chen, H.; Cai, D.; Chen, C.; Wang, J.; Qin, P.; Tan, T. Novel distillation process for effective and stable separation of high-concentration acetone–butanol–ethanol mixture from fermentation–pervaporation integration process. *Biotechnol. Biofuels* **2018**, *11*, 286. [CrossRef]
25. Haigh, K.F.; Petersen, A.M.; Gottumukkala, L.; Mandegari, M.; Naleli, K.; Görgens, J.F. Simulation and comparison of processes for biobutanol production from lignocellulose via ABE fermentation. *Biofuels Bioprod. Bioref.* **2018**, *12*, 1023–1036. [CrossRef]
26. Qureshi, N.; Lin, X.; Liu, S.; Saha, B.C.; Mariano, A.P.; Polaina, J.; Ezeji, T.C.; Friedl, A.; Maddox, I.S.; Klasson, K.T.; et al. Global view of biofuel butanol and economics of its production by fermentation from sweet sorghum bagasse, food waste, and yellow top presscake: Application of novel technologies. *Fermentation* **2020**, *6*, 58. [CrossRef]
27. Pudjiastuti, L.; Widjaja, T.; Iskandar, K.K.; Sahid, F.; Nurkhamidah, S.; Altway, A.; Putra, A.P. Modelling and simulation of multicomponent acetone-butanol-ethanol distillation process in a sieve tray column. *Heliyon* **2021**, *7*, e06641. [CrossRef] [PubMed]
28. Jones, D.T.; Woods, D.R. Acetone-butanol fermentation revisited. *Microbiol. Rev.* **1986**, *50*, 484–524. [CrossRef]
29. Datta, R.; Zeikus, J.G. Modulation of acetone-butanol-ethanol fermentation by carbon monoxide and organic acids. *Appl. Environ. Microbiol.* **1985**, *49*, 522–529. [CrossRef]
30. Yerushalmi, L.; Volesky, B.; Szczesny, T. Effect of increased hydrogen partial pressure on the acetone-butanol fermentation by *Clostridium acetobutylicum*. *Appl. Microbiol. Biotechnol.* **1985**, *22*, 103–107. [CrossRef]
31. Anbarasan, P.; Baer, Z.C.; Sreekumar, S.; Gross, E.; Binder, J.B.; Blanch, H.W.; Clark, D.S.; Dean Toste, F. Integration of chemical catalysis with extractive fermentation to produce fuels. *Nature* **2012**, *491*, 235–239. [CrossRef]
32. Klasson, K.T.; Qureshi, N.; Heckemeyer, M.; Eggleston, G. Biobutanol production from sweet sorghum biorefinery byproducts. In Proceedings of the Advances in Sugar Crop Processing and Conversion, New Orleans, LA, USA, 15–18 March 2018; pp. 274–281.
33. Corsi, A.K. A biochemist's guide to *Caenorhabditis elegans*. *Anal. Biochem.* **2006**, *359*, 1–17. [CrossRef]
34. Day, R.W.; Quinn, G.P. Comparisons of treatments after an analysis of variance in ecology. *Ecol. Monogr.* **1989**, *59*, 433–463. [CrossRef]
35. Kearn, J.; Ludlow, E.; Dillon, J.; O'Connor, V.; Holden-Dye, L. Fluensulfone is a nematicide with a mode of action distinct from anticholinesterases and macrocyclic lactones. *Pestic. Biochem. Physiol.* **2014**, *109*, 44–57. [CrossRef]
36. DiLegge, M.J.; Manter, D.K.; Vivanco, J.M. A novel approach to determine generalist nematophagous microbes reveals *Mortierella globalpina* as a new biocontrol agent against *Meloidogyne* spp. nematodes. *Sci. Rep.* **2019**, *9*, 7521. [CrossRef]
37. Porta-de-la-Riva, M.; Fontrodona, L.; Villanueva, A.; Cerón, J. Basic *Caenorhabditis elegans* methods: Synchronization and observation. *J. Visual. Exp.* **2012**, *64*, e4019. [CrossRef]
38. Stiernagle, T. Maintenance of *C. elegans* (February 11, 2006). In *WormBook*, ed.; The C. elegans Research Community. Available online: http://www.wormbook.org (accessed on 19 January 2023). [CrossRef]
39. Watson, T.T. Sensitivity of *Meloidogyne enterolobii* and *M. incognita* to fluorinated nematicides. *Pest Manage. Sci.* **2022**, *78*, 1398–1406. [CrossRef]
40. Klasson, K.T.; Cole, M.R.; Pancio, B.T.; Heckemeyer, M. Development of an enzyme cocktail to bioconvert untapped starch in sweet sorghum processing by-products: Part II. Application and economic potential. *Ind. Crops Prod.* **2022**, *176*, 114370. [CrossRef]

41. Cheng, M.H.; Huang, H.; Dien, B.S.; Singh, V. The costs of sugar production from different feedstocks and processing technologies. *Biofuels Bioprod. Bioref.* **2019**, *13*, 723–739. [CrossRef]
42. Intratec. Commodity Production Costs. Available online: https://www.intratec.us/products/commodity-production-costs (accessed on 15 December 2022).
43. Almodares, A.; Ranjbar, M.; Hadi, M.R. Effects of nitrogen treatments and harvesting stages on the aconitic acid, invert sugar and fiber in sweet sorghum cultivars. *J. Environ. Biol.* **2010**, *31*, 1001–1005. [PubMed]
44. Amorim, H.V. Challenges to produce ethanol from sweet sorghum in Brazil. In Proceedings of the Sweet Sorghum Association 2015 Annual Conference, Orlando, FL, USA, 27–29 January 2015.
45. Klasson, K.T. The inhibitory effects of aconitic acid on bioethanol production. *Sugar Tech* **2018**, *20*, 88–94. [CrossRef]
46. Piper, P.; Calderon, C.O.; Hatzixanthis, K.; Mollapour, M. Weak acid adaptation: The stress response that confers yeasts with resistance to organic acid food preservatives. *Microbiology* **2001**, *147*, 2635–2642. [CrossRef]
47. Khanna, N.; Cressman Iii, C.P.; Tatara, C.P.; Williams, P.L. Tolerance of the nematode *Caenorhabditis elegans* to pH, salinity, and hardness in aquatic media. *Arch. Environ. Contam. Toxicol.* **1997**, *32*, 110–114. [CrossRef]
48. Melakeberhan, H.; Dey, J.; Baligar, V.C.; Carter, T.E., Jr. Effect of soil pH on the pathogenesis of *Heterodera glycines* and *Meloidogyne incognita* on *Glycine max* genotypes. *Nematology* **2004**, *6*, 585–592. [CrossRef]

life

Article

Genome-Wide Identification of *LeBAHDs* in *Lithospermum erythrorhizon* and In Vivo Transgenic Studies Confirm the Critical Roles of *LeBAHD1/LeSAT1* in the Conversion of Shikonin to Acetylshikonin

Xuan Wang [1,2], Zhuoyu He [1], Huan Yang [1], Cong He [1], Changyi Wang [1], Aliya Fazal [1], Xiaohui Lai [1], Liangjie Yang [3], Zhongling Wen [1,2], Minkai Yang [1,2], Shenglin Ma [1], Wencai Jie [1], Jinfeng Cai [2], Tongming Yin [2], Bao Liu [4], Yonghua Yang [1,2,*] and Jinliang Qi [1,2,*]

1 State Key Laboratory of Pharmaceutical Biotechnology, Institute for Plant Molecular Biology, School of Life Sciences, Nanjing University, Nanjing 210023, China
2 Co-Innovation Center for Sustainable Forestry in Southern China, Nanjing Forestry University, Nanjing 210037, China
3 Yili Key Laboratory of Applied Research and Development on Active Ingredients of Chinese Herbal Medicine, Yili National Agricultural Science and Technology Park at Xinjiang, Yili 835600, China
4 Key Laboratory of Molecular Epigenetics of the Ministry of Education (MOE), Northeast Normal University, Changchun 130024, China
* Correspondence: yangyh@nju.edu.cn (Y.Y.); qijl@nju.edu.cn (J.Q.)

Abstract: The BAHD acyltransferase family is a unique class of plant proteins that acylates plant metabolites and participates in plant secondary metabolic processes. However, the BAHD members in *Lithospermum erythrorhizon* remain unknown and uncharacterized. Although the heterologously expressed *L. erythrorhizon* BAHD family member LeSAT1 in *Escherichia coli* has been shown to catalyze the conversion of shikonin to acetylshikonin in vitro, its in vivo role remains unknown. In this study, the characterization, evolution, expression patterns, and gene function of *LeBAHDs* in *L. erythrorhizon* were explored by bioinformatics and transgenic analysis. We totally identified 73 *LeBAHDs* in the reference genome of *L. erythrorhizon*. All *LeBAHDs* were phylogenetically classified into five clades likely to perform different functions, and were mainly expanded by dispersed and WGD/segmental duplication. The in vivo functional investigation of the key member *LeBAHD1/LeSAT1* revealed that overexpression of *LeBAHD1* in hairy roots significantly increased the content of acetylshikonin as well as the conversion rate of shikonin to acetylshikonin, whereas the CRISPR/Cas9-based knockout of *LeBAHD1* in hairy roots displayed the opposite trend. Our results not only confirm the in vivo function of *LeBAHD1/LeSAT1* in the biosynthesis of acetylshikonin, but also provide new insights for the biosynthetic pathway of shikonin and its derivatives.

Keywords: *Lithospermum erythrorhizon*; acetylshikonin; shikonin; BAHD; LeSAT1; hairy root

check for updates

Citation: Wang, X.; He, Z.; Yang, H.; He, C.; Wang, C.; Fazal, A.; Lai, X.; Yang, L.; Wen, Z.; Yang, M.; et al. Genome-Wide Identification of *LeBAHDs* in *Lithospermum erythrorhizon* and In Vivo Transgenic Studies Confirm the Critical Roles of *LeBAHD1/LeSAT1* in the Conversion of Shikonin to Acetylshikonin. *Life* **2022**, *12*, 1775. https://doi.org/10.3390/life12111775

Academic Editor: Jianfeng Xu

Received: 20 September 2022
Accepted: 30 October 2022
Published: 3 November 2022

1. Introduction

Red naphthoquinones are the main secondary metabolites accumulated in the roots of a few *Boraginaceae* plants such as *Lithospermum erythrorhizon* and *Arnebia euchroma*, among which shikonin and its derivatives such as acetylshikonin, isobutyrylshikonin, β,β-dimethylacrylshikonin have been shown to have broad medicinal value and industrial applications [1–3]. Acetylshikonin is the main naphthoquinone in the root periderm of the traditional Chinese medicine plant *L. erythrorhizon*, accounting for about 50% of the total [4]. Acetylshikonin has attracted much attention as a potential anticancer drug because of its ability to inhibit cell proliferation [5] and induce cell autophagy [6], and can inhibit the growth of colorectal cancer [7], liver cancer [8], oral cancer [9], melanoma [10], etc., and is less toxic to normal human cells [11,12]. Therefore, it has become urgent and crucial to dissect the biosynthetic mechanism of acetylshikonin and improve its yield in plant-derived materials.

A two stage culture system, where: (1) cell cultures [13] or hairy roots [14] of *L. erythrorhizon* are cultured in a growth medium (B5 medium) under light for proliferation; (2) the proliferated hairy roots or cells are transferred to the production medium (M9 medium) in darkness to induce the production of shikonin and its derivatives, has been successfully developed to study the biochemical and molecular mechanisms of the biosynthesis of shikonin and its derivatives, as well as the yield improvement of these useful compounds [15]. Based on this culture system, many key genes, such as *LePGT* [16], *LeC4H* [17], *LePAL* [18], *Le4CL* [19], hydroxylase *CYP76B74* [20], *CYP76B101* [21] *LeDSH* [22], etc., have been excavated, and the pathway for shikonin biosynthesis was well characterized. However, the key enzymes that catalyze the conversion of shikonin to its derivatives have not been fully elucidated.

Most shikonin derivatives are produced by the acylation of different types of acyl groups from CoA thioester acyl donors to shikonin side chains catalyzed by acyltransferases [23]. The plant BAHD acyltransferases family, which was named after the initials of the four characterized members (benzylalcohol O-acetyltransferase from *Clarkia breweri*, BEAT; anthocyanin O-hydroxycinnamoyltransferases from *Petunia*, *Senecio*, *Gentiana*, *Perilla*, and *Lavandula*, AHCTs; anthranilate N-hydroxycinnamoyl/benzoyltransferase from *Dianthus caryophyllus*, HCBT; deacetylvindoline 4-O-acetyltransferase from *Catharanthus roseus*, DAT) [24,25], has been shown to catalyze the transfer of acyl moieties to a variety of acceptor molecules and were involved in the biosynthesis of various natural secondary metabolites such as flavonoids, phenols, alkaloids, anthocyanins, and volatile esters [25,26]. Recently, a shikonin O-acyltransferases gene (*LeSAT1*) belonging to the BAHD gene family was identified in *L. erythrorhizon*, its function of catalyzing acetyl-CoA, isobutyryl-CoA, and isovaleryl-CoA as acyl donors to generate acetylshikonin, isobutyrylshikonin, and isovalerylshikonin was verified by in vitro enzyme activity, and the acetylation activity of *LeSAT1* was the strongest [27]. However, the in vitro catalytic experiments of *LeSAT1's* functions need to be further confirmed by in vivo transgenic studies. After that, Tang [28] conducted a preliminary screening and evolutionary process exploration of the alkannin/shikonin O-acyltransferase gene (i.e., AAT/SAT)-like superfamily members in *L. erythrorhizon* with AAT/SAT's amino acid sequences (i.e., BBV14785.1 and BBV14786.1) as the queries. However, the genome-wide identification and characterization of the *LeBAHD* gene family in *L. erythrorhizon* still remains elusive.

Here, we conducted genome-wide identification of *LeBAHD* members in *L. erythrorhizon*, fully characterized their regulatory elements and expression patterns, and predicted gene functions based on their evolutionary status; moreover, we confirmed the in vivo effect of LeBAHD1/LeSAT1 in catalyzing the conversion of shikonin to acetylshikonin in plants using overexpression and CRISPR/Cas9 strategies based on the two-stage culture system of *L. erythrorhizon* transgenic hairy roots. Our study not only pioneers the use of the CRISPR/Cas9 gene knockout system on the hairy roots of *L. erythrorhizon*, laying a practical foundation for more precisely revealing the biochemical and molecular mechanisms underlying the biosynthesis and regulation of shikonin and its derivatives, but also provides a method for the efficient biosynthesis of acetylshikonin through genetic engineering of *L. erythrorhizon*.

2. Materials and Methods

2.1. Plant Materials and Growth Conditions

The seeds of *Lithospermum erythrorhizon* Sieb. et Zucc collected in Tuoyaozi Town, Huanan County, Jiamusi City, Heilongjiang Province, China (130°78′81″ E/46°23′95″ N), were dispersed in wet sand for about one month at 4 °C in the dark. After germination, the seeds were transplanted into the soil in the greenhouse for cultivation at 25 °C with a 16 h day/8 h night photoperiod and 100 μmol m^{-2} s^{-1} light intensity until the seedlings developed 8 leaves, which were used as explants for hairy root induction.

2.2. Identification of LeBAHD Superfamily

To identify the putative *LeBAHD* genes from *L. erythrorhizon*, several approaches were synergistically employed. The BAHD family characteristic domain (Pfam domain: PF02458) and HMMER 3.0 were used to search the genome of *L. erythrorhizon* for candidate *LeBAHD* genes with E-values less than e^{-10} [29]; in addition, 55 AtBAHD protein sequences of *Arabidopsis thaliana* were used to blastp the *L. erythrorhizon* genome; then, redundant sequences and abnormal sequences (including incomplete PF02458 domain, lacking initiation codon and/or termination codon, and lack of one or two characteristic motifs: HXXXD and DFGWG) identified by Batch CD-Search (https://www.ncbi.nlm.nih.gov/Structure/bwrpsb/bwrpsb.cgi, accessed on 10 September 2022) and Motif analysis (https://meme-suite.org/meme/tools/meme, accessed on 10 September 2022) were then removed.

2.3. Bioinformatic Analysis

The ExPASy website (https://web.expasy.org/protparam/, accessed on 10 September 2022) was used to analyze the chemical characteristics of LeBAHDs, and the subcellular localization prediction was carried out using Cell-PLoc 2.0 (http://www.csbio.sjtu.edu.cn/bioinf/Cell-PLoc-2/, accessed on 10 September 2022). The full-length amino acid sequence of LeBAHDs was subjected to a conserved motif analysis using the MEME program (https://meme-suite.org/meme/tools/meme, accessed on 10 September 2022) with the default setting of 15 motifs. The gene structure was analyzed using the Gene Structure Display Server 2.0 (http://gsds.gao-lab.org/, accessed on 10 September 2022). Gene duplication types of *LeBAHDs* were analyzed using the 'MCScanX' and 'duplicate_gene_classifier' programs implemented in the MCScanX package. The cis-acting elements of the promoter region (2000 bp sequence upstream of the start codon) of the *LeBAHD* genes were analyzed using the PlantCARE program (http://bioinformatics.psb.ugent.be/webtools/plantcare/html/, accessed on 10 September 2022).

The phylogenetic trees were constructed using the following three steps: (1) the amino acid sequences of LeBAHD family members were aligned via prank v17042751 (key parameter: –F–codon); (2) the preliminary alignment was trimmed using trimAL v.1.2.rev59 (key parameter: –gt 0.50); (3) the trimmed alignment was then used to construct the phylogenetic tree via iqtree-2.2.0-Windows according to the Maximum Likelihood (ML) method with 1000 bootstrap replications. The constructed phylogenetic tree was embellished using iTOL (https://itol.embl.de/, accessed on 10 September 2022). Protein sequences of the published BAHD family members (Table S1) used for LeBAHDs phylogenetic analysis were retrieved from the Genbank database.

To explore the specific expression patterns of *LeBAHDs* in different *L. erythrorhizon* tissues, RNA-seq data from *L. erythrorhizon* mature roots (MR), root periderm (PD), root cortex (CT), root stele (SE), mature leaves + stems (ML) and flowers (FL) were downloaded from the NCBI SRA (accession ID: SRP141330) and used for gene expression analysis. The heatmap was drawn in Rstudio using pheatmap.

2.4. cDNA Cloning of LeBAHD1/SAT1, Plasmid Construction and Transformation

Total RNA of the red roots of *L. erythrorhizon* seedlings was extracted using the FastPure Plant Total RNA Isolation Kit (Vazyme, #RC401, Nanjing, China) and cDNA was synthesized by reverse transcription with the HiScript III 1st Strand cDNA Synthesis Kit (+gDNA wiper) (Vazyme, #R312, Nanjing, China). With the specific primers (Table S2) designed by Primer Premier 5.0, the full-length open-reading frame (ORF) of *LeBAHD1*/*LeSAT1* (1317 bp, GenBank number: LC520137.1) was amplified with a high-fidelity enzyme (Vazyme, #P515, Nanjing, China) using the above cDNA as the template. The PCR parameters were as follows: 95 °C for 3 min, followed by 34 cycles at 95 °C for 15 s, 58 °C for 15 s, and 72 °C for 1 min, with a final extension at 72 °C for 10 min.

For the construction of *LeBAHD1*/*LeSAT1* overexpression vector (pBI121–*LeBAHD1*/*LeSAT1*–*eGFP*), the above amplified full-length ORF of *LeSAT1* was subcloned into the *Xba* I/*Bam* HI sites of the plant expression vector pBI121–eGFP by homologous recombination. The *LeBAHD1/LeSAT1* knockout vector (pYLCRISPR/Cas9–*LeBAHD1/LeSAT1* was constructed using the CRISPR/Cas9 vector system (pYLCRISPR/Cas9P_{ubi}-H, sgRNA promoter of *A. thaliana* AtU6-29). The sgRNA target was designed on the sense and antisense strand of its first exon according to the *LeSAT1* gene sequence, and the target sequence was imported into the sgRNA expression box by overlapping PCR. The PCR parameters were as follows: 95 °C for 3 min, followed by 30 cycles at 95 °C for 15 s, 58 °C for 15 s, and 68 °C for 20 s, with a final extension at 68 °C for 2 min. Then, the sgRNA expression box was added to the skeleton vector through the digestion at *Bsa* I site.

All the recombined expression vectors were respectively transformed into competent cells of *Escherichia coli* strain Top10 by heat shock and verified before being introduced into competent cells of *Agrobacterium rhizogenes* strain ATCC15834 by electroporation. All the primers used for the vector construction, as well as the verification and identification of *A. rhizogenes* strain ATCC15834 harboring the constructs, are listed in Table S2.

2.5. Hairy Root Induction, Culture, and Validation

For transgenic hairy root induction and culture, the explants cut from seedling leaves (1.0–1.5 cm) were placed in MS cultures containing 0.2 mg/L 6-benzylaminopurine and 2 mg/L 2,4-dichlorophenoxyacetic acid and incubated in the dark at 25 °C for 2 days. Meanwhile, the strain ATCC15834 was successfully transformed with pBI121–eGFP (EV), pBI121–*LeBAHD1/LeSAT1* (OE), pYLCRISPR/Cas9Pubi-H (MH), or pYLCRISPR/Cas9–*LeBAHD1/LeSAT1* (MH-K) plasmid, and incubated in liquid YEB medium with 50 mg/L kanamycin on a rotary shaker at 120 rpm at 26–28 °C to an OD_{600} of 0.6. Then acetosyringone was added to the above bacterial culture medium to a final concentration of 0.1 mM to prepare an infection solution. The precultured leaf explants were placed in the infection solution and treated in the dark at 28 °C for 30 min. Then, the explants were incubated in MS solid medium at 26 °C for 2 days in the dark. Afterwards, the infected explants were washed three times with sterilized water, and then the explants were put into MS solid medium supplemented with 500 mg of cefotaxime and cultured at 26 °C in the dark. Hairy roots appeared after 2–3 weeks of cultivation. The developed hairy roots of approximately 2 cm in length were excised from the infection sites and then subcultured on solid 1/2 B5 medium supplemented with 500 mg/L cefotaxime at 26–28 °C for about 1 week to eliminate *Agrobacterium*. The concentration of cefotaxime was constantly reduced to eliminate *Agrobacterium* completely. Finally, the hairy roots of EV, MH, OE or MH-K were transferred into antibiotic-free and hormone-free 1/2 B5 solid medium for continuous growth at 26–28 °C. Then, the *rolC* of *A. rhizogenes* or hygromycin resistance gene (*HPT*) gene was amplified in the genomic DNA extracted from transgenic hairy roots to confirm that the hairy roots were transgenic hairy roots, not aerial roots. The PCR parameters were as follows: 95 °C for 3 min, followed by 34 cycles at 95 °C for 15 s, 54 °C for 15 s, and 72 °C for 45 s, with a final extension at 72 °C for 10 min. The transgenic efficiency of the OE hairy roots was verified by quantitative real-time PCR (RT-qPCR) using gene specific primers. The RT-qPCR parameters were as follows: 95 °C for 5 min, followed by 40 cycles at 95 °C for 10 s and 60 °C for 30 s, with a final extension at amplification conditions of a default dissolution curve. Primers were designed on 200–300 bp sequences on both sides of the sgRNA target, and the target sequence was amplified from genomic DNA extracted from MH-K hairy roots to detect its gene editing effect; the PCR parameters were as follows: 95 °C for 3 min, followed by 34 cycles at 95 °C for 15 s, 56 °C for 15 s, and 72 °C for 40 s, with a final extension at 72 °C for 10 min. All primers used for transgenic hairy root validation are listed in Table S2.

The subcultured hairy roots were transferred into a 50 mL Erlenmeyer flask containing 20 mL of 1/2 B5 liquid medium at 28 °C under light with shaking at 100 rpm for rapid proliferation. Then, these proliferated hairy roots were transferred from 1/2 B5 proliferation medium into 20 mL of M9 production medium and incubated in the dark at 28 °C and 100 rpm to induce the production of shikonin and its derivatives, as described in a previous report [30].

2.6. eGFP Fluorescence Detection of Hairy Roots

The transgenic hairy roots expressing 35S: eGFP or 35S: *LeBAHD1/LeSAT1*–eGFP were directly put on a confocal laser scanning fluorescence microscope (LSFM, FV10-ASW, Olympus, Japan) for the observation of eGFP fluorescence. The fluorescence signal of GFP was excited at 488 nm and the emission wavelength was detected at 510 nm as previously reported [31].

2.7. RNA Extraction and RT-qPCR Analysis

Total RNA was extracted from different transgenic hairy root lines (EV, OE, MH, or MH-K), or from different tissues of *L. erythrorhizon* (root, stem, or leaf) using the FastPure Plant Total RNA Isolation Kit (Vazyme, #RC401, Nanjing, China). The RNA purity and integrity were assessed based on the A_{260}/A_{280} absorbance ratio and 1.0% agarose gel electrophoresis. Approximately 1 µg of total RNA was used to synthesize cDNA by reverse transcription with the HiScript III 1st Strand cDNA Synthesis Kit (+gDNA wiper) (Vazyme, #R312, Nanjing, China), and RT-qPCR was performed using ChamQ Universal SYBR qPCR Master Mix (Vazyme, #Q711, Nanjing, China) with gene-specific primers (Table S2) on an Applied Biosystems 7500 Real-Time PCR System and StepOnePlus™ Real-Time PCR System. The RT-qPCR parameters were as follows: 95 °C for 5 min, followed by 40 cycles at 95 °C for 10 s and 60 °C for 30 s, with a final extension at amplification conditions of a default dissolution curve. Gene expression levels of each sample were normalized relative to the glyceraldehyde-3-phosphate dehydrogenase encoding gene (*GAPDH*) mRNA as an internal standard, and were calculated using the $2^{-\Delta\Delta Ct}$ method [32]. At least three independent experiments were performed for each analysis.

2.8. Extraction and Quantitative Determination of Shikonin and Acetylshikonin

Shikonin and its derivatives were extracted and quantified from hairy roots and the M9 medium as previously reported [30]. In brief, both fresh hairy roots (approximately 0.5 g) and the M9 production medium (20 mL) were extracted with petroleum ether and then dissolved in 1 mL of methanol following rotary evaporation. Using shikonin and its derivatives, including shikonin and acetylshikonin, as standards, a total of 2 µL of methanolic extract was analyzed using high performance liquid chromatography (HPLC) [1].

2.9. Statistical Analysis

All data are presented as means with standard deviations (SDs). The means ± SD values were calculated for each material using Microsoft Excel 2019. Statistical analysis using the Student's *t*-test was performed using GraphPad Prism 8. * $p < 0.05$, ** $p < 0.01$, *** $p < 0.001$, **** $p < 0.0001$.

3. Results

3.1. Identification and Characterization of LeBAHD Genes in L. erythrorhizon Genome

In order to have a comprehensive understanding of the characteristics and functions of BAHD family genes in *L. erythrorhizon*, 104 members were initially screened out from the *L. erythrorhizon* genome based on sequences homology and the conserved BAHD superfamily feature domain (Pfam: PF02458). A total of 73 genes were then identified following a series of de-redundant and abnormal sequence exclusions (Tables S3 and S4). Analysis of the basic physicochemical properties of putative LeBAHDs showed that the full length of the coding region is between 1068 and 1464 bp, the molecular weight of LeBAHD proteins is about 39.3–54.17 kDa, 47.9% of LeBAHDs were unstable proteins, and the theoretical isoelectric points (pI) of proteins varied widely from 4.86 to 9. In addition, subcellular localization predictions suggested all the proteins likely function in the cytoplasm. The relevant information of the putative LeBAHDs is listed in Table S4.

The distribution of protein motifs may provide insight into the functional diversity of a gene family's members. MEME detected fifteen motifs in the predicted full-length protein sequences of LeBAHDs (Figure 1A and Figure S1), and all LeBAHDs contained both motif 1 (FYDVDFGWGKP) and motif 4 (HKVGDGTSLSNFLNAWAEJAR), which correspond to the conserved domains DFGWG and HXXXD, respectively, associated with enzymatic activity. Moreover, the HXXXD domain is conserved in all potential genes, whereas the DFGWG domain demonstrated heterogeneity (Figure 1B). Consistent with the classification of LeBAHDs into five clades in their phylogenetic tree, the type and distribution of conserved motifs within the same subclade are identical (Figure 1A). Gene structure analysis using the GSDS tool revealed that the number of exons in *LeBAHDs* ranged from one to six, and that 31 proteins lacked introns (Figure 1A).

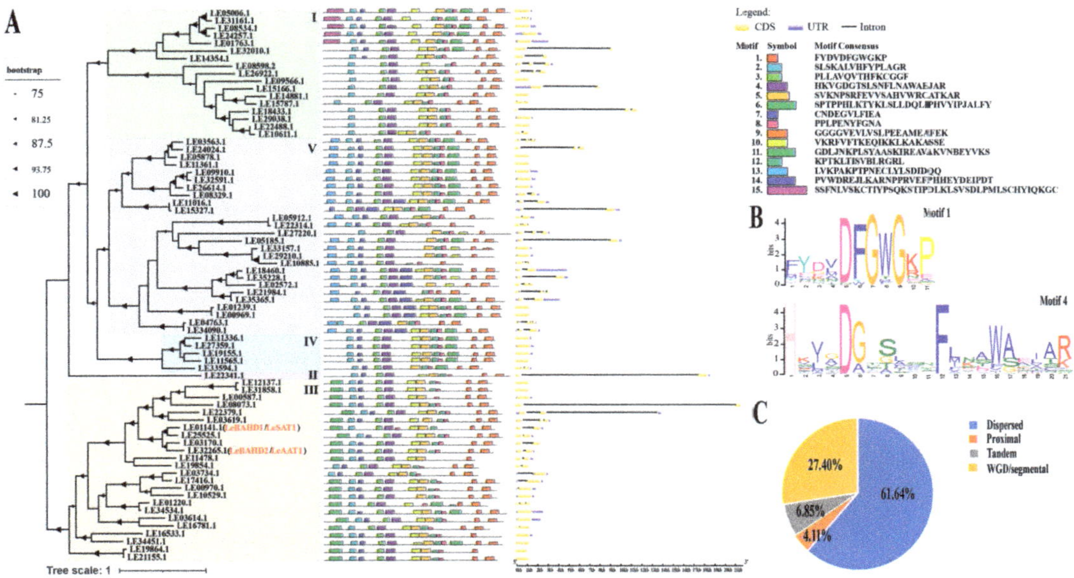

Figure 1. Phylogenetic relationships, conserved motifs, gene structures, and the expansion events of *LeBAHDs*. (**A**) The phylogenetic tree, conserved motifs, and gene structures of 73 *LeBAHDs* identified in *L. erythrorhizon*. I–V represent different clades in the phylogenetic tree. (**B**) The logos of Motif 1 and Motif 4 of LeBAHD proteins were identified using the MEME search tool, and Motif 1 and Motif 4 have conserved domains of DFGWG and HXXXD, respectively. (**C**) Gene duplication types and their proportions in 73 *LeBAHDs*.

To better comprehend the evolutionary history of LeBAHDs, we analyzed and counted the duplication type of all *LeBAHDs*. The results showed that more than half of *LeBAHDs* were derived from dispersed duplication, 27.39% of the members may have evolved through WGD/segmental, and the remaining genes were generated from tandem and proximal duplications (Figure 1C, Table S4).

3.2. Phylogenetic Classes and Function Analysis of LeBAHD Proteins

Similar evolutionary constraints apply to proteins with similar functions [33]. To further speculate on the evolutionary classification and functions of LeBAHDs, we constructed a phylogenetic tree of 119 BAHD amino acid sequences using the maximum likelihood method (Figure 2), including 73 LeBAHDs and 46 canonical sequences of BAHDs (Tables S1 and S4). LeBAHDs were divided into five clades, which was in accordance with the evolutionary relationships described by D'Auria [26]. Based on the phylogenetic tree, 17 LeBAHDs belonging to clade I are closely related to Dv3MAT in *Dahlia variabilis*, Gt5AT in *Gentiana triflora*; LE22341.1, closely related to Glossy2 in maize and CER2 in *A. thaliana* is the only one member classified in clade II. There are 24 LeBAHDs in clade III, which are closely related to proteins with acetylation function, such as Vinorine synthase identified in *Rauvolfia serpentina* and DAT identified in *C. roseus*; LeBAHD1/LeSAT1 is closely related to Le25525.1, and they converge with LeAAT1 and LE03170.1 to form a small branch in clade III. The five LeBAHDs in clade IV are closest to ACT in *Hordeum vulgare*; clade V contains the highest number of LeBAHDs, which are closely related to various proteins with different functions, such as BanAAT in *Musa sapientum* and NtBEBT in *Nicotiana tabacum*.

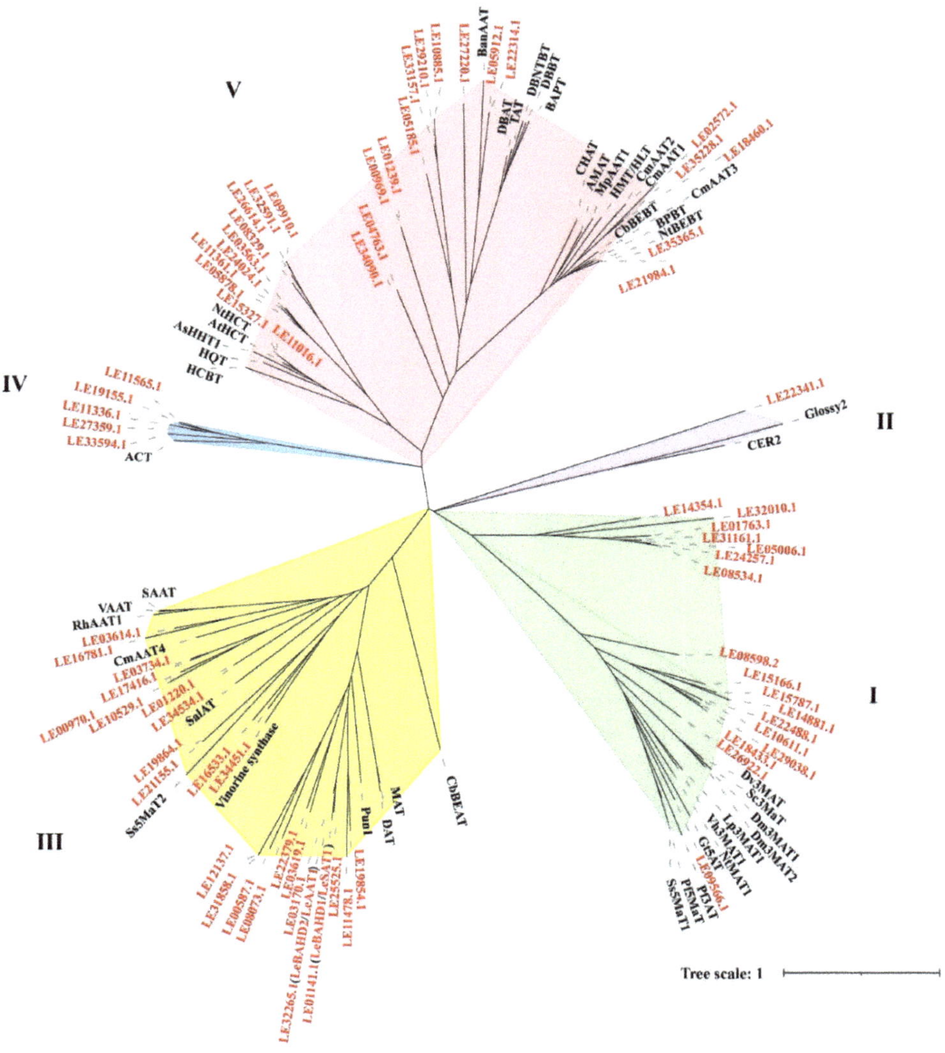

Figure 2. The phylogenetic tree of 119 BAHDs from 28 species including 73 LeBAHDs identified from *L. erythrorhizon* and 46 BAHD used in D'Auria's study [27]. The phylogenetic tree was constructed based on full-length protein sequences using the maximum likelihood method with 1000 replicates. I–V represent different clades in the phylogenetic tree.

3.3. Expression Patterns of LeBAHD Genes in Different Tissues of L. erythrorhizon

To further explore the functions of LeBAHD members in *L. erythrorhizon*, we analyzed the expression patterns of 73 *LeBAHDs* in different tissues of *L. erythrorhizon* (ML:

mature stems + leaves; FL: mature flower; PD: root periderm; MR: mature root; CT: root cortex; SE: root stele) using the published transcriptome data (NCBI SRA accession ID: SRP141330). The heat map results showed that the expression of *LeBAHDs* was distributed in almost every tissue (Figure 3). Among the expressed genes, the members LE17416.1, LE00587.1, LE16781.1, LE12137.1, and LE31858.1 in clade III and LE18460.1 in clade V had significantly higher expression levels in leaves than in other tissues; 12 *LeBAHD* genes showed high transcription levels in flowers, including LE03614.1, LE05912.1 and LE11478.1; LE29038.1, LE09566.1, LE02572.1 and LE21155.1 were preferentially expressed in root cortices or root steles; Importantly, some *LeBAHDs*, especially *LeBAHD1/LeSAT1* and LE25525.1, showed elevated transcription levels in mature root and/or root periderm where shikonin and its derivatives are biosynthesized (Figure 3).

In order to verify the root specific expression pattern of *LeBAHD1/LeSAT1*, the relative expression levels of this gene in roots, stems and leaves of *L. erythrorhizon* seedlings were determined by RT-qPCR. Consistent with the transcriptome data, the results show that *LeBAHD1/LeSAT1* is preferentially expressed in the roots, showing that the expression level in roots was 738 times and 119 times higher than that in leaves and stems, respectively (Figure S2).

3.4. Analysis of Cis-Acting Elements in the Promoters of LeBAHD Genes

To assess putative cis-acting elements in the promoter regions of *LeBAHDs* that regulate their expression, the PlantCARE tool was used to investigate the 2000 bp sequence upstream of the start codon of 73 *LeBAHD* genes. In all *LeBAHDs* promoters, 12,138 cis-acting elements of 102 types were predicted (Table S5). The most significant environmental regulator of *LeBAHDs* among them was light signals, followed by phytohormones (Figure 4 and Figure S3). A total of 27 types of light-responsive elements were identified in the *LeBAHD* promoter sequences, including Box 4, GA-motif, G-box, TCT-motif, etc. (Figure 4). Meanwhile, 19 different types of phytohormone-related cis-elements were identified, with the majority of *LeBAHD* promoters containing phytohormone-responsive binding sites: abscisic acid-responsive elements (ABRE motif and AAGAA-motif), MeJA-responsive elements (CGTCA-motif and TGACG-motif), ethylene-responsive element (ERE), and auxin-responsive element (TGA-element) (Figure 4). In addition, stress-stimulating elements such as anaerobic element (ARE), low temperature element (LTR), stress response element (STRE) and drought element (MSB), plant development-related elements, and myb and myc transcription binding sites exist in *LeBAHDs* promoters, which may possibly regulate their expression (Figure 4). Overall, certain elements such as Box 4, G-Box, ERE, and STRE were identified as high-frequency elements in the promoters of *LeBAHDs* (Figure S3). Importantly, the promoter region of the acyltransferase gene *LeBAHD1/LeSAT1* was enriched with cis-acting elements related to the light signal, ethylene, fungal initiators, etc., for the biosynthesis of shikonin and its derivatives (Figure 4).

3.5. Induction and Identification of the Transgenic Hairy Roots

The function of *LeBAHD1/LeSAT1* in catalyzing the conversion of shikonin to acetylshikonin has been verified in vitro using the heterologously expressed protein in *E. coli*. In order to further confirm the function of *LeBAHD1/LeSAT1* in vivo in *L. erythrorhizon*, we constructed *A. rhizogenes* ATCC15834 strains transformed with overexpression and knockout (CRISPR-Cas9 system) plasmids of *LeBAHD1/LeSAT1* (Figure 5A), then a series of transgenic hairy root lines for EV (pBI121 empty vector), OE, MH (CRISPR/Cas9 empty vector), and MH-K were generated using the method we reported previously [34] (Figure 5B). Furthermore, the marker *rolC* gene of ATCC15834 and the marker *HPT* of CRISPR/Cas9 vector were amplified from the obtained series of EV, OE, MH, and MH-K hairy roots, confirming the effective acquisition of transgenic hairy roots (Figure 5C).

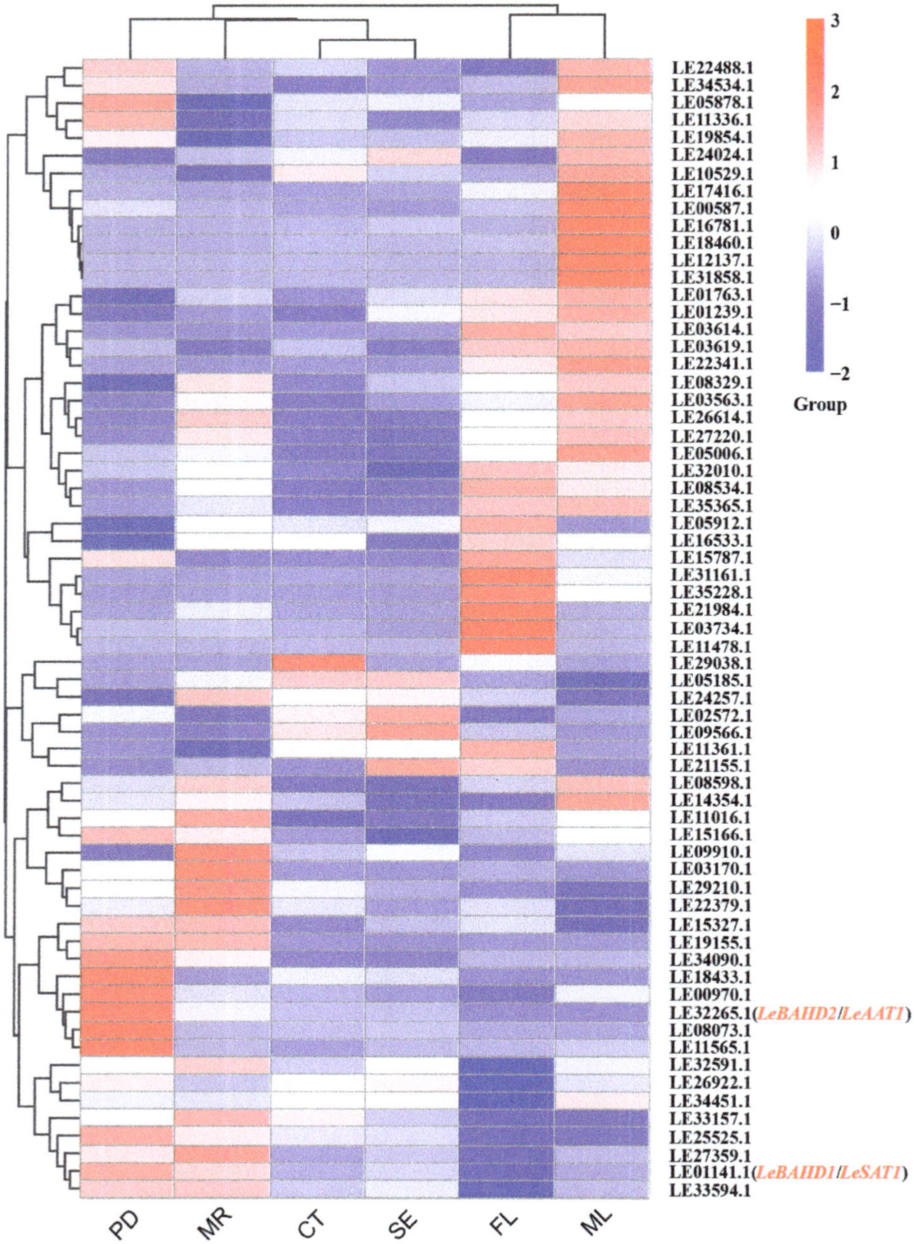

Figure 3. Expression patterns of *LeBAHDs* in different tissues of *L. erythrorhizon* seedlings based on transcriptome data. MR: mature roots; PD: root periderm; FL: flowers; ML: mature leaves + stems; CT: root cortex; SE: root stele.

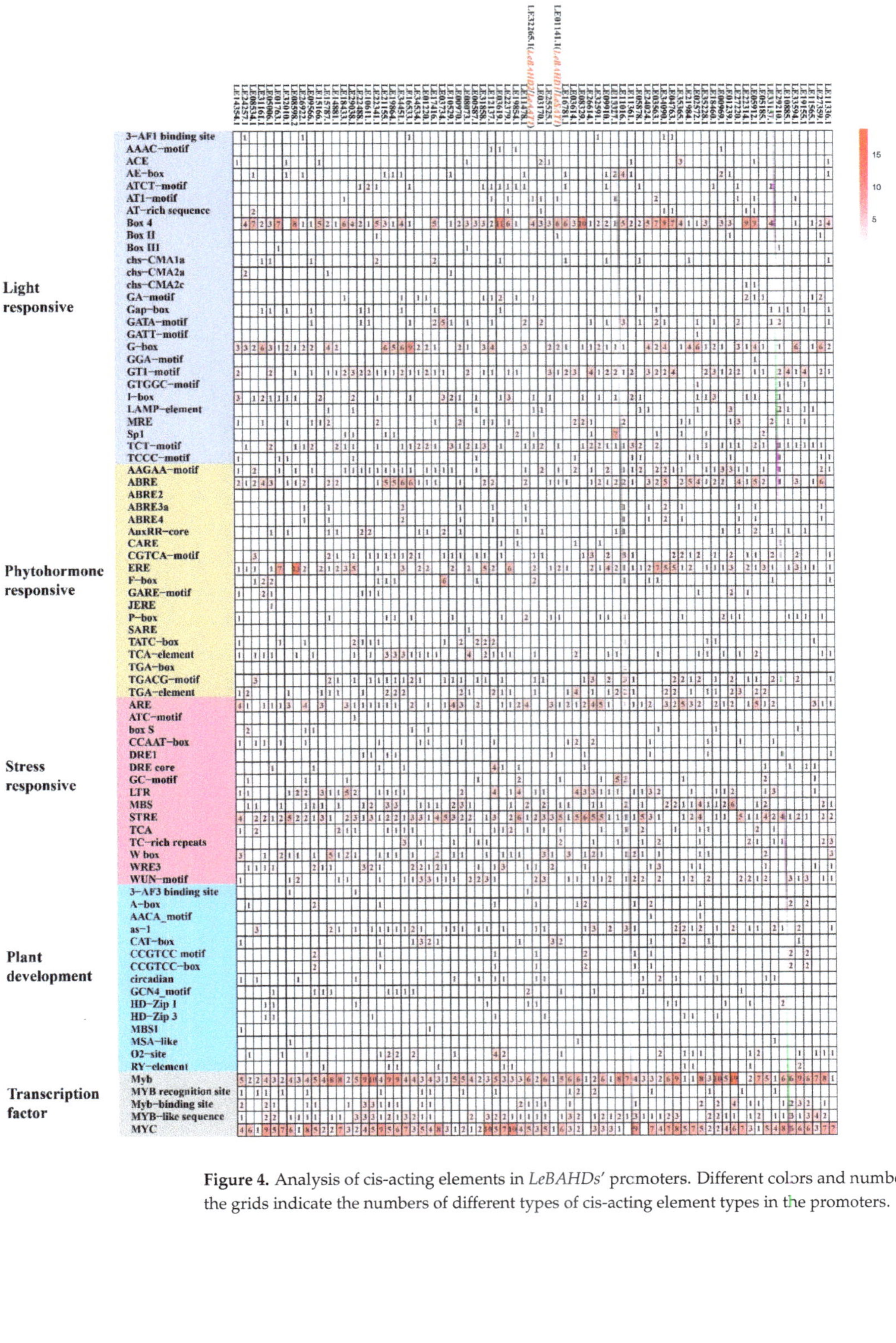

Figure 4. Analysis of cis-acting elements in *LeBAHDs'* promoters. Different colors and numbers in the grids indicate the numbers of different types of cis-acting element types in the promoters.

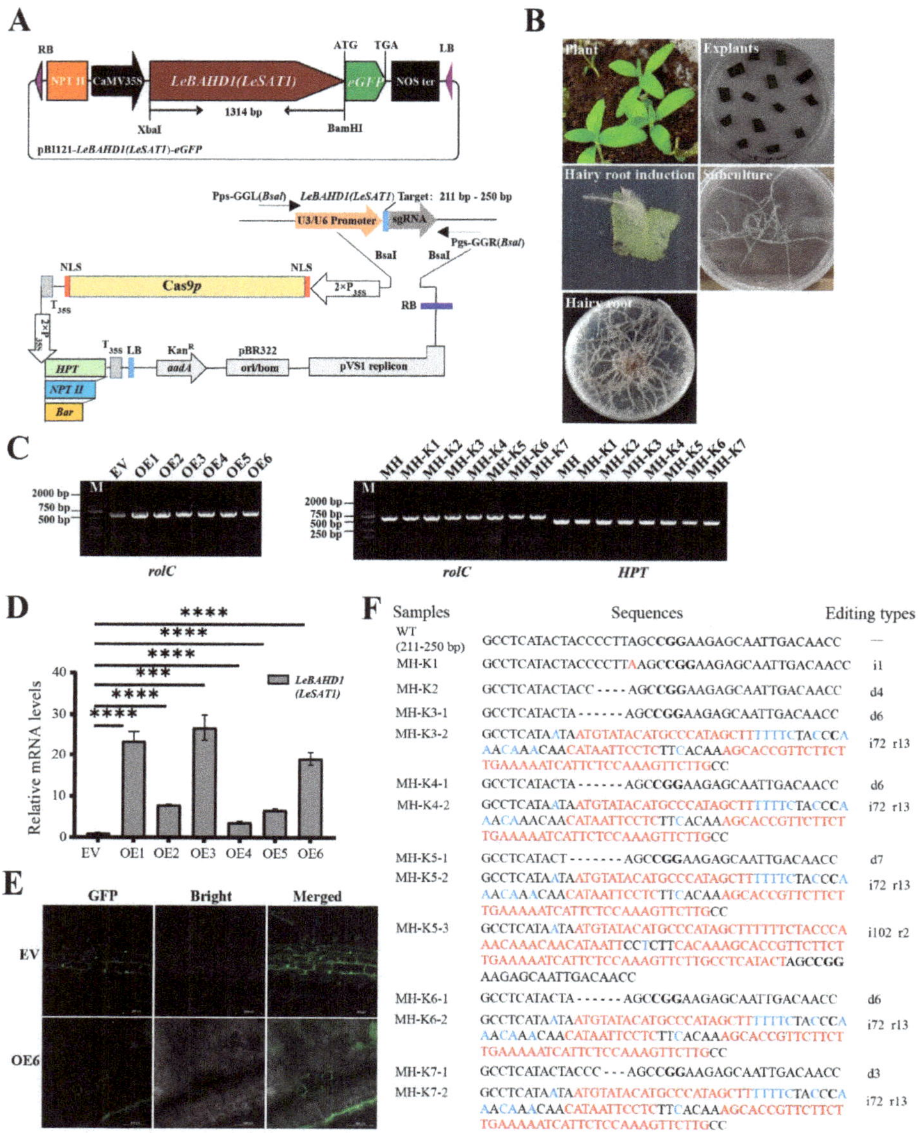

Figure 5. Generation and identification of transgenic hairy roots. (**A**) The structural maps of pBI121–*LeBAHD1*/*LeSAT1*–eGFP overexpression vector and pYLCRISPR/Cas9–*LeBAHD1*/*LeSAT1* knockout vectors. (**B**) Induction and subculture of EV, OE, MH and MH-K in *L. erythrorhizon* hairy roots. (**C**) PCR verification of the *rolC* gene in the *L. erythrorhizon* hairy root lines EV, OE1-6, MH, and MH-K1-7, and the *HPT* gene in MH, MH-K1-7. (**D**) Transcript levels of *LeBAHD1*/*LeSAT1* in the overexpression hairy roots cultured in 1/2 B5 multiplication medium under light at 26–28 °C. Asterisks indicate significant differences between each OE lines and EVs by the Student's *t*-test. *** $p < 0.001$, **** $p < 0.0001$. All data are means ± SD (n = 3). (**E**) Subcellular localization of LeBAHD1/LeSAT1 in EV and OE6 *L. erythrorhizon* hairy roots. Scale bar = 100 µm. (**F**) Gene editing types of each of MH-K knockout hairy root lines, i: number of inserted bases, marked with red; d: number of missing bases, marked with a dash; r: number of replaced bases, marked with blue.

The results showed that the expression levels of *LeBAHD1/LeSAT1* were up-regulated to varied degrees in the six OE hairy root lines compared to EV, with the expression levels in the OE1, OE3 and OE6 lines being significantly raised by 23-, 27- and 19-fold, respectively (Figure 5D). As a result, these three lines were used as subsequent experimental materials. In addition, eGFP fluorescence detection was performed to identify the successful induction of EV and OE hairy roots as well as the subcellular localization of *LeBAHD1/LeSAT1*. In contrast to the apparent triple distribution of GFP in the PM, cytoplasmic space, and nucleus of EV hairy roots, GFP fluorescence was primarily found in the cytoplasm of OE hairy roots, which was consistent with the cytoplasm localization prediction of LeBAHD1/LeSAT1 (Figure 5E, Table S4), suggesting that the LeBAHD1–eGFP fusion protein was successfully expressed under the control of the 35S promoter.

Meanwhile, to investigate the impact of gene editing, *LeBAHD1/LeSAT1* sequences from the MH and MH-k hairy roots were amplified and sequenced. The findings verified the generation of knockout hairy root lines and the first effective use of the CRISPR-Cas9 system for gene editing in *L. erythrorhizon* hairy roots. The nucleotide sequences of *LeBAHD1/LeSAT1* in the seven knockout hairy root lines (MH-K1~MH-K7) were edited with nucleotide insertion, deletion, and/or substitution (Figure 5F). Among them, MH-K1 and MH-K2 were homozygotes, inserting one base and deleting four bases, respectively, in 211–250 bp of the *LeBAHD1/LeSAT1* coding sequence (CDS); MH-K5 contained three editing types in 211–250 bp of the *LeBAHD1/LeSAT1* CDS: (1) 7 bases are missing, (2) 72 bases are inserted and 13 bases are replaced, (3) 102 bases are inserted and 2 bases are replaced (Figure 5F). These three lines were then selected for the identification of acyltransferase function of LeBAHD1/LeSAT1 in *L. erythrorhizon*.

3.6. *LeBAHD1/LeSAT1 Confers the Conversion of Shikonin to Acetylshikonin In Vivo in L. erythrorhizon Hairy Roots*

To investigate the role of LeBAHD1/LeSAT1 in enhancing acetylshikonin production in vivo in *L. erythrorhizon*, we measured the amount of shikonin and its derivatives produced by EV, MH, OE and MH-K hairy roots incubated in M9 medium for 9 days in the dark. Observing the color of the cultured transgenic hairy roots revealed that the color of three OE lines (OE1, OE3 and OE6) was significantly more red than that of the control EV and MH, whereas the color of the three knockout hairy root lines appeared yellow (Figure 6A). The pigment extracted from each hairy roots' culture was then analyzed using HPLC. According to the peak diagram of each sample, the content of shikonin at 4.061 min retention time in all hairy roots was significantly lower than acetylshikonin at 4.792 retention time. The relative quantitative results showed that the content of shikonin in OE1, OE3, and OE6 was not significantly different from that of EV (Figure 6B); however, the amount of acetylshikonin produced from the three OE lines was significantly higher than that of EV by 3.58-, 3.92-, and 2.72-fold, respectively (Figure 6C). Statistically, the ratio of acetylshikonin to shikonin in the OE1, OE3, and OE6 was significantly increased by 3.5-, 4.6-, and 3.1-fold, respectively, when compared to the relative control (Figure 6D). Shikonin and acetylshikonin content, as well as the ratio of these two metabolites, was significantly lower in MH-K2 than in the control line MH, whereas there was no significant difference in pigment content or ratio in the other lines (Figure 6E–G).

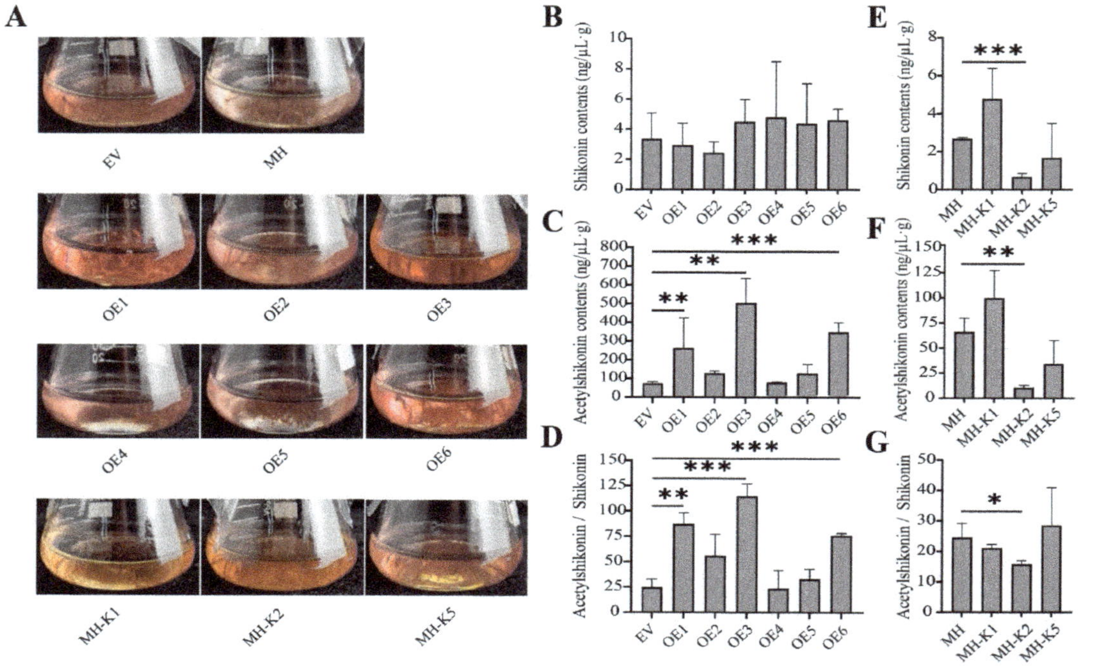

Figure 6. Color observation of the hairy root culture medium and HPLC analysis of shikonin and its derivatives in control (EV, or MH), overexpression (OE), knockout (MH-K) hairy roots of *L. erythrorhizon*. (**A**) Color observation of control (EV and MH) and transgenic hairy roots (OE and MH-K) cultured in M9 in darkness for 9 days. (**B**) Content analysis of shikonin in EV and OE hairy roots cultured in M9 in darkness for 9 days using HPLC analysis. (**C**) Content analysis of acetylshikonin in EV and OE hairy roots cultured in M9 in darkness for 9 days using HPLC analysis. (**D**) The ratio of acetylshikonin to shikonin in EV and OE hairy roots cultured in M9 in darkness for 9 days. (**E**) Content analysis of shikonin in MH and MH-K hairy roots cultured in M9 in darkness for 9 days using HPLC analysis. (**F**) Content analysis of acetylshikonin in MH and MH-K hairy roots cultured in M9 in darkness for 9 days using HPLC analysis. (**G**) The ratio of acetylshikonin to shikonin in MH and MH-K hairy roots cultured in M9 in darkness for 9 days. Asterisks indicate significant differences by Student's *t*-test. * $p < 0.05$, ** $p < 0.01$, *** $p < 0.001$. All data are means ± SD (n = 3).

4. Discussion

Traditional Chinese medicinal herbs, such as *L. erythrorhizon* and *A. euchroma*, get their significance from the secondary metabolites accumulating in their roots, primarily shikonin and its derivatives, which have a variety of pharmacological activities [35]. After paclitaxel and camptothecin, shikonin and its dervatives are regarded as promising natural antitumor agents due to their excellent anticancer activity [35]. Acetylshikonin is converted from shikonin by acyltransferase, and its content is 15 times higher than shikonin in the red roots of *L. erythrorhizon*. Moreover, acetylshikonin is less toxic to normal cells while also being anti-cancerous, and it has greater medicinal potential than shikonin [11,12]. Therefore, it is of importance to demonstrate how shikonin is converted to acetylshikonin and to increase the content of acetylshikonin in the callus cells or hairy roots of *Boraginaceae* plants.

The plant BAHD gene family members have been proven to have acyltransferase activity and participate in the biosynthesis of flavonoids, anthocyanins, and other secondary metabolites [25]. The BAHD gene family members in genomes of *A. thaliana*, *Populus tomentosa*, *Oryza sativa*, *Vitis vinifera* and other species have also been fully identified, with 55, 100, 84 and 52 members, respectively [36]. However, no results have been reported about the genome-wide identification of *BAHDs* in *Boraginaceae* plants. In the present study, we made a successful identification and characterization of 73 BAHD family members in the genome of *L. erythrorhizon*, and performed functional analysis for these genes based on evolutionary status and tissue expression patterns. In particular, the positive regulatory effect of the candidate gene *LeSAT1* in catalyzing the conversion of shikonin to acetylshikonin was verified by a in vivo transgenic strategy.

The 73 LeBAHD family members we identified included the 54 AAT/SAT's members previously identified by Tang from the *L. erythrorhizon* genome [28]. Tang's research firstly used LeSAT1/LeAAT1 amino acid sequences as queries to blast the *L. erythrorhizon* genome, the initial sequence was obtained based on sequence homology, and 54 members were then screened out by removing the sequences with redundancy or lack of conserved functional domains or conserved motifs [28]. However, not all members of the BAHD acyltransferase gene family could be screened out from the *L. erythrorhizon* genome by only using AAT/SAT amino acid sequences as a query. In the present study, the query Pfam domain PF02458 was firstly used to search the genome, and the initial 104 sequences were obtained based on the presence in the sequences of characteristic domains. After that 73 members were screened out by removing the sequences with redundancy or lack of conserved functional domains or conserved motifs. Although all of the 54 family members in Tang's study were represented by our 73 members, the purpose and gene family definitions of the two studies differed: Tang focused on an alkannin/shikonin O-acyltransferase gene family and its evolutionary history [28], while we focused on the BAHD acyltransferase family and its characterization and function.

A lot of evidence shows that the BAHD family members are involved in a variety of biological reactions: Dv3MAT in *D. variabilis* and Sc3MaT in *Pericullis cruenta* are responsible for the modification of anthocyanins [37,38], while NtMAT1 in *N. tabacum* is responsible for the modification of flavonoid and napthol glucosides [39]; based on the evolutionary relationship, 17 LeBAHDs were clustered into the clade I with these three proteins (Figure 2), implying they might have similar functions. LE22341.1 is a member in clade II and may have similar functions to Glossy2 in maize and CER2 in *Arabidopsis*, which mainly regulates cuticle wax extension and prevents pathogen invasion [26,40] (Figure 2); five LeBAHDs may be able to acylate a nitrogen to generate the corresponding amide, similar to ACT in clade IV [26,41] (Figure 2). The 26 LeBAHDs belonging to clade V are closely related to AtHCT in *Arabidopsis* and BanAAT in *M. sapientum*, which may be involved in the biosynthesis of volatile esters, hydroxyl cinnamyl quate/oxalate [26,42–44] (Figure 2). The substrate of most members of BAHD III subfamily is alcohols, and most of these enzymes use acetyl-CoA as the main acyl donor [26,45,46]. For example, *DAT* isolated from *C. roseus* can catalyze the acetylation of alcohol substance deacetylvindoline to produce an anti-cancer alkaloid drug—ventolin [24,26] (Figure 2). Since the catalytic process from shikonin to acetylshikonin is an acetylation modification reaction of alcohol hydroxyl, we speculate that LeBAHD1/LeSAT1 in Branch III could acetylate shikonin to acetylshikonin.

The tissue expression pattern of genes also could reflect their possible functions in plants. According to the transcriptome analysis of six tissues of *L. erythrorhizon* (Figure 3), LE22341.1, a member in clade II, is predominantly highly expressed in ML and might be responsible for cuticle wax extension in the leaves and stems of *L. erythrorhizon* in order to prevent water loss and microbial infestation. Twelve *LeBAHD* genes displayed higher transcription levels in flowers, which may be related to the synthesis of flower-specific metabolites. *LeBAHD1/LeSAT1* and *LeBAHD56*(LE25525.1) are mostly strongly expressed in MR and/or PD, the principle biosynthetic tissues of shikonin and acetylshikonin in *L. erythrorhizon*, suggesting that they may be involved in the acylation of shikonin.

Recently, Oshikiri et al. identified LeBAHD1/LeSAT1 using comparative transcriptome and proteomic analysis of *L. erythrorhizon*, and its function of catalyzing acetyl-CoA as acyl donors to generate acetylshikonin was verified by heterologous expression systems and in vitro enzyme activity [27]. However, heterologous in vitro experiments cannot fully imitate the complex biosynthetic and regulatory mechanisms of secondary metabolites in plants. It is therefore necessary to confirm the in vivo role of *LeBAHD1/LeSAT1* in *L. erythrorhizon*. As a result, the function of the *LeBAHD1/LeSAT1* enzyme in catalyzing the acetylation of shikonin was studied further in this study by constructing the overexpression and knockout lines of transgenic hairy roots of *LeBAHD1/LeSAT1*. The results showed that the yields of acetylshikonin, as well as the ratio of acetylshikonin to shikonin, were significantly higher in the overexpression lines OE1, OE3, and OE6, compared to the control lines, confirming that *LeBAHD1/LeSAT1* positively regulates shikonin conversion to acetylshikonin. However, only the MH-K2 line out of the three knockout hairy root lines had significantly lower shikonin yield, acetylshikonin yield, and the ratio of acetylshikonin to shikonin than the MH line. As a result, this finding is insufficient to establish the specific acyl transferase function of LeBAHD1/LeSAT1 in *L. erythrorhizon*. It is likely that other important genes with functional redundancy, such as *LeBAHD56* (LE25525.1), which clusters in the same clade as LeBAHD1/LeSAT1 with high sequence homology and that has greater transcript levels in roots where shikonin and its derivatives are biosynthesized, may account for the phenotypic variability in knockout hairy root lines. Therefore, more single-gene knockout hairy root lines of *LeBAHD1/LeSAT1* and *LeBAHD56* (LE25525.1) and double-gene knockout hairy root lines of these two genes are required to clarify their contribution ratio in regulating shikonin's conversion to acetylshikonin in *L. erythrorhizon* using a CRISPR/Cas9-based knockout system. After demonstrating that *LeBAHD56* (LE25525.1) acetylates shikonin, we can generate double-gene overexpression hairy roots to determine if they can produce more acetylshikonin than single-gene overexpression hairy roots.

Due to its high efficiency of gene editing, CRISPR/Cas9-based knockout technology has been widely utilized to study gene function in numerous plant species, including the model plants *A. thaliana* [47] and *O. sativa* [48], woody plants such as *Populus* [49] and *Malus pumila* [50], and medicinal plants such as *Salvia miltiorrhiza* [51] and *Dendrobium officinale* [52]. However, previous studies have not reported the application of CRISPR/Cas9 technology in *Boraginaceae* plants. In this study, in addition to constructing *LeBAHD1/LeSAT1*-overexpressing transgenic hairy root lines to study their acyltransferase function, we also successfully introduced the CRISPR/Cas9 technology into the hairy root system of *L. erythrorhizon* and produced seven knockout lines with a total of seven gene-editing types, of which MH-K1 and MH-K2 were insertion homozygous and deletion homozygous, respectively, and MH-K5 contained three different editing types. This investigation of the practical feasibility of CRISPR/Cas9 technology in *L. erythrorhizon* can provide valuable research experience and methods for more precise characterization of the biosynthesis and regulatory pathways of shikonin and its derivatives.

In conclusion, our study employed a successful identification and characterization of *LeBAHD* family members from *L. erythrorhizon*. We confirmed *LeBAHD1/LeSAT1*'s function in the biosynthesis of acetylshikonin by converting shikonin in the transgenic hairy root system in vivo utilizing overexpression and CRISPR/Cas9-based knockout transgenic experiments. We also provided some potential candidate genes with functional redundancies to *LeBAHD1/LeSAT1* in the same phylogenetical branch with similar expression patterns, such as *LeBAHD56*(LE25525.1). Our findings not only contribute to a better understanding of the regulatory mechanism governing the biosynthesis of shikonin and its derivatives in *L. erythrorhizon* or other secondary metabolites in non-model medicinal plants, but also offer compelling evidence that there is a possibility to produce high yields of acetylshikonin in in vivo in *L. erythrorhizon* by manipulating *LeBAHD1/LeSAT1* through genetic engineering.

Supplementary Materials: The following supporting information can be downloaded at: https://www.mdpi.com/article/10.3390/life12111775/s1, Figure S1. Logos of 15 conserved motifs were identified in 73 LeBAHDs from *L. erythrorhizon* using the MEME search tool. Figure S2. The expression level of *LeBAHD1/LeSAT1* in the leaf, stem, and root of *L. erythrorhizon* seedlings were analyzed using RT-qPCR. Figure S3. Cloud word graph of cis-acting elements in *LeBAHDs'* promoters. Table S1. Basic information about published BAHD reference genes in D'Auria's study. Table S2. Primers used for experiment of *LeBAHD1/LeSAT1* on *L. erythrorhizon*. Table S3. Identification of BAHDs from the genome of *L. erythrorhizon*. Table S4. The basic information of *LeBAHDs* in *L. erythrorhizon*. Table S5. The number of cis-acting elements in *LeBAHDs'* promoters.

Author Contributions: Y.Y. and J.Q. conceived and designed the experiments. X.W., Z.H., H.Y. and C.H. performed the experiments. X.W., C.W., X.L., L.Y., Z.W. and M.Y. analyzed the data. S.M., W.J., J.C., T.Y. and B.L. contributed to resources. X.W., Z.H., H.Y. and C.H. wrote the draft of the manuscript. Y.Y., J.Q. and A.F. contributed to review and edit the manuscript. All authors have read and agreed to the published version of the manuscript.

Funding: This research was funded by the National Natural Science Foundation of China (31970321, 31670298, U1903201, 31771413), the Natural Science Foundation of Jiangsu Bureau of Science and Technology (BK20191254), the Open Project Program from the MOE Key Laboratory of Molecular Epigenetics of China, and the Program for Changjiang Scholars and Innovative Research Team in University from the Ministry of Education of China (IRT_14R27).

Institutional Review Board Statement: Not applicable.

Informed Consent Statement: Not applicable.

Data Availability Statement: All supporting data can be found within the manuscript and its supplementary files.

Acknowledgments: We thank Yaoguang Liu from South China Agricultural University for providing the CRISPR/Cas9 vector system.

Conflicts of Interest: The authors declare that they have no conflict of interest

References

1. Albreht, A.; Vovk, I.; Simonovska, B.; Srbinoska, M. Identification of shikonin and its ester derivatives from the roots of *Echium italicum* L. *J. Chromatogr. A* **2009**, *1216*, 3156–3162. [CrossRef]
2. Kumar, A.; Shashni, S.; Kumar, P.; Pant, D.; Singh, A.; Verma, R.K. Phytochemical constituents, distributions and traditional usages of *Arnebia euchroma*: A review. *J. Ethnopharmacol.* **2021**, *271*, 113896. [CrossRef]
3. Wang, R.B.; Yin, R.T.; Zhou, W.; Xu, D.F.; Li, S.S. Shikonin and its derivatives: A patent review. *Expert Opin. Ther. Pat.* **2012**, *22*, 977–997. [CrossRef] [PubMed]
4. Hu, Y.; Jiang, Z.H.; Leung, K.S.Y.; Zhao, Z.Z. Simultaneous determination of naphthoquinone derivatives in Boraginaceous herbs by high-performance liquid chromatography. *Anal. Chim. Acta* **2006**, *577*, 26–31. [CrossRef]
5. Cho, S.C.; Choi, B.Y. Acetylshikonin inhibits human pancreatic PANC-1 cancer cell proliferation by suppressing the NF-kB activity. *Biomol. Ther.* **2015**, *23*, 428–433. [CrossRef]
6. Wu, M.D.; Zhang, Y.Y.; Yi, S.Y.; Sun, B.B.; Lan, J.; Jiang, H.M.; Hao, G.P. Acetylshikonin induces autophagy-dependent apoptosis through the key LKB1-AMPK and PI3K/Akt-regulated mTOR signalling pathways in HL-60 cells. *J. Cell. Mol. Med.* **2022**, *26*, 1606–1620. [CrossRef]
7. Zhao, R.; Choi, B.Y.; Wei, L.; Fredimoses, M.; Yin, F.; Fu, X.; Chen, H.; Liu, K.; Kundu, J.K.; Dong, Z.; et al. Acetylshikonin suppressed growth of colorectal tumour tissue and cells by inhibiting the intracellular kinase, T-lymphokine-activated killer cell-originated protein kinase. *Br. J. Pharmacol.* **2020**, *177*, 2303–2319. [CrossRef] [PubMed]
8. Park, S.H.; Phuc, N.M.; Lee, J.; Wu, Z.; Kim, J.; Kim, H.; Kim, N.D.; Lee, T.; Song, K.S.; Liu, K.H. Identification of acetylshikonin as the novel CYP2J2 inhibitor with anti-cancer activity in HepG2 cells. *Phytomedicine* **2017**, *24*, 134–140. [CrossRef]
9. Cho, B.H.; Jung, Y.H.; Kim, D.J.; Woo, B.H.; Jung, J.E.; Lee, J.H.; Choi, Y.W.; Park, H.R. Acetylshikonin suppresses invasion of porphyromonas gingivalis-infected YD10B oral cancer cells by modulating the interleukin-8/matrix metalloproteinase axis. *Mol. Med. Rep.* **2018**, *17*, 2327–2334. [CrossRef] [PubMed]
10. Rajasekar, S.; Park, D.J.; Park, C.; Park, S.; Park, Y.H.; Kim, S.T.; Choi, Y.H.; Choi, Y.W. In vitro and in vivo anticancer effects of *Lithospermum erythrorhizon* extract on B16F10 murine melanoma. *J. Ethnopharmacol.* **2012**, *144*, 335–345. [CrossRef] [PubMed]
11. Figat, R.; Zgadzaj, A.; Geschke, S.; Sieczka, P.; Pietrosiuk, A.; Sommer, S.; Skrzypczak, A. Cytotoxicity and antigenotoxicity evaluation of acetylshikonin and shikonin. *Drug Chem. Toxicol.* **2018**, *44*, 140–147. [CrossRef] [PubMed]
12. Kim, D.; Lee, J.H.; Park, H.R.; Choi, Y.W. Acetylshikonin inhibits growth of oral squamous cell carcinoma by inducing apoptosis. *Arch. Oral Biol.* **2016**, *70*, 149–157. [CrossRef] [PubMed]
13. Fujita, Y.; Tabata, M.; Nishi, A.; Yamada, Y. New medium and production of secondary compounds with the two-staged culture method. In Proceedings of the Plant Tissue Culture 1982: Proceedings, 5th International Congress of Plant Tissue and Cell Culture, Tokyo, Japan, 11–16 July 1982.
14. Shimomura, K.; Sudo, H.; Saga, H.; Kamada, H. Shikonin production and secretion by hairy root cultures of *Lithospermum-erythrorhizon*. *Plant Cell Rep.* **1991**, *10*, 282–285. [CrossRef]

15. Fang, R.J.; Zou, A.L.; Zhao, H.; Wu, F.Y.; Zhu, Y.; Zhao, H.; Liao, Y.H.; Tang, R.J.; Pang, Y.J.; Yang, R.W.; et al. Transgenic studies reveal the positive role of *LeEIL-1* in regulating shikonin biosynthesis in *Lithospermum erythrorhizon* hairy roots. *BMC Plant Biol.* **2016**, *16*, 121. [CrossRef] [PubMed]
16. Yazaki, K.; Kunihisa, M.; Fujisaki, T.; Sato, F. Geranyl diphosphate: 4-hydroxybenzoate geranyltransferase from *Lithospermum erythrorhizon*—Cloning and characterization of a key enzyme in shikonin biosynthesis. *J. Biol. Chem.* **2002**, *277*, 6240–6246. [CrossRef]
17. Yamamura, Y.; Ogihara, Y.; Mizukami, H. Cinnamic acid 4-hydroxylase from *Lithospermum erythrorhizon*: cDNA cloning and gene expression. *Plant Cell Rep.* **2001**, *20*, 655–662. [CrossRef]
18. Yazaki, K.; Kataoka, M.; Honda, G.; Severin, K.; Heide, L. cDNA cloning and gene expression of phenylalanine ammonia-lyase in *Lithospermum erythrorhizon*. *Biosci. Biotechnol. Biochem.* **1997**, *61*, 1995–2003. [CrossRef]
19. Yazaki, K.; Ogawa, A.; Tabata, M. Isolation and characterization of two cDNAs encoding 4-coumarate:CoA ligase in *Lithospermum* cell cultures. *Plant Cell Physiol.* **1995**, *36*, 1319–1329. [CrossRef]
20. Wang, S.; Wang, R.S.; Liu, T.; Lv, C.G.; Liang, J.W.; Kang, C.Z.; Zhou, L.Y.; Guo, J.; Cui, G.H.; Zhang, Y.; et al. CYP76B74 catalyzes the 3″-hydroxylation of geranylhydroquinone in shikonin biosynthesis. *Plant Physiol.* **2019**, *179*, 402–414. [CrossRef]
21. Song, W.; Zhuang, Y.B.; Liu, T. Potential role of two cytochrome P450s obtained from *Lithospermum erythrorhizon* in catalyzing the oxidation of geranylhydroquinone during shikonin biosynthesis. *Phytochemistry* **2020**, *175*, 112375. [CrossRef]
22. Song, W.; Zhuang, Y.B.; Liu, T. CYP82AR subfamily proteins catalyze C-1′ hydroxylations of deoxyshikonin in the biosynthesis of shikonin and alkannin. *Org. Lett.* **2021**, *23*, 2455–2459. [CrossRef] [PubMed]
23. Takanashi, K.; Nakagawa, Y.; Aburaya, S.; Kaminade, K.; Aoki, W.; Saida-Munakata, Y.; Sugiyama, A.; Ueda, M.; Yazaki, K. Comparative proteomic analysis of *Lithospermum erythrorhizon* reveals regulation of a variety of metabolic enzymes leading to comprehensive understanding of the shikonin biosynthetic pathway. *Plant Cell Physiol.* **2019**, *60*, 19–28. [CrossRef]
24. St-Pierre, B.; Laflamme, P.; Alarco, A.M.; De Luca, V. The terminal O-acetyltransferase involved in vindoline biosynthesis defines a new class of proteins responsible for coenzyme A-dependent acyl transfer. *Plant J.* **1998**, *14*, 703–713. [CrossRef]
25. St-Pierre, B.; Luca, V.D. Evolution of acyltransferase genes: Origin and diversification of the BAHD superfamily of acyltransferases involved in secondary metabolism. *Recent Adv. Phytochem.* **2000**, *34*, 285–315. [CrossRef]
26. D'Auria, J.C. Acyltransferases in plants: A good time to be BAHD. *Curr. Opin. Plant Biol.* **2006**, *9*, 331–340. [CrossRef]
27. Oshikiri, H.; Watanabe, B.; Yamamoto, H.; Yazaki, K.; Takanashi, K. Two BAHD acyltransferases catalyze the last step in the shikonin/alkannin biosynthetic pathway. *Plant Physiol.* **2020**, *184*, 753–761. [CrossRef] [PubMed]
28. Tang, C.Y. Exploring the evolutionary process of alkannin/shikonin O-acyltransferases by a reliable *Lithospermum erythrorhizon* genome. *DNA Res.* **2021**, *28*, dsab015. [CrossRef] [PubMed]
29. Yuan, Z.; Yang, H.L.; Pan, L.W.; Zhao, W.H.; Liang, L.P.; Gatera, A.; Tucker, M.R.; Xu, D.W. Systematic identification and expression profiles of the BAHD superfamily acyltransferases in barley (*Hordeum vulgare*). *Sci. Rep.* **2022**, *12*, 5063. [CrossRef]
30. Zhao, H.; Chang, Q.S.; Zhang, D.X.; Fang, R.J.; Zhao, H.; Wu, F.Y.; Wang, X.M.; Lu, G.H.; Qi, J.L.; Yang, Y.H. Overexpression of *LeMYB1* enhances shikonin formation by up-regulating key shikonin biosynthesis-related genes in *Lithospermum erythrorhizon*. *Biol. Plant.* **2015**, *59*, 429–435. [CrossRef]
31. Yu, P.L.; Yuan, J.H.; Deng, X.; Ma, M.; Zhang, H.Y. Subcellular targeting of bacterial CusF enhances Cu accumulation and alters root to shoot Cu translocation in *Arabidopsis*. *Plant Cell Physiol.* **2014**, *55*, 1568–1581. [CrossRef]
32. Kenneth, S.T.L. Analysis of relative gene expression data using real-time quantitative PCR and the $2^{\Delta\Delta}$CT method. *Methods* **2001**, *25*, 402–408. [CrossRef]
33. Durand, P.M.; Hazelhurst, S.; Coetzer, T.L. Evolutionary rates at codon sites may be used to align sequences and infer protein domain function. *BMC Bioinform.* **2010**, *11*, 151. [CrossRef] [PubMed]
34. Zhu, Y.; Lu, G.H.; Bian, Z.W.; Wu, F.Y.; Pang, Y.J.; Wang, X.M.; Yang, R.W.; Tang, C.Y.; Qi, J.L.; Yang, Y.H. Involvement of *LeMDR*, an ATP-binding cassette protein gene, in shikonin transport and biosynthesis in *Lithospermum erythrorhizon*. *BMC Plant Biol.* **2017**, *17*, 198. [CrossRef] [PubMed]
35. Guo, C.J.; He, J.L.; Song, X.M.T.; Tan, L.; Wang, M.; Jiang, P.D.; Li, Y.Z.; Cao, Z.X.; Peng, C. Pharmacological properties and derivatives of shikonin-a review in recent years. *Pharmacol. Res.* **2019**, *149*, 104463. [CrossRef] [PubMed]
36. Tuominen, L.K.; Johnson, V.E.; Tsai, C.J. Differential phylogenetic expansions in BAHD acyltransferases across five angiosperm taxa and evidence of divergent expression among *Populus paralogues*. *BMC Genom.* **2011**, *12*, 236. [CrossRef]
37. Suzuki, H.; Nakayama, T.; Yonekura-Sakakibara, K.; Fukui, Y.; Nakamura, N.; Yamaguchi, M.A.; Tanaka, Y.; Kusumi, T.; Nishino, T. cDNA cloning, heterologous expressions, and functional characterization of malonyl-coenzyme a:anthocyanidin 3-o-glucoside-6″-o-malonyltransferase from dahlia flowers. *Plant Physiol.* **2002**, *130*, 2142–2151. [CrossRef]
38. Suzuki, H.; Sawada, S.Y.; Yonekura-Sakakibara, K.; Nakayama, T.; Yamaguchi, M.A. Identification of a cDNA encoding malonyl—Coenzyme a: Anthocyanidin 3-O-glucoside 6″-O-malonyltransferase from cineraria (*Senecio cruentus*) flowers. *Plant Biotechnol.* **2003**, *20*, 229–234. [CrossRef]
39. Taguchi, G.; Shitchi, Y.; Shirasawa, S.; Yamamoto, H.; Hayashida, N. Molecular cloning, characterization, and downregulation of an acyltransferase that catalyzes the malonylation of flavonoid and naphthol glucosides in tobacco cells. *Plant J.* **2005**, *42*, 481–491. [CrossRef]
40. Costaglioli, P.; Joubes, K.; Garcia, C.; Stef, M.; Arveiler, B.; Lessire, R.; Garbay, B. Profiling candidate genes involved in wax biosynthesis in *Arabidopsis thaliana* by microarray analysis. *BBA-Mol. Cell Biol. Lipids* **2005**, *1734*, 247–258. [CrossRef]

41. Burhenne, K.; Kristensen, B.K.; Rasmussen, S.K. A new class of N-hydroxycinnamoyltransferases. Purification, cloning, and expression of a barley agmatine coumaroyltransferase (EC 2.3.1.64). *J. Biol. Chem.* **2003**, *278*, 13919–13927. [CrossRef]
42. D'Auria, J.C.; Chen, F.; Pichersky, E. Characterization of an acyltransferase capable of synthesizing benzylbenzoate and other volatile esters in flowers and damaged leaves of *Clarkia breweri*. *Plant Physiol.* **2002**, *130*, 466–476. [CrossRef]
43. Hoffmann, L.; Besseau, S.; Geoffroy, P.; Ritzenthaler, C.; Meyer, D.; Lapierre, C.; Pollet, B.; Legrand, M. Acyltransferase-catalysed p-coumarate ester formation is a committed step of lignin biosynthesis. *Plant Biosyst.* **2005**, *139*, 50–53. [CrossRef]
44. Beekwilder, J.; Alvarez-Huerta, M.; Neef, E.; Verstappen, F.W.; Bouwmeester, H.J.; Aharoni, A. Functional characterization of enzymes forming volatile esters from strawberry and banana. *Plant Physiol.* **2004**, *135*, 1865–1878. [CrossRef] [PubMed]
45. El-Sharkawy, I.; Manriquez, D.; Flores, F.B.; Regad, F.; Bouzayen, M.; Latche, A.; Pech, J.C. Functional characterization of a melon alcohol acyl-transferase gene family involved in the biosynthesis of ester volatiles. Identification of the crucial role of a threonine residue for enzyme activity. *Plant Mol. Biol.* **2005**, *59*, 345–362. [CrossRef] [PubMed]
46. Grothe, T.; Lenz, R.; Kutchan, T.M. Molecular characterization of the salutaridinol 7-O-acetyltransferase involved in morphine biosynthesis in opium poppy *Papaver somniferum*. *J. Biol. Chem.* **2001**, *276*, 30717–30723. [CrossRef]
47. Li, P.; Li, X.W.; Jiang, M. CRISPR/Cas9-mediated mutagenesis of WRKY3 and WRKY4 function decreases salt and Me-JA stress tolerance in *Arabidopsis thaliana*. *Mol. Biol. Rep.* **2021**, *48*, 5821–5832. [CrossRef]
48. Nguyen, T.M.; Lu, C.A.; Huang, L.F. Applications of CRISPR/Cas9 in a rice protein expression system via an intron-targeted insertion approach. *Plant Sci.* **2022**, *315*, 111132. [CrossRef]
49. Liu, T.T.; Fan, D.; Ran, L.Y.; Jiang, Y.Z.; Liu, R.; Luo, K.M. Highly efficient CRISPR/Cas9-mediated targeted mutagenesis of multiple genes in *Populus*. *Yi Chuan Hered.* **2015**, *37*, 1044–1052. [CrossRef]
50. Nishitani, C.; Hirai, N.; Komori, S.; Wada, M.; Okada, K.; Osakabe, K.; Yamamoto, T.; Osakabe, Y. Efficient genome editing in apple using a CRISPR/Cas9 system. *Sci. Rep.* **2016**, *6*, 31481. [CrossRef]
51. Li, B.; Li, J.W.; Chai, Y.Q.; Huang, Y.Y.; Li, L.; Wang, D.H.; Wang, Z.Z. Targeted mutagenesis of CYP76AK2 and CYP76AK3 in *Salvia miltiorrhiza* reveals their roles in tanshinones biosynthetic pathway. *Int. J. Biol. Macromol.* **2021**, *189*, 455–463. [CrossRef]
52. Kui, L.; Chen, H.; Zhang, W.; He, S.; Xiong, Z.; Zhang, Y.; Yan, L.; Zhong, C.; He, F.; Chen, J.; et al. Building a genetic manipulation tool box for orchid biology: Identification of constitutive promoters and application of CRISPR/Cas9 in the orchid, *Dendrobium officinale*. *Front. Plant Sci.* **2016**, *7*, 2036. [CrossRef] [PubMed]

life

Article

Addition of Arbuscular Mycorrhizal Fungi Enhances Terpene Synthase Expression in *Salvia rosmarinus* Cultivars

Emily Leggatt [1], Alistair Griffiths [2], Simon Budge [3], Anthony D. Stead [1], Alan C. Gange [1] and Paul F. Devlin [1,*]

[1] Department of Biological Sciences, Royal Holloway University of London, Egham TW20 0EX, UK
[2] Science Department, Royal Horticultural Society, Woking GU23 6QB, UK
[3] Vitacress Herbs, Chichester PO20 1LJ, UK
* Correspondence: paul.devlin@rhul.ac.uk

Abstract: Culinary herbs are commercially cultivated for their wide range of volatile compounds that give characteristic aromas and tastes. Rosemary (*Salvia rosmarinus* Spenn.) is an excellent model for assessment of methods improvement of volatile production as cultivars offer a wide variety of aromatic profiles due to the large family of terpene synthase genes. Arbuscular mycorrhizal fungi (AMF) associations have been shown to improve essential oil production in aromatic plants and offer one approach to enhance aroma in commercial herb production. Changes in the expression of seven different terpene synthases were compared in six rosemary cultivars in response to addition of AMF to a peat substrate. Addition of AMF profoundly influenced terpene synthase expression in all cultivars and did so without impacting the optimised plant size and uniformity achieved in these conditions. In addition, two methods for AMF application, developed with the horticultural industry in mind, were tested in this study. Uniform incorporation of AMF mixed into the growing substrate prior to planting of a root plug produced the most consistent root colonisation. Overall, our findings demonstrate the potential for the use of AMF in the improvement of aroma in culinary herbs within a commercial setting but show that outcomes are likely to greatly vary depending on variety.

Keywords: arbuscular mycorrhizal fungi; commercial herb cultivation; gene expression; rosemary; terpene synthase; volatile production

Citation: Leggatt, E.; Griffiths, A.; Budge, S.; Stead, A.D.; Gange, A.C.; Devlin, P.F. Addition of Arbuscular Mycorrhizal Fungi Enhances Terpene Synthase Expression in *Salvia rosmarinus* Cultivars. *Life* **2023**, *13*, 315. https://doi.org/10.3390/life13020315

Academic Editor: Jianfeng Xu

Received: 30 December 2022
Revised: 17 January 2023
Accepted: 20 January 2023
Published: 23 January 2023

1. Introduction

Culinary herbs are commercially grown across the world for their wide range of volatile compounds that give characteristic aromas and tastes. Leading commercial growers are located globally and freshly grown herbs represent 5% of the global horticulture market with a predicted revenue of USD 1183.2 million between 2018 and 2025 [1]. There is a growing demand by consumers for high quality fresh herbs in cooking, leading to increasing potted herbs sales with consumers favouring plants with strong aromas and taste. As a result, improvements to the growth, aroma and taste are sought after by horticultural growers to meet with consumer demands.

Aromatic volatiles are phenolic terpenoid compounds synthesised in plants through two main cellular pathways: the plastidial methylerythritol 4-phosphate (MEP) pathway and the cytosolic mevalonate (MVA) pathway. All terpenoids are derived from the precursors dimethylallyl pyrophosphate (DMAPP) and isoprene diphosphate (IPP) formed during the later catalytic steps of both pathways [2]. IPP and DMAPP are further condensed by prenyltransferases, resulting in intermediates of different chain lengths and cyclization [3]. These intermediates are then used by terpene synthases to produce aromatic volatiles. The diversity of terpenoids produced is attributed to the enzymatic activity of these pathways but also to the large family of multi-substrate terpene synthases [4]. Altering the expression of terpene synthases in planta would likely alter the aromatic profile.

Rosemary (*Salvia rosmarinus* Spenn.) is a Mediterranean shrub belonging to the *Lamiaceae* Martinov family and is commonly cultivated for the active properties of its essential oils. It is an excellent model for assessment of methods looking to improve volatile production in potted culinary herbs. The antioxidant, antimicrobial, aromatic and preservative properties obtained from the leaves make rosemary valuable to both the pharmaceutical and food industries. Rosemary is widely cultivated for the fresh herb industry as a potted plant for culinary use. The main volatile constituents of rosemary that contribute to aroma are α-pinene, linalool, cymene, and eucalyptol [5] among others, and are all synthesised by terpene synthases. In the family *Lamiaceae*, there is a large chemo-diversity of terpene synthases, as studied in sweet basil (*Ocimum basilicum* L.), sweet marjoram (*Origanum majorana* L.), oregano (*Origanum vulgare* L.) and rosemary [6]. This diversity in terpene synthases provides the basis for breeding rosemary cultivars with desirable aromas by using gene markers for selected terpene synthases. The horticultural industry also seeks improvements to current rosemary cultivars on the market without the need for extensive breeding programmes. The following trials have been conducted to investigate the effects arbuscular mycorrhizal fungi (AMF) have on the volatile production in current horticultural rosemary cultivars.

The addition of AMF has been shown to improve the volatile production in rosemary [7]. Previous studies have generally investigated the effect of AMF bio-stimulants on rosemary growth and aroma while under environmental stressors, such as high salinity, drought, or nutrient deficiencies, as these are the challenges that face field grown rosemary crops [8,9]. In glasshouse herbs, such as basil, the addition of fungal bio-stimulants to the substrate mitigated the effects of salinity stress and improved the production of aromatic products in the plant [10]. However, another study conducted by Saia et al. [10] investigated the effect of AMF bio-stimulants in plants that were not saline stressed. They found the antioxidant content in basil increased with AMF addition through increases in rosmarinic acid production [10].

The addition of fertilizers (inorganic and organic) to the growth substrate of rosemary revealed a positive correlation between total terpene compounds and the nitrogen and phosphorus content in leaves [11]. It was suggested that the increased availability of nitrogen for terpene synthase activity, and the availability of phosphorus for precursors in the MEP pathway increased aromatic volatile production. Bustamante et al. [11] also reported that the fertilisation effects are a direct nitrogen trade-off between growth and the terpenoid pathways by which volatiles are synthesised. Therefore, increasing fertilization throughout the growth of the crop may not directly benefit the terpenoid pathway. Mycorrhizal associations in rosemary have previously been found to enhance growth and antioxidant properties of the leaves through increases in carnosol, ferulic acid, asiatic acid and vanillin [12]. It is known that mycorrhizal associations improve growth in plants and can improve production of specialised metabolites [13]. It is less well understood how AMF influence gene expression in specialised metabolism, and, in particular, terpene synthase expression.

One of the main challenges facing the horticultural industry is to find solutions that improve the growing conditions and the aromatic profiles of the plants. In addition, there is a need in horticulture for environmentally conscious and sustainable solutions whereby there can be a reduced use of chemical sprays and fertilizers. Mycorrhizal fungal associations with the plant can provide part of a solution to sustainable horticulture as they enhance metabolite production by increasing plant nutrient uptake. The aim of this investigation was to assess the effect of beneficial mycorrhizae on gene expression of terpene synthases for improved volatile production in horticultural rosemary. The effect of AMF on physiology and terpene synthase expression among different rosemary cultivars was investigated to better understand how cultivars respond to AMF in terms of growth and aroma production.

2. Materials and Methods

2.1. Propagation of Rosemary Cuttings

Semi-hard wood cuttings 8 cm long were taken from six cultivars of rosemary, namely, 'Bolham Blue'; 'Blue Boy'; 'Logee Blue'; and 'Roman Beauty', sourced from Downderry Nursery, Tonbridge, UK; 'Perigord' obtained from Vitacress Herbs, Runcton, UK; and 'Vatican Blue' obtained from Jekka's, Bristol, UK. The top leaves were left on the cutting and a sharp scalpel was used to remove excess leaves and cut below an internode. The cuttings were rooted in TPS peat substrate mix (Jiffy Products International, Moerdijk, The Netherlands). They were maintained in propagating boxes at 18 °C with 12 h warm white fluorescent light (120 mmol m^{-2} s^{-1}, 12 h dark cycles of for 2–3 weeks until roots were produced. The cuttings were kept moist during root growth using a spray bottle. The cuttings were re-potted once rooted in 0.4 L pots filled with TPS peat substrate mix.

2.2. AMF Trials Design

The trial was conducted at the Vitacress commercial glasshouse in Runcton, UK, under controlled ambient conditions, an average temperature of 20.2 °C (±0.8 in the day and average temperature of 18.3 °C (±0.9) during the night over nine weeks. The experiment was carried out from February to April 2020, with natural light supplemented by SON-T high pressure sodium lamps (6000 lux) when the natural light was less than 8000 lux to maintain a light schedule of 12 h light/dark cycle. Pots were irrigated with potable water as required to keep the substrate moist. The AMF mixture used was the RGPRO HORTI 2 (PlantWorks Ltd., Sittingbourne, UK) mixture composed of five AMF species: *Funneliformis mosseae, Funneliformis geosporus, Claroideoglomus claroideum, Rhizophagus intraradices,* and *Glomus microaggregatum*. Two AMF application methods were used. The direct application method (Method 1) involved 5 g of AMF mixture added directly to the planting hole of each pot at the repotting stage, then the rosemary root plug was planted directly into the hole containing AMF. The AMF mixture application method (Method 2) involved 25 g of AMF mixture mixed into the peat-based compost, and dispersed evenly, per pot prior to planting the root plug. The control condition was peat substrate without the AMF mixture. In Trial 1, the cultivar Perigord was grown with both AMF application methods. Twelve pots were used as a control of just peat substrate, 12 pots were treated with the direct application (Method 1), and 26 pots were treated with an AMF mixture (Method 2). Plants were grown under controlled conditions for 64 days. In Trial 2, five rooted plugs of each cultivar were potted in peat substrate containing the AMF mixture and were grown for 64 days in controlled conditions. At the end of the trials, random sampling of six plants was performed to take measurements of plant height, plant width, fresh weight of the leaves and stems, dry weight of leaves and stems, total phenolic and total antioxidant content of the leaves. Root samples from five plants were taken from each substrate condition to assess AMF colonization.

2.3. Plant Growth Measurements

Plant growth was assessed to investigate any changes after AMF addition. The fresh and dry weight of the plants was recorded to assess any growth difference by measuring the total biomass of the stems and leaves of the plants. Height and width measurements were taken in triplicate from images of plants from the experimental conditions of each cultivar using automated measuring via image analysis software, ImageJ [14]. Six plant replicates of 'Perigord' were used to measure the fresh and dry weight of the stems and leaves. The fresh weight was measured after separating the stems from the roots and the stems were weighed. The dry weight was measured after air drying the stems at 50 °C in a drying cabinet for three days.

2.4. Total Antioxidant Content and Phenolic Content Assays

Total phenolic content was determined by the Folin-Ciocalteu (F-C) assay, as modified by Sánchez-Rangel et al. [15]. Plant extracts were prepared by grinding 0.2 g of rosemary leaf tissue in 2 mL of Acetate buffer (1.6% Acetic acid and 6% Sodium Acetate Trihydrate) with a pestle and mortar. The extract was transferred to a 1.5 mL microcentrifuge tube and centrifuged at 8000× *g* for five minutes. The supernatant was then used either directly in the assay or diluted with acetate buffer. The dilution factor was accounted for during data analysis. The assay solution consisted of 17.84 mM F-C reagent (Sigma-Aldrich, Gillingham, UK) and 71.39 mM sodium carbonate in distilled water. A total of 300 µL of reagent was added to 15 µL of the extracted sample into a well of a 96 well microtiter plate and incubated for 2 h. Gallic acid half serial dilutions from 1 mM stock were used as a standard. The absorbance was read using a plate-reading spectrophotometer (Spectramax Plus 384 Microplate Reader, Molecular Devices LLC, Wokingham, UK) at 765 nm. The Gallic Acid Equivalent (GAE) was calculated from the absorbance and expressed in mg of GAE/g of fresh weight tissue.

A ferric reducing antioxidant power assay was used to determine the antioxidant capacity of rosemary extracts in ascorbic acid equivalents. Each reagent was prepared fresh. Reagents included 25 mL of 50 mmol/L acetate buffer at pH 3.6, 10 mmol/L of 2,4,6-tripyridyl-s-triazine (Sigma-Aldrich, UK) dissolved in 2.56 mL 40 mmol/L HCl and 20 mmol/L of Ferric Chloride Hexahydrate dissolved in 25 mL distilled water. A total of 300 µL of this solution was then pipetted into a 96-well microtiter plate containing 30 µL of the extracted samples. The absorbance was read immediately using a plate-reading spectrophotometer (Spectramax Plus 384 Microplate Reader, Molecular Devices LLC, UK) at 595 nm. The Ascorbic Acid Equivalent (AAE) was then calculated in mg of AAE/g of fresh weight leaf tissue.

2.5. Root Staining and Microscopy

The roots from the rosemary plants treated with the two AMF application methods were stained for fungal structures and observed under a microscope to estimate percentage colonisation. The following protocol was adapted to clear the roots of rosemary and stain AMF structures within the roots of the AMF treated plants. The roots of un-supplemented plants were also stained for AMF presence as a control. A total of 5 g of root material was taken from the plant and washed in water to remove the substrate. Roots were placed in a 50 mL Falcon tube with 50 mL of 10% *w/v* KOH and given a heat pre-treatment of 60 °C in a water bath for 1 h. The roots were left to clear at room temperature for 24 h. The cleared roots were rinsed and 5% *v/v* HCl was added to the roots for 1 min. The HCl solution was removed and the stain trypan blue was added (0.01% trypan blue (Sigma-Aldrich, UK), 2.5% acetic acid, 50% glycerol). Whole roots were measured then mounted on microscope slides and observed under a compound microscope. Colonization was quantified using the cross-hair eyepiece method [16,17].

2.6. RNA Extraction from Rosemary Leaves

RNA extraction was performed using the Qiagen RNeasy Plant Mini Kit (Qiagen, Germantown, MD, USA). Initial leaf tissue disruption was carried out by grinding leaf samples in liquid nitrogen with a pestle and mortar. An alternative lysis buffer was used to improve RNA yield. The lysis buffer was developed by [18] for plant tissues with high levels of phenolics and polysaccharides. The lysis buffer consisted of 4 M guanidine isothiocyanate, 0.2 M sodium acetate at pH 5.0, 25 mM PVP-40 (polyvinylpyrollidone, Sigma-Aldrich, UK) and 1% final concentration of β-mercaptoethanol (Sigma-Aldrich, UK) was added immediately before use. A total of 20% sarkosyl (Sigma-Aldrich, UK) was added to the sample lysate before incubating at 70 °C for 10 min. The Qiagen protocol was followed as instructed in the kit. The extracted RNA was quantified using a NanoDrop spectrophotometer (Thermo Scientific, Waltham, MA, USA). RNA integrity was checked by gel electrophoresis, with RNA gel loading dye (Thermo Scientific, MA, USA) and an agarose gel with TAE as the buffer and stained with ethidium bromide. Bands were visualised under UV light with the NuGenius (Syngene, Cambridge, UK) imaging system.

2.7. Reverse Transcription and Quantitative PCR

Reverse transcription of extracted RNA was performed with the QuantiTect Reverse Transcription Kit (Qiagen, Maryland). gDNA was removed as per instructions, then 50 µL per sample was reverse transcribed at a final concentration of 1 µg of RNA. The cDNA synthesis was performed in a Techne 5-prime thermocycler (Cole-Parmer Ltd., Saint Neots, UK) with the following temperatures: 42 °C for 60 min, followed by 70 °C for 10 min and a final hold of 4 °C. qPCR was performed on a Rotorgene 6000 (Qiagen, MD, USA) using the primer sequences in Table 1 and using GAPDH as the housekeeping gene. Three biological replicates, chosen at random from the trials, and two technical replicates were used. 0.5 µg of cDNA per sample was pipetted along with 1X SYBR Green master mix (QuantiTect SYBR Green RT-PCR kit from Qiagen, MD, USA) with each primer set at a final concentration of 200 nM. A QIAgility robot (Qiagen, MD, USA) automated pipette was used for pipetting accuracy. The following temperature programme was used for qPCR: an initial denaturing of 94 °C for 2 min, followed by a cycle of 94 °C for 15 s, 58 °C for 45 s and 72 °C s for 30 cycles. Cycle threshold values were used to calculate relative expression using the ΔΔCt method [19].

Table 1. qPCR primers for selected terpene synthases, showing sequence of Forward and Reverse primers, their temperature ranges, G-C% content and product length in bp.

Gene Name	Arabidopsis Ortholog	Forward Primer	Reverse Primer
Glyceraldehyde 3-phosphate dehydrogenase	GAPDH	AAGCATCGGAGACCAAGCTC	CGCGAGAACTGTAACCCCAT
Ocimene synthase	TPS03	GGTACCACACGGGGCATAAA	CAAGATCATCTGCAAGCCGC
β-caryophyllene synthase	TPS12	AGACTGGCCGTAGCAAACTC	CCGATTGTTCAGGCAACACG
Cineole synthase	TPS27	CAGGCATCCTTGCCACATGA	GCCAAACGTTGAGAAAGCCC
Linalool synthase	TPS14	GCCAAATTCAGAGAGGCCCTT	TTGTCCGAGAAGGAAGCACG
Myrcene synthase	TPS24	TGACGCGAACCCTATTCTGG	CAAACCCCAACTTTTCCGGC
Lupeol synthase	LUP2	CTGGCTCTTCCCTTCCGTTT	TAAAACGACGTCGGTGAGGG
Terpene synthase 07	TPS07	CGATGTTCGTGTTCTTGCCC	CCTTCAAATCTCCTCCCCCG

2.8. Statistical Analysis

For comparison of the two application methods of AMF, statistical analysis was conducted by one-way ANOVA in R Studio version 1.4.1106 [20]. The data were tested for normality of residuals and Levene's test for equal variances were $p \leq 0.05$. Two-way ANOVA with Tukey multiple comparisons of means post hoc was performed on the cultivar comparisons. The dependent factors were height of stems, width of stems, total antioxidant content, total phenolic content, dry weight of stems, fresh weight of stems and relative gene expression. Each dependent factor was analysed separately, the independent factors in each case were the presence/absence of AMF and the rosemary cultivars.

3. Results

3.1. AMF Addition to the Substrate Increases Phenolic Content and Alters Gene Expression of Terpene Synthases without Altering the Biomass or Morphology of Rosemary 'Perigord'

The addition of AMF to the substrate of the current commercial cultivar, 'Perigord', was trialled to assess the change in specialised metabolite production and aromatic content. Two different AMF application methods were trialled, one was a direct application to the root plug, the second method was a mixture of AMF into the substrate. Colonisation was observed with both application methods while no mycorrhizal colonisation was detected in any of the control plants. Microscopic observation of AMF root colonisation demonstrated no statistical difference between the levels of colonisation in each application method; however, the plug application method showed more variable colonisation (Figure 1).

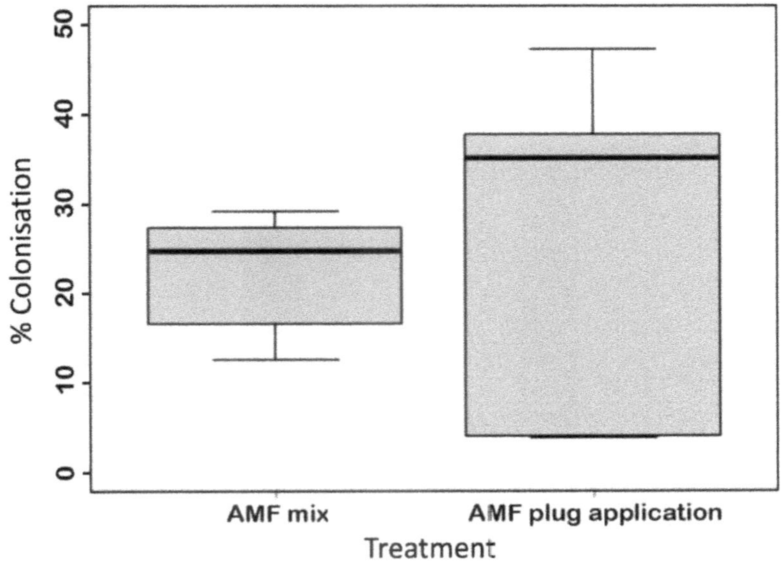

Figure 1. Observations of Arbuscular mycorrhizal fungi colonization in the roots of rosemary 'Perigord'. Mean percentage colonization of two AMF treatments; an AMF mixture added to the substrate and a direct application to the root plug.

The addition of AMF with either method had no significant effect on fresh and dry weight of rosemary plants (Figure 2A) or on the height and width of the plants (Figure 2B and Supplementary Figure S1).

Both application methods had no effect on the total antioxidant content of the plants ($F_{2,6} = 0.21$, $p > 0.05$, Figure 3A). However, a FRAP assay showed a clear increase in phenolic content by 33% for 'Perigord' treated with AMF mix and by 54% for 'Perigord' treated with the AMF root plug (Figure 3B). Statistical analysis, however, showed this was only significant in the case of the plug application method ($F_{2,6} = 3.76$, $p < 0.05$). The AMF mixture showed no significant difference in the phenolic content (Figure 3B).

Terpene synthases play a key role in volatile and essential oil synthesis and their expression contributes to the aroma and taste of the plant. A range of terpene synthase genes were selected for analysis based on sequences previously identified in the rosemary genome sequence [6], which showed detectable transcripts when tested by qPCR. These represent a mix of mono-sesqi-and tri-terpene synthases responsible for terminal steps in synthesis of terpenes associated with rosemary volatiles and essential oil. The AMF root plug application showed statistically significant upregulation of expression of the genes, *β-caryophyllene synthase* and *Linalool synthase*, compared to the control condition (Figure 4). *β-caryophyllene synthase* showed a 32-fold increase in expression in rosemary 'Perigord' compared to the control ($F = 151.3$, $p < 0.001$). *Linalool synthase* also showed increased gene expression levels, by 4-fold, in the plug application method ($F_{2,10} = 151.3$, $p < 0.01$). The other four terpene synthases exhibited no significant change in expression with the AMF plug application method (*Ocimene synthase, Myrcene synthase, Lupeol synthase* and *Cineole synthase*).

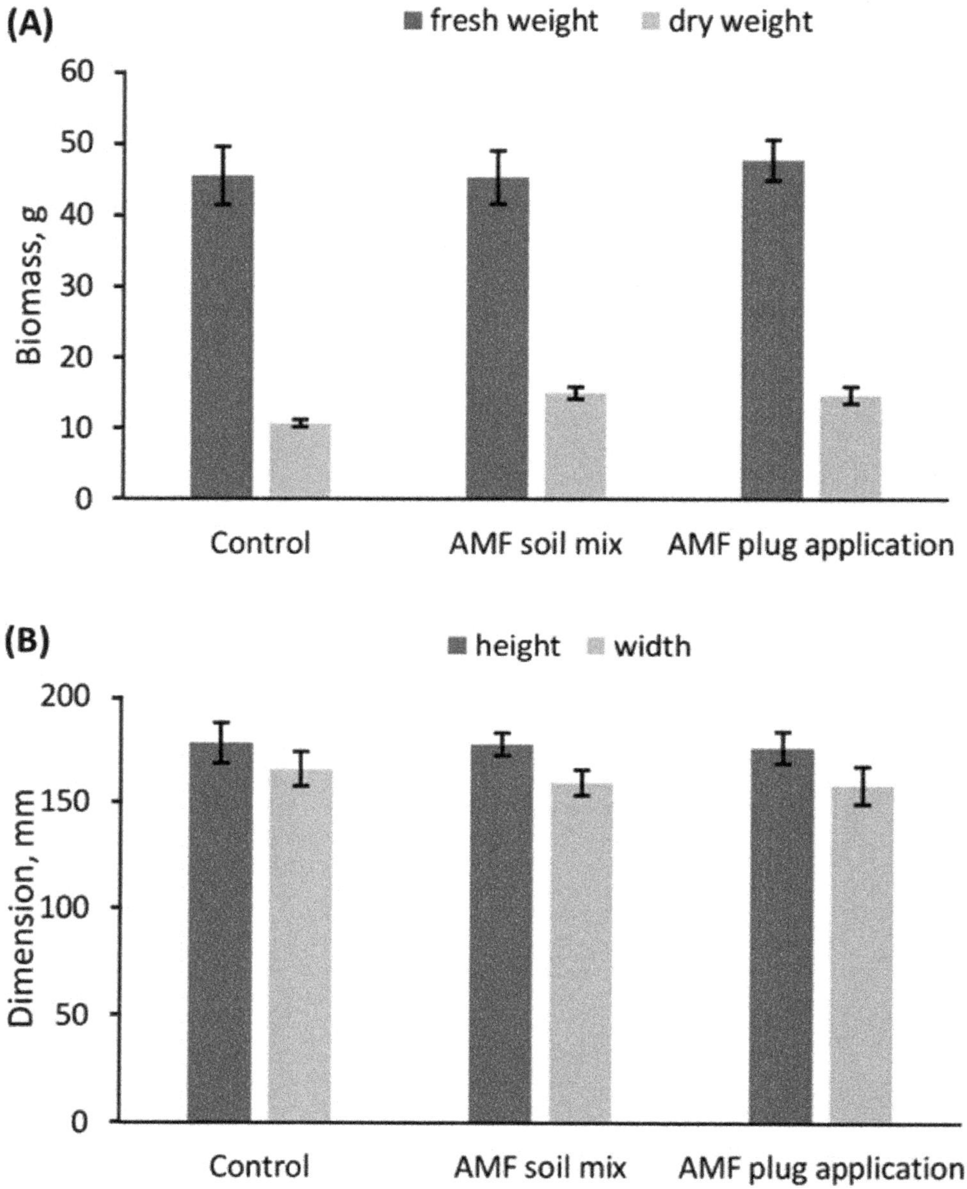

Figure 2. Growth assessment of rosemary 'Perigord' grown in peat substrate with additional Arbuscular Mycorrhizal Fungi applied via two different application methods. (**A**) Fresh weight and dry weight of the stems. (**B**) Height and width measurements of the plants. Error bars are SEM for each AMF condition.

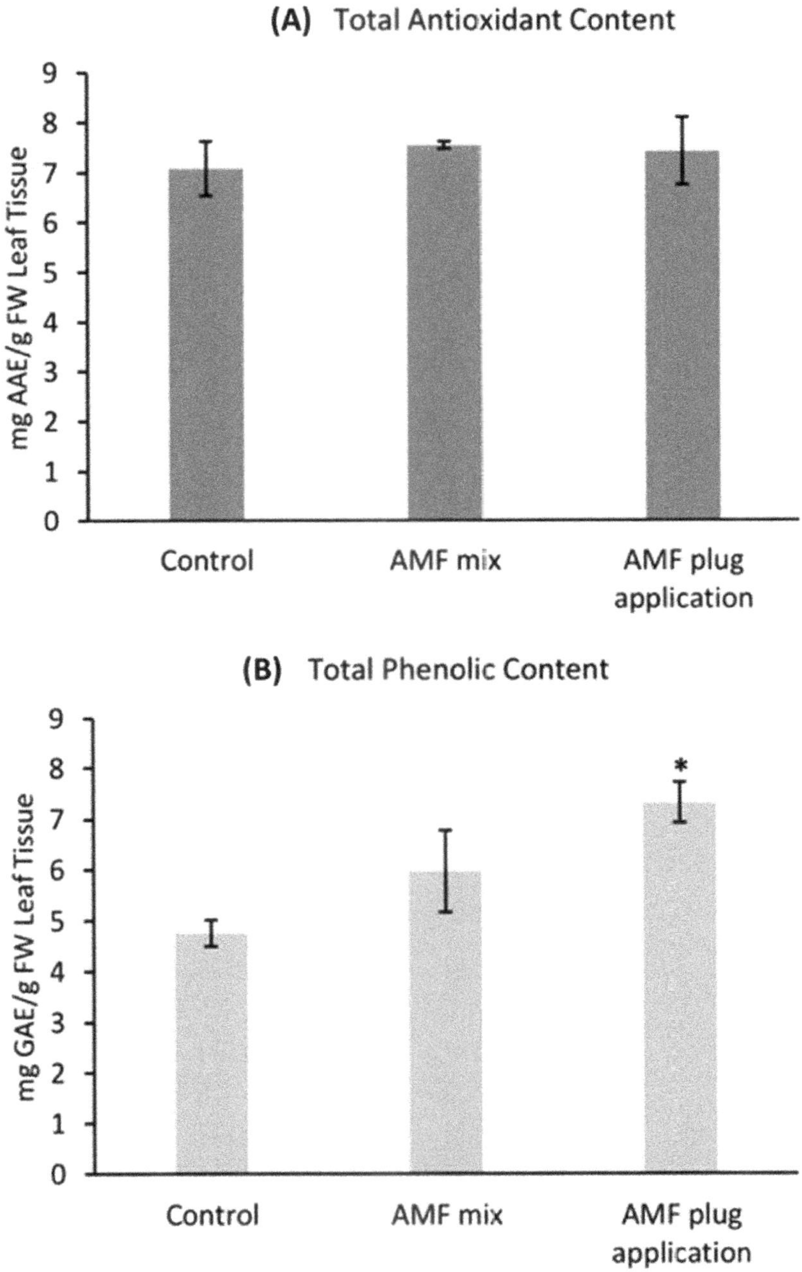

Figure 3. Specialised metabolite production in rosemary 'Perigord' treated with two Arbuscular mycorrhizal fungi AMF application methods. (**A**) Antioxidant content expressed as ascorbic acid equivalent (AAE). (**B**) Phenolic content expressed as gallic acid equivalent (GAE). Mean of treated plants with * above are significantly different to the control ($p < 0.05$ based on the Tukey HSD-test). Error bars are SEM of each treatment group.

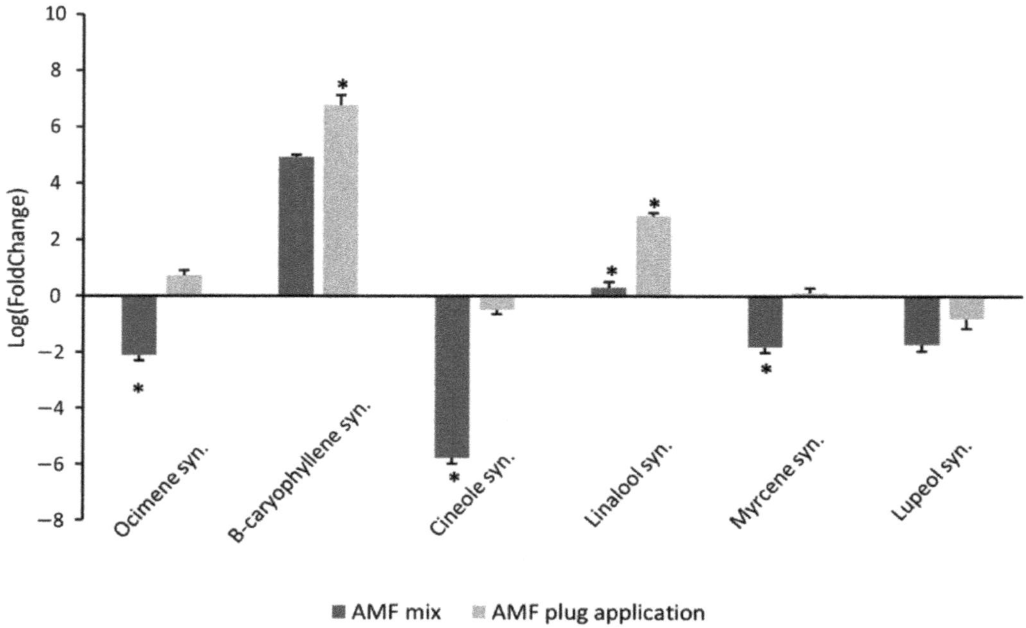

Figure 4. Change in gene expression of six terpene synthases in Rosemary 'Perigord' treated with different additions of AMF to the substrate. Fold change was calculated using 'Perigord' without AMF addition in peat substrate as control condition. Error bars are SEM of samples for each gene. * indicates a significant change versus untreated ($p < 0.05$ based on the Tukey HSD-test).

Meanwhile, for the AMF treatment mixed into the substrate, there was an increased expression of *β-caryophyllene synthase*, by 9-fold, compared to the control plants ($F_{5,10} = 151.3$, $p < 0.001$, Figure 4). However, this change in levels of expression was significantly lower than in the direct application method with levels for the AMF mixture of only 0.28-fold of those observed for the plug application method ($F_{5,10} = 151.3$, $p < 0.001$). Other terpene synthases were downregulated with the addition of the AMF substrate mixture. *Cineole synthase* showed a large decrease with levels of 0.02-fold ($F_{5,10} = 151.3$, $p < 0.001$) compared to the control. *Ocimene synthase* and *Myrcene synthase* were both downregulated with levels of 0.29 and 0.17-fold, respectively, compared to the control ($F_{5,10} = 151.3$, $p < 0.001$).

3.2. AMF Substrate Mixture Alters Gene Expression of Terpene Synthases in Six Rosemary Cultivars

The mixture application method was then selected for further analysis of the impact of AMF supplementation on terpene synthase gene expression in additional cultivars. While both application methods showed good colonisation, the mixture application method was both less variable and provided a much more scalable approach, easily integrated with current practices within the horticultural industry and so this was chosen as the focus of our subsequent analysis. We assessed the effect of AMF addition to five additional rosemary cultivars, while also characterising variation between the cultivars themselves. There were considerable differences in height and width between the various cultivars, 'Blueboy' and 'Vatican Blue' had the tallest plants and were wider than 'Perigord'. The shortest cultivar with a broader width was 'Bolham Blue' (Supplementary Figure S2). The addition of AMF did not affect the height or width of these cultivars, however, apart from 'Logee Blue' which showed an increase in plant height by an average 34.3% ($F_{1,4} = 9.61$, $p < 0.05$). Specialised metabolite production was assessed by measuring antioxidant content as an indicator of increased activity of specialised metabolism and potentially the improved production of volatile compounds. The commercial cultivar 'Perigord' had the highest antioxidant content of all the cultivars when grown in control conditions of peat substrate with antioxidant levels on average two-fold those observed in other cultivars ($F_{1,5} = 35.11$, $p < 0.05$, Figure 5). All other cultivars showed a consistent level of antioxidants with no substantial differences to each other. Overall, there was no significant difference between the antioxidant content of rosemary cultivars grown in peat substrate and those grown with addition of AMF. 'Logee Blue' treated with AMF showed 42.85% of the antioxidant content of control plants; however, this was not significant ($F_{1,5} = 35.11$, $p > 0.05$) based on the Tukey HSD-test.

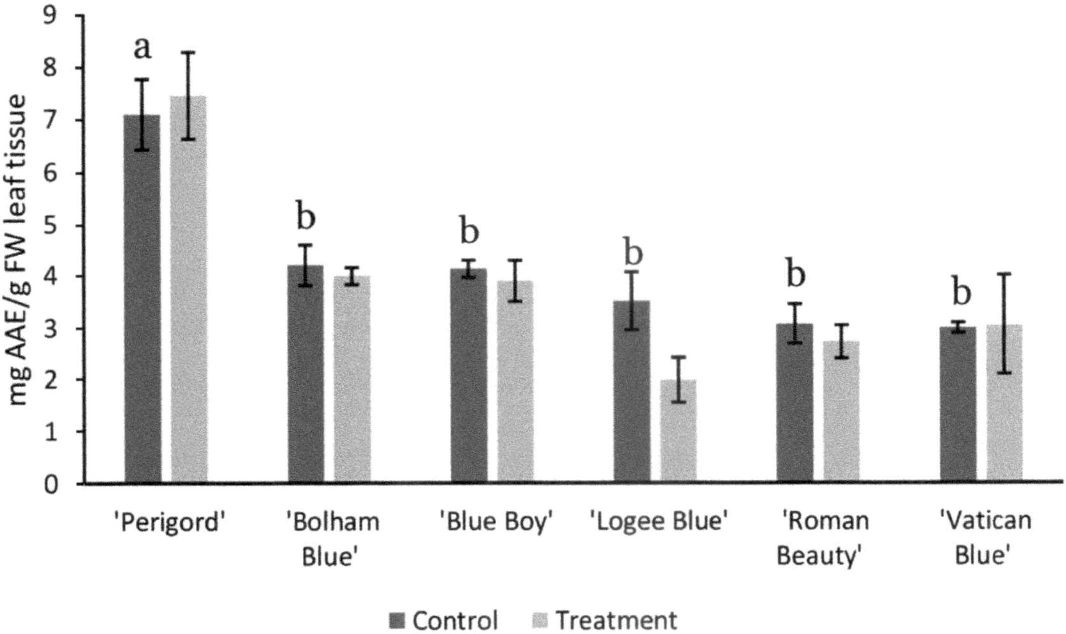

Figure 5. Antioxidant content (ascorbic acid equivalent, AAE) of the leaves of six different rosemary cultivars grown in peat substrate or treated with the addition of AMF. Error bars are SEM. Means in control conditions not sharing any letter are significantly different ($p < 0.05$ based on the Tukey HSD-test.

The total phenolic content was also measured in the six cultivars. Under control conditions, there were several small but significant differences between cultivars. 'Perigord' had the highest phenolic content, while the cultivar 'Bolham Blue' also had a significantly higher phenolic content than 'Blue Boy', 'Roman Beauty' or 'Vatican Blue' ($F_{1,5} = 2.21$, $p < 0.05$, Figure 6). In addition, 'Perigord' showed a statistically significant (F1,5 = 2.21, $p < 0.001$) 1.5-fold increase in phenolic content when treated with AMF mix. However, the other rosemary cultivars showed no differences.

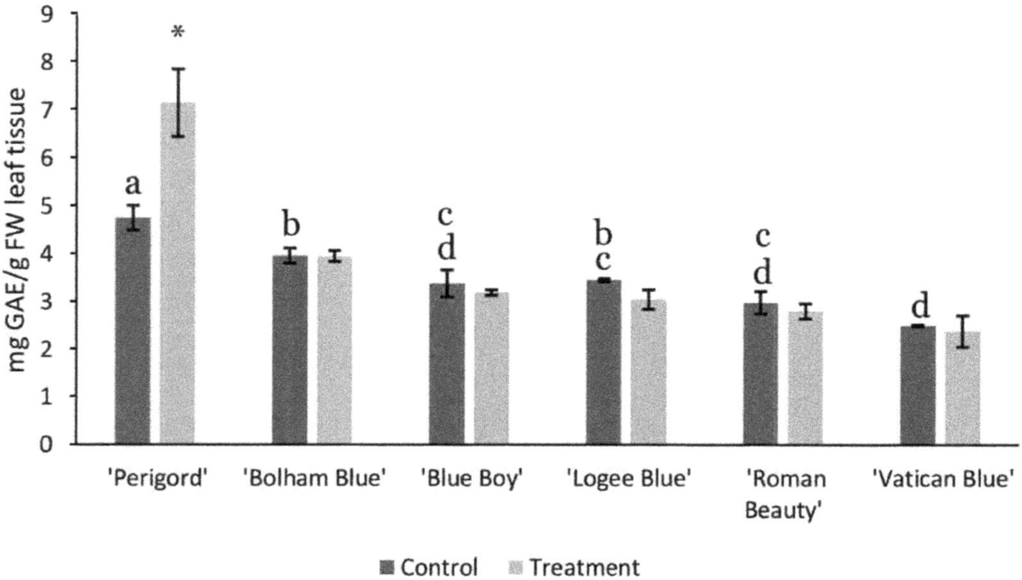

Figure 6. Total phenolic content (gallic acid equivalent, AAE) of six different rosemary cultivars grown in peat substrate containing AMF mixed in, compared with a control of peat substrate. Error bars are SEM. Means in control conditions not sharing any letter are significantly different; * represents significant difference between treated and control; ($p < 0.05$ based on the Tukey HSD-test).

The gene expression of seven terpene synthases was evaluated in the six cultivars of rosemary with and without the addition of AMF. The plants grown in peat substrate showed different relative expression levels between cultivars (Figure 7). 'Logee Blue' showed the highest overall levels of terpene synthase gene expression, while 'Blue Boy' and 'Vatican Blue' showed relatively low levels of all synthases. However, each cultivar had a unique gene expression profile in the control conditions. Notably, 'Perigord' showed relatively high levels of *Linalool synthase*; 'Bolham Blue' showed relatively high levels of *Ocimene* and *Cineole synthases*; 'Logee Blue' showed relatively high levels of *Cineole* and *Lupeol synthases*; while Roman Beauty showed relatively high levels of *Myrcene synthase*.

The addition of the AMF mixture had varying effects on the expression of terpene synthases (Figure 8). 'Bolham Blue' showed small but significant increases in the expression of *Ocimene, Cineole, Linalool* and *Myrcene synthases* ($F_{5,24} = 171.19$, $p < 0.001$). 'Blueboy' showed a significant upregulation of *Cineole* and *Lupeol synthases* as well as an upregulation of *Terpene synthase 7* but a downregulation in *β-caryophyllene Linalool,* and *Myrcene synthases* when treated with AMF, while 'Logee Blue' showed significant increases in *β-caryophyllene synthase* and *Terpene synthase 7* but a downregulation in *Myrcene, Linalool* and *Lupeol synthases* ($F_{5,24} = 171.19$, $p < 0.001$). When 'Roman Beauty' was treated with AMF, the cultivar showed an upregulation of *Ocimene, Cineole* and *Lupeol synthases* but downregulation of *Linalool synthase* ($F_{5,24} = 171.19$, $p < 0.001$). 'Vatican Blue' showed significant upregulation of *β-caryophyllene, Linalool* and *Myrcene synthases* and downregulation of *Ocimene, Cineole Lupeol synthases* ($F_{5,24} = 171.19$, $p < 0.001$). In all, this suggests that selection of variety and addition of AMF offer opportunities for significant modulation of aroma and flavour in rosemary.

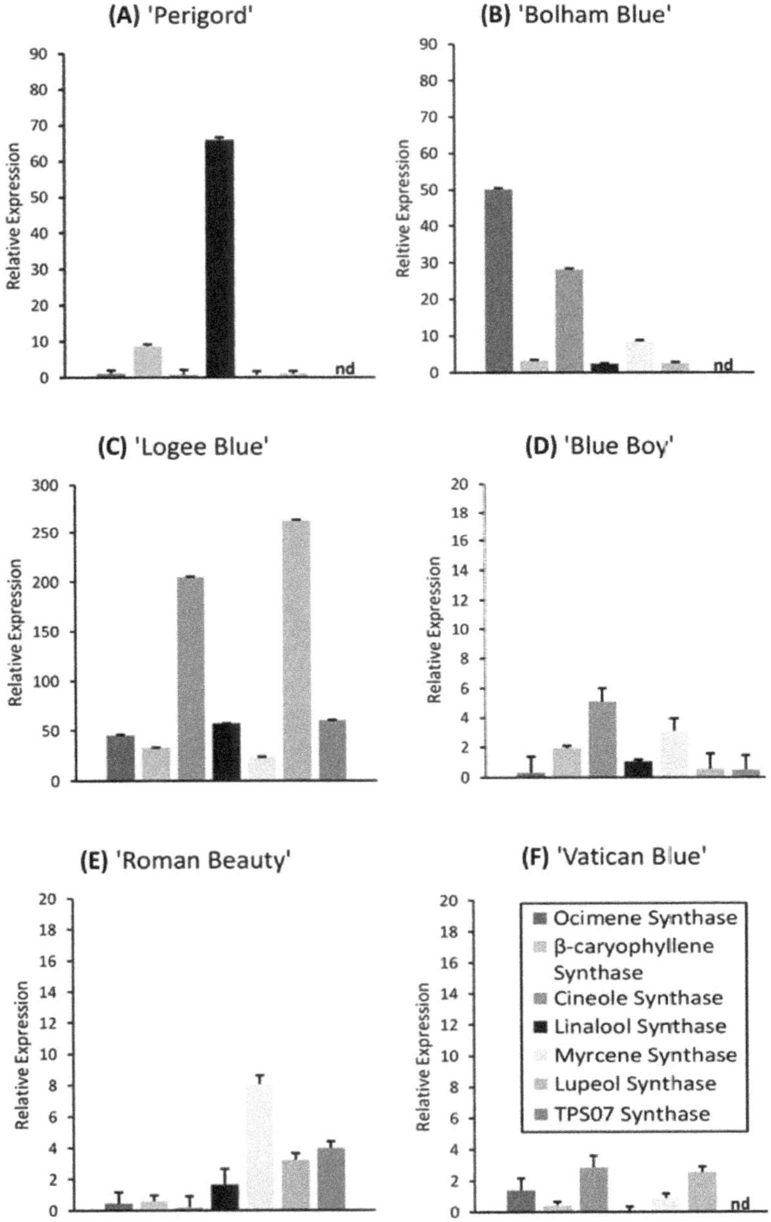

Figure 7. Cultivar comparison of relative expression levels of seven terpene synthases in different rosemary cultivars grown in un-supplemented peat substrate. The housekeeping gene GAPDH was used to calculate relative expression in each cultivar. No detection of expression represented by 'nd'. Error bars are SEM.

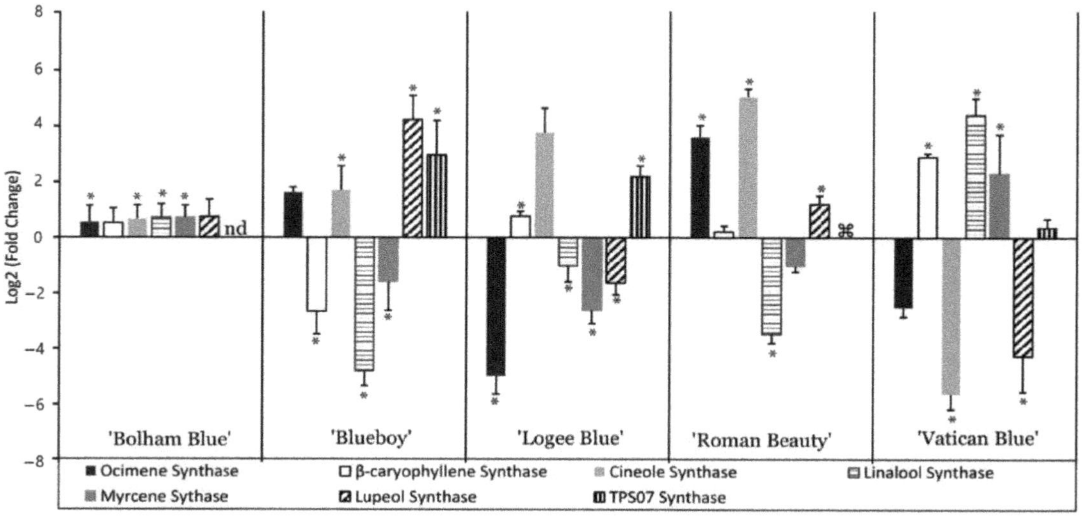

Figure 8. Change in gene expression of seven terpene synthases in five cultivars treated with AMF mixed into the substrate. Fold change was calculated by comparison of gene expression in each cultivar with controls in untreated peat substrate. * indicates a significant difference in gene expression between control and AMF conditions ($p < 0.01$ based on a Tukey HSD after ANOVA). nd: not detected in cultivar. ⌘: not detected following AMF treatment. Error bars are SEM.

4. Discussion

4.1. Root Colonisation Patterns Differed between the Two AMF Application Methods

The two AMF application methods trialled here both resulted in successful colonisation of the roots of rosemary plant cultivar, 'Perigord'. In horticultural production, reducing the number of steps and resources to produce a high-quality plant product is highly desirable and the mixing of AMF into the substrate prior to planting plugs is more favourable in this respect than direct addition to the plug before planting. Fewer steps are involved in the AMF mixture method and so it would be easier to implement into current horticultural practices, making this method more favourable for commercial growers. A large batch of AMF substrate mixture can be easily prepared prior to transplanting. In contrast, with the direct application method a small, measured quantity of AMF is added to the location of every transplant before the plant is added to the pot. The latter would require an additional step to be integrated into the production line, reducing the efficiency of current horticultural practices.

None the less, there are some potential advantages to the plug application method. Rosemary plugs are young cuttings. When transplanted, their roots will take some time to grow and establish in the new substrate. With the direct application method to the root plug, the AMF is in higher concentration close to growing roots and, possibly, the AMF associate with the roots earlier as a result. This proposal is supported by the work of Scagel et al. [21] with Hick's yew (*Taxus media*), a conifer which is also propagated by semi-hard wood cuttings. Arbuscular mycorrhizal fungi have been shown to improve the establishment of cuttings in Hick's yew. The speed of root establishment is increased with the addition of AMF, indicating that mycorrhizal associations with young roots can significantly boost root growth. As a consequence, this will increase nutrients available to the plant in the early stages of growth [21]. Crucially, Scagel et al. [21] demonstrated that roots established more quickly, and the fungal-root cell interaction happened earlier when AMF was added directly to roots of young cuttings.

After root staining, differences in colonisation of the roots were observed between the AMF application methods. The percentage root length colonisation was higher in the direct application method than using the AMF mix method. However, the AMF plug application method had a more variable colonisation rate (as seen in Figure 1). The more uniform colonisation rates in plants treated with the AMF mixture will likely mean more consistent effects on the plants. Such consistency is desirable in pot grown herbs and would represent a further advantage of using this inoculation method over the plug application method in horticultural production.

4.2. AMF Application Enhanced Specialised Metabolism but Did Not Change Biomass or Morphology

Both application methods showed an increase in total phenolics in the commercial cultivar, 'Perigord'. This could indicate that AMF stimulates the specialised metabolism including the MEP and MVA pathways for volatile production. The addition of arbuscular mycorrhizal fungi has been shown to boost the production of essential oils in rosemary cultivars [7,12]. Increasing the content of phenolic chemicals likely includes those that are associated with aroma and taste. As consumers choose fresh herbs based on a strong aroma, these changes to plant specialised metabolism would be beneficial. However, for cultivars other than 'Perigord', the addition of AMF made no significant difference to the quantity of phenolic compounds in the leaf extracts in our assay.

It is also important to note there was no morphological change after the AMF treatment for most of the rosemary cultivars. This could be due to conditions in horticultural glasshouses which are already fully optimised for growth. Other studies which have shown that AMF addition can provide a considerable boost to plant growth have generally studied growth in stressed conditions where growth is not already optimised [8–10,22]. This lack of a morphological response to AMF addition is, however, beneficial for commercial cultivars, which already have a desirable morphology for horticultural production and so no change to its morphology with treatment is desirable. This suggests that AMF addition is desirable for rosemary producers who wish to boost the specialised metabolism of the plants whilst retaining consistent morphological characteristics.

4.3. AMF Enhances Gene Expression of Terpene Synthases in Rosemary Cultivars

The gene expression analysis showed that there were considerable differences in terpene expression between control and AMF treated rosemary plants. It also showed that the cultivars had different patterns of terpene synthase expression without the AMF addition. 'Logee Blue' is reported to be a particularly aromatic cultivar compared to the others [23], which may be due to the high expression levels of seven terpene synthases observed in these plants. Responses to the AMF mixture also vary among cultivars. A general upregulation of terpene synthases was seen in 'Bolham Blue' which had increases in all six of the terpene synthases detected. In other cultivars, responses varied, with large increases often seen in just one or two terpene synthases. Patterns of up and downregulation varied greatly between cultivars; though, there was some commonality between certain cultivars. For example, both 'Blue Boy' and 'Roman Beauty' both showed increases in *Cineole* and *Lupeol synthases* and downregulation of *Linalool synthase*. Likewise, both 'Logee Blue' and 'Vatican Blue' showed upregulation of *β-caryophyllene synthase* and downregulation of *Lupeol synthase*. For the grower, these results show that cultivar selection is relevant not only when selecting for increased aromatics but also when considering application of AMF. As rosemary cultivars have demonstrated here a wide variation in responses in terpene synthase gene expression, this implies that a similar variation might be expected in terms of aroma.

Prediction of details of aroma changes based purely on gene expression analysis is quite unreliable as the final aroma is a result of the blend of volatiles produced [24]. However, some predictions based on individual pathway end products may be attempted. For 'Perigord', the AMF addition enhanced the gene expression of some key terpene synthases. *β-caryophyllene synthase* was upregulated, along with *Linalool synthase*. The upregulation of *β-caryophyllene synthase* could lead to an increase in synthesis of two sesquiterpene products, β-caryophyllene and α-humulene [25]. These are associated with a peppery clove smell and a woody, slightly bitter odour, respectively. The upregulation of *Linalool synthase* would also be likely to alter the aroma profile of 'Perigord' since Linalool is a main constituent of rosemary essential oil. It contributes to the aroma profile with an aroma of pine and floral notes. This may be a beneficial enhancement of the aroma profile of 'Perigord' as consumers may find this rosemary more fragrant. *Ocimene synthase* is responsible for the synthesis of β-ocimene, terpinolene, β-myrcene, and β-pinene, all constituents of rosemary volatiles and essential oils. This synthase was upregulated in 'Bolham Blue', 'Blue Boy' and 'Roman Beauty', indicating that AMF may be enhancing the quality of these aromatics in these cultivars. However, there were some notable decreases that should be considered. For example, 'Blue Boy' and 'Logee Blue' showed decreases in *Myrcene synthase*, with addition of AMF. This is responsible for the synthesis of myrcene, which has a peppery aroma [24].

Most frequently, addition of AMF led to an increase in expression of *Cineole synthase*, a phenomenon observed in cultivars 'Bolham Blue', 'Blueboy', 'Logee Blue' and 'Roman Beauty'. That would likely result in an increase in monoterpenes with strong woody, spicey, and floral scents.

Control of gene expression in metabolic pathways often involves complex regulatory systems. It is possible that specific changes in gene expression may be triggered by chemical communication between the AMF and the plant [26]. Equally, they may be triggered by changes in levels of an enzyme's substrate leading to increased enzyme production [27]. In the case of AMF, such changes in substrate could result from improved availability of nutrients resulting in a change in the flux of metabolites through other connected biosynthetic pathways. For example, since sesquiterpenes are synthesised using the MVA pathway [28], it is possible that AMF addition is altering specialised metabolite production at an earlier stage of the MVA pathway. A possible mechanism for this is the AMF providing additional substrates and nutrients, such as phosphorus [29], for the MVA pathway, thereby increasing the availability of the precursors, DMAPP and IPP, for terpene synthases. In support of this, a previous study has shown that phosphate fertilizers improve essential oil yield and linalool quantities in lavender [30]. Therefore, AMF could be enhancing terpene synthesis through the additional nutrient uptake provided to the plant. It has also been demonstrated that microbial supplementation, specifically, using plant growth-promoting bacteria, can reduce the impact of environmental stress on plants, altering antioxidant levels [31]. It is possible that AMF addition could also be acting in this case via alteration of plant stress responses which may influence terpene synthase expression. Finally, it is important to note that the exact impact of AMF addition may vary depending on the initial microbial content of the substrate. The magnitude and even direction of the effect may be impacted by interaction of the initial substrate microbial community with the AMF. However, our key conclusion is that AMF addition offers a treatment which can be used with potted rosemary, grown in short term cultivation in a horticultural setting, to potentially alter volatile production and, therefore, aroma.

5. Conclusions

We have used rosemary as a model to investigate the potential for AMF to modify pathways involved in synthesis of volatile aromatic compounds in potted herbs. We investigated the impact of AMF addition on expression of several genes encoding key enzymes within volatile biosynthetic pathways in rosemary. Our findings show that a more consistent colonisation of rosemary roots was achieved when AMF was mixed into the substrate prior to planting of root plugs as opposed to being locally applied as the plug was planted. AMF application did not significantly affect the biomass or morphology of plants in a commercial setting but did alter gene expression values of terpene synthases in all cultivars tested. Significant varietal differences were observed between cultivars both in terms of basal gene expression values and in terms of response to AMF addition. The absence of an effect on biomass and morphology is of benefit to the commercial potted rosemary industry, where parameters of uniformity and height are already optimised for large scale cultivation and shipping to supermarkets. However, by revealing an extensive reprogramming of terpene synthase gene expression in response to AMF, in a range of rosemary cultivars, we demonstrate the potential for AMF addition in a commercial setting to significantly alter the aromatic profile of the plants. Overall, our findings demonstrate that AMF addition to commercial pot-grown herbs has the potential to enhance the aroma and taste, while maintaining consistency of plant shape and visual aspects.

Supplementary Materials: The following supporting information can be downloaded at: https://www.mdpi.com/article/10.3390/life13020315/s1, Figure S1: Rosemary cultivar, Perigord, after 9 weeks growth in a glasshouse following transplant of an initial the root plug (A) into untreated substrate, (B) into substrate in which AMF mixture was mixed, and (C) into untreated substrate following AMF addition to the surface of the initial root plug; Figure S2: Height and width measurements of potted rosemary varieties treated with AMF.

Author Contributions: Conceptualization, A.G., S.B., A.D.S., A.C.G. and P.F.D.; investigation, E.L.; writing—original draft preparation, E.L.; writing—review and editing, A.G., A.C.G. and P.F.D.; supervision, A.D.S., A.C.G. and P.F.D.; project administration, P.F.D. All authors have read and agreed to the published version of the manuscript.

Funding: This research was funded by a studentship to Emily Leggatt from the Royal Horticultural Society, Vitacress Herbs Ltd. and Royal Holloway University of London.

Institutional Review Board Statement: Not applicable.

Informed Consent Statement: Not applicable.

Data Availability Statement: Not applicable.

Acknowledgments: The authors would like to thank Stacey Vincent, Royal Holloway University of London, for help with training in qPCR analysis; Vitacress for help with provision of plant material and care of plants; and Chongboi Haokip, Royal Holloway University of London, for care of plants.

Conflicts of Interest: Emily Leggatt received a studentship from the RHS, Vitacress Herbs and Royal Holloway University of London. Alistair Griffiths is employed by the Royal Horticultural Society. Simon Budge is employed by Vitacress Herbs. Anthony D. Stead is employed by Royal Holloway University of London. Alan C. Gange is employed by Royal Holloway University of London. Paul F. Devlin is employed by Royal Holloway University of London.

References

1. TechNavio. *Global Greenhouse Horticulture Market 2021–2025*; TechNavio (Infiniti Research Ltd.): London, UK, 2021.
2. Tetali, S.D. Terpenes and isoprenoids: A wealth of compounds for global use. *Planta* **2018**, *249*, 1–8. [CrossRef] [PubMed]
3. Yazaki, K.; Sasaki, K.; Tsurumaru, Y. Prenylation of aromatic compounds, a key diversification of plant secondary metabolites. *Phytochemistry* **2009**, *70*, 1739–1745. [CrossRef] [PubMed]
4. Pazouki, L.; Niinemets, U. Multi-Substrate Terpene Synthases: Their Occurrence and Physiological Significance. *Front. Plant Sci.* **2016**, *7*, 1019. [CrossRef] [PubMed]
5. Özcan, M.M.; Chalchat, J.-C. Chemical composition and antifungal activity of rosemary (*Rosmarinus officinalis* L.) oil from Turkey. *Int. J. Food Sci. Nutr.* **2008**, *59*, 691–698. [CrossRef]
6. Bornowski, N.; Hamilton, J.P.; Liao, P.; Wood, J.C.; Dudareva, N.; Buell, C.R. Genome sequencing of four culinary herbs reveals terpenoid genes underlying chemodiversity in the Nepetoideae. *DNA Res.* **2020**, *27*, dsaa016. [CrossRef]
7. Tarraf, W.; Ruta, C.; De Cillis, F.; Tagarelli, A.; Tedone, L.; De Mastro, G. Effects of mycorrhiza on growth and essential oil production in selected aromatic plants. *Ital. J. Agron.* **2015**, *10*, 160–162. [CrossRef]
8. Shenata, A.; Nosir, W.; Ahmed, A. Using some biofertilizers treatments to promote growth and oil yield of rosemary plant (*Rosmarinus officinalis* L.) grown in sandy calcareous soil. *Future J. Biol.* **2019**, *3*, 26–33.
9. Begum, N.; Akhtar, K.; Ahanger, M.A.; Iqbal, M.; Wang, P.; Mustafa, N.S.; Zhang, L Arbuscular mycorrhizal fungi improve growth, essential oil, secondary metabolism, and yield of tobacco (*Nicotiana tabacum* L.) under drought stress conditions. *Environ. Sci. Pollut. Res.* **2021**, *28*, 45276–45295. [CrossRef]
10. Saia, S.; Corrado, G.; Vitaglione, P.; Colla, G.; Bonini, P.; Giordano, M.; Stasio, E.; Raimondi, G.; Sacchi, R.; Rouphael, Y. An Endophytic Fungi-Based Biostimulant Modulates Volatile and Non-Volatile Secondary Metabolites and Yield of Greenhouse Basil (*Ocimum basilicum* L.) through Variable Mechanisms Dependent on Salinity Stress Level. *Pathogens* **2021**, *10*, 797. [CrossRef]

11. Bustamante, M.; Michelozzi, M.; Caracciolo, A.B.; Grenni, P.; Verbokkem, J.; Geerdink, P.; Safi, C.; Nogues, I. Effects of Soil Fertilization on Terpenoids and Other Carbon-Based Secondary Metabolites in *Rosmarinus officinalis* Plants: A Comparative Study. *Plants* **2020**, *9*, 830. [CrossRef]
12. Seró, R.; Núñez, N.; Núñez, O.; Camprubí, A.; Grases, J.M.; Saurina, J.; Moyano, E.; Calvet, C. Modified distribution in the polyphenolic profile of rosemary leaves induced by plant inoculation with an arbuscular mycorrhizal fungus. *J. Sci. Food Agric.* **2018**, *99*, 2966–2973. [CrossRef] [PubMed]
13. Emmanuel, O.C.; Babalola, O.O. Productivity and quality of horticultural crops through co-inoculation of arbuscular mycorrhizal fungi and plant growth promoting bacteria. *Microbiol. Res.* **2020**, *239*, 126569. [CrossRef] [PubMed]
14. Schneider, C.A.; Rasband, W.S.; Eliceiri, K.W. NIH Image to ImageJ: 25 Years of image analysis. *Nat. Methods* **2012**, *9*, 671–675. [CrossRef] [PubMed]
15. Sánchez-Rangel, J.C.; Benavides, J.; Heredia, J.B.; Cisneros-Zevallos, L.; Jacobo-Velázquez, D.A. The Folin–Ciocalteu assay revisited: Improvement of its specificity for total phenolic content determination. *Anal. Methods* **2013**, *5*, 5990–5999. [CrossRef]
16. McGONIGLE, T.P.; Miller, M.H.; Evans, D.G.; Fairchild, G.L.; Swan, J.A. A new method which gives an objective measure of colonization of roots by vesicular—Arbuscular mycorrhizal fungi. *New Phytol.* **1990**, *115*, 495–501. [CrossRef] [PubMed]
17. Giovannetti, M.; Mosse, B. An evaluation of techniques for measuring vesicular arbuscular mycorrhizal infection in roots. *New Phytol.* **1980**, *84*, 489–500. [CrossRef]
18. MacKenzie, D.J.; McLean, M.A.; Mukerji, S.; Green, M. Improved RNA Extraction from Woody Plants for the Detection of Viral Pathogens by Reverse Transcription-Polymerase Chain Reaction. *Plant Dis.* **1997**, *81*, 222–226. [CrossRef]
19. Rao, X.; Huang, X.; Zhou, Z.; Lin, X. An improvement of the 2 (-delta delta CT) method for quantitative re-al-time polymerase chain reaction data analysis. *Biostat. Bioinform. Biomath.* **2013**, *3*, 71–85.
20. RStudio Team. *RStudio: Integrated Development Environment for R*; RStudio PBC: Boston, MA, USA, 2021.
21. Scagel, C.; Reddy, K.; Armstrong, J. Mycorrhizal Fungi in Rooting Substrate Influences the Quantity and Quality of Roots on Stem Cuttings of Hick's Yew. *Horttechnology* **2003**, *13*, 62–66. [CrossRef]
22. Begum, N.; Qin, C.; Ahanger, M.A.; Raza, S.; Khan, M.I.; Ashraf, M.; Ahmed, N.; Zhang, L. Role of Arbuscular Mycorrhizal Fungi in Plant Growth Regulation: Implications in Abiotic Stress Tolerance. *Front. Plant Sci.* **2019**, *10*, 1068. [CrossRef]
23. The Sandy Mush Herb Nursery. *The Sandy Mush Herb Nursery Handbook*, 9th ed.; Sandy Mush Herb Nursery: Leicester, NC, USA, 2004.
24. Blerot, B.; Martinelli, L.; Prunier, C.; Saint-Marcoux, D.; Legrand, S.; Bony, A.; Sarrabère, L.; Gros, F.; Boyer, N.; Caissard, J.-C.; et al. Functional Analysis of Four Terpene Synthases in Rose-Scented Pelargonium Cultivars (*Pelargonium* × *hybridum*) and Evolution of Scent in the Pelargonium Genus. *Front. Plant Sci.* **2018**, *9*, 1435. [CrossRef] [PubMed]
25. Li, F.-Q.; Fu, N.-N.; Zhou, J.-J.; Wang, G.-R. Functional characterization of (E)-β-caryophyllene synthase from lima bean and its up-regulation by spider mites and alamethicin. *J. Integr. Agric.* **2017**, *16*, 2231–2238. [CrossRef]
26. Duc, N.H.; Vo, A.T.; Haddidi, I.; Daood, H.; Posta, K. Arbuscular Mycorrhizal Fungi Improve Tolerance of the Medicinal Plant *Eclipta prostrata* (L.) and Induce Major Changes in Polyphenol Profiles Under Salt Stresses. *Front. Plant Sci.* **2021**, *11*, 612299. [CrossRef] [PubMed]
27. Mandal, S.; Upadhyay, S.; Wajid, S.; Ram, M.; Jain, D.C.; Singh, V.P.; Abdin, M.Z.; Kapoor, R. Arbuscular mycorrhiza increase artemisinin accumulation in Artemisia annua by higher expression of key biosynthesis genes via enhanced jasmonic acid levels. *Mycorrhiza* **2014**, *25*, 345–357. [CrossRef] [PubMed]
28. Maffei, M. Sites of synthesis, biochemistry and functional role of plant volatiles. *South Afr. J. Bot.* **2010**, *76*, 612–631. [CrossRef]
29. Kobae, Y. Dynamic Phosphate Uptake in Arbuscular Mycorrhizal Roots Under Field Conditions. *Front. Environ. Sci.* **2019**, *6*, 159. [CrossRef]
30. Peçanha, D.A.; Freitas, M.S.M.; Vieira, M.E.; Cunha, J.M.; de Jesus, A.C. Phosphorus fertilization affects growth, essential oil yield and quality of true lavender in Brazil. *Ind. Crop. Prod.* **2021**, *170*, 113803. [CrossRef]
31. Motamedi, M.; Zahedi, M.; Karimmojeni, H.; Motamedi, H.; Mastinu, A. Effect of rhizosphere bacteria on antioxidant enzymes and some biochemical characteristics of *Medicago sativa* L. subjected to herbicide stress. *Acta Physiol. Plant.* **2022**, *44*, 1–12. [CrossRef]

life

MDPI

Article

Properties of Human Gastric Lipase Produced by Plant Roots

François Guerineau

BioEcoAgro Research Unit, Université de Picardie Jules Verne, 33 Rue St Leu, 80039 Amiens, France; francois.guerineau@u-picardie.fr

Abstract: The properties of recombinant human gastric lipase produced in *Arabidopsis thaliana* roots have been investigated with the goal of determining the potential of the enzyme. This enzyme is stably bound to roots and can be extracted using a buffer at pH 2.2. This enzyme retains over 75% of its activity after two weeks at room temperature when stored in a pH 2.2 buffer. Some of this activity loss was due to the adsorption of the enzyme to the surface of the container. There was no loss of lipase activity in dehydrated roots stored at room temperature for 27 months. The half-life of the enzyme was approximately 15 min when stored in solution at 60 °C whereas dried roots retained 90% lipase activity after one hour at 80 °C. In vitro binding assays using different root cell wall extracts suggested that the lipase was bound to pectin in the roots. Lipase released from the root powder hydrolyzed tributyrin. The high stability of the recombinant human gastric lipase makes this enzyme a good candidate to be tested as a catalyst, whether in solution or bound to roots.

Keywords: *Arabidopsis*; hairy roots; lipase; protein production

check for updates

Citation: Guerineau, F. Properties of Human Gastric Lipase Produced by Plant Roots. *Life* **2022**, *12*, 1249. https://doi.org/10.3390/life12081249

Academic Editor: Jianfeng Xu

Received: 15 July 2022
Accepted: 15 August 2022
Published: 16 August 2022

1. Introduction

Lipases (EC 3.1.1.3) are enzymes of the hydrolase family that: (i) act on triacyl glycerides as part of the catabolic process required for fatty acid absorption and (ii) exhibit esterase activity. The two main lipases involved in digestion in humans are produced by pancreas and stomach cells [1]. Unlike pancreatic lipase, gastric lipase (GL) acts at acidic pH and does not require a colipase for activity [2]. Gastric lipase is a soluble enzyme found at the interface of lipid droplets and aqueous solutions [3]. The hydrophobic active center inside the enzyme structure is covered by a lid that opens upon contact of the enzyme with a substrate [4].

The low pH tolerance of gastric lipase makes it a candidate for enzyme supplementation therapies for pancreatitis or cystic fibrosis patients [5]. Lipase supplementation is currently done using enzymes extracted from livestock. A human recombinant enzyme would be an interesting alternative to lipases of animal origin [6]. Lipases, especially of microbial origin, have been used as biocatalysts in numerous biosynthetic reactions [7]. They can act in various solvents or ionic liquids to produce innovative molecules of therapeutic interest [8]. Human gastric lipase may also be used as a catalyst in esterification or transesterification reactions.

Plants are now considered valuable hosts for the production of heterologous proteins [9]. As a consequence of the ban on transgenic crops in some parts of the world, in vitro plant systems have been developed for the production of heterologous proteins with the first approved plant-produced protein for therapeutic use being glucocerebrosidase, produced by transgenic carrot cells grown in bioreactors [10]. Hairy roots are another in vitro system that can be used for the production of heterologous proteins [11]. Hairy roots emerge following infection of plant tissue by the bacteria *Rhizobium rhizogenes*. Hairy roots have higher stability than alternative cell culture systems, likely because they can grow without the need to add phytohormones which can induce genetic instability [12–14]. The Brassicaceae plant family can be used as an alternative to tobacco species for the production of heterologous proteins [15–17]. In particular, *Arabidopsis thaliana* hairy roots have been established for the production of human gastric lipase [18]. Although the enzyme, in this system, was targeted for secretion, it did not diffuse from the hairy roots to the culture medium and had to be extracted from the roots, which suggested some interactions between the enzyme and the roots producing it. In this work, I further investigated the interaction of gastric lipase with the roots. Given that the stability of an enzyme produced in a heterologous host was sometimes found to be lower than that of the native enzyme [19], I investigated if this was similarly the case in recombinant human gastric lipase produced by *Arabidopsis thaliana* hairy roots. This investigation paves the way for using this recombinant human gastric lipase enzyme as a biocatalyst or for lipase supplementation in certain patients.

2. Materials and Methods

2.1. Materials

The *Arabidopsis thaliana* hairy root line used in this work was the GL28 line described in [18]. It was obtained by hypocotyl transformation of the *Arabidopsis sgs3-12* co-suppression mutant line with the *Rhizobium rhizogenes* 15834 strain containing the cDNA encoding the mature human gastric lipase, fused to a plant signal peptide-encoding sequence, under control of the 2×35 S cauliflower mosaic virus promoter. The lipase substrates 4-methylumbelliferyloleate (MUO) (ref. 75164) and tributyrin (ref. W222305) were from Sigma-Aldrich, St-Quentin-Fallavier, France.

2.2. Root Culture and Storage

Hairy roots were grown at 21 °C on an orbital shaker at 60 RPM in 55 mm diameter Petri dishes in Gamborg B5 medium containing sucrose at 5% and 2,4-D at 0.5 mg/L. They were harvested after 18 to 21 days, briefly washed in distilled water, and freeze-dried overnight. The dehydrated roots were kept in Eppendorf tubes having their lids punctured. The tubes were placed in closed containers containing silica gel and stored in the dark at room temperature.

2.3. Lipase Extraction and Assay

Lipase was extracted by grinding 5 to 10 mg of dry roots in 200 to 400 μL of lipase extraction buffer (LEB: 0.1 M glycine-NaOH pH 2.2, 0.15 M NaCl). The mixture was incubated for 15 min at 37 °C in an Eppendorf Thermomixer at 1000 RPM and then centrifuged at 13,000× *g* for 2 min. Fluorometric assays were done on the supernatants diluted 1/100 or 1/200 in LEB in Eppendorf Protein LoBind® tubes. As previous work has established linearity of lipase activity on tributyrin and esterase activity on MUO [18], for convenience, the fluorometric assay using MUO was used to quantify hGL activity. The assays were done in MUO substrate solution (10 mM acetate buffer pH 5, 0.15 M NaCl, 7 ppm Triton X-100, 0.15 mM MUO) as described before [18]. The assays were done in duplicates. The activity was expressed in μmoles of 4-methylumbelliferone (MU) produced per min per ml of undiluted extract or per mg of roots. It was then converted to units/mg of roots, knowing that one unit of enzyme catalyzed 0.219 μmoles of MUO per min under our assay conditions [18]. For the reaction on tributyrin, 100 μL of tributyrin was emulsified in 10 mL of 10 mM acetate buffer pH 5, 0.15 M NaCl, 1% molten agarose and the mixture was poured into a 90 mm diameter Petri dish. After setting, root powder was sprinkled on the plate. Photographs were taken after 1, 4, and 24 h.

2.4. Lipase Stability and Tube Binding Assays

Lipase was extracted from different batches of dry roots. Extracts were diluted 1/200 either in 50 mM glycine buffer at pH 2.2, 0.15 M NaCl or in 50 mM acetate buffer at pH 5, 0.15 M NaCl. The diluted extracts were kept in Protein LoBind® Eppendorf tubes at room temperature and assayed for lipase activity after 1, 3, 6, 10, and 14 days. After 14 days, the solutions were transferred into new tubes and the lipase activity was assayed 24 h later. The activities in stored extracts were expressed as a percentage of activity related to freshly diluted extracts. To test for lipase binding to tubes, 100 µL of lipase diluted 1/200 in one or the other above buffer were placed in Protein LoBind® Eppendorf tubes for 24 h. The solution was removed and the tubes were rinsed with 100 µL of buffer of the same composition as the one used to dilute the extracts. One hundred µL of MUO substrate solution was placed for one min in the empty tubes. One hundred µL of stop solution was added and the fluorescence of the MU (4-methylumbelliferone) was measured, indicating the activity of lipase bound to the tubes.

2.5. Cell Wall Polysaccharide Extraction

Alcohol-insoluble matter (AIM) was prepared using a method described in [20]. Two hundred mg of freeze-dried untransformed roots were ground in 2 mL of 70% ethanol. The suspension was vortexed for 20 s and centrifuged at 13,000× *g* for 30 s. The pellet was further extracted twice with 2 mL of 70% ethanol, twice with 100% ethanol, twice with chloroform/methanol (vol/vol), and once with acetone. The dry pellet was AIM. For pectin removal, AIM was extracted twice with 20 mM Tris-HCl pH 6.8, 50 mM EDTA for 15 min in an Eppendorf Thermomixer set at 60 °C and 1000 RPM, and then twice with 50 mM Na_2CO_3 under the same conditions. The pectin-deprived AIM pellet was then washed in 1 mL of acetone and dried at 65 °C.

2.6. In Vitro Lipase Binding Assays

Lipase was extracted from GL28 roots using LEB as described above (40 µL LEB/mg roots). The proteins were precipitated by 70% ammonium sulfate for 5 min at room temperature. After centrifugation for 5 min at 13,000× *g*, the pellets were dissolved in the same volume of 10 mM acetate buffer, pH 5. Two hundred µL of the solution was incubated for 15 min with 5 mg of either root powder or AIM or pectin-deprived AIM in an Eppendorf Thermomixer set at 30 °C and 900 RPM. The suspension was centrifuged at 13,000× *g* for 2 min. The supernatants were collected for the quantification of unbound lipase. The pellets were washed once with 10 mM acetate buffer pH 5 and resuspended in 200 µL of LEB for lipase extraction for 15 min in an Eppendorf Thermomixer set at 37 °C and 1000 RPM. After centrifugation at 13,000× *g* for 2 min, the lipase activity in the supernatant was assayed, indicating the amount of bound lipase.

2.7. Statistical Analysis

For each experiment, n indicates the number of lipase extractions done on dry roots. Two enzyme assays were done on each extract. The two values were averaged and the mean activities in the different extracts were calculated. The confidence intervals of the means (CI; $p = 0.95$) were calculated on vassarstats.net and indicated as error bars on the graphs. The number of replicates (n) is given for each experiment in the figures or after the mean values and CI in the text.

3. Results

3.1. Extraction of Lipase from Roots

Dehydrated roots were ground in various extraction buffers and the lipase activities in the extracts were measured. As found before [18], much less lipase was released in acetate buffer at pH 5 than in glycine buffer at pH 2.2 (LEB). However, the addition of 1 M NaCl or of 0.2 M $CaCl_2$ to acetate buffer resulted in the release of 2.7 and 8 times more lipase activity than by acetate alone, respectively (Figure 1).

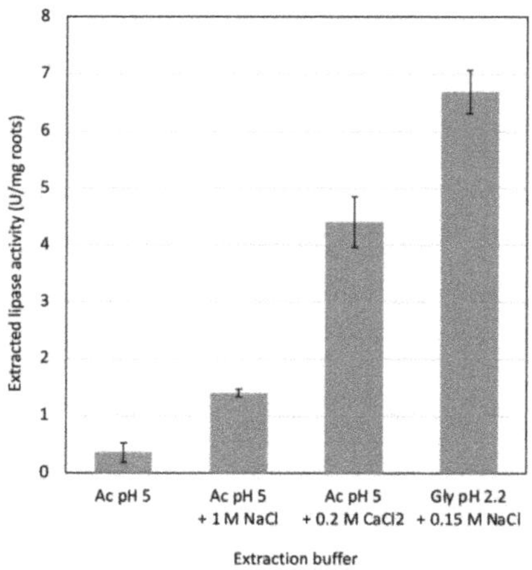

Figure 1. Release of gastric lipase from dry root powder in different solutions. Ac is 10 mM acetate buffer. Gly is 0.1 M glycine buffer. Bars are confidence intervals (0.95; n = 5).

3.2. Stability of Lipase in Solution at Room Temperature

A time course of lipase activity in solution at pH 2.2 or at pH 5 kept at room temperature was performed. These pHs were chosen because they are the pH of the extraction buffer and of the assay buffer, respectively. Whereas extracts at pH 2.2 retained approximately 75% of lipase activity after 14 days, those at pH 5 lost over 50% activity over the same time span (Figure 2). Transferring the diluted solutions of lipase to new tubes on day 14 resulted in a sharp drop of lipase activity after 24 h, especially at pH 5. To test for adsorption on tubes, lipase diluted in solutions at pH 2.2 or at pH 5 were placed in tubes for 24 h and removed. The tubes were then rinsed, and the activity of lipase adsorbed onto the tube was measured. The lipase activity per tube was 0.095 ± 0.012 mU (0.95 CI, n = 6) or 1.5 ± 0.39 mU (0.95 CI, n = 6) for lipase in buffer at pH 2.2 or at pH 5, respectively.

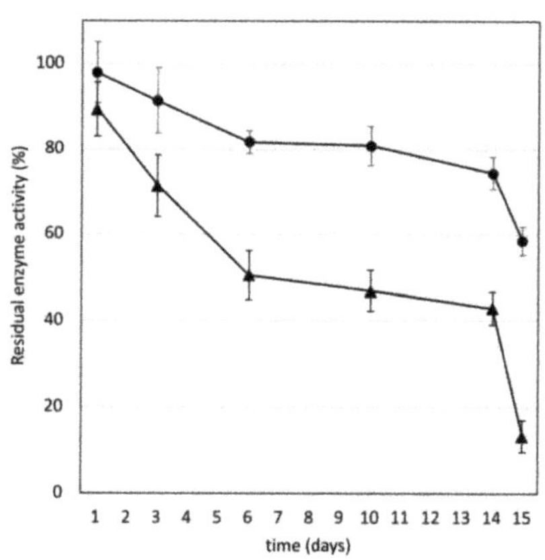

Figure 2. Time course of gastric lipase activity in extracts kept at 20 °C, at pH 2.2 (circles) or at pH 5 (triangles). The extracts were placed in new tubes on day 14. Bars are confidence intervals (0.95; n = 6).

3.3. Long-Term Storage of Lipase in Roots

Extracts from roots that had been freeze-dried the day before had a lipase activity of 6.59 ± 0.26 U/mg roots (0.95 CI, n = 10). Extracts from roots that were kept in silica gel-containing pots for 27 months had a lipase activity of 6.63 ± 0.3 U/mg roots (0.95 CI, *n* = 10). Similar lipase activity was observed in extracts from dry roots that were stored for shorter amounts of time (Table S1).

3.4. Temperature Stability of Lipase

To investigate whether the association of gastric lipase to roots translated into higher temperature stability, the stability at 60 °C of lipase in solution or in dehydrated roots was compared. The half-life of lipase in lipase extraction buffer at 60 °C was approximately 15 min whereas there was no loss of lipase activity extracted from dry roots kept at 60 °C for one hour (Figure 3). Incubating lipase in extraction buffer for 5 min at 80 °C resulted in a total loss of lipase activity (data not shown). In contrast, the activity of lipase extracted from dry roots incubated at 80 °C for one hour was 92.7 ± 9.4% (0.95 CI, *n* = 4) of that of lipase extracted from the control roots kept at room temperature. Similarly, 56.4 ± 17.1% (0.95 CI, *n* = 4) activity could be recovered from roots incubated for one hour at 100 °C. To assess the relative contributions of the association with roots or of the dehydration in the thermal stability of dry root-associated lipase, dehydrated roots were ground and hydrated in 20 mM acetate buffer pH 5, 0.15 M NaCl. The mixes were incubated either at room temperature or at 60 °C for 30 min. The lipase was then extracted from the roots and assayed for lipase activity. The activity extracted from heat-treated hydrated roots was 33.4 ± 5.4% (0.95 CI, *n* = 4) of the activity extracted from hydrated roots kept at room temperature.

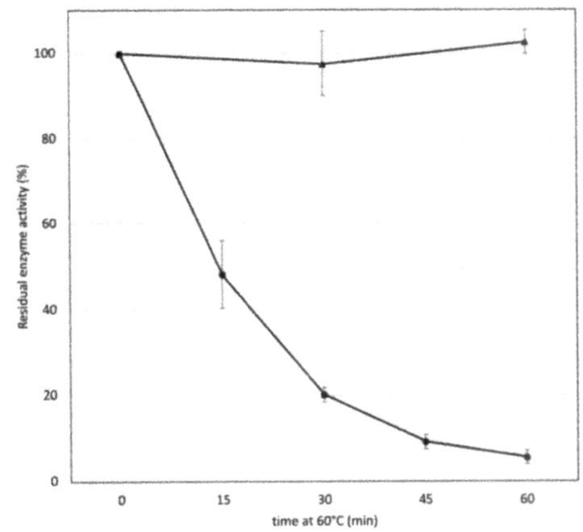

Figure 3. Stability at 60 °C of gastric lipase in solution in lipase extraction buffer (circles) or in dry roots (triangles). Bars are confidence intervals (0.95; *n* = 4).

3.5. In Vitro Binding Assays

To investigate the interaction of human gastric lipase with plant cell wall polysaccharides, in vitro binding assays were undertaken. Lipase in 10 mM acetate buffer at pH 5 was incubated with either untransformed root powder, cell wall powder, or cell wall powder from which pectin had been extracted. The unbound and bound fractions of lipase were extracted and assayed as indicated in the materials and methods section above. The binding of lipase to each powder was expressed as a percentage of bound lipase activity related to the total lipase activity. The binding of hGL was much more extensive on root powder and cell wall powder than on pectin-deprived cell wall powder (Figure 4).

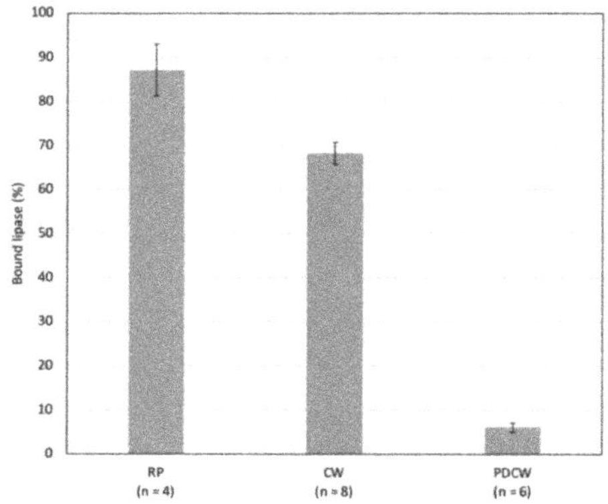

Figure 4. In vitro binding of gastric lipase to root powder (RP), cell wall powder (CW), or pectin-deprived cell wall powder (PDCW). Bars are confidence intervals (0.95, *n* indicated on graph).

3.6. Enzyme Release from Root Powder

To see whether gastric lipase could be released from root powder to hydrolyze a substrate, root powder was sprinkled on a tributyrin-containing agarose plate. Untransformed root powder was used as a control. The formation of a clear halo around each grain of transgenic root indicated tributyrin hydrolysis, with the size of the halos increasing with time (Figure 5).

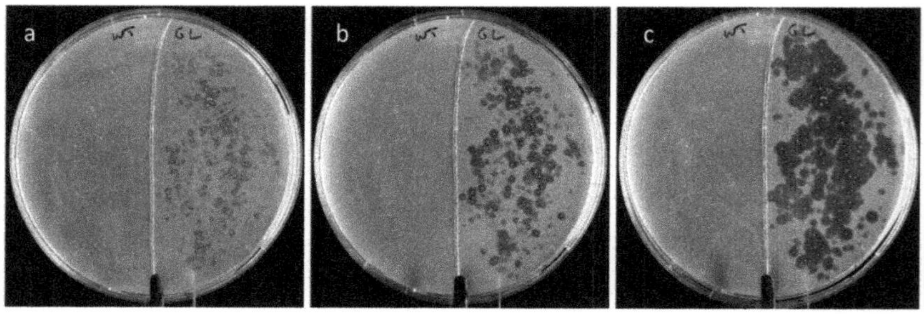

Figure 5. Degradation of tributyrin by gastric lipase diffusing from transgenic root powder on an agarose plate. (**a**) after 1 h; (**b**) after 4 h; (**c**) after 24 h. In each plate: left, untransformed root powder; right, transgenic root powder.

4. Discussion

Previous work has shown that unlike EGFP (enhanced green fluorescent protein), the human gastric lipase targeted for secretion failed to diffuse from transgenic hairy roots into the culture medium [18]. The enzyme, which had likely accumulated in the cell walls, had to be extracted from the roots. Similarly, gastric lipase produced in *Pichia pastoris* also remained in the cell wall [21]. The efficiency of the extraction of human gastric lipase from hairy roots was shown to be pH-dependent, with markedly more lipase being released at pH 2.2 than at pH 5 [18]. Here, the effect of salts on lipase release was investigated. The addition of 1 M NaCl or of 0.2 M $CaCl_2$ enhanced the release of lipase from root powder at pH 5 (Figure 1). The effect of the 0.2 M $CaCl_2$ solution on lipase release was found to be three times higher than that of the 1 M NaCl solution, although the ionic strength of the $CaCl_2$ solution was lower than that of the NaCl solution. This result suggests a specific action of calcium ions. In plant cell walls, calcium ions interact with pectin by forming bridges between homogalacturonan chains [22]. The extensive effect of $CaCl_2$ on lipase release from roots suggests involvement of pectin in lipase retention in the cell wall at pH 5.

To investigate the interaction of lipase with cell wall components, gastric lipase was added to either root powder or crude cell wall extracts or cell wall extracts from which soluble pectin had been removed. The binding of lipase to root powder and to cell wall powder was very effective whereas it was markedly lower on cell wall powder from which pectin had been extracted (Figure 4). This strengthened the hypothesis of lipase interaction with pectin in root cell walls. The negative charges of pectin at pH 5 might be responsible for the association with lipase, expected to be positively charged below its pI of 6.9. In contrast, at pH 2.2, homogalacturonans are expected not to be charged, which would explain the release of lipase at this pH. Similarly, the addition of $CaCl_2$ might trigger lipase release at pH 5 by competing for the negative charges of pectin.

The time course analysis of lipase activity in solution at pH 2.2 revealed a slow decrease down to approximately 75% of the initial activity after 14 days (Figure 2). There was more activity loss at pH 5, down to approximately 45%. Although this could indicate higher stability of lipase at pH 2.2 than at pH 5, the following observation points towards another mechanism. The transfer of lipase solutions into new tubes after 14 days resulted in a sharp drop in activity after 24 h. This strongly suggested that lipase adsorption on tubes was occurring. This hypothesis was confirmed by a tube-binding assay that revealed lipase activity on the surface of the tubes. The binding was 15 times more extensive at pH 5 than at pH 2.2. The adsorption of various lipases on polypropylene beads has been reported, with the immobilization causing a loss of enzyme activity in some cases [23]. The propensity of gastric lipase to adsorb on surfaces has to be taken into account when using this enzyme.

Human gastric lipase was found to be more stable in gastric juice than when purified [24]. It was found that the half-life of purified gastric lipase was only 25 min at pH 2 and pH 3 and 101 min at pH 5 whereas it was over 24 h in gastric juice at pH 3 and pH 5 and 510 min at pH 2. The stability of recombinant lipase extracted from plant roots here was higher than that of native lipase in gastric fluid. Some plant co-extracted proteins or other molecules may have a stabilizing effect on lipase. Alternatively, higher stability might result from differences in the protein structures of the native and the recombinant enzymes, due to differences in the glycan structures deposited on proteins by animal and plant cells [16]. In any case, the higher stability of the recombinant enzyme extracted from plant roots is a valuable feature for its use as a biocatalyst.

The effect of the association of lipase with root cell wall on the lipase thermal tolerance was evaluated. The half-life of lipase in solution in LEB at 60 °C was approximately 15 min (Figure 3). In contrast, the half-life of lipase in dry roots at 100 °C was over an hour, revealing much higher temperature stability of the lipase in dehydrated roots than in solution. When hydrated roots were incubated at 60 °C for 30 min, the lipase activity dropped by 67% whereas lipase in solution under similar treatment lost 80% of its activity (Figure 3). The thermal tolerance of lipase in dehydrated roots is therefore mainly due to dehydration.

The protective effect of dehydration was also confirmed by long-term stability measurements. It has been previously shown that approximately 17% of lipase activity was lost in roots kept at room temperature for two months [18]. In that experiment, no silica gel was used for the storage of the roots, which resulted in the roots being exposed to atmospheric moisture. Here, the storage of roots in the presence of silica gel ensured more extensive dehydration and long-term preservation of lipase activity. There was no loss of extractable lipase activity in dehydrated roots over a period of 27 months. Although freeze-drying is a well-established method of tissue or protein preservation, long-term studies are lacking [25]. In order to maintain the activity of proteins upon freeze-drying, a protecting agent must often be added to them [26]. The high stability of human gastric lipase in lyophilized plant roots revealed a high tolerance of the enzyme to dehydration.

Tributyrin was hydrolyzed around root powder specked on an agarose plate (Figure 5), which indicated that some gastric lipase was released from the roots and diffused onto the agarose to hydrolyze the substrate. The slow release of lipase at pH 5 initiated the production of butyric acid, which decreased the pH around the roots, causing the release of more enzymes. This induction mechanism likely explains the diffusion of the enzyme on the agar plate, indicated by the enlargement of the halos around the specks, over 24 h.

5. Conclusions

This work has revealed interesting properties of a plant-produced human gastric lipase, such as high stability at room temperature and high tolerance to dehydration, that could be leveraged in the design of supplement formulations for the treatment of certain patients. The propensity of gastric lipase to bind pectin or to adsorb on surfaces suggests that the immobilization of the enzyme may be easy. The stability of the enzyme favors its use as a biocatalyst in solvents or in ionic liquids. The strong association of lipase with roots provided a slow-release form of the enzyme that can be utilized in biochemical reactions, whether in aqueous or non-aqueous solvents. The ease of production of this extremophilic human enzyme by plant roots enables the production of structural variants of the enzyme that will improve its properties and its biosynthesis capacity.

Supplementary Materials: The following supporting information can be downloaded at: https://www.mdpi.com/article/10.3390/life12081249/s1, Table S1: Lipase activity extracted from roots kept at room temperature for various amounts of time.

Funding: This research was funded by MESRI (Ministère de l'Enseignement Supérieur, de la Recherche et de l'Innovation), grant number EA3900 and by the SFR (Structure Fédérative de Recherche) Condorcet, LIPACT project.

Institutional Review Board Statement: Not applicable.

Informed Consent Statement: Not applicable.

Data Availability Statement: Data is contained within the article or Supplementary Materials.

Acknowledgments: I am very grateful to Catherine Sarazin for helpful discussions and critical reading of the manuscript.

Conflicts of Interest: The author declares no conflict of interest.

References

1. Roussel, A.; Canaan, S.; Egloff, M.P.; Rivière, M.; Dupuis, L.; Verger, R.; Cambillau, C. Crystal structure of human gastric lipase and model of lysosomal acid lipase, two lipolytic enzymes of medical interest. *J. Biol. Chem.* **1999**, *274*, 16995–17002. [CrossRef] [PubMed]
2. Chahinian, H.; Snabe, T.; Attias, C.; Fojan, P.; Petersen, S.B.; Carrière, F. How gastric lipase, an interfacial enzyme with a Ser-His-Asp catalytic triad, acts optimally at acidic pH. *Biochemistry* **2006**, *45*, 993–1001. [CrossRef] [PubMed]
3. Miled, N.; Canaan, S.; Dupuis, L.; Roussel, A.; Rivière, M.; Carrière, F.; de Caro, A.; Cambillau, C.; Verger, R. Digestive lipases: From three-dimensional structure to physiology. *Biochimie* **2000**, *82*, 973–986. [CrossRef]
4. Canaan, S.; Roussel, A.; Verger, R.; Cambillau, C. Gastric lipase: Crystal structure and activity. *Biochim. Biophys. Acta* **1999**, *1441*, 197–204. [CrossRef]
5. Fieker, A.; Philpott, J.; Armand, M. Enzyme replacement therapy for pancreatic insufficiency: Present and future. *Clin. Exp. Gastroenterol.* **2011**, *4*, 55–73.
6. Aloulou, A.; Carrière, F. Gastric lipase: An extremophilic interfacial enzyme with medical applications. *Cell. Mol. Life Sci.* **2008**, *65*, 851–854. [CrossRef]
7. De Godoy Daiha, K.; Angeli, R.; de Oliveira, S.D.; Almeida, R.V. Are lipases still important biocatalysts? A study of scientific publications and patents for technological forecasting. *PLoS ONE* **2015**, *10*, e0131624.
8. Elgharbawy, A.A.; Riyadi, F.A.; Alam, M.Z.; Moniruzzaman, M. Ionic liquids as a potential solvent for lipase-catalysed reactions: A review. *J. Mol. Liq.* **2018**, *251*, 150–166. [CrossRef]
9. Schillberg, S.; Raven, N.; Spiegel, H.; Rasche, S.; Buntru, M. Critical analysis of the commercial potential of plants for the production of recombinant proteins. *Front. Plant Sci.* **2019**, *10*, 720. [CrossRef]
10. Tekoah, Y.; Shulman, A.; Kizhner, T.; Ruderfer, I.; Fux, L.; Nataf, Y.; Bartfeld, D.; Ariel, T.; Gingis-Velitski, S.; Hanania, U.; et al. Large-scale production of pharmaceutical proteins in plant cell culture-the protalix experience. *Plant Biotechnol. J.* **2015**, *13*, 1199–1208. [CrossRef]
11. Ono, N.N.; Tian, L. The multiplicity of hairy root cultures: Prolific possibilities. *Plant Sci.* **2011**, *180*, 439–446. [CrossRef]
12. Aird, E.L.N.; Hamill, J.D.; Rhodes, M.J.C. Cytogenetic analysis of hairy root cultures from a number of plant species transformed by *Agrobacterium rhizogenes*. *Plant Cell Tissue Organ Cult.* **1988**, *15*, 47–57. [CrossRef]

13. Häkkinen, S.T.; Moyano, E.; Cusido, R.M.; Oksman-Caldentey, K.M. Exploring the metabolic stability of engineered hairy roots after 16 years maintenance. *Front. Plant Sci.* **2016**, *7*, 1486. [CrossRef]
14. Sun, J.; Ma, L.; San, K.Y.; Peebles, C.A.M. Still stable after 11 years: A *Catharanthus roseus* hairy root line maintains inducible expression of anthranilate synthase. *Biotechnol. Prog.* **2017**, *33*, 66–69. [CrossRef]
15. Huet, Y.; Ele Ekouna, J.P.; Caron, A.; Mezreb, K.; Boitel-Conti, M.; Guerineau, F. Production and secretion of a heterologous protein by turnip hairy roots with superiority over tobacco hairy roots. *Biotechnol. Lett.* **2014**, *36*, 181–190. [CrossRef]
16. Ele Ekouna, J.P.; Boitel-Conti, M.; Lerouge, P.; Bardor, M.; Guerineau, F. Enhanced production of recombinant human gastric lipase in turnip hairy roots. *Plant Cell Tissue Organ Cult.* **2017**, *131*, 601–610. [CrossRef]
17. Mai, N.T.P.; Boitel-Conti, M.; Guerineau, F. *Arabidopsis thaliana* hairy roots for the production of heterologous proteins. *Plant Cell Tissue Organ Cult.* **2016**, *127*, 489–496. [CrossRef]
18. Guerineau, F.; Mai, N.T.P.; Boitel-Conti, M. *Arabidopsis* hairy roots producing high level of active human gastric lipase. *Mol. Biotechnol.* **2020**, *62*, 168–176. [CrossRef]
19. Rai, M.; Padh, H. Expession systems for production of heterologous proteins. *Curr. Sci.* **2001**, *80*, 1121–1128.
20. Fangel, J.U.; Jones, C.Y.; Ulvskov, P.; Harholt, J.; Willats, W.G.T. Analytical implications of different methods for preparing plant cell wall material. *Carbohydr. Polym.* **2021**, *261*, 117866. [CrossRef]
21. Sams, L.; Amara, S.; Chakroun, A.; Coudre, S.; Paume, J.; Giallo, J.; Carrière, F. Constitutive expression of human gastric lipase in *Pichia pastoris* and site-directed mutagenesis of key lid-stabilizing residues. *BBA-Mol. Cell Biol. Lipids* **2017**, *1862*, 1025–1034. [CrossRef]
22. Cao, L.; Lu, W.; Mata, A.; Nishinari, K.; Fang, Y. Egg-box model-based gelation of alginate and pectin: A review. *Carbohydr. Polym.* **2020**, *242*, 116389. [CrossRef]
23. Gitlesen, T.; Bauer, M.; Adlercreutz, P. Adsorption of lipase on polypropylene powder. *Biochim. Biophys. Acta* **1997**, *1345*, 188–196. [CrossRef]
24. Ville, E.; Carrière, F.; Renou, C.; Laugier, R. Gastric disorders physiological study of pH stability and sensitivity to pepsin of human gastric lipase. *Digestion* **2002**, *65*, 73–81. [CrossRef]
25. Molnar, A.; Lakat, T.; Hosszu, A.; Szebeni, B.; Balogh, A.; Orfi, L.; Szabo, A.J.; Fekete, A.; Hodrea, J. Lyophilization and homogenization of biological samples improves reproducibility and reduces standard deviation in molecular biology techniques. *Amino Acids* **2021**, *53*, 917–928. [CrossRef]
26. Roy, I.; Gupta, M.N. Freeze-drying of proteins: Some emerging concerns. *Biotechnol. Appl. Biochem.* **2004**, *177*, 165–177. [CrossRef]

life

Review

Naringin and Naringenin Polyphenols in Neurological Diseases: Understandings from a Therapeutic Viewpoint

Talha Bin Emran [1,2,*], Fahadul Islam [2], Nikhil Nath [3], Hriday Sutradhar [4], Rajib Das [4], Saikat Mitra [4], Mohammed Merae Alshahrani [5], Abdulaziz Hassan Alhasaniah [5] and Rohit Sharma [6]

1 Department of Pharmacy, BGC Trust University Bangladesh, Chittagong 4381, Bangladesh
2 Department of Pharmacy, Faculty of Allied Health Sciences, Daffodil International University, Dhaka 1207, Bangladesh
3 Department of Pharmacy, International Islamic University Chittagong, Chittagong 4318, Bangladesh
4 Department of Pharmacy, Faculty of Pharmacy, University of Dhaka, Dhaka 1000, Bangladesh
5 Department of Clinical Laboratory Sciences, Faculty of Applied Medical Sciences, Najran University, P.O. Box 1988, Najran 61441, Saudi Arabia
6 Department of Rasa Shastra and Bhaishajya Kalpana, Faculty of Ayurveda, Institute of Medical Sciences, Banaras Hindu University, Varanasi 221005, Uttar Pradesh, India
* Correspondence: talhabmb@bgctub.ac.bd

Abstract: The glycosides of two flavonoids, naringin and naringenin, are found in various citrus fruits, bergamots, tomatoes, and other fruits. These phytochemicals are associated with multiple biological functions, including neuroprotective, antioxidant, anticancer, antiviral, antibacterial, anti-inflammatory, antiadipogenic, and cardioprotective effects. The higher glutathione/oxidized glutathione ratio in 3-NP-induced rats is attributed to the ability of naringin to reduce hydroxyl radical, hydroperoxide, and nitrite. However, although progress has been made in treating these diseases, there are still global concerns about how to obtain a solution. Thus, natural compounds can provide a promising strategy for treating many neurological conditions. Possible therapeutics for neurodegenerative disorders include naringin and naringenin polyphenols. New experimental evidence shows that these polyphenols exert a wide range of pharmacological activity; particular attention was paid to neurodegenerative diseases such as Alzheimer's and Parkinson's diseases, as well as other neurological conditions such as anxiety, depression, schizophrenia, and chronic hyperglycemic peripheral neuropathy. Several preliminary investigations have shown promising evidence of neuroprotection. The main objective of this review was to reflect on developments in understanding the molecular mechanisms underlying the development of naringin and naringenin as potential neuroprotective medications. Furthermore, the configuration relationships between naringin and naringenin are discussed, as well as their plant sources and extraction methods.

Keywords: naringin; naringenin; polyphenols; neurological disease; dietary interventions

check for updates

Citation: Emran, T.B.; Islam, F.; Nath, N.; Sutradhar, H.; Das, R.; Mitra, S.; Alshahrani, M.M.; Alhasaniah, A.H.; Sharma, R. Naringin and Naringenin Polyphenols in Neurological Diseases: Understandings from a Therapeutic Viewpoint. *Life* **2023**, *13*, 99. https://doi.org/10.3390/life13010099

Academic Editor: Jianfeng Xu

Received: 9 November 2022
Revised: 21 December 2022
Accepted: 22 December 2022
Published: 29 December 2022

1. Introduction

The term "neurological disease" is often used to refer to anything that affects the nervous system. In the brain, spinal cord, or other nerves, structural, metabolic, or electrical dysfunctions may cause a broad range of symptoms. These symptoms include altered states of consciousness, convulsions, muscular weakness, poor coordination, a loss of feeling, and a lack of sensation. Other symptoms include disorientation, pain, and discomfort. Extensive study has led to the discovery of several neurological abnormalities, some of which are somewhat common while others are relatively uncommon [1,2]. Some of these problems are inherited while others are not. Neurology and clinical neuropsychology can identify and treat these illnesses using the appropriate diagnostic and therapeutic methods. Although encased in the skull and spinal vertebrae and chemically isolated by the blood–brain barrier, the brain and spinal cord are susceptible organs [3]. Because of their placement just beneath the epidermis, nerves are nevertheless vulnerable to harm despite the protective layers of tissue around them. Individual neurons and the neural circuits and nerves they form are susceptible to damage to their electrical and structural integrity [4].

Phytochemicals may be broken down into several different classes, such as flavanols, flavan-3-ols, isoflavones, flavanones, anthocyanidins, and flavones [5]. It has been shown that flavonoids promote apoptosis and decrease metastasis, angiogenesis, and proliferation in the setting of carcinogenesis by interfering with various cell signaling pathways. The flavanone-7-O-glycoside naringin is found in many plants, while citrus fruits are the most common place to find it. A flavanone called naringenin is its main component [6,7]. Many different signaling pathways and signaling molecules are affected by this chemical. Its pharmacological properties include but are not limited to an antioxidant, an anti-inflammatory, an anti-apoptotic, an anti-tumor, and an anti-viral, as well as effects on metabolic syndrome, bone regeneration, neurological illnesses, cardiovascular disease, and genetic damage [8]. Naringin may cause drug interactions due to its ability to inhibit cytochrome P450 enzymes, including CYP3A4 and CYP1A2 [9].

Neurological disorders may be alleviated with the help of naringenin (4′,5,7-trihydroxy flavanone), a flavonoid abundant in the peels of citrus fruits (particularly grapes and tomatoes) [10]. Among the top dietary sources of this noteworthy flavonoid are grapefruits and oranges. A few of the many medical advantages of this drug include preventing or reversing weight gain, enhancing metabolic health, and restoring typical lipid profiles in patients with dyslipidemia [11]. Among naringenin's many biological properties is an antioxidant effect. It has been shown that they may help with pain, inflammation, kidney health, and even nerve function. Research has revealed that naringenin may help reduce pain and inflammation associated with several medical disorders [12]. Mechanistically, naringenin is a pleiotropic molecule that inhibits leukocyte recruitment, pro-inflammatory cytokine release, and cytokine-induced analgesia by modulating transient receptor potential (TRP) channels and triggering the nitric oxide (NO)/cyclic guanosine monophosphate (cGMP)/protein kinase G(PKG)/adenosine triphosphate (ATP The compound's low bioavailability and limited brain access are significant roadblocks to broader use despite naringenin's promise in treating neurological diseases [13,14]. The therapeutic potential of naringin and naringenin for various symptoms related to neurological diseases has been shown by encouraging results from preclinical investigations [15].

The principal objective of this review was to report on the progress made toward understanding the molecular mechanisms behind the development of naringin and naringenin as potential neuroprotective therapeutics. Additionally, the configuration links between naringin and naringenin, as well as their plant sources and extraction methods, are described.

2. Methodology

The keywords naringin and naringenin, neurological diseases, botanical sources, and neuroprotective action were used to search the following databases: PubMed, Scopus, and Web of Science. Up to 2022, English-language research reports, reviews, and original research articles were considered and studied. An algorithm was used that followed the flowchart in Figure 1 and had all of the steps and requirements for choosing the required literature, in accordance with the recommendations of Page et al. [16].

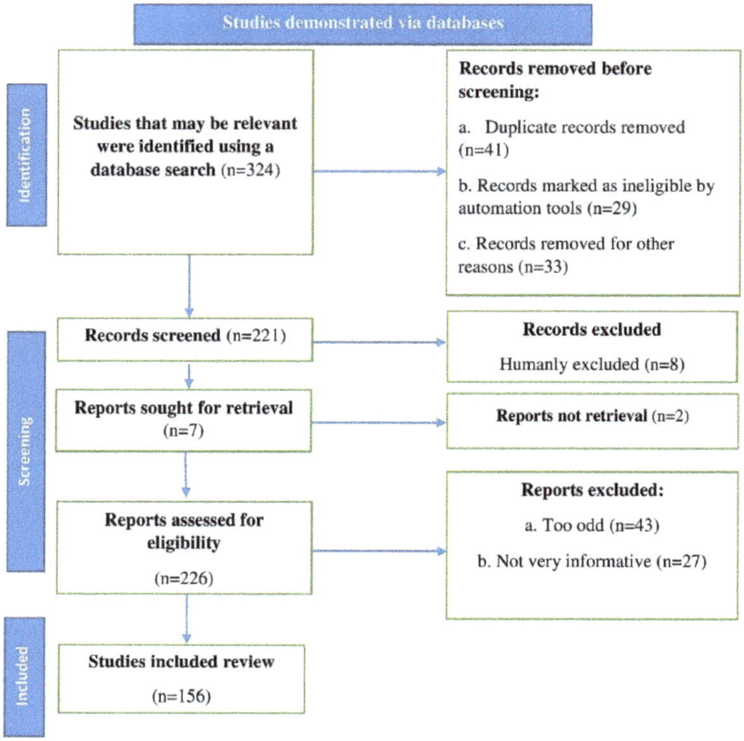

Figure 1. Flowchart depicting the steps to be followed when selecting published data for the current study; n stands for the number of literature reports.

3. Naringin and Naringenin

The flavanone naringenin and the glycosylated derivative naringin are in high concentrations in grapefruit and other citrus fruits. The antioxidant and anti-inflammatory properties of flavonoids have gained widespread recognition [17]. When naringenin is added to SH-SY5Y cells or when 6-OHDA is injected into mice, the cells are protected in an Nrf2-dependent manner, just like the dopaminergic neurons [18]. This is due to the complete elimination of protective effects and expression of Nrf2-dependent cytoprotective genes after treatment with Nrf2 short interfering RNA [19]. Naringenin prevented rotenone-induced structural alterations in muscle and motor impairment in rat models when given after the administration of the medication. The expression of DJ-1 and chaperone-associated E3 ligase was increased in the striatum and SN after treatment with naringenin [20]. Target proteins are ubiquitinated by DJ-1 and a chaperone-associated E3 ligase and then sent on to be degraded by the proteasome. In a rotenone model of Parkinson's disease, naringenin was found to have neuroprotective effects [21,22]. Naringin may aid dopaminergic neuron recovery after injury if given soon after the damage occurs. Dopaminergic neurons were preserved, GDNF levels in the SN were restored, and the number of ionized calcium-binding adaptor molecule 1 (Iba-1) and tumor necrosis factor-alpha (TNF-α) immunoreactive neurons in the striatum were reduced after pretreatment with naringin in rats with a unilateral MPP+-lesion. Eukaryotic initiation factor 4E-binding protein 1 (4E-BP1) and growth differentiation and neurotrophic factor (GDNF) were upregulated in the SN after a single injection of naringin [23].

On top of that, it is an effective antioxidant. During fasting and stimulated states, rapid glucose uptake is impaired in BC. Insulin stimulates phosphoinositide 3-kinase PIP3/Akt and mitogen-activated protein kinase (MAPK) activity, and naringenin prevents this MAPK. It decreases TNF-α and COX-2 levels and raises the transcription factor Nrf2 [24,25].

4. Botanical Sources

Flavonoids are phenolic compounds associated with a wide range of biological functions. There are more than 4000 different flavonoids known to science, most of which are found in their natural, unaltered plant-based forms. Flavonoids can be a dietary supplement [26,27]. Grapefruit and other citrus fruits get their distinctive bitter taste from flavonoid naringin. Although the number of flavonoids taken from food may be large and the flavonoids exhibit potential biological action, they have attracted significantly less attention than flavanols and isoflavones [28,29].

In most cases, the researchers focused on flavanols and isoflavones—intracellular cycling of naringenin, hesperidin, and its glycosylated derivatives, naringenin, hesperidin, and rutin. Grapefruit, bergamot, sour orange, tart cherry, tomato, chocolate, Greek oregano, water mint, and beans are all foods and plants that contain norepinephrine or its glycosides [29,30].

5. Neuroprotective Action

5.1. Alzheimer's Disease

Alzheimer's disease (AD) is the most common type of dementia in the industrialized world and is a severe public health problem [31]. Several different molecular processes caused AD, but its exact pathophysiology is still poorly understood. In several pre-clinical studies, naringin and some of its derivatives, such as naringenin, changed these pathways in ways that could be used to treat AD [32,33]. The disease is caused by the death of cholinergic neurons in the frontal lobe and the formation of Amyloid-β (Aβ) plaques outside the body [34].

Memantine, an antagonist of the NMDA glutamate receptor, breaks down at a much slower rate than acetylcholinesterase (AChE) inhibitors, such as donepezil aricept, which is often used to treat the symptoms of AD [35,36]. Synaptic dysfunction, in which synapses are damaged, cells are killed, and mental impairments occur, results from consuming too much. This dysfunction can be fixed entirely with the proper treatment. It is essential to understand how Aβ is currently in charge of producing and storing memories and how this affects synaptic plasticity in the brain network. Evidence suggests that calcium/calmodulin-dependent protein kinase II (CaMKII) is a critical synaptic target for Aβ-induced synaptic depression [37,38]. Several plant species high in flavonoids have been used in traditional medicine for hundreds of years. Epidemiological and dietary studies on both people and animals have shown that these flavonoids protect against and slow down neurodegeneration, especially regarding the cognitive decline that accompanies aging [39]. The flavonoid glycoside naringin, found in citrus fruits in large amounts, is effective against many diseases and conditions, such as cancer, inflammation, ulcers, osteoporosis, and apoptosis. Naringin has been shown to improve behavior and thinking in animal models of epilepsy caused by kainic acid and Huntington's disease caused by 3-nitropropionic acid [40]. The effects of colchicine and D-galactose on learning and memory are also undone by treatment with naringin. Naringenin has been shown to improve insulin signaling and cognitive ability in the brain and reduce the effects of intracerebroventricular-streptozotocin on neurodegeneration caused by AD (Table 1 and Figure 2) [41].

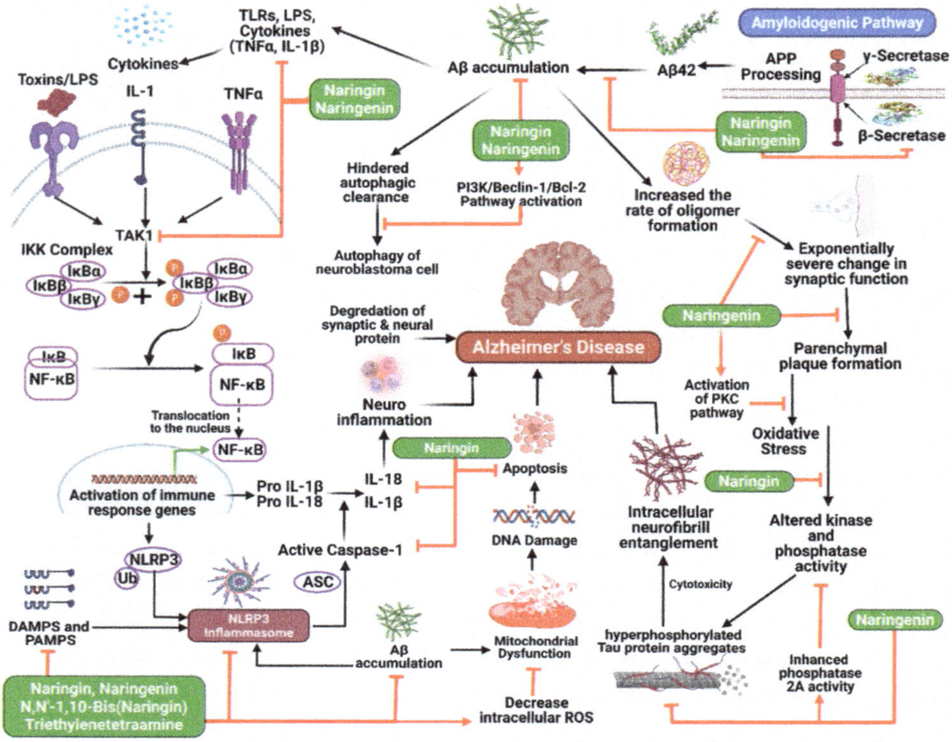

Figure 2. Illustration representing the site of action of naringin and naringenin in Alzheimer's disease pathway.

5.2. Parkinson's Disease

Parkinson's disease (PD) is a degenerative neurological disease that results in impaired motor function due to dopaminergic neuropathy in the substantia nigra [42–44]. Genetic predisposition may play a role in developing mitochondrial damage and oxidative stress; however, additional molecular routes exist. In recent years, oxidative stress has been investigated as a potential mechanism in neurodegeneration, which is only one example of many similarities between AD and PD. The antioxidant properties of flavonoids, alkaloids, and other polyphenols are coming to the fore. Plant chemicals may modulate enzymes and metabolic signaling pathways to reduce ROS production. Bioactive metabolite flavonoids have anti-oxidative actions on the liver's metabolic pathway [45–47].

The antioxidant and neuroprotective properties of naringin have been studied. Asahina and Inubuse found the blueprints for naringin in 1928 [48,49]. Naringenin is linked to naringin through the C-7 hydroxyl group. Metabolites of naringenin may be found in phases I and II of the drug's metabolism. Glucose is the source of naringin's bitter flavor. It reacts with potassium hydroxide or other essential substances to form 1,3-diphenylpropan-1-one, which has a scent reminiscent of menthol [50–52]. It has been shown that naringin may neutralize ROS, scavenge superoxide, suppress xanthine oxide, decrease lipid peroxidation, and decrease the permeability of oxygen-stimulated K+ erythrocytes. Naringin's antioxidant potentials may aid in treating neurology and diabetes. Degeneration of nerve cells in the striatum and substantia nigra pars compacta kills dopamine-producing brain cells, leading to Parkinson's disease (Table 1 and Figure 3) [53]. Neurons in the substantia nigra are affected as microglia are activated, and protein clumps form. Causes of neurodegeneration include oxidative stress, dopamine depletion, and neuroinflammation. In addition to boosting dopamine, naringenin also reduces inflammation [54]. By activating Nrf2/ARE and its downstream target genes, including HO-1 and glutathione cysteine ligase regulatory subunit, naringenin protects mice against 6-hydroxydopamine-induced dopaminergic neurodegeneration and oxidative damage. By inhibiting and caspase 3, naringenin was able to prevent apoptosis. Grapefruit suppresses CYP3A4 much more so than naringin. With less naringin than grapefruit juice, orange juice also stops CYP3A4. Preclinical tests have shown that naringin is not transformed into naringenin in cultivated cells [55–57].

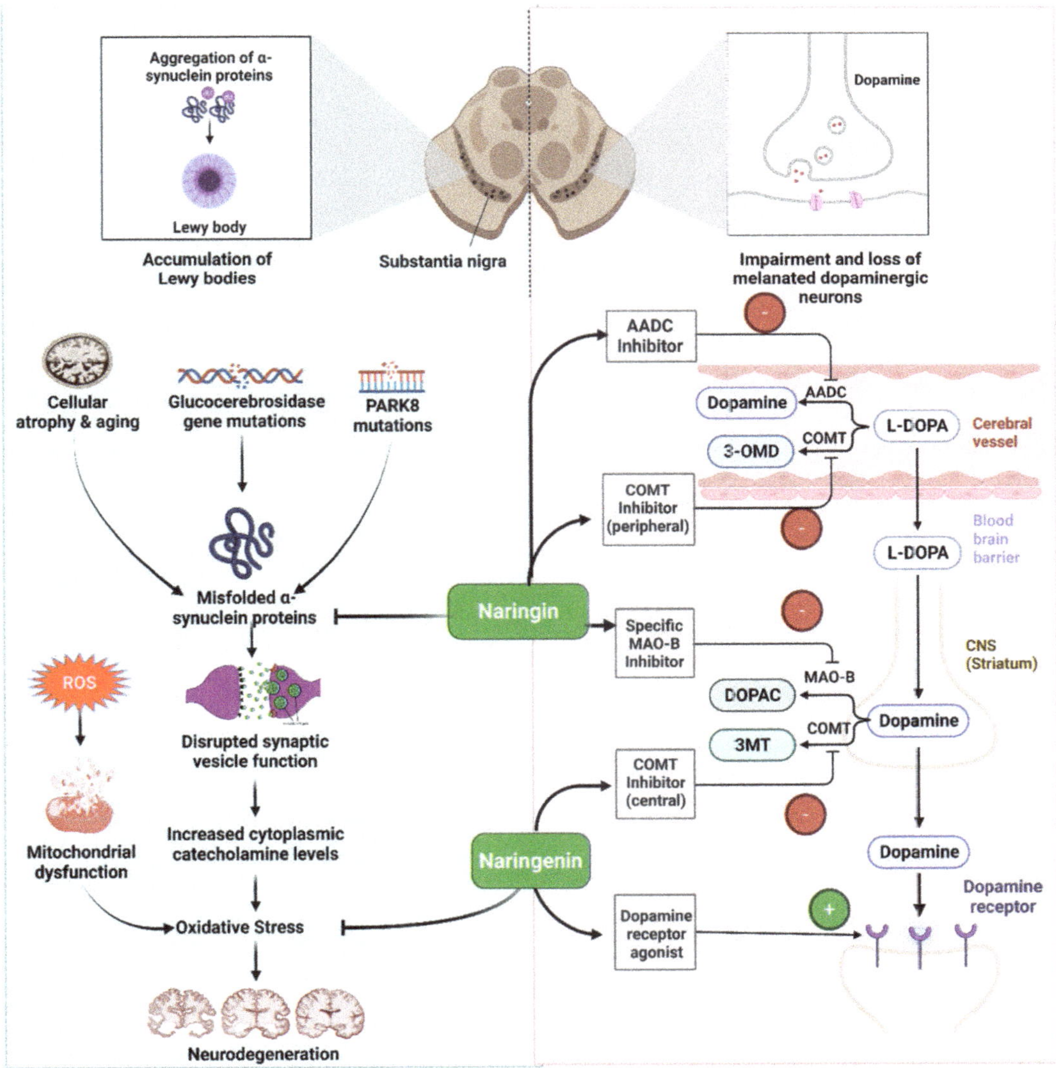

Figure 3. Illustration representing the site of action of naringin and naringenin in Parkinson's disease.

Table 1. Preclinical findings on the use of naringin and naringenin polyphenols in neurological disorders.

Disease	Compound	Dose/Conc.	Study Model	Findings	References
Alzheimer Disease	Naringin	50, 100 and 200 mg/kg; PO) for twentyone days	ICV-STZ rats	Restoration of cognitive deficits in ICV-STZ rat along with mitigation of mitochondrial dysfunction mediated oxido-nitrosative stress and cytokine release	[58]
	Naringin	50 or 100 mg/kg/day	APPswe/PSΔE9 transgenic mice	Reduction in plaque burden and an increase in glucose uptake through the inhibition of GSK-3β activity	[59]
	Naringin	40 and 80 mg/kg	Wistar rats	Protection against ICV β-A1–42 and intranasal manganese induced memory dysfunction possibly due to its antioxidant, anti-inflammatory, anti-amyloidogenesis	[60]
	Naringin	100 mg/kg/day	Mice	Neuroprotective effects through a variety of mechanisms, including amyloid β metabolism, Tau protein hyperphosphorylation, acetyl cholinergic system, glutamate receptor system, oxidative stress and cell apoptosis	[61]
	Naringenin	70–210 μg/mL	PC12 cells	Inhibition of AChE activity	[62]
	Naringenin	50 mg/kg	Male albino Wistar rats	Reduced oxidative stress markers: 4-HNE, MDA, TBARS, H_2O_2, PC, GSH in the hippocampus; Increase antioxidant level: GPx, GR, GST, SOD, CAT and Na+/K+-ATPase in the hippocampus	[63]
	Naringenin	25, 50 and 100 mg/kg	Male Sprague-Dawley rats	Increased the mRNA expression of INS and INSR in cerebral cortex and hippocampus. In addition, NAR reversed ICV-STZ induced Tau hyper-phosphorylation in both hippocampus and cerebral cortex through downregulation of GSK-3β activity	[41]
	Naringenin	25, 50 and 100 μM and 1.5, 3.0 and 4.5 mg/kg	PC12 cells and male ICR mice	Decreased ROS level and LDH activity	[64]
	Naringenin	25 and 50 mg/kg	Male Sprague-Dawley rats	Decreased oxidative stress by depleting elevated lipid peroxide and nitric oxide and elevating reduced glutathione levels and exert cholinergic function through the inhibition of elevated ChE activity	[65]
	Naringin dihydrochalcone	100 mg/kg	APPswe/PS1ΔE9 (APP/PS1) transgenic mice	Reduction in amyloid plaque burden and Aβ levels, suppression of neuroinflammation and promotion of neurogenesis	[66]

Table 1. *Cont.*

Disease	Compound	Dose/Conc.	Study Model	Findings	References
Alzheimer Disease	Naringin	40 and 80 mg/kg, PO	Male Wistar rats	Improvement in the cognitive performance and attenuated oxidative damage, as evidenced by lowering of malondialdehyde level and nitrite concentration and restoration of superoxide dismutase, catalase, glutathione S-transferase, and reduced glutathione levels, and acetylcholinesterase activity	[67]
	Naringin	30 or 60 mg/kg/day	NMRI male mice	Reduction of Aβ plaque numbers in CA1, CA3, and DG areas of the hippocampus	[68]
	Naringin	25, 50 and 100 mg/kg PO	Westar rats	Reduced lipid peroxidation, restored reduced superoxide dismutase and catalase) and acetylcholine esterase activity were significantly decreased	[69]
	N,N′-1,10-Bis(Naringin) Triethylenete-traamine	10–200 μM	PC12 cells	Deceased the level of ROS in Cu^{2+}-$A\beta_{1\text{-}42}$-treated PC12 cells and elevate the SOD activity in Cu^{2+}-$A\beta_{1\text{-}42}$-treated PC12 cells	[70]
	Naringin	2.5, 5 and 10 mg/kg	Swiss mice	Increased the activities of superoxide dismutase and catalase, and glutathione and decreased malondialdehyde and nitrite contents, and reduced brain acetylcholinesterase activity in mice brains	[71]
	Naringin	80 mg/kg	Wistar albino rats	Improvement of the Aβ-induced cholinergic dysfunction and increase in the activity of AChE in rat hippocampus, prefrontal cortex, and amygdala. Furthermore, naringin attenuated Aβ-induced decrease in mitochondrial function, integrity, and bioenergetics as well as mitochondrial and cytosolic calcium level in all the brain regions. Moreover, reversal of Aβ-induced increase in apoptosis and level of mitochondrial calcium uniporter and decrease in the level of hemeoxygenase-1	[72]
	Naringenin	100 mg/kg, orally	male Wistar rats	Lowered hippocampal MDA content	[73]
Parkinson Disease	Naringin	80 mg/kg	rat model	Protection of the nigrostriatal DA projection by increasing glial cell line-derived neurotrophic factor expression and decreasing TNF-a expression in DA neurons and microglia	[74]
	Naringin	8 or 80 mg/kg per day	Female Sprague Dawley (SD) rats	Increased the level of GDNF in DA neurons, contributing to neuroprotection in the MPP+ rat model of PD, with activation of mammalian target of rapamycin complex 1 and pre-treatment with naringin could attenuate the level of TNF-α in the substantia nigra of MPP+-treated brains	[75]

Table 1. *Cont.*

Disease	Compound	Dose/Conc.	Study Model	Findings	References
Parkinson Disease	Naringin	80 mg/kg	male Wistar albino rats	Neuroprotective activity against rotenone-induced toxicity in the animals possibly through Nrf2-mediated pathway	[76]
	Naringin	50, 100 and 200 mg/kg	Swiss albino mice	Reduction in haloperidol-induced cataleptic scores in both bar test and block test	[77]
	Naringenin	25, 50, 100 mg/kg/b.w, PO	male C57BL/6J mice	Reversed the toxic effects of MPTP by reducing LPO levels and increasing the activities of glutathione reductase and catalase along with improved behavioral performance	[78]
	Naringenin	50 mg/kg, orally	albino Wister rats	Improved oxidative stress status by decreasing MDA and increasing glutathione content	[79]
	Naringenin	50, 100 mg/kg	male Sprague-Dawley rats BV-2 and MN9D cell lines	Inhibition of microglia-induced neuroinflammation via NLRP3 inflammasome inactivation	[80]
	Naringenin	25, 50, and 100 mM	SH-SY5Y Cell Line	Reduction of the ROS production by decreasing oxidative stress markers such as LPO and NO and increasing SOD level. In addition, pretreatment with NGN decreased the inflammatory markers such as TNF-α and NF-κβ in MPP+-treated SH-SY5Y cells. Further, NGN decreased the pro-apoptotic marker—Bax—and increased the anti-apoptotic marker—Bcl-2—in MPP+-induced SH-SY5Y cells	[81]
	Naringenin	40 μM	Primary rat mesencephalic cultures	Decreased TH-positive neurons and TUNEL positive neurons	[82]
	Naringenin	50 mg/kg	Male Sprague-Dawley rats	Restoration of dopamine concentrations due to neuroprotective effects rather than compensatory effects by remaining TH-positive cells after 6-OHDA lesioning	[83]
	Naringenin	20, 40 and 80 mM (in vitro) 70 mg/kg, orally (in vivo)	Human neuroblastoma SH-SY5Y cells and male C57BL/6 mice	Activated Nrf2/ARE pathway in dopaminergic (in vitro) Up regulated protein levels of Nrf2/ARE genes (in vivo) Reduced striatal oxidative stress and subsequent apoptotic signalling cascades in striatum (in vivo)	[84]
	Naringenin	12.5 μM and 25 μM	SH-SY5Y Human Neuroblastoma cell line	Downregulation of the expression of some Parkinsonian genes such as casp9, lrrk2, and polg and upregulate pink1	[85]
	Naringenin	25, 50, and 100 mg/ kg/p.o	Male C57BL/6J mice	Reduced NO content and restored SOD activity, also downregulated TNF-α and IL-1β expression	[86]

Table 1. *Cont.*

Disease	Compound	Dose/Conc.	Study Model	Findings	References
Parkinson Disease	Naringenin	10 and 50 μM	Female Wistar rats	Enhanced astroglial neurotrophic effects on DA neurons through the regulation of Nrf2 activation,	[87]
Anxiety and depression	Naringenin	5, 10 and 20 mg/kg	Male ICR mice	Increased hippocampal 5-HT, NE and GR levels, and reduced serum corticosterone levels	[88]
	Naringenin	5, 10 and 20 mg/kg	Male ICR mice	Up-regulation of BDNF	[89]
	Naringenin	10, 25 and 50 mg/kg	Swiss mice	Naringenin (25–50 mg/ kg) ameliorated the hypolocomotion, depressive- and anxiety-like behaviors in hypoxic mice Naringenin (10 mg/kg) increases BDNF expression but did not significantly ($p <$ 0.05) alter corticosterone and catalase contents. The increased expressions of iNOS and NF-kB as well as loss of amygdala neuronal cells were reduced by naringenin (10 mg/kg)	[90]
	Naringin	10 mg/kg	Adult male Swiss mice	Alleviation of the depressive and anxiogenic behaviors evidenced by the increased preference to sucrose and open arm entries and duration in SPT and EPM respectively	[91]
	Naringin	25–100 mg/kg, i.p	Swiss mice	Increased the levels of GAD67, glutathione and decrease AChE activities, pro-inflammatory cytokines (TNF-α, IL-6), malondialdehyde, nitrite concentrations	[92]
	Naringenin	50 mg/kg/day	Adult male Wistar rats	Mitigation of morphological anomalies in the hippocampal CA1 region and cortex and upregulation of BDNF, Shh, GLI1, NKX2.2, and PAX6	[93]
Huntington's disease	Naringenin	0.2, 0.4 mM	C3H10T1/2 cells	Suppression of the protein aggregation caused by EGFP-polyQ97 in mammalian cells.	[94]
	Naringin	50, 100 mg/kg	Male Wistar rats	Protection against 3-nitropropionic acid induced neurotoxicity via nitric oxide mechanism	[95]
	Naringenin	50 mg/kg b.w, PO	Albino Wistar rats	Improvement of the behavioral function and restored the activity of MAO and 5-HT levels and reduction of the activation of astrocytes against 3-NP induced neurotoxicity	[96]
	Naringin	40, and 80 mg/kg	Adult male Sprague-Dawley rats	Modulation of oxido-nitrosative stress, neuroinflammatory, apoptotic markers and mitochondrial complex activity	[97]
	Naringin	(80 mg/kg b.w/day, orally)	Male Wistar rats	Enhancement of phase II and antioxidant gene expressions via Nrf2 activation	[98]
	Naringin	10 μM	PC12 cells	Modulation in expressions of B-cell lymphoma 2 and Bcl-2-associated X protein and enhancement of the nuclear translocation of Nrf2	[99]

Table 1. *Cont.*

Disease	Compound	Dose/Conc.	Study Model	Findings	References
Ischemic brain injury	Naringin	50 and 100 mg/kg	Male Wistar rats	Restoration of reduced glutathione and catalase activity and mitochondrial enzyme activities in cortex, striatum, cerebellum	[100]
	Naringin	106 mg/kg/day	male C57BL/6 strain mice	Suppression of neuronal cell death, reversed the reduction in the level of phosphorylated calcium-calmodulin-dependent protein kinase II, had the tendency to reverse the reduction in the level of glutathione, and blockade of excessive activation of microglia and astrocytes	[101]
	Naringin	40, 80 mg/kg	Male Sprague-Dawley rats	Improvement of early brain injury (EBI), including subarachnoid hemorrhage (SAH) severity, neurologic deficits, brain edema and blood-brain barrier integrity by attenuating SAH-induced oxidative stress and apoptosis, and reduction of the oxidative damage and apoptosis by inhibiting the activation of MAPK signaling pathway	[102]
	Naringenin	50 and 100 mg/kg	Male Sprague–Dawley rats	Down-regulation of NOD2, RIP2, NF-κB, MMP-9 and up-regulation of claudin-5 expression	[103]
	Naringin	80, 120, or 160 mg/kg/ day	SH-SY5Y cells	Reduced 3-nitrotyrosine formation, NADPH oxidase, and iNOS expression. Increased nNOS, p47, and p67 expression. Decreased mitophagy	[104]
	Naringin	100 mg/kg/day	Adult Wistar male rats	Continual treatment increased SOD activity, decreased MDA, NO, iNOS, and IL-1β. It also improved rats' behavioral performance	[105]
Spinal cord injury	Naringin	20, 40 mg/kg	Female Sprague-Dawley rat	Upregulation of the expression of NKx2.2 and 2′3′-cyclic nucleotide 3′-phosphodiesterase, and inhibition of β-catenin expression and GSK-3β phosphorylation	[106]
	Naringenin	5, 10, 15 mM	Male Wistar rats	Suppression of MMP-9 activity and upregulation of GSH, catalase and MMP-2 activation	[107]
	Naringenin	50–100 mg/kg	Female Wistar rats	Repression of miR-223	[108]
	Naringin	25, 50, and 100 mg/kg	Adult Sprague Dawley rats	Reduction of TNF-α, IL-8 as well as MDAcontent and elevation of IL-10 as well as SOD activity	[109]

Table 1. *Cont.*

Disease	Compound	Dose/Conc.	Study Model	Findings	References
Chronic hyper-glycemic peripheral neuropathy	Naringenin	50, 100 and 200 mg/kg	Male Sprague Dawley rats	Inhibition of upregulated expression of TNF-α, IL-1β and MCP-1 level; GFAP and Mac-1 mRNA expression	[110]
	Naringenin	25 and 50 mg/kg	Male Sprague Dawley rats	Increase GSH level and decrease MDA and NO level	[111]
	Naringin	50 and 100 mg/kg, b.w	Rat model of OXL-induced peripheral neuropathy	Improved the level of superoxide dismutase, catalase, glutathione peroxidase, nuclear factor erythroid 2-related factor 2, Heme oxygenase-1, nuclear factor-κ B, tumor necrosis factor-α, interleukin-1β, Bax, Bcl-2, caspase-3, paraoxonase, mitogen-activated protein kinase 14, neuronal nitric oxide synthase (nNOS), acetylcholinesterase, and arginase 2	[112]

4-HNE: 4-hydroxynonenal, TBARS: Thiobarbituric reactive substances, H_2O_2: Hydrogen peroxide, PC: Protein carbonyl, GSH: Reduced glutathione, GPx: Glutathione peroxidase, GR: Glutathione reductase, GST: Glutathione-S-transferase, SOD: Superoxide dismutase, CAT: catalase, GSK-3β: Glycogen synthase kinase-3β, ROS: Reactive oxygen species, AchE: Acetylcholine esterase, DA: Dopaminergic, LPO: Lipid peroxidation, MPTP: 1-methyl-4-phenyl-1,2,3,6-tetrahydropyridine, 5-HT: Serotonin, NE: Norepinephrine, BDNF: brain-derived neurotrophic factor, MDA: Malondialdehyde.

5.3. Cerebral Ischemia

Cerebral ischemia is a disorder that may trigger a cascade of unfavorable biochemical responses in the brain, leading to malfunction of key brain regions and, commonly, neuropathy [113,114]. An inflammatory reaction and the production of ROS following an ischemia event may damage brain tissue and lead to neuronal death [115]. Ischemia-induced damage involves several kinases, including mitogen-activated protein kinases, extracellular signal-regulated kinases, signal transducers and activators of transcription 1, calcium/calmodulin-dependent kinases, etc. ROS initiate the caspase cascade and encourage the synthesis of pro-inflammatory cytokines including interleukin (IL)-1, IL-6, and tumor necrosis factor-alpha, all of which contribute to cell death. Although progress has been made, a full understanding of the molecular mechanisms behind post-ischemic neuronal damage remains elusive. However, naringin and naringenin have been shown to have a neuroprotective effect after ischemia [116–119].

Naringin lowers cholesterol, prevents blood clots, and improves blood circulation and nutrient supply [100]. In addition, naringin protects against central nervous and cardiovascular diseases. To investigate the role of NFKB1 in OGD/R + injured PC12 cells, the previous study measured the components of the HIF-1α/AKT/mTOR signal path. HIF-1α, phosphorylated AKT, and mTOR were all higher in the OGD/R + naringin, OGD/R + si-NFKB1, and OGD/R + naringin + si-NFKB1 groups than in the OGD/R group. Significantly higher levels of HIF-1α, activated kinase AKT, and mTOR signals were expressed in the OGD/R + si-NFKB1 group ($p < 0.01$). In conclusion, naringin targets NFKB1 and modifies PC12 cell proliferation and apoptosis by affecting HIF-1α, p-AKT, and p-mTOR levels [120–123].

There is evidence that naringenin may help reduce the adverse effects of oxidative stress on the body, making it a potentially beneficial treatment option for various chronic illnesses. In addition to modulating the activity of antioxidant enzymes and regulating the expression of antioxidant genes, the flavonoid may have direct antioxidant effects, such as scavenging reactive species and reducing oxidative damage [124,125]. When Nrf2 is activated, antioxidant defenses are enhanced by maintaining redox equilibrium and decreasing oxidative stress and inflammation. TNF-α and IL-1β, two key inflammatory activators, are elevated in response to oxidative stress, whereas naringenin suppresses mRNA expression in the substantia nigra, hippocampus, and BV2 microglial cells [126,127]. For example, studies on rat hippocampus and BV cells have revealed that naringenin reduces NF-kB activation. Naringenin inhibited NF-kB, COX-2, iNOS, and their immunoreactivity, preventing the inflammatory cascade that leads to neurological diseases [51,128].

5.4. Anxiety and Depression

Anxiety and depression are two of the most common mental illnesses, both of which have complex origins at the intersection of several biological systems. Li et al. performed the first studies on the antidepressant effects of naringin and naringenin. The chemicals were tested on mice models of depression brought on by chronic unpredictable mild stress (CUMS) [129]. Anxiety may cause a variety of uncomfortable physical and emotional symptoms, including but not limited to: irritation, impatience, weariness, difficulty concentrating, a racing heart, chest pain, and an upset stomach. Anxiety comes in a variety of forms, and each is treated differently [130]. The serotonergic and noradrenergic systems have been connected to mood disorders such as depression and anxiety. The serotonergic system has far-reaching effects on cognitive processes in the brain, in addition to its role in regulating mood and appetite. Memory and focus are only two of the cognitive functions that are controlled by the noradrenergic system. Increases in serotonin (5-HT) and norepinephrine (NE) receptors, activation of brain-derived neurotrophic factor (BDNF), and decreased blood corticosterone are hypothesized to underlie NRG's antidepressant-like effects [131,132]. It inhibits monoamine oxidase, which may also help those who are depressed. Increased rearing activity, decreased immobility, and increased social communication was seen in mice administered NRG intraperitoneally (at doses of 2.5, 5, and 10 mg/kg), which is consistent with anti-depressant-like and anxiolytic-like effects. Reductions were seen in nitrosative stress, lipid peroxidation, and cholinergic transmission. Mental diseases, often known as mental illnesses or psychiatric disorders, are characterized by persistent patterns of thinking or behavior that significantly impair an individual's capacity to function in everyday life. Both the frequency and length of time that these symptoms will persist are unknown at this time. Various diseases and disorders have been identified, and each has its signs. In some instances, seeking the assistance of a clinical psychologist or psychiatrist specializing in evaluating and managing mental health conditions may be beneficial [133–136].

Neuronal inflammatory mediator release exacerbates tissue damage and reactive oxygen and nitrogen species (ROS/RNS) production, perpetuating the neuronal degeneration in stress-induced neuropsychiatric diseases, including depression and cognitive loss. Decreased antioxidant defenses in neurons promote neuroinflammation and neurodegeneration [137]. Several mediators and intracellular signaling molecules have been connected to neuroinflammatory responses to hypoxia damage. When hypoxia occurs, inflammatory transcription factors, including the NF-κB pathway, are activated, increasing pro-inflammatory cytokine production [137]. Physical changes in the brain's dendritic arbourization and synaptic architecture have been linked to psychological and neurological issues, including depression, anxiety, and memory loss. Naringenin, a dietary flavanone, may be abundant in various foods, including citrus fruits, vegetables, and other berries and nuts [138]. Chronic illnesses and ailments may benefit from consuming a diet high in NG-rich fruits and vegetables. Animal pharmacokinetic studies have shown that NG rapidly undergoes intermediate glucuronide metabolism in the liver and readily crosses the blood–brain barrier (BBB). NG's high permeability across the BBB has been attributed to its association with a broad range of CNS effects. However, the oral bioavailability of naringenin is limited by its metabolism in the liver and its degradation by bacterial enzymes in the colon. Lowered levels of inflammatory mediators were seen in rats, including TNF-α, cyclooxygenase-2, and inducible nitric oxide synthase (iNOS). Studies of naringenin's effects on the brain and spinal cord show it may help treat various neurological disorders [139–141].

5.5. Schizophrenia

Schizophrenia is a disabling brain illness characterized by a wide range of neurotic symptoms including but not limited to hallucinations, delusions, cognitive impairment, disorganized speech, and aberrant motor activities [142]. Many different causes contribute to the development of schizophrenia. These include structural brain abnormalities, impaired neurotransmission, and stress-induced signaling cascades. Disruptions in epidermal growth factor (EGF) signaling and abnormalities in the processing or expression of the EGF receptors ErbB1 and ErbB2 are common in all schizophrenias [143]. As the pathophysiology of schizophrenia worsens, oxidative stress has been suspected to be a contributing factor. Several signs of oxidative stress, including ROS, reduced antioxidant enzyme activity (catalase), depleted glutathione, and oxidized lipids, have been associated with schizophrenia [144].

Schizophrenia and apoptosis have been connected via both intrinsic (mitochondrial death) and extrinsic (death receptor) pathways [145]. Cytochrome C interacts with proapoptotic and anti-apoptotic proteins to trigger the release of activated caspase-3 (primarily Bax and Bcl-2). Diabetes was prevented in streptozotocin-treated rats by administration of naringin, which inhibited the production of inflammatory and oxidative stress mediators [146]. Reduced free radical production, decreased release of proinflammatory cytokines (such as interleukin-6 and TNF-α), and down-regulation of inflammatory proteins such as NF-κB have all been linked to its anti-inflammatory effects in diabetic, chronic bronchitis, and walker carcinosarcoma rats. To evaluate if naringenin protects interendothelial tight junctions, we analyzed the expression and localization of ZO-1, occludin, claudin-1, and claudin-2 across experimental groups. ZO-1 protein expression was significantly reduced in the TNF-α treated group compared to the control group ($p < 0.05$) [147–151].

However, in TNF-α induced RIMVECs ($p < 0.05$), treatment with naringin dramatically increased ZO-1 protein expression. Immunofluorescence's structured cell death also showed ZO-1 distribution. As proposed here, the medication that acts as a positive allosteric modulator of GABA neurotransmission from chandelier neurons is thought to improve the function of dorsolateral prefrontal cortex circuitry in people with schizophrenia by increasing gamma-band synchronization of pyramidal neuron activity [152–155].

6. Concluding Remarks and Future Directions

Although technological advances have substantially sped up research on phytochemicals, we still have a long way to go before we gather more definitive evidence regarding the neurotherapeutic benefits of herbal medicines. Our data and other researchers' data lead us to believe that naringin and naringenin may be beneficial as neurotherapeutic medications because of their ability to alter several signaling pathways. The outcomes thus far are in line with this theory.

Despite the limitations of ongoing clinical studies, naringenin and naringin are promising therapies for various neurological conditions, including AD, PD, cerebral ischemia, anxiety, depression, schizophrenia, and chronic hyperglycemic peripheral neuropathy. Given these obstacles, it is essential that pharmacokinetic research on naringin and naringenin administration be performed, that more accurate dosage designs for different illnesses be developed, and that innovative drug delivery strategies be developed to boost bioavailability in healthcare situations.

Author Contributions: Conceptualization, T.B.E., F.I., R.D. and S.M.; data curation, T.B.E., F.I., R.D. and S.M.; formal analysis, T.B.E., F.I., N.N., H.S., R.D. and S.M.; investigation, T.B.E., F.I., N.N., H.S., R.D. and S.M.; methodology, T.B.E., F.I., N.N., H.S., R.D. and S.M.; project administration, M.M.A., A.H.A. and R.S.; resources, M.M.A., A.H.A., R.S. and T.B.E.; supervision, T.B.E.; validation, M.M.A., A.H.A., R.S. and T.B.E.; visualization, T.B.E., F.I., R.D. and S.M.; writing—original draft, T.B.E., F.I., R.D. and S.M.; writing—review and editing, M.M.A., A.H.A., R.S. and T.B.E. All authors have read and agreed to the published version of the manuscript.

Funding: This research received no external funding.

Institutional Review Board Statement: Not applicable.

Informed Consent Statement: Not applicable.

Data Availability Statement: All data used to support the findings of this study are included within the article.

Conflicts of Interest: The authors declare that they have no conflict of interest.

Abbreviations

PD	Parkinson's disease
TRP	Transient receptor potential
NO	Nitric oxide
cGMP	Cyclic guanosine monophosphate
PKG	Protein kinase G
MAPK	Mitogen-activated protein kinase
GDNF	Growth differentiation and neurotrophic factor
AD	Alzheimer's disease
Aβ	Amyloid-β
AChE	Acetylcholinesterase
CaMKII	Calcium/calmodulin-dependent protein kinase II
GSH	Reduced glutathione
GPx	Glutathione peroxidase
GST	Glutathione-S-transferase
SOD	Superoxide dismutase
ROS	Reactive oxygen species
LPO	Lipid peroxidation
MDA	Malondialdehyde
5-HT	Serotonin
NE	Norepinephrine
BBB	Blood–brain barrier
iNOS	Inducible nitric oxide synthase

References

1. Patel, V.; Chisholm, D.; Dua, T.; Laxminarayan, R.; Medina-Mora, M.E. *Mental, Neurological, and Substance Use Disorders: Disease Control Priorities*; The International Bank for Reconstruction and Development/The World Bank: Washington, DC, USA, 2016; Volume 4.
2. Zis, P.; Hadjivassiliou, M. Treatment of Neurological Manifestations of Gluten Sensitivity and Coeliac Disease. *Curr. Treat. Options Neurol.* **2019**, *21*, 10. [CrossRef] [PubMed]
3. Ludvigsson, J.F.; Bai, J.C.; Biagi, F.; Card, T.R.; Ciacci, C.; Ciclitira, P.J.; Green, P.H.R.; Hadjivassiliou, M.; Holdoway, A.; Van Heel, D.A. Diagnosis and Management of Adult Coeliac Disease: Guidelines from the British Society of Gastroenterology. *Gut* **2014**, *63*, 1210–1228. [CrossRef] [PubMed]
4. Sapone, A.; Bai, J.C.; Ciacci, C.; Dolinsek, J.; Green, P.H.R.; Hadjivassiliou, M.; Kaukinen, K.; Rostami, K.; Sanders, D.S.; Schumann, M. Spectrum of Gluten-Related Disorders: Consensus on New Nomenclature and Classification. *BMC Med.* **2012**, *10*, 13. [CrossRef] [PubMed]
5. Fuhr, U.; Kummert, A.L. The Fate of Naringin in Humans: A Key to Grapefruit Juice-drug Interactions? *Clin. Pharmacol. Ther.* **1995**, *58*, 365–373. [CrossRef] [PubMed]
6. Alam, M.A.; Subhan, N.; Rahman, M.M.; Uddin, S.J.; Reza, H.M.; Sarker, S.D. Effect of Citrus Flavonoids, Naringin and Naringenin, on Metabolic Syndrome and Their Mechanisms of Action. *Adv. Nutr.* **2014**, *5*, 404–417. [CrossRef]
7. Zhang, J.; Gao, W.; Liu, Z.; Zhang, Z.; Liu, C. Systematic Analysis of Main Constituents in Rat Biological Samples after Oral Administration of the Methanol Extract of *Fructus aurantii* by HPLC-ESI-MS/MS. *Iran. J. Pharm. Res.* **2014**, *13*, 493–503. [CrossRef]
8. Ramesh, E.; Alshatwi, A.A. Naringin Induces Death Receptor and Mitochondria-Mediated Apoptosis in Human Cervical Cancer (SiHa) Cells. *Food Chem. Toxicol.* **2013**, *51*, 97–105. [CrossRef]
9. Romagnolo, D.F.; Selmin, O.I. Flavonoids and Cancer Prevention: A Review of the Evidence. *J. Nutr. Gerontol. Geriatr.* **2012**, *31*, 206–238. [CrossRef]
10. Bharti, S.; Rani, N.; Krishnamurthy, B.; Arya, D.S. Preclinical Evidence for the Pharmacological Actions of Naringin: A Review. *Planta Med.* **2014**, *80*, 437–451. [CrossRef]
11. Esaki, S.; Nishiyama, K.; Sugiyama, N.; Nakajima, R.; Takao, Y.; Kamiya, S. Preparation and Taste of Certain Glycosides of Flavanones and of Dihydrochalcones. *Biosci. Biotechnol. Biochem.* **1994**, *58*, 1479–1485. [CrossRef]

12. Felgines, C.; Texier, O.; Morand, C.; Manach, C.; Scalbert, A.; Régerat, F.; Remesy, C.; Felgines, C.; Texier, O.; Morand, C.; et al. Bioavailability of the Flavanone Naringenin and Its Glycosides in Rats. *Am. J. Physiol. Gastrointest. Liver Physiol.* **2000**, *279*, G1148–G1154. [CrossRef] [PubMed]
13. Jadeja, R.N.; Devkar, R.V. Chapter 47—Polyphenols and Flavonoids in Controlling Non-Alcoholic Steatohepatitis. In *Polyphenols in Human Health and Disease*; Watson, R.R., Preedy, V.R., Zibadi, S., Eds.; Academic Press: San Diego, CA, USA, 2014; pp. 615–623, ISBN 978-0-12-398456-2.
14. Manchope, M.F.; Ferraz, C.R.; Borghi, S.M.; Rasquel-Oliveira, F.S.; Franciosi, A.; Bagatim-Souza, J.; Dionisio, A.M.; Casagrande, R.; Verri, W.A. Chapter 38—Therapeutic Role of Naringenin to Alleviate Inflammatory Pain. In *Treatments, Mechanisms, and Adverse Reactions of Anesthetics and Analgesics*; Rajendram, R., Patel, V.B., Preedy, V.R., Martin, C.R., Eds.; Academic Press: Cambridge, MA, USA, 2022; pp. 443–455, ISBN 978-0-12-820237-1.
15. Li, Z.-X.; Cao, Y.; Yan, P. Topological Insulators and Semimetals in Classical Magnetic Systems. *Phys. Rep.* **2021**, *915*, 1–64. [CrossRef]
16. Page, M.J.; McKenzie, J.E.; Bossuyt, P.M.; Boutron, I.; Hoffmann, T.C.; Mulrow, C.D.; Shamseer, L.; Tetzlaff, J.M.; Akl, E.A.; Brennan, S.E.; et al. The PRISMA 2020 Statement: An Updated Guideline for Reporting Systematic Reviews. *Syst. Rev.* **2021**, *10*, 89. [CrossRef] [PubMed]
17. Abaza, M.S.I.; Orabi, K.Y.; Al-Quattan, E.; Al-Attiyah, R.J. Growth Inhibitory and Chemo-Sensitization Effects of Naringenin, a Natural Flavanone Purified from Thymus Vulgaris, on Human Breast and Colorectal Cancer. *Cancer Cell Int.* **2015**, *15*, 1–19. [CrossRef] [PubMed]
18. Ademosun, A.O.; Oboh, G.; Passamonti, S.; Tramer, F.; Ziberna, L.; Boligon, A.A.; Athayde, M.L. Inhibition of Metalloproteinase and Proteasome Activities in Colon Cancer Cells by Citrus Peel Extracts. *J. Basic Clin. Physiol. Pharmacol.* **2015**, *26*, 471–477. [CrossRef] [PubMed]
19. Hsiu, S.-L.; Huang, T.-Y.; Hou, Y.-C.; Chin, D.-H.; Chao, P.-D.L. Comparison of Metabolic Pharmacokinetics of Naringin and Naringenin in Rabbits. *Life Sci.* **2002**, *70*, 1481–1489. [CrossRef]
20. Moghaddam, R.H.; Samimi, Z.; Moradi, S.Z.; Little, P.J.; Xu, S.; Farzaei, M.H. Naringenin and Naringin in Cardiovascular Disease Prevention: A Preclinical Review. *Eur. J. Pharmacol.* **2020**, *887*, 173535. [CrossRef]
21. Zhang, L.; Song, L.; Zhang, P.; Liu, T.; Zhou, L.; Yang, G.; Lin, R.; Zhang, J. Solubilities of Naringin and Naringenin in Different Solvents and Dissociation Constants of Naringenin. *J. Chem. Eng. Data* **2015**, *60*, 932–940. [CrossRef]
22. Fujioka, K.; Greenway, F.; Sheard, J.; Ying, Y. The Effects of Grapefruit on Weight and Insulin Resistance: Relationship to the Metabolic Syndrome. *J. Med. Food* **2006**, *9*, 49–54. [CrossRef]
23. Ishii, K.; Furuta, T.; Kasuya, Y. Determination of Naringin and Naringenin in Human Plasma by High-Performance Liquid Chromatography. *J. Chromatogr. B Biomed. Sci. Appl.* **1996**, *683*, 225–229. [CrossRef]
24. Lee, C.-H.; Jeong, T.-S.; Choi, Y.-K.; Hyun, B.-H.; Oh, G.-T.; Kim, E.-H.; Kim, J.-R.; Han, J.-I.; Bok, S.-H. Anti-Atherogenic Effect of Citrus Flavonoids, Naringin and Naringenin, Associated with Hepatic ACAT and Aortic VCAM-1 and MCP-1 in High Cholesterol-Fed Rabbits. *Biochem. Biophys. Res. Commun.* **2001**, *284*, 681–688. [CrossRef] [PubMed]
25. Renugadevi, J.; Prabu, S.M. Naringenin Protects against Cadmium-Induced Oxidative Renal Dysfunction in Rats. *Toxicology* **2009**, *256*, 128–134. [CrossRef] [PubMed]
26. Cook, N.C.; Samman, S. Flavonoids—Chemistry, Metabolism, Cardioprotective Effects, and Dietary Sources. *J. Nutr. Biochem.* **1996**, *7*, 66–76. [CrossRef]
27. Choudhury, R.; Chowrimootoo, G.; Srai, K.; Debnam, E.; Rice-Evans, C.A. Interactions of the Flavonoid Naringenin in the Gastrointestinal Tract and the Influence of Glycosylation. *Biochem. Biophys. Res. Commun.* **1999**, *265*, 410–415. [CrossRef] [PubMed]
28. Croft, K.D. The Chemistry and Biological Effects of Flavonoids and Phenolic Acids A. *Ann. N. Y. Acad. Sci.* **1998**, *854*, 435–442. [CrossRef]
29. Scalbert, A.; Williamson, G. Dietary Intake and Bioavailability of Polyphenols. *J. Nutr.* **2000**, *130*, 2073S–2085S. [CrossRef]
30. Pietta, P.; Minoggio, M.; Bramati, L. Plant Polyphenols: Structure, Occurrence and Bioactivity. *Stud. Nat. Prod. Chem.* **2003**, *28*, 257–312. [CrossRef]
31. Navipour, E.; Neamatshahi, M.; Barabadi, Z.; Neamatshahi, M.; Keykhosravi, A. Epidemiology and Risk Factors of Alzheimer's Disease in Iran: A Systematic Review. *Iran. J. Public Health* **2019**, *48*, 2133–2139. [CrossRef]
32. Pedersen, W.A.; McMillan, P.J.; Kulstad, J.J.; Leverenz, J.B.; Craft, S.; Haynatzki, G.R. Rosiglitazone Attenuates Learning and Memory Deficits in Tg2576 Alzheimer Mice. *Exp. Neurol.* **2006**, *199*, 265–273. [CrossRef]
33. Rajmohan, R.; Reddy, P.H. Amyloid-Beta and Phosphorylated Tau Accumulations Cause Abnormalities at Synapses of Alzheimer's Disease Neurons. *J. Alzheimer's Dis.* **2017**, *57*, 975–999. [CrossRef]
34. Jin, H.; Wang, W.; Zhao, S.; Yang, W.; Qian, Y.; Jia, N.; Feng, G. Aβ-HBc Virus-like Particles Immunization without Additional Adjuvant Ameliorates the Learning and Memory and Reduces Aβ Deposit in PDAPP Mice. *Vaccine* **2014**, *32*, 4450–4456. [CrossRef] [PubMed]
35. Gauthier, S.; Scheltens, P.; Cummings, J. *Alzheimer's Disease and Related Disorders*; CRC Press: Boca Raton, FL, USA, 2005; ISBN 0203931742.
36. Mimura, M.; Yano, M. Memory Impairment and Awareness of Memory Deficits in Early-Stage Alzheimer's Disease. *Rev. Neurosci.* **2006**, *17*, 253–266. [CrossRef] [PubMed]

37. Fakhri, S.; Abbaszadeh, F.; Dargahi, L.; Jorjani, M. Astaxanthin: A Mechanistic Review on Its Biological Activities and Health Benefits. *Pharmacol. Res.* **2018**, *136*, 1–20. [CrossRef] [PubMed]
38. Bao, X.-Q.; Li, N.; Wang, T.; Kong, X.-C.; Tai, W.-J.; Sun, H.; Zhang, D. FLZ Alleviates the Memory Deficits in Transgenic Mouse Model of Alzheimer's Disease via Decreasing Beta-Amyloid Production and Tau Hyperphosphorylation. *PLoS ONE* **2013**, *8*, e78033. [CrossRef]
39. Mimura, M. Memory Impairment and Awareness of Memory Deficits in Early-Stage Alzheimer's Disease. *Tohoku J. Exp. Med.* **2008**, *215*, 133–140. [CrossRef]
40. Obulesu, M.; Jhansilakshmi, M. Neuroinflammation in Alzheimer's Disease: An Understanding of Physiology and Pathology. *Int. J. Neurosci.* **2014**, *124*, 227–235. [CrossRef]
41. Yang, W.; Ma, J.; Liu, Z.; Lu, Y.; Hu, B.; Yu, H. Effect of Naringenin on Brain Insulin Signaling and Cognitive Functions in ICV-STZ Induced Dementia Model of Rats. *Neurol. Sci.* **2014**, *35*, 741–751. [CrossRef]
42. Burke, R.E.; O'Malley, K. Axon Degeneration in Parkinson's Disease. *Exp. Neurol.* **2013**, *246*, 72–83. [CrossRef]
43. Savitt, J.M.; Dawson, V.L.; Dawson, T.M. Diagnosis and Treatment of Parkinson Disease: Molecules to Medicine. *J. Clin. Investig.* **2006**, *116*, 1744–1754. [CrossRef]
44. Shulman, J.M.; De Jager, P.L.; Feany, M.B. Parkinson's Disease: Genetics and Pathogenesis. *Annu. Rev. Pathol. Mech. Dis.* **2011**, *6*, 193–222. [CrossRef]
45. Dexter, D.T.; Jenner, P. Parkinson Disease: From Pathology to Molecular Disease Mechanisms. *Free Radic. Biol. Med.* **2013**, *62*, 132–144. [CrossRef] [PubMed]
46. AlDakheel, A.; Kalia, L.V.; Lang, A.E. Pathogenesis-Targeted, Disease-Modifying Therapies in Parkinson Disease. *Neurotherapeutics* **2014**, *11*, 6–23. [CrossRef] [PubMed]
47. Olanow, C.W.; Tatton, W.G. Etiology and Pathogenesis of Parkinson's Disease. *Annu. Rev. Neurosci.* **1999**, *22*, 123. [CrossRef] [PubMed]
48. Zbarsky, V.; Datla, K.P.; Parkar, S.; Rai, D.K.; Aruoma, O.I.; Dexter, D.T. Neuroprotective Properties of the Natural Phenolic Antioxidants Curcumin and Naringenin but Not Quercetin and Fisetin in a 6-OHDA Model of Parkinson's Disease. *Free Radic. Res.* **2005**, *39*, 1119–1125. [CrossRef] [PubMed]
49. Golechha, M.; Chaudhry, U.; Bhatia, J.; Saluja, D.; Arya, D.S. Naringin Protects against Kainic Acid-Induced Status Epilepticus in Rats: Evidence for an Antioxidant, Anti-Inflammatory and Neuroprotective Intervention. *Biol. Pharm. Bull.* **2011**, *34*, 360–365. [CrossRef]
50. Ahmed, S.; Khan, H.; Aschner, M.; Hasan, M.M.; Hassan, S.T.S. Therapeutic Potential of Naringin in Neurological Disorders. *Food Chem. Toxicol.* **2019**, *132*, 110646. [CrossRef]
51. Raza, S.S.; Khan, M.M.; Ahmad, A.; Ashafaq, M.; Islam, F.; Wagner, A.P.; Safhi, M.M. Neuroprotective Effect of Naringenin Is Mediated through Suppression of NF-KB Signaling Pathway in Experimental Stroke. *Neuroscience* **2013**, *230*, 157–171. [CrossRef]
52. Harrison, W.T.A.; Yathirajan, H.S.; Sarojini, B.K.; Narayana, B.; Anilkumar, H.G. Do C—H$\cdots$ O and C—H$\cdots$ π Interactions Help to Stabilize a Non-Centrosymmetric Structure for Racemic 2, 3-Dibromo-1, 3-Diphenylpropan-1-One? *Acta Crystallogr. Sect. C Cryst. Struct. Commun.* **2005**, *61*, o728–o730. [CrossRef]
53. Gopinathan, G.; Teravainen, H.; Dambrosia, J.M.; Ward, C.D.; Sanes, J.N.; Stuart, W.K.; Evarts, E.V.; Calne, D.B. Lisuride in Parkinsonism. *Neurology* **1981**, *31*, 371. [CrossRef]
54. Kim, H.D.; Jeong, K.H.; Jung, U.J.; Kim, S.R. Naringin Treatment Induces Neuroprotective Effects in a Mouse Model of Parkinson's Disease in Vivo, but Not Enough to Restore the Lesioned Dopaminergic System. *J. Nutr. Biochem.* **2016**, *28*, 140–146. [CrossRef]
55. Kensler, T.W.; Wakabayashi, N.; Biswal, S. Cell Survival Responses to Environmental Stresses via the Keap1-Nrf2-ARE Pathway. *Annu. Rev. Pharmacol. Toxicol.* **2007**, *47*, 89–116. [CrossRef] [PubMed]
56. Lee, J.-M.; Johnson, J.A. An Important Role of Nrf2-ARE Pathway in the Cellular Defense Mechanism. *BMB Rep.* **2004**, *37*, 139–143. [CrossRef] [PubMed]
57. Eagling, V.A.; Profit, L.; Back, D.J. Inhibition of the CYP3A4-Mediated Metabolism and P-Glycoprotein-Mediated Transport of the HIV-1 Protease Inhibitor Saquinavir by Grapefruit Juice Components. *Br. J. Clin. Pharmacol.* **1999**, *48*, 543. [CrossRef] [PubMed]
58. Sachdeva, A.K.; Kuhad, A.; Chopra, K. Naringin Ameliorates Memory Deficits in Experimental Paradigm of Alzheimer's Disease by Attenuating Mitochondrial Dysfunction. *Pharmacol. Biochem. Behav.* **2014**, *127*, 101–110. [CrossRef] [PubMed]
59. Wang, D.; Gao, K.; Li, X.; Shen, X.; Zhang, X.; Ma, C.; Qin, C.; Zhang, L. Long-Term Naringin Consumption Reverses a Glucose Uptake Defect and Improves Cognitive Deficits in a Mouse Model of Alzheimer's Disease. *Pharmacol. Biochem. Behav.* **2012**, *102*, 13–20. [CrossRef]
60. Kaur, G.; Prakash, A. Involvement of the Nitric Oxide Signaling in Modulation of Naringin against Intranasal Manganese and Intracerbroventricular β-Amyloid Induced Neurotoxicity in Rats. *J. Nutr. Biochem.* **2020**, *76*, 108255. [CrossRef]
61. Meng, X.; Fu, M.; Wang, S.; Chen, W.; Wang, J.; Zhang, N. Naringin Ameliorates Memory Deficits and Exerts Neuroprotective Effects in a Mouse Model of Alzheimer's Disease by Regulating Multiple Metabolic Pathways. *Mol. Med. Rep.* **2021**, *23*, 332. [CrossRef]
62. Heo, H.J.; Kim, M.J.; Lee, J.M.; Choi, S.J.; Cho, H.Y.; Hong, B.; Kim, H.K.; Kim, E.; Shin, D.H. Naringenin from Citrus Junos Has an Inhibitory Effect on Acetylcholinesterase and a Mitigating Effect on Amnesia. *Dement. Geriatr. Cogn. Disord.* **2004**, *17*, 151–157. [CrossRef]

63. Khan, M.B.; Khan, M.M.; Khan, A.; Ahmed, M.E.; Ishrat, T.; Tabassum, R.; Vaibhav, K.; Ahmad, A.; Islam, F. Naringenin Ameliorates Alzheimer's Disease (AD)-Type Neurodegeneration with Cognitive Impairment (AD-TNDCI) Caused by the Intracerebroventricular- Streptozotocin in Rat Model. *Neurochem. Int.* **2012**, *61*, 1081–1093. [CrossRef]
64. Heo, H.J.; Kim, D.O.; Shin, S.C.; Kim, M.J.; Kim, B.G.; Shin, D.H. Effect of Antioxidant Flavanone, Naringenin, from Citrus Junos on Neuroprotection. *J. Agric. Food Chem.* **2004**, *52*, 1520–1525. [CrossRef]
65. Rahigude, A.; Bhutada, P.; Kaulaskar, S.; Aswar, M.; Otari, K. Participation of Antioxidant and Cholinergic System in Protective Effect of Naringenin against Type-2 Diabetes-Induced Memory Dysfunction in Rats. *Neuroscience* **2012**, *226*, 62–72. [CrossRef] [PubMed]
66. Yang, W.; Zhou, K.; Zhou, Y.; An, Y.; Hu, T.; Lu, J.; Huang, S.; Pei, G. Naringin Dihydrochalcone Ameliorates Cognitive Deficits and Neuropathology in APP/PS1 Transgenic Mice. *Front. Aging Neurosci.* **2018**, *10*, 169. [CrossRef] [PubMed]
67. Kumar, A.; Dogra, S.; Prakash, A. Protective Effect of Naringin, a Citrus Flavonoid, against Colchicine-Induced Cognitive Dysfunction and Oxidative Damage in Rats. *J. Med. Food* **2010**, *13*, 976–984. [CrossRef] [PubMed]
68. Jahanshahi, M.; Khalili, M.; Margedari, A. Naringin Chelates Excessive Iron and Prevents the Formation of Amyloid-Beta Plaques in the Hippocampus of Iron-Overloaded Mice. *Front. Pharmacol.* **2021**, *12*, 518. [CrossRef]
69. Prabhakar, O. Naringin Attenuates Aluminum Induced Cognitive Deficits in Rats. *Indian J. Pharm. Educ. Res.* **2020**, *54*, 674–681. [CrossRef]
70. Guo, L.; Sun, B. N,N'-1,10-Bis(Naringin) Triethylenetetraamine, Synthesis and as a Cu(II) Chelator for Alzheimer's Disease Therapy. *Biol. Pharm. Bull.* **2021**, *44*, 51–56. [CrossRef]
71. Ben-Azu, B.; Nwoke, E.; Aderibigbe, A.; Omogbiya, I.; Ajayi, A.; Olonode, E.; Umukoro, S.; Iwalewa, E. Possible Neuroprotective Mechanisms of Action Involved in the Neurobehavioral Property of Naringin in Mice. *Biomed. Pharmacother.* **2019**, *109*, 536–546. [CrossRef]
72. Varshney, V.; Garabadu, D. Naringin Exhibits Mas Receptor-Mediated Neuroprotection Against Amyloid Beta-Induced Cognitive Deficits and Mitochondrial Toxicity in Rat Brain. *Neurotox. Res.* **2021**, *39*, 1023–1043. [CrossRef]
73. Ghofrani, S.; Joghataei, M.; Mohseni, S.; Baluchnejadmojarad, T.; Bagheri, M.; Khamse, S.; Roghani, M. Naringenin Improves Learning and Memory in an Alzheimer's Disease Rat Model: Insights into the Underlying Mechanisms. *Eur. J. Pharmacol.* **2015**, *764*, 195–201. [CrossRef]
74. Jung, U.; Kim, S. Effects of Naringin, a Flavanone Glycoside in Grapefruits and Citrus Fruits, on the Nigrostriatal Dopaminergic Projection in the Adult Brain. *Neural Regen. Res.* **2014**, *9*, 1514.
75. Leem, E.; Nam, J.; Jeon, M.; Shin, W.; Won, S.; Park, S.; Choi, M.; Jin, B.; Jung, U.; Kim, S. Naringin Protects the Nigrostriatal Dopaminergic Projection through Induction of GDNF in a Neurotoxin Model of Parkinson's Disease. *J. Nutr. Biochem.* **2014**, *25*, 801–806. [CrossRef]
76. Garabadu, D.; Agrawal, N. Naringin Exhibits Neuroprotection Against Rotenone-Induced Neurotoxicity in Experimental Rodents. *NeuroMolecular Med.* **2020**, *22*, 314–330. [CrossRef] [PubMed]
77. Kumar, N.; James, R.; Sinha, S.; Kinra, M.; Anuranjana, P.V.; Nandakumar, K. Naringin Exhibited Anti-Parkinsonian like Effect against Haloperidol-Induced Catalepsy in Mice. *Res. J. Pharm. Technol.* **2021**, *14*, 662–666. [CrossRef]
78. Sugumar, M.; Sevanan, M.; Sekar, S. Neuroprotective Effect of Naringenin against MPTP-Induced Oxidative Stress. *Int. J. Neurosci.* **2018**, *129*, 534–539. [CrossRef] [PubMed]
79. Madani, M.; Attia, A.; ElSalalm, R.; El-Shenawy, S.; Arbid, M. Neuropharmacological Effects of Naringenin, Harmine and Adenosine on Parkinsonism Induced in Rats. *Der Pharm. Lett.* **2016**, *8*, 45–57.
80. Chen, C.; Wei, Y.; He, X.; Li, D.; Wang, G.; Li, J. Naringenin Produces Neuroprotection Against LPS-Induced Dopamine Neurotoxicity via the Inhibition of Microglial NLRP3 Inflammasome Activation. *Front. Immunol.* **2019**, *10*, 936. [CrossRef]
81. Mani, S.; Sekar, S.; Chidambaram, S.B.; Sevanan, M. Naringenin Protects against 1-Methyl-4-Phenylpyridinium-Induced Neuroinflammation and Resulting Reactive Oxygen Species Production in SH - SY5Y Cell Line: An In Vitro Model of Parkinson's Disease. *Pharmacogn. Mag.* **2018**, 458–464. [CrossRef]
82. Mercer, L.; Kelly, B.; Horne, M.; Beart, P. Dietary Polyphenols Protect Dopamine Neurons from Oxidative Insults and Apoptosis: Investigations in Primary Rat Mesencephalic Cultures. *Biochem. Pharmacol.* **2005**, *69*, 339–345. [CrossRef]
83. Kabir, M.S.H.; Hossain, M.M.; Kabir, M.I.; Rahman, M.M.; Hasanat, A.; Emran, T.B.; Rahman, M.A. Phytochemical screening, Antioxidant, Thrombolytic, alpha-amylase inhibition and cytotoxic activities of ethanol extract of *Steudnera colocasiifolia* K. Koch leaves. *J. Young Pharm.* **2016**, *8*, 391. [CrossRef]
84. Lou, H.; Jing, X.; Wei, X.; Shi, H.; Ren, D.; Zhang, X. Naringenin Protects against 6-OHDA-Induced Neurotoxicity via Activation of the Nrf2/ARE Signaling Pathway. *Neuropharmacology* **2014**, *79*, 380–388. [CrossRef]
85. Kesh, S.; Rajesh, R.; Balakrishnan, A. Naringenin Alleviates 6-Hydroxydopamine Induced Parkinsonism in SHSY5Y Cells and Zebrafish Model. *Comp. Biochem. Physiol. Part C Toxicol. Pharmacol.* **2021**, *239*, 108893. [CrossRef] [PubMed]
86. Mani, S.; Sekar, S.; Barathidasan, R.; Manivasagam, T.; Thenmozhi, A.J.; Sevanan, M.; Chidambaram, S.B.; Essa, M.M.; Guillemin, G.J.; Sakharkar, M.K. Naringenin Decreases α -Synuclein Expression and Neuroinflammation in MPTP-Induced Parkinson's Disease Model in Mice. *Neurotox. Res.* **2018**, *33*, 656–670. [CrossRef] [PubMed]
87. Wang, G.; Zhang, B.; He, X.; Li, D.; Shi, J.; Zhang, F. Naringenin Targets on Astroglial Nrf2 to Support Dopaminergic Neurons. *Pharmacol. Res.* **2018**, *139*, 452–459. [CrossRef] [PubMed]

88. Yi, L.; Li, J.; Li, H.; Su, D.; Quan, X.; He, X.; Wang, X. Antidepressant-like Behavioral, Neurochemical and Neuroendocrine Effects of Naringenin in the Mouse Repeated Tail Suspension Test. *Prog. Neuropsychopharmacol. Biol. Psychiatry* **2012**, *39*, 175–181. [CrossRef]
89. Yi, L.; Liu, B.; Li, J.; Luo, L.; Liu, Q.; Geng, D.; Tang, Y.; Xia, Y.; Wu, D. BDNF Signaling Is Necessary for the Antidepressant-like Effect of Naringenin. *Prog. Neuro-Psychopharmacol. Biol. Psychiatry* **2014**, *48*, 135–141. [CrossRef]
90. Olugbemide, A.S.; Ben-Azu, B.; Bakre, A.G.; Ajayi, A.M.; Femi-Akinlosotu, O.; Umukoro, S. Naringenin Improves Depressive- and Anxiety-like Behaviors in Mice Exposed to Repeated Hypoxic Stress through Modulation of Oxido-Inflammatory Mediators and NF-KB/BDNF Expressions. *Brain Res. Bull.* **2021**, *169*, 214–227. [CrossRef]
91. Anyanwu, G.E.; Atuadu, V.O.; Esom, E.A.; Nto, J.N.; Katchy, A.U. Putative Role of Monoaminergic Systems in Antidepressant and Anxiolytic Effects of Naringin in Mice: An Interaction Study with Receptor Antagonists. *J. Pharm. Res. Int.* **2021**, *33*, 661–676. [CrossRef]
92. Oladapo, O.M.; Ben-azu, B.; Ajayi, A.M.; Emokpae, O.; Eneni, A.O. Naringin Confers Protection against Psychosocial Defeat Stress-Induced Neurobehavioral Deficits in Mice: Involvement of Glutamic Acid Decarboxylase Isoform-67, Oxido-Nitrergic Stress, and Neuroinflammatory Mechanisms. *J. Mol. Neurosci.* **2020**, *71*, 431–445. [CrossRef]
93. Tayyab, M.; Farheen, S.; Mariyath, M.; Khanam, P.M.N.; Hossain, M.M. Antidepressant and Neuroprotective Effects of Naringenin via Sonic Hedgehog-GLI1 Cell Signaling Pathway in a Rat Model of Chronic Unpredictable Mild Stress. *Neuromol. Med.* **2019**, *21*, 250–261. [CrossRef]
94. Yamagishi, N.; Yamamoto, Y.; Noda, C.; Hatayama, T. Naringenin Inhibits the Aggregation of Expanded Polyglutamine Tract-Containing Protein through the Induction of Endoplasmic Reticulum Chaperone GRP78. *Biol. Pharm. Bull.* **2012**, *35*, 1836–1840. [CrossRef]
95. Kumar, P.; Kumar, A. Protective Effect of Hesperidin and Naringin against 3-Nitropropionic Acid Induced Huntington' s like Symptoms in Rats: Possible Role of Nitric Oxide. *Behav. Brain Res.* **2010**, *206*, 38–46. [CrossRef] [PubMed]
96. Salman, M.; Sharma, P.; Alam, I.; Tabassum, H. Naringenin Mitigates Behavioral Alterations and Provides Neuroprotection against 3-Nitropropinoic Acid-Induced Huntington's Disease like Symptoms in Rats. *Nutr. Neurosci.* **2021**, *25*, 1–11. [CrossRef]
97. Cui, J.; Wang, G.; Kandhare, A.D.; Mukherjee, A.A.; Bodhankar, S.L. Neuroprotective Effect of Naringin, a Flavone Glycoside in Quinolinic Acid-Induced Neurotoxicity: Possible Role of PPAR-γ, Bax/Bcl-2, and Caspase-3. *Food Chem. Toxicol.* **2018**, *121*, 95–108. [CrossRef] [PubMed]
98. Gopinath, K.; Sudhandiran, G. Naringin modulates oxidative stress and inflammation in 3-nitropropionic acid-induced neurodegeneration through the activation of nuclear factor-erythroid 2-related factor-2 signalling pathway. *Neuroscience* **2012**, *227*, 134–143. [CrossRef]
99. Kulasekaran, G.; Ganapasam, S. Neuroprotective Efficacy of Naringin on 3-Nitropropionic Acid- Induced Mitochondrial Dysfunction through the Modulation of Nrf2 Signaling Pathway in PC12 Cells. *Mol. Cell. Biochem.* **2015**, *409*, 199–211. [CrossRef] [PubMed]
100. Gaur, V.; Aggarwal, A.; Kumar, A. Protective Effect of Naringin against Ischemic Reperfusion Cerebral Injury: Possible Neurobehavioral, Biochemical and Cellular Alterations in Rat Brain. *Eur. J. Pharmacol.* **2009**, *616*, 147–154. [CrossRef] [PubMed]
101. Okuyama, S.; Yamamoto, K.; Mori, H.; Sawamoto, A.; Amakura, Y. Neuroprotective Effect of Citrus Kawachiensis (Kawachi Bankan) Peels, a Rich Source of Naringin, against Global Cerebral Ischemia/Reperfusion Injury in Mice. *Biosci. Biotechnol. Biochem.* **2018**, *82*, 1216–1224. [CrossRef]
102. Han, Y.; Su, J.; Liu, X.; Zhao, Y.; Wang, C.; Li, X. Naringin Alleviates Early Brain Injury after Experimental Subarachnoid Hemorrhage by Reducing Oxidative Stress and Inhibiting Apoptosis. *Brain Res. Bull.* **2016**, *133*, 42–50. [CrossRef]
103. Bai, X.; Zhang, X.; Chen, L. Protective Effect of Naringenin in Experimental Ischemic Stroke: Down-Regulated NOD2, RIP2, NF-KappaB, MMP-9 and Up-Regulated Claudin-5 Expression. *Neurochem. Res.* **2014**, *39*, 1405–1415. [CrossRef]
104. Feng, J.; Chen, X.; Lu, S.; Li, W.; Yang, D.; Su, W.; Wang, X. Naringin Attenuates Cerebral Ischemia-Reperfusion Injury Through Inhibiting Peroxynitrite-Mediated Mitophagy Activation. *Mol. Neurobiol.* **2018**, *55*, 9029–9042. [CrossRef]
105. Cui, Q.; Wang, L.; Wei, Z.; Qu, W. Continual Naringin Treatment Benefits the Recovery of Traumatic Brain Injury in Rats Through Reducing Oxidative and Inflammatory Alterations. *Neurochem. Res.* **2014**, *39*, 1254–1262. [CrossRef] [PubMed]
106. Rong, W.; Pan, Y.; Cai, X.; Song, F.; Zhao, Z.; Xiao, S.; Zhang, C. The Mechanism of Naringin-Enhanced Remyelination after Spinal Cord Injury. *Neural Regen. Res.* **2017**, *12*, 470. [PubMed]
107. Fakhri, S.; Sabouri, S.; Kiani, A.; Farzaei, M.H.; Rashidi, K. Intrathecal Administration of Naringenin Improves Motor Dysfunction and Neuropathic Pain Following Compression Spinal Cord Injury in Rats: Relevance to Its Antioxidant and Anti-Inflammatory Activities. *Korean J. Pain* **2022**, *35*, 291–302. [CrossRef] [PubMed]
108. Shi, L.; Tang, P.; Zhang, W.; Zhao, Y.; Zhang, L.; Zhang, H. Naringenin Inhibits Spinal Cord Injury-Induced Activation of Neutrophils through MiR-223. *Gene* **2016**, *592*, 128–133. [CrossRef] [PubMed]
109. Wang, L.; Zhang, Z.; Wang, H. Naringin Attenuates Cerebral Ischemia-Reperfusion Injury in Rats by Inhibiting Endoplasmic Reticulum Stress. *Transl. Neurosci.* **2021**, *12*, 190–197. [CrossRef]
110. Hu, C.; Zhao, Y. Analgesic Effects of Naringenin in Rats with Spinal Nerve Ligation-Induced Neuropathic Pain. *Biomed. Rep.* **2014**, *2*, 569–573. [CrossRef]
111. Kaulaskar, S.; Bhutada, P.; Rahigude, A.; Jain, D.; Harle, U. Effects of Naringenin on Allodynia and Hyperalgesia in Rats with Chronic Constriction Injury-Induced Neuropathic Pain. *J. Chin. Integr. Med.* **2012**, *10*, 1482–1489. [CrossRef]

112. Semis, H.; Kandemir, F.; Caglayan, C.; Kaynar, O.; Genc, A.; Arıkan, S. Protective Effect of Naringin against Oxaliplatin-induced Peripheral Neuropathy in Rats: A Behavioral and Molecular Study. *J. Biochem. Mol. Toxicol.* **2022**, *36*, e23121. [CrossRef]
113. Yang, J.; Yuan, L.; Wen, Y.; Zhou, H.; Jiang, W.; Xu, D.; Wang, M. Protective Effects of Naringin in Cerebral Infarction and Its Molecular Mechanism. *Med. Sci. Monit. Int. Med. J. Exp. Clin. Res.* **2020**, *26*, e918772. [CrossRef]
114. Sveinsson, O.A.; Kjartansson, O.; Valdimarsson, E.M. Cerebral Ischemia/Infarction-Diagnosis and Treatment. *Laeknabladid* **2014**, *100*, 393–401.
115. Olmez, I.; Ozyurt, H. Reactive Oxygen Species and Ischemic Cerebrovascular Disease. *Neurochem. Int.* **2012**, *60*, 208–212. [CrossRef] [PubMed]
116. Tsung, A.; Klune, J.R.; Zhang, X.; Jeyabalan, G.; Cao, Z.; Peng, X.; Stolz, D.B.; Geller, D.A.; Rosengart, M.R.; Billiar, T.R. HMGB1 Release Induced by Liver Ischemia Involves Toll-like Receptor 4–Dependent Reactive Oxygen Species Production and Calcium-Mediated Signaling. *J. Exp. Med.* **2007**, *204*, 2913–2923. [CrossRef] [PubMed]
117. Ma, L.L.; Song, L.; Yu, X.D.; Yu, T.X.; Liang, H.; Qiu, J.X. The Clinical Study on the Treatment for Acute Cerebral Infarction by Intra-Arterial Thrombolysis Combined with Mild Hypothermia. *Eur. Rev. Med. Pharmacol. Sci.* **2017**, *21*, 1999–2006.
118. Jeong, H.S.; Song, H.-J.; Kim, S.-B.; Lee, J.; Kang, C.W.; Koh, H.-S.; Shin, J.E.; Lee, S.H.; Kwon, H.J.; Kim, J. A Comparison of Stent-Assisted Mechanical Thrombectomy and Conventional Intra-Arterial Thrombolysis for Acute Cerebral Infarction. *J. Clin. Neurol.* **2013**, *9*, 91–96. [CrossRef] [PubMed]
119. Bailey, D.G.; Arnold, J.M.O.; Strong, H.A.; Munoz, C.; Spence, J.D. Effect of Grapefruit Juice and Naringin on Nisoldipine Pharmacokinetics. *Clin. Pharmacol. Ther.* **1993**, *54*, 589–594. [CrossRef] [PubMed]
120. Cao, W.; Feng, S.-J.; Kan, M.-C. Naringin Targets NFKB1 to Alleviate Oxygen-Glucose Deprivation/Reoxygenation–Induced Injury in PC12 Cells via Modulating HIF-1α/AKT/MTOR-Signaling Pathway. *J. Mol. Neurosci.* **2021**, *71*, 101–111. [CrossRef] [PubMed]
121. Zhang, Y.; Zhang, J.; Wu, C.; Guo, S.; Su, J.; Zhao, W.; Xing, H. Higenamine Protects Neuronal Cells from Oxygen-glucose Deprivation/Reoxygenation-induced Injury. *J. Cell. Biochem.* **2019**, *120*, 3757–3764. [CrossRef] [PubMed]
122. Shu, K.; Zhang, Y. Protodioscin Protects PC12 Cells against Oxygen and Glucose Deprivation-Induced Injury through MiR-124/AKT/Nrf2 Pathway. *Cell Stress Chaperones* **2019**, *24*, 1091–1099. [CrossRef]
123. Wang, C.; Yang, Y.-H.; Zhou, L.; Ding, X.-L.; Meng, Y.-C.; Han, K. Curcumin Alleviates OGD/R-Induced PC12 Cell Damage via Repressing CCL3 and Inactivating TLR4/MyD88/MAPK/NF-KB to Suppress Inflammation and Apoptosis. *J. Pharm. Pharmacol.* **2020**, *72*, 1176–1185. [CrossRef]
124. Rahman, J.; Tareq, A.M.; Hossain, M.; Sakib, S.A.; Islam, M.N.; Ali, M.H.; Uddin, A.B.M.N.; Hoque, M.; Nasrin, M.S.; Emran, T.B.; et al. Biological evaluation, DFT calculations and molecular docking studies on the antidepressant and cytotoxicity activities of *Cycas pectinata* Buch. - Ham. Compounds. *Pharmaceuticals* **2020**, *13*, 232. [CrossRef]
125. Tejera, D.; Heneka, M.T. Microglia in Neurodegenerative Disorders. In *Microglia*; Springer: New York, NY, USA, 2019; pp. 57–67.
126. Abdel-Magied, N.; Shedid, S.M. The Effect of Naringenin on the Role of Nuclear Factor (Erythroid-derived 2)-like2 (Nrf2) and Haem Oxygenase 1 (HO-1) in Reducing the Risk of Oxidative Stress-related Radiotoxicity in the Spleen of Rats. *Environ. Toxicol.* **2019**, *34*, 788–795. [CrossRef] [PubMed]
127. Drishya, S.; Dhanisha, S.S.; Raghukumar, P.; Guruvayoorappan, C. Amomum Subulatum Mitigates Total Body Irradiation-Induced Oxidative Stress and Associated Inflammatory Responses by Enhancing the Antioxidant Status and Regulating pro-Inflammatory Cytokines. *J. Nutr. Biochem.* **2022**, *107*, 109064. [CrossRef] [PubMed]
128. Saha, S.; Buttari, B.; Panieri, E.; Profumo, E.; Saso, L. An Overview of Nrf2 Signaling Pathway and Its Role in Inflammation. *Molecules* **2020**, *25*, 5474. [CrossRef] [PubMed]
129. Li, M.; Fu, Q.; Li, Y.; Li, S.; Xue, J.; Ma, S. Emodin Opposes Chronic Unpredictable Mild Stress Induced Depressive-like Behavior in Mice by Upregulating the Levels of Hippocampal Glucocorticoid Receptor and Brain-Derived Neurotrophic Factor. *Fitoterapia* **2014**, *98*, 1–10. [CrossRef] [PubMed]
130. Mitra, S.; Anjum, J.; Muni, M.; Das, R.; Rauf, A.; Islam, F.; Emran, T.B.; Semwal, P.; Hemeg, H.A.; Alhumaydhi, F.A. Exploring the Journey of Emodin as a Potential Neuroprotective Agent: Novel Therapeutic Insights with Molecular Mechanism of Action. *Biomed. Pharmacother.* **2022**, *149*, 112877. [CrossRef]
131. Mitra, S.; Lami, M.S.; Uddin, T.M.; Das, R.; Islam, F.; Anjum, J.; Hossain, M.J.; Emran, T.B. Prospective multifunctional roles and pharmacological potential of dietary flavonoid narirutin. *Biomed. Pharmacother.* **2022**, *150*, 112932. [CrossRef]
132. Halbreich, U.; Kahn, L.S. Role of Estrogen in the Aetiology and Treatment of Mood Disorders. *CNS Drugs* **2001**, *15*, 797–817. [CrossRef]
133. Bruce, M.L.; Raue, P.J. Mental Illness as Psychiatric Disorder. In *Handbook of the Sociology of Mental Health*; Springer: Berlin/Heidelberg, Germany, 2013; pp. 41–59.
134. Evans, D.L.; Foa, E.B.; Gur, R.E.; Hendin, H.; O'Brien, C.P.; Seligman, M.E.P.; Walsh, B.T. *Treating and Preventing Adolescent Mental Health Disorders: What We Know and What We Don't Know*; Oxford University Press: Oxford, UK, 2005; ISBN 0198038712.
135. Chtourou, Y.; Gargouri, B.; Kebieche, M.; Fetoui, H. Naringin Abrogates Cisplatin-Induced Cognitive Deficits and Cholinergic Dysfunction through the down-Regulation of AChE Expression and INOS Signaling Pathways in Hippocampus of Aged Rats. *J. Mol. Neurosci.* **2015**, *56*, 349–362. [CrossRef]

136. Kwatra, M.; Jangra, A.; Mishra, M.; Sharma, Y.; Ahmed, S.; Ghosh, P.; Kumar, V.; Vohora, D.; Khanam, R. Naringin and Sertraline Ameliorate Doxorubicin-Induced Behavioral Deficits through Modulation of Serotonin Level and Mitochondrial Complexes Protection Pathway in Rat Hippocampus. *Neurochem. Res.* **2016**, *41*, 2352–2366. [CrossRef]
137. Tallei, T.E.; Niode, N.J.; Idroes, R.; Zidan, B.M.R.M.; Mitra, S.; Celik, I.; Nainu, F.; Ağagündüz, D.; Emran, T.B.; Capasso, R.A. Comprehensive Review of the Potential Use of Green Tea Polyphenols in the Management of COVID-19 Evid-based Complement. *Altern. Med.* **2021**, *2021*, 7170736. [CrossRef]
138. D'Angelo, S.; Mele, E.; Di Filippo, F.; Viggiano, A.; Meccariello, R. Sirt1 Activity in the Brain: Simultaneous Effects on Energy Homeostasis and Reproduction. *Int. J. Environ. Res. Public Health* **2021**, *18*, 1243. [CrossRef] [PubMed]
139. Bhattacharya, I.; Domínguez, A.P.; Drägert, K.; Haas, E.; Battegay, E.J. Hypoxia Potentiates Tumor Necrosis Factor-α Induced Expression of Inducible Nitric Oxide Synthase and Cyclooxygenase-2 in White and Brown Adipocytes. *Biochem. Biophys. Res. Commun.* **2015**, *461*, 287–292. [CrossRef] [PubMed]
140. Matheus, M.E.; de Almeida Violante, F.; Garden, S.J.; Pinto, A.C.; Fernandes, P.D. Isatins Inhibit Cyclooxygenase-2 and Inducible Nitric Oxide Synthase in a Mouse Macrophage Cell Line. *Eur. J. Pharmacol.* **2007**, *556*, 200–206. [CrossRef] [PubMed]
141. Shin, J.-S.; Park, Y.M.; Choi, J.-H.; Park, H.-J.; Shin, M.C.; Lee, Y.S.; Lee, K.-T. Sulfuretin Isolated from Heartwood of Rhus Verniciflua Inhibits LPS-Induced Inducible Nitric Oxide Synthase, Cyclooxygenase-2, and pro-Inflammatory Cytokines Expression via the down-Regulation of NF-KB in RAW 264.7 Murine Macrophage Cells. *Int. Immunopharmacol.* **2010**, *10*, 943–950. [CrossRef]
142. Feifel, D.; Shilling, P.D.; MacDonald, K. A Review of Oxytocin's Effects on the Positive, Negative, and Cognitive Domains of Schizophrenia. *Biol. Psychiatry* **2016**, *79*, 222–233. [CrossRef]
143. Lisman, J.E.; Coyle, J.T.; Green, R.W.; Javitt, D.C.; Benes, F.M.; Heckers, S.; Grace, A.A. Circuit-Based Framework for Understanding Neurotransmitter and Risk Gene Interactions in Schizophrenia. *Trends Neurosci.* **2008**, *31*, 234–242. [CrossRef]
144. Obrosova, I.G. Diabetes and the Peripheral Nerve. *Biochim. Biophys. Acta Mol. Basis Dis.* **2009**, *1792*, 931–940. [CrossRef]
145. Howes, O.; McCutcheon, R.; Stone, J. Glutamate and Dopamine in Schizophrenia: An Update for the 21st Century. *J. Psychopharmacol.* **2015**, *29*, 97–115. [CrossRef]
146. Li, P.-F.; Dietz, R.; von Harsdorf, R. P53 Regulates Mitochondrial Membrane Potential through Reactive Oxygen Species and Induces Cytochrome C-Independent Apoptosis Blocked by Bcl-2. *EMBO J.* **1999**, *18*, 6027–6036. [CrossRef]
147. Zou, W.; Xiao, Z.; Wen, X.; Luo, J.; Chen, S.; Cheng, Z.; Xiang, D.; Hu, J.; He, J. The Anti-Inflammatory Effect of *Andrographis paniculata* (Burm. f.) Nees on Pelvic Inflammatory Disease in Rats through down-Regulation of the NF-KB Pathway. *BMC Complement. Altern. Med.* **2016**, *16*, 483. [CrossRef]
148. Stenvinkel, P.; Ketteler, M.; Johnson, R.J.; Lindholm, B.; Pecoits-Filho, R.; Riella, M.; Heimbürger, O.; Cederholm, T.; Girndt, M. IL-10, IL-6, and TNF-α: Central Factors in the Altered Cytokine Network of Uremia—The Good, the Bad, and the Ugly. *Kidney Int.* **2005**, *67*, 1216–1233. [CrossRef] [PubMed]
149. Yao, Q.; Pecoits-Filho, R.; Lindholm, B.; Stenvinkel, P. Traditional and Non-Traditional Risk Factors as Contributors to Atherosclerotic Cardiovascular Disease in End-Stage Renal Disease. *Scand. J. Urol. Nephrol.* **2004**, *38*, 405–416. [CrossRef] [PubMed]
150. Higley, M.J.; Picciotto, M.R. Neuromodulation by Acetylcholine: Examples from Schizophrenia and Depression. *Curr. Opin. Neurobiol.* **2014**, *29*, 88–95. [CrossRef] [PubMed]
151. Futamura, T.; Toyooka, K.; Iritani, S.; Niizato, K.; Nakamura, R.; Tsuchiya, K.; Someya, T.; Kakita, A.; Takahashi, H.; Nawa, H. Abnormal Expression of Epidermal Growth Factor and Its Receptor in the Forebrain and Serum of Schizophrenic Patients. *Mol. Psychiatry* **2002**, *7*, 673–682. [CrossRef] [PubMed]
152. Zhong, J.; Yu, R.; Zhou, Q.; Liu, P.; Liu, Z.; Bian, Y. Naringenin Prevents TNF-α-Induced Gut-Vascular Barrier Disruption Associated with Inhibiting the NF-KB-Mediated MLCK/p-MLC and NLRP3 Pathways. *Food Funct.* **2021**, *12*, 2715–2725. [CrossRef]
153. Bitanihirwe, B.K.Y.; Woo, T.-U.W. Oxidative Stress in Schizophrenia: An Integrated Approach. *Neurosci. Biobehav. Rev.* **2011**, *35*, 878–893. [CrossRef]
154. Jarskog, L.F.; Glantz, L.A.; Gilmore, J.H.; Lieberman, J.A. Apoptotic Mechanisms in the Pathophysiology of Schizophrenia. *Prog. Neuro-Psychopharmacology Biol. Psychiatry* **2005**, *29*, 846–858. [CrossRef]
155. Missonnier, P.; Hasler, R.; Perroud, N.; Herrmann, F.R.; Millet, P.; Richiardi, J.; Malafosse, A.; Giannakopoulos, P.; Baud, P. EEG Anomalies in Adult ADHD Subjects Performing a Working Memory Task. *Neuroscience* **2013**, *241*, 135–146. [CrossRef]

life

Review

Applications and Pharmacological Properties of Cactus Pear (*Opuntia* spp.) Peel: A Review

Salvador Manzur-Valdespino [1], José Arias-Rico [2], Esther Ramírez-Moreno [1], María de Cortes Sánchez-Mata [3], Osmar Antonio Jaramillo-Morales [4], Julieta Angel-García [2], Quinatzin Yadira Zafra-Rojas [1], Rosario Barrera-Gálvez [2] and Nelly del Socorro Cruz-Cansino [1,*]

1 Área Académica de Nutrición, Instituto de Ciencias de la Salud, Universidad Autónoma del Estado Hidalgo, Circuito Ex Hacienda La Concepción S/N, Carretera Pachuca-Actopan, San Agustín Tlaxiaca 42160, Mexico
2 Área Académica de Enfermería, Instituto de Ciencias de la Salud, Universidad Autónoma del Estado Hidalgo, Circuito Ex Hacienda La Concepción S/N, Carretera Pachuca-Actopan, San Agustín Tlaxiaca 42160, Mexico
3 Department of Nutrition and Food Sciences, Pharmacy Faculty, Universidad Complutense de Madrid, Plaza de Ramón y Cajal s/n, E-28040 Madrid, Spain
4 Nursing and Obstetrics Department, Life Sciences Division, Campus Irapuato-Salamanca, University of Guanajuato, Ex Hacienda El Copal, Km. 9 Carretera Irapuato-Silao, A.P 311, Irapuato 36500, Guanajuato, Mexico
* Correspondence: ncruz@uaeh.edu.mx

Abstract: Nowadays, there is a growing interest in the exploitation of by-products from fruits and vegetables, generated from industrial processing or human feeding. Residues of popularly consumed fruits such as orange, lemon, banana, pomegranate, among others, have been widely described and studied; however, cactus pear (*Opuntia* spp.) residues, as a locally consumed product, have been forgotten. The whole fruit can be divided into the edible portion (pulp) and the non-edible portion (seeds and peel). Several studies mainly focus on the characteristics of the edible portion or in the whole fruit, ignoring by-products such as peels, which are rich in compounds such as phenols, flavonoids and dietary fiber; they have also been proposed as an alternative source of lipids, carbohydrates and natural colorants. Some uses of the peel have been reported as a food additives, food supplements, as a source of pectins and for wastewater treatment; however, there have not been any deep investigations of the characteristics and potential uses of the cactus pear peel (CPP). The aim of the present paper is to provide an overview of the current research on CPP. CPP has many bio-active compounds that may provide health benefits and may also be useful in pharmaceutical, food and manufacturing industries; however, greater research is needed in order to gain thorough knowledge of the possibilities of this by-product.

Keywords: waste; by-product; composition; bioactive compounds; health benefits

check for updates

Citation: Manzur-Valdespino, S.; Arias-Rico, J.; Ramírez-Moreno, E.; Sánchez-Mata, M.d.C.; Jaramillo-Morales, O.A.; Angel-García, J.; Zafra-Rojas, Q.Y.; Barrera-Gálvez, R.; Cruz-Cansino, N.d.S. Applications and Pharmacological Properties of Cactus Pear (*Opuntia* spp.) Peel: A Review. *Life* **2022**, *12*, 1903. https://doi.org/10.3390/life12111903

Academic Editor: Jianfeng Xu

Received: 1 November 2022
Accepted: 14 November 2022
Published: 16 November 2022

1. Introduction

Cactus pear is the fruit of the nopal cactus, is native to the arid and semi-arid regions of Mexico and Mesoamerica and has spread to many regions [1,2] This fruit is commonly known as cactus pera fruit, prickly pear, tuna (Mexico), higo (Colombia) higo chumbo (Spain), fico d'India, figue de barbarie (France), among others. There are some countries where there is significant production of this fruit, such as Italy, which is the most important producer in the Mediterranean area, and on the African continent, it is produced in the Cape region and in South Africa; in countries such as Israel, Chile and Argentina, it is produced on a small scale and it is also possible to find it in some plantations in Brazil, Colombia, Peru, Spain, Greece and Turkey [3]. However, Mexico is the main producer worldwide and cactus pear cultivation is considered highly profitable, because in optimal conditions, the production is 40 tons per hectare by year [2,4]. In Mexico, an area of 48,000 ha is dedicated to its cultivation, in which 352,000 tons by year are produced, through the participation of around 20,000 producers [2,5]. The main producing regions in Mexico are the southern region (Puebla and Oaxaca), the central region (Estado de Mexico and Hidalgo) and the north-central region (Guanajuato, Jalisco, Aguascalientes, San Luis Potosí and Zacatecas) [2,6].

It is estimated that the consumption per capita is only around 3.7 kg/year, since it is a seasonal fruit and is only available a few months per year [7]. The fruit is very popular, and it is mainly consumed fresh, although it is also processed in products manufactured on a small scale or in an artisanal way, and can be found in jams, yogurts, juices or candies [8]. Because of how it is consumed and processed, only the edible portion is used, generating a large amount of residue between peels and seeds. The non-edible portion known as the peel comprises two fractions, the mesocarp and the pericarp, and depending on the variety, it may represent between 33 to 55% of the total weight. It is usually discarded as by-product and may represent a problem due to waste management issues [9–11]; therefore, different alternatives have been sought for its use, revealing that is an inexpensive source of many nutrients, such as minerals [12], aminoacids [13], polyunsaturated fatty acids [14] and carbohydrates (which have been applied as a source of fiber, sweeteners and pectins for food applications) [15].

In addition, several studies have shown that cactus pear by-products are rich in bioactive compounds [16] such as phenolics, flavonoids, pigments, fibers, polysaccharides and fatty acids [8]. They can provide many health benefits such as inhibition and protection against free radicals [17], cytotoxic activity against some cancer cell lines [18] as well as the reduction of atherosclerosis and glycaemia [19]. However, many studies have been performed by using the pear peel for multiple purposes such as for pigment extraction [20] a preservative for margarine [21], as snacks [22], a dietary supplement with hypoglycemic properties [23] and in wastewater treatment [24].

The present review describes the nutritional characterization of cactus pear peel as a promising source in pharmaceutical, food, textile and wastewater industries.

2. Information Sources and Search Strategies

Scientific databases such as PubMed, Research Gate and Science Direct were used in the literature search, using the following search key words: *Opuntia* spp., Cactus pear and peel; as filters to search all fields, the words: composition, bioactive compounds, applications and health benefits were used. From the articles resulting from the search, the abstracts were carefully read, and relevant studies were selected and reviewed.

Study Eligibility Criteria

Original articles written in English and Spanish were included, which the pericarp and endocarp of the fruit were used or described; those in which the properties or effects of other by-products, such as the pulp or the seeds, were excluded. The process was carried out following the recommendations of the PRISMA Flow Diagram [25] as presented in Figure 1.

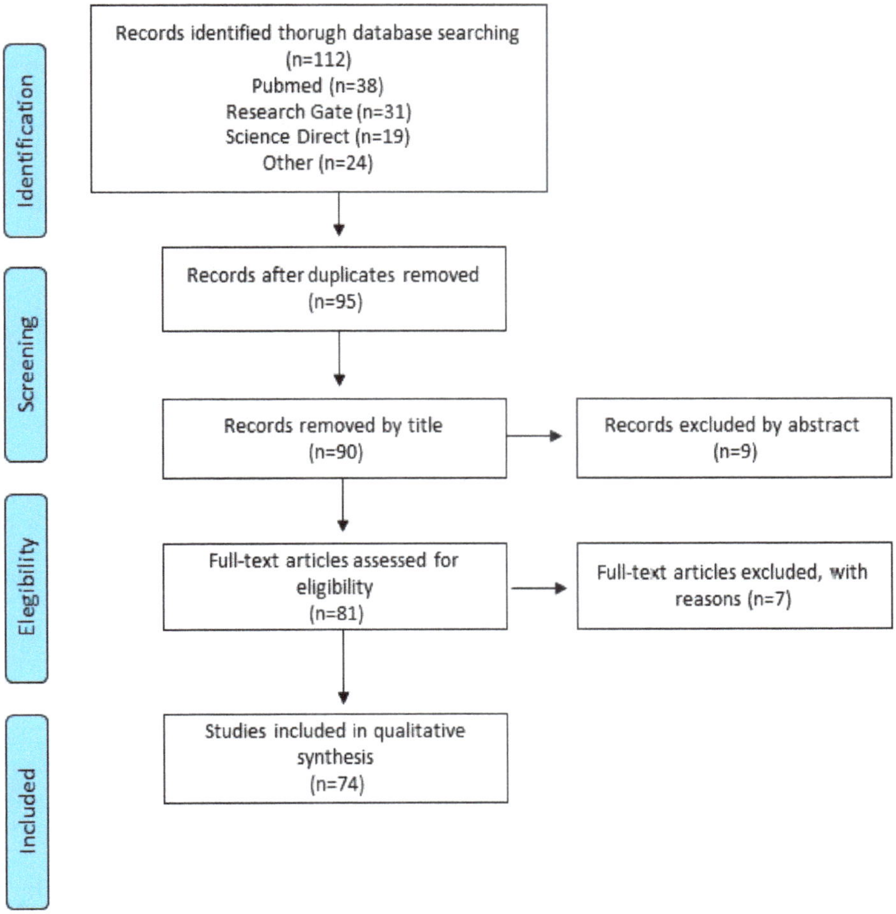

Figure 1. PRISMA flow-chart.

3. Cactus Pear Description and Structure

Cactus pear is the fruit of the *Opuntia* spp., which is a shrubby plant of the *Opuntia* genus and belongs to the Cactaceae family [26]. The fruit usually has a spherical or turbine shape, with elliptical or cylindrical variations. The color combinations such as bright yellow, green, red or purple are provided by pigments such as betanins and betaxanthins. The size is approximately 7 to 9 cm long and 5 to 6 cm wide, whereas the weight varies from 86 to 146 g. The peel has the same color of the pulp and has an umbilical zone from 50 to 70 "areolas", with small spines from 3 to 10 mm [27] and the peel thickness is approximately 0.65 cm [28]. At the beginning of fruit development, the peel or skin predominates over the locular tissue, whereas as the fruit grows, the pulp proportion increases in comparison to the peel [29], and when the fruit matures, the peel still comprises the highest percentage [30].

The first function of the fruit peel is to protect the pulp from weather and the sun [31], and the peel also indicates the moment when the fruit is ripe through physical parameters, such as the change of color, shape, firmness, diameter and volume, which allow for knowing the optimal time for harvesting [29]. In the whole fresh fruit, the amount of peel may be of approximately 40%, and on a dry weight basis, it may represent 25% of the total weight [32].

Cactus pear peel (CPP) has been studied mainly for the extraction of starches, pectins and fiber. The peel cells are mainly composed of collenchyma and parenchyma cells, which are rich in pectic polysaccharides [33]. The organization and morphology of the different cell types that conform to the peel from the outside to the inside can be described as follows: the chlorenchyma cells are thin-walled epidermal cells that are rich in non-cellulosic components and have thick cell walls, and on the other hand, parenchyma cells show thin walls. Inside the tissues of the collenchyma and parenchyma, there are mucilaginous cells that act as a polysaccharide storage, which is common in succulent plants, and is related to the osmotic function for water molecule retention [32].

3.1. Physical Chemical Characteristics and Nutrient Composition

The peel of the fruit shares some properties with the fruit; its physical characteristics give it very specific organoleptic properties, in addition to allowing for knowing the ripeness of the fruit. The acidity of CPP is 0.02 to 0.12% of citric acid, which gives the characteristic flavor of the fruit and is related to the pH, which is an indicator of the maturity of the fruit. CPP can reach values of 4.5–5.9, and in this sense, it could be considered a low acid by-product (pH > 4.5) [12,34,35]. Soluble solids content (°Brix) is a parameter used for the screening of the evolution and ripening of the fruit, correlating with the content of sugars through the refraction properties of the total soluble solids [36]. The peel °Brix varies from 6.16 to 15.00 depending on the state of maturity of the fruit at harvest time [12,35].

CPP also has nutritional properties of interest, which are summarized in Table 1. CPP is notorious for its high percentage of moisture and the low amount of lipids and proteins; however, it also has essential amino acids, which are described later.

Carbohydrates are the major component of CPP, which are rich in fibers and polysaccharides. It may change depending on the variety and color of the peel; for example, the peels of red cactus pear present high humidity and high soluble fiber content, whereas green fruit peels present a higher content of fat and insoluble fiber [11].

Since the demand for novel sources of quality protein have increased, the use of vegetables as an alternative source of protein has been proposed, especially those obtained from agro-industrial by-products. The use of cactus pear residue is available at low costs and can contribute to the generation of value-added protein, leading to environmental sustainability [39].

It is well known that the amino acids of CPP are linked to the betaxanthins, one of the main pigments of the cactus pear, which result from the conjugation of betalamic acid with protein or non-protein amino acids and biogenic amines. In CPP, sixteen betaxanthins have been identified, which include amino acids in their structure; they are arginine-betaxanthin, aspartic acid-betaxanthin, lysine-betaxanthin, proline-betaxanthin, serine-betaxanthin, tyramine-betaxanthin and threonine-betaxanthin [13].

Fruit peels could be considered as a promising source of essential fatty acids and fat-soluble antioxidants. According with Ramadan and Mörsel [14], around 36.8 g/kg (dry weight) of lipids have been found [14], and depending on the variety, the fatty acids that may be present in the peels are Lauric acid (C12:0), Myristic acid (C14:0),Palmitic acid (C16:0), Palmitoleic acid (C16:1), Stearic acid (C18:0), Oleic acid (C18:1), Linoleic acid (C18:2), Linolenic acid (C18:3) and Arachidonic acid (C20:0) [40]. From the recovered lipids, the unsaponifiable comprises 12.8% [14]. However, the amount of lipids is not only influenced by the color or variety, but is also affected by processing and storage conditions [12], since during the storage of oils and fats, lipid peroxidation takes place, affecting the nutritional and organoleptic properties, as the unsaturated nature of the fatty acids from cactus pears makes them highly susceptible to oxidation [9,11]. The fatty acids quantifications that are specifically found in the peel of the fruits of *Opuntia* spp. L. Mill are shown in Table 1.

Table 1. Nutritional composition of cactus pear peel.

	Ranges	Refs.
Moisture (%)	80.17–90.33	[12,37]
Ash (%)	1.60–3.05	[12,37]
Minerals		
Mg (mg/100 g)	0.987	[12]
Ca (mg/100 g)	0.951	
Na (mg/100 g)	0.925	
K (mg/100 g)	0.320	
Fe (mg/100 g)	0.129	
Mn (mg/100 g)	0.090	
Zn (mg/100 g)	0.090	
Protein (%)	0.90–4.14	[12,35]
Lipids (%)	0.94–2.43	[4,11]
Palmitic (%)	23.71	[12]
Stearic (%)	3.93	
Arachidonic (%)	5.52	
Palmitoleic (%)	2.46	
Oleic (%)	19.73	
Linoleic (%)	28.96	
Linolenic (%)	15.68	
Carbohydrates	27.60	[4]
Total sugars (%)	3.53	[35]
Reducing sugars (%)	2.07	[35]
Saccharose (mg/100 g)	0.00225	[37]
Glucose (mg/100 g)	0.014	[37]
Fructose (mg/100 g)	0.0029	[37]
Galacturonic acid (mg/100 g)	2.23	[12]
Stachyose (mg/100 g)	1.81	[12]
Mannitol (mg/100 g)	1.48	[12]
Sorbitol (mg/100 g)	0.71	[12]
Arabinose (mg/100 g)	0.05	[12]
Starch (mg/100 g)	7.12	[4]
Fiber		
Total fiber	40.80	[4]
Crude fiber (%)	0.96	[12]
Insoluble fiber (%)	7.98–8.12	[38]
Soluble fiber (%)	19.39–34.95	[38]
Hemicellulose (%)	20.80	[4]
Cellulose (%)	27.0	[11]
Lignin (%)	2.4	[11]
Pectin (%)	7.71	[4]
Mucilage (%)	4.10	[11]
Vitamins		
Ascorbic acid (mg/100 g)	27.3	[38]
Niacin (B_3) (mg/100 g)	0.26	
Pyridoxine (B_6) (mg/100 g)	0.19	
Folic acid (B_9) (mg/100 g)	0.11	

Mineral composition is highly influenced by the soil where the plant is grown and may vary from place to place, together with the variety of the fruit and the climate of that region. CPP is characterized by a high content of Mg and Ca [12] (Table 1). In this sense, the consumption of only 20 g of peel would cover 90% of the recommended daily intake (RDI) of magnesium and 20% of calcium for the general population [41]. Although, it is important to mention that calcium is not bio-accessible [10,11,42]. In addition, the presence of potassium is very important, since it helps to mitigate the negative effects of high sodium consumption on blood pressure [43], and manganese, zinc and copper are also present in the peels, being relevant because they are used for bone mineralization, muscle contraction, nerve stimulus transmission and act as a cofactor of many enzymes involved human metabolism [41,44,45].

As shown in Table 1, the peel contains mainly vitamin C, B3, B6 and B9 [46], whereas other vitamins, such as, thiamine or riboflavin, are found in trace amounts [47,48].

Ascorbic acid is one of the antioxidant agents found in abundance in the peels in comparison with the pulp [49], ranging from 46.40 to 86.28 mg AAE (Ascorbic Acid Equivalent)/100 g [50], and there are variations in the content depending on the color of the fruit; the highest content of ascorbic acid has been found in pink peels, followed by the orange ones and, lastly, the red varieties [50].

3.2. *Bioactive Compounds*

In some varieties of cactus pear fruit, the total phenolic content (TPC) as well as ascorbic acid is higher in the peel [51–53], mainly in red peel fruits [54], and a high amount is associated with the matrix of the dietary fiber [55]. Amounts up to 1534 mg GAE (Gallic Acid Equivalent)/100 g (fresh weight) can be found in the peels. The amount found depends on the cultivar, season and soil properties [49,53], fruit maturity and climate [53,56], as seen in Table 2.

Table 2. Bioactive compounds and antioxidant activity of cactus pear peels.

Bioactive Compounds and Antioxidant Activity	Ranges	Refs.
Phenolics (mg GAE/100 g)	14–376	[50,57]
Flavonoids (mg RE/100 g)	8–66	[57]
Tannins (mg CE/100 g)	23–144	[57]
Carotenoids (μg/g)	1.79–6.06	[50]
Betaxanthins (mg/100 g)	83.4	[53]
Betacyanins (mg/100 g)	13,468	[53]
DPPH (%)	90.9—96.8	[50]
ABTS (μM TE/100 g)	529	[50]
Chelating activity (%)	69–97	[50]
Reducing power (EC 50 mg/mL)	2.08–2.65	[57]
ORAC (μM TE/100 g)	37.4	[53]
Β-Carotene bleaching inhibition (EC 50 mg/mL)	3.87–6.49	[57]

GAE: Gallic Acid Equivalent; RE: Rutin Equivalent; CE: Catechin Equivalent; TE: Trolox Equivalent; EC: Extinction coefficient.

Purple peels have high concentrations of betalains, which are water-soluble, natural pigment derivatives, which yield a variety of colors, from red-violet (betacyanins) to yellow-orange (betaxanthins) [58,59], whereas green peel varieties have the lowest concentrations of these pigments.

The lipids from CPP present high levels of β-carotene [14]; the main carotenoids of the peel (representing about 80%) are lutein, β-carotene and violaxanthin [50]. In the different varieties of cactus pear (red, yellow, orange and green), the presence of carotenoids varies; however, the presence of norbixin, antheraxanthin, astaxanthin, canthaxanthin and ζ-carotene is consistent [57,60–63].

There is little research about the safety of oral ingestion of fruit peels, especially in the presence of pesticides, and there are minimal amounts of pesticides (such as malathion, chlorpyrifos, permethrin, diazinon, dimethoate, spinosad and abamectin) and heavy metals (copper, chromium, arsenic, cadmium, lead, and selenium). However, these are under the maximum limits of toxic residues established by the North American Free Trade Agreement (NAFTA), so that phytotoxic elements do not trigger health risks [8,11].

3.3. Potential Uses, Applications and Health Benefits

Many authors have focused their interest on this fruit, valuing it as a functional food because of its high fiber and secondary metabolites (polyphenols, betaxanthins, organic acids, among others) that confer properties such as having an anti-inflammatory effect, are lipid lowering, have a hypoglycemic effect, among others [64–73]. In these diseases, the chronic use of conventional anti-inflammatory drugs may lead to some adverse effects. For that reason, the anti-inflammatory activity of some natural compounds present in food products or herbal drugs may be valued as adjuvants to relieve the symptoms of these diseases [67].

3.3.1. Digestive System

In an animal model of colonic inflammation caused by irradiation, it was observed that by applying a pretreatment with *Opuntia* spp. peel extract in rats, a prophylactic effect against the damage is produced in the colonic tissue, decreasing inflammation markers, as well as increasing intrinsic anti-inflammatory agents [73] because of the high content of phenolic and flavonoids that are linked to the by-products of the cactus pear [11]. Therefore, *Opuntia* spp. fruit peel extract could have some potential to improve colonic inflammation processes [73].

There are also other digestive issues such as constipation that is related to a deficient intake of fiber. This component plays an important role in human health since it is associated with prevention and the treatment of diseases such as colitis, colon cancer and high cholesterol levels (36). *Opuntia* ssp.peel is considered a good source of dietary fiber because 40.8% of the dry weight is fiber [33], and the recommended daily intake of dietary fiber in adults must be >25 g/day [74]; therefore, the daily intake could be easily covered with 62 g of dried CPP [31]. The beneficial effects of fibers in human health are widely known through the effects on the digestive system, helping to relieve constipation, increasing the bulk and softness of the feces on the intestine, accelerating the pass through the bowel and easing the evacuation [74]. In addition, dietary fibers have some functional properties such as an oil and water retention capacity, and which have lipid-lowering and anti-constipation effects, respectively [75].

It has been demonstrated in studies in vivo with rats that the consumption of cactus pear residues favors the growth of beneficial bacteria in the gut. In a study with rats fed with CPP flour, it was observed that the growth of lactic acid bacteria and bifidobacteria is promoted [76] because of the combination of non-digestible sugars with different dietary fibers [77].

3.3.2. Antimicrobial, Antifungal, Antiviral and Insecticidal Activities

Opuntia spp. peel extract is considered a promising source of new natural antibacterial agents against some microbes, and peel extracts have been proven to have significant antimicrobial activity, which can vary according to the type of extract. The ethanolic extract of the peel has high activity against some microbes, even higher than the activity presented by the pulp extracts, showing an increase in the inhibition zone against *Staphylococcus aureus* and *Escherichia coli*. The peel extract also has antifungal activity against *Candida albicans* and has antibacterial activity against Gram-positive and Gram-negative bacteria because of the presence of several potent bioactive components such as sterols, tannins, alkaloids and other phenolic compounds [46].

Different solvents were used on the CPP in order to extract the bioactive compounds, and to evaluate the inhibitory effect against pathogens that cause pneumonia. Extractions with ethyl acetate (EtOAc) show the highest effect against microbes such as *Streptococcus pneumoniae*, *Stenotrophomonas maltophilia* and *Klebsiella pneumoniae*, among others. Furthermore, the isolated compound with the highest antimicrobial activity is Quercetin 5,4′-dimethyl ether [78]. Extracts of CPP have shown greater antimicrobial activity against Gram-positive bacteria than Gram-negative bacteria and antimicrobial activity against S. typhimurium and Bacillus subtilis [79] has been demonstrated. In addition, *Opuntia* ssp. peel extracts showed potential antiviral activity against H5N1 and rotavirus [80]. Concerning insecticidal activity, prickly pear peel waste has shown larvicidal activity as well as a decrease in the fecundity and hatchability of the *C. pipiens* mosquito [81]; however, more studies are needed to assess the antimicrobial activity of CPP extracts.

3.3.3. Hypolipidemic Effect

Dyslipidemias are a group of asymptomatic diseases originating in abnormal concentrations of blood lipoproteins [82]. The hypolipidemic effect of the peel extract was demonstrated in a study with hamsters fed a diet containing CPP extract, and after five weeks, the plasmatic and hepatic cholesterol reduced at 35% in comparison to the control diet [83]. In a study using rats, LDL cholesterol decreased significantly through the treatment with CPP [12], owing to its high content of ascorbic acid, which protects the essential fatty acids (omega-3, omega-6, α-linolenic acid and linolenic acid) from oxidation [80,81]. Its high fiber content that helps to the lower cholesterol was also noted [11]. The CPP extract is rich in phytosterols such as lanosterol, campesteryl β-D-glucoside, stigmasteryl β-D-glucoside and sitosteryl β-D-glucoside [37], which have a hypocholesterolemic effect by a competitive mechanism with cholesterol absorption. This evidence suggests that this by-product has some potential to be employed as an ingredient or a supplement to low cholesterol and prevent cardiovascular diseases [83].

3.3.4. Cytotoxic and Anticancer Activity

CPP extracts possess cytotoxic activity in human liver cancer cell lines (Hep G2), colorectal adenocarcinoma (Caco-2) and breast cells (MCF-7), decreasing the viability of cancer cells, by increasing the concentrations of bioactive compounds of an ethanolic extract. The highest concentrations cause a reduction in the viability of cancer cells, especially in the human liver cancer cell line (Hep G2) [84–86]. The anticancer effect may be due to the presence of polyphenols that play an important role in antioxidant activity and show antiproliferative activity or cytotoxicity in human cancer cells [87]. In addition, the extract contains gallic acid, which also shows cytotoxic activity against tumor cells, as in the case of lung cancer, prostate or leukemia [88]. Furthermore, sterols inhibit tumor promotion in carcinogenesis in mice, altering the expression of certain genes related to cell growth and apoptosis. Furthermore, the presence of quercetin could be one of the active compounds responsible for the anticancer and apoptosis-inducing effects of the extracts [89]. CPP contains large amounts of isorhamnetin (3′-methoxy-3,4′,5,7-tetrahydroxyflavone) that exerts anticancer action by the inhibition of epidermal growth factor (EGF); it also improves the skin barrier function through activation of the peroxisome proliferator-activated receptor (PPAR)-α and the suppression of inflammatory cytokines production. In a model of gastric cancer cells and, in combination with chemotherapeutic drugs, isorhamnetin also has strong antiproliferative effects and causes cytotoxicity [90].

3.3.5. Hypoglycemic Effect

Regarding the benefits in glucose levels, in a study with a group of rats fed with CPP flour, glycaemia was lower in comparison with a group fed with flour with apple residue, and the weight gain was lower compared to the control (inulin) [91] due to the low soluble sugars, the fructans content of fructans and high fiber content in the CPP [92].

Different studies aim at developing nutraceuticals or dietary supplements that may provide health benefits. Although no clinical trials have been developed exclusively with CPP, the effects of the product "OpunDia^{TM}", made with cactus cladode and CPP extracts (70:30 *w*/*w*), has potential hypoglycemic activity. After administration of 400 mg/day for 16 weeks, in obese pre-diabetic men and women, there was a significant decrease in blood glucose concentrations in an oral glucose tolerance test [93]. Furthermore, in athletes, this dietary supplement stimulates insulin secretion before and after exercise, lowering blood glucose levels and lowering the area under the blood glucose curve by 30%, a reduction of 10% in the glucose levels and greater insulin concentrations after ingestion [94].

3.4. Application of CPP in Industry

3.4.1. Food Industry

A high amount of peel waste is generated from fruit and vegetable-based industries and has led to an economical and nutritional losses. Processing of fruits and vegetables generates a significant amount of residue, among 25–30% of the total product, which have many bioactive compounds and have many applications in some industries such as food additives or ingredients, to develop films, for probiotics development, among others. The utilization of these low-cost horticultural wastes as a value-added product is a novel step for sustainable production [95], since the presence of complex polysaccharides composed of arabinose, galactose, rhamnose and galacturonic acid may influence the pleasant flavor. These characteristics make the CPP a suitable option as a sweetener in foods [12]. The great amount of carbohydrate polymers makes the peels a good source of fibers; therefore, their use is relevant to the food industry as a viscosity agent in food components [33].

The carbohydrates of CPP provide techno-functional properties such as water holding capacity (WHC) and lipid holding capacity (LHC), which range from 3.20 to 4.60 g/g, 1.73–1.90 g/g, respectively. Another technological property is the swelling capacity (SC), which varies from 9.82 to 12.33 mL/g. These functional properties are relevant because they may help to improve the sensory characteristics of some foods such as sausages or bakery products [37].

Several trials have included the CPP in different food products in order to obtain different sensorial attributes or improve nutritional value or presence of bioactive compounds.

An assay was done adding freeze-dried CPP in a snack of rice flour produced by extrusion cooking, in order to improve the nutritional and organoleptic properties where the cooking process does not significantly affect the content of bioactive compounds [96]. In addition, a snack from CPP using instant pressure drop texturing (a new technology applying high pressure at high temperatures) was applied to develop a "healthy snack", with a high content of phenolic compounds and β-carotene [23]. In another study, 10% CPP flour was added to a snack of amaranth and rice flour; as a result, the fat content decreased mainly due to the fiber content, which also provided better sensorial acceptance, better attributes of color, texture and oiliness [97].

CPP are used to produce biscuits rich in phenolic compounds, with greater sweetness and stable attractive colors, taking advantage of the natural colorants of the peels [98]. It has also been incorporated in the preparation of muffins mixed with wheat flour; the product presents a high fiber and moisture content and a lower fat content with great acceptability in the sensory analysis in comparison with commercial muffins [99].

The addition of CPP powder as a source of carbon to produce baker's yeast, using Saccharomyces cerevisiae, has also been assayed. The maximum cell mass production is reached at 24 h of inoculum time, with a temperature of 30 °C, agitation at 200 rpm and an inoculum size of 10%. Therefore, CPP has a great potential in the production of baker's yeast for industrial bakery applications [100].

3.4.2. Animal Fodder

CPP is commonly used as animal fodder due to their nutritional properties, such as the moderate content of sugars, starch, ether extract, crude protein, amino acids, fiber, and for providing a good amount of the animal requirements for vitamins and calcium, representing a better feed for ruminants than commercial feeds [101]. In rabbits fed with diets with 50% of CPP, giblets, liver and heart were heavier, and abdominal fat, triglycerides and LDL cholesterol were reduced, while the concentration of HDL cholesterol increased [102]. Adding a 15% of CPP to the traditional corn diets for commercial Cobb chicken, the weight gain improves in 5.78%, as well as the total protein and globulin in blood serum, resulting in superior nutritional status, greater daily weight gain, and better sensorial characteristics of the meat, including taste, color, odor, texture, and general acceptability [103]. The addition of CPP to the traditional diets, lead to a greater economic efficiency by minimizing the costs of the expensive yellow corn grain which is the food base in the poultry diets [104]. Farmers consider the fruit peel as an excellent supplemental feed and usually offered animals such as draught oxen, pregnant and milking animals [105].

3.4.3. Colorants

Synthetic colorants have been used in different types of industries because they present good stability and are cheap. However, the trend for using natural colorants is increasing, and the market derived from natural sources such as fruits, vegetables, insects or minerals represents a promising industry [106]. The peels of the cactus pear, mainly the red ones, are an important source of betanins, one of the most valued red natural colorants [107]. These pigments are of great importance in the industry because of their ecological value and non-toxicity. Betalains from CPP are colorants with a potential to be applied in functional foods, not only for their action as colorants also for their as antioxidants, antimicrobial, anti-proliferative and hypolipidemic properties [37,61], and are considered a permitted colorant for foods (USDA) [60].

Cactus pear by-products can be used in a more profitable way, by extracting the colorants before being used for animal feed [108], either using solvents, or novel clean technologies, such as the application of microwaves or pulsed electrical fields or ultrasound, which provide better recovery of the intracellular compounds with less impurity [107,109]. The textile fibers dyed with the extract of *Opuntia* spp. peels yield pink colors with great solidity, and by adding lemon juice (widely used in popular tradition as a natural mordant), depth tones are reached, achieving an environmentally friendly staining process [108].

3.4.4. Other Applications

The aqueous extract and the powder of CPP have been added in biofilms to improve the physical and antioxidant characteristics of edible carboxymethylcellulose films, obtaining a formulation with a high content of betalains and phenolic compounds by adding 1.7% of peel powder and 3.3% of the aqueous extract [110]. In addition, the mucilage from CPP is extracted to create a biopolymer with good solubility in water, foam and emulsion capacity and with a thermal stability of up to 250 °C, which could see this biopolymer applied in biodegradable containers [111]. The raw CPP has been successfully used as an agent for the decontamination of wastewaters that contain dyes, pesticides, high levels of turbidity, chemical oxygen demand and heavy metal ions. By using only 0.5 g of CPP at a particle size of 10 mm, it has highly efficient decontamination power, produced by the high biosorption capacity of the CPP [25].

4. Conclusions

The functional benefits and bio-active compounds that CPP provides generate great scientific interest in the exploration of this by-product as a inexpensive source of fibers, antioxidants, fatty acids and colorants, in addition to the proven health benefits, such as its hypolipidemic, hypocholesterolemic and cytotoxic activity. In this sense, cactus pear residues may be used on a wide spectrum of applications either as an ingredient for functional foods, as food supplements or to improve the sensorial characteristics of food and pharmaceutical products with the use of natural ingredients, which is generally better accepted in contrast to synthetic colorants or other additives. There remains great potential for future studies on the recovery of the compounds and their utilization. The health benefits that cactus pear peel could provide needs deeper research, mainly through clinical studies, in order to apply the current research. Furthermore, the use and exploitation of the by-products leads to more sustainable and environmentally friendly processes.

Author Contributions: Conceptualization, N.d.S.C.-C.; methodology, S.M.-V.; software, R.B.-G.; formal analysis, M.d.C.S.-M.; investigation, J.A.-G.; resources, Q.Y.Z.-R.; writing—original draft preparation, S.M.-V. and O.A.J.-M.; writing—review and editing, J.A.-R., E.R.-M., O.A.J.-M. and N.d.S.C.-C. All authors have read and agreed to the published version of the manuscript.

Funding: This research received no external funding.

Institutional Review Board Statement: Not applicable.

Informed Consent Statement: Not applicable.

Data Availability Statement: Not applicable.

Acknowledgments: The first author would like to thank the Consejo Nacional de Ciencia y Tecnología (CONACyT) for the fellowship assigned (CVU 708817).

Conflicts of Interest: The authors declare no conflict of interest.

References

1. Inglese, P. Cactus pear: Gift of the new world. *Chron. Hort.* **2009**, *49*, 15–19. [CrossRef]
2. National Germplasm Resources Laboratory. National Genetic Resources Program. Germplasm Resources Information Network: USDA, ARS. 2005. Available online: http://www.ars-grin.gov/cgi-bin/npgs/html/tax_search.pl (accessed on 10 November 2022).
3. Boyle, T.H.A.E. Biodiversity and conservation. In *Cacti Biology and Uses*; Univ of California Press: Los Angeles, CA, USA, 2002; pp. 125–141. [CrossRef]
4. El-Kossori, R.L.; Villaume, C.; El Boustani, E.; Sauvaire, Y.M.L. Composition of pulp, skin and seeds of prickly pears fruit (*Opuntia ficus-indica* sp.). *Plant Food Hum. Nutr.* **1998**, *52*, 263–270. [CrossRef] [PubMed]
5. SIAP. Servicio de Información y Estadística Agroalimentaria y Pesquera. In *Anuario Estadístico de la Producción Agrícola de los Estados Unidos Mexicanos*; SIAP: Mexico City, Mexico, 2017. Available online: http://www.siap.gob.mx (accessed on 25 October 2022).
6. SIACON. Sistema de Información Agropecuaria de Consulta. In *Anuarios Estadísticos*; SIACON: Mexico City, Mexico, 2007.
7. Inglese, P.; Mondragon, C.; Nefzaoui, A.; Saenz, C. *Crop Ecology, Cultivation and Uses of Cactus Pear*; Food and Agriculture Organization of the United Nations (FAO): Rome, Italy, 2017. Available online: http://www.fao.org/3/a-i7628e.pdf (accessed on 25 October 2022).
8. Ochoa, C.E.; Guerrero, J.A. Efecto del almacenamiento a diferentes temperaturas sobre la calidad de tuna roja (*Opuntia ficus indica* (L. Miller). *Rev. Inf. Tecnol.* **2012**, *23*, 117–128. [CrossRef]
9. Barba, F.J.; Putnik, P.; Kovačević, D.B.; Poojary, M.M.; Roohinejad, S.; Lorenzo, J.M. Impact of conventional and nonconventional processing on prickly pear (*Opuntia* spp.) and their derived products: From preservation of beverages to valorization of by-products. *Trends Food Sci. Technol.* **2017**, *67*, 260–270. [CrossRef]
10. Piga, A. Cactus pear: A fruit of nutraceutical and functional importance. *J. Prof. Assoc. Cactus Dev.* **2004**, *6*, 9–22.
11. Bensadón, S.; Hervert-Hernández, D.; Sáyago-Ayerdi, S.G. By-products of *Opuntia ficus-indica* as a source of antioxidant dietary fiber. *Plant Food Hum. Nutr.* **2010**, *65*, 210–216. [CrossRef]
12. El-Said, N.M.; Nagib, A.I.; Rahman, Z.A.; Deraz, S.F. Prickly pear [*Opuntia ficus-indica* (L.) Mill] peels: Chemical composition, nutritional value, and protective effects on liver and kidney functions and cholesterol in rats. *Funct. Plant Sci. Biotechnol.* **2011**, *5*, 30–35.
13. Delgado-Vargas, F.; Jiménez, A.R.; Paredes-López, O. Natural pigments: Carotenoids, anthocyanins, and betalains–characteristics, biosynthesis, processing, and stability. *Crit. Rev. Food Sci. Nutr.* **2000**, *40*, 173–289. [CrossRef]
14. Ramadan, M.F.; Mörsel, J.T. Recovered lipids from prickly pear [*Opuntia ficus-indica* (L.) Mill] peel: A good source of polyunsaturated fatty acids, natural antioxidant vitamins and sterols. *Food Chem.* **2003**, *83*, 447–456. [CrossRef]
15. Cerezal, P.; Duarte, G. Use of skin in the elaboration of concentrated products of cactus pear (*Opuntia ficus-indica* (L.) Miller). *J. Prof. Assoc. Cactus Dev.* **2005**, *7*, 61–82.
16. Basile, F. Economic aspects of Italian cactus pear production and market. *J. Prof. Assoc. Cactus Dev.* **2001**, *4*, 31–46.
17. Villacís-Chiriboga, J.; Elst, K.; Van Camp, J.; Vera, E.; Ruales, J. Valorization of byproducts from tropical fruits: Extraction methodologies, applications, environmental, and economic assessment: A review (Part 1: General overview of the byproducts, traditional biorefinery practices, and possible applications). *Compr. Rev. Food Sci. Food Saf.* **2020**, *19*, 405–447. [CrossRef] [PubMed]
18. Hernández-Carranza, P.; Rivadeneyra-Mata, M.; Ramos-Cassellis, M.E.; Aparicio-Fernández, X.; Navarro-Cruz, A.R.; Ávila-Sosa, R.; Ochoa-Velasco, C.E. Characterization of red prickly pear peel (*Opuntia ficus-indica* L.) and its mucilage obtained by traditional and novel methodologies. *J. Food Meas. Charact.* **2019**, *13*, 1111–1119. [CrossRef]
19. Jimenez-Aguilar, D.M.; Mújica-Paz, H.; Welti-Chanes, J. Phytochemical characterization of prickly pear (*Opuntia* spp.) and of its nutritional and functional properties: A review. *Curr. Nutr. Food Sci.* **2014**, *10*, 57–69. [CrossRef]
20. Stintzing, F.C.; Herbach, K.M.; Mosshammer, M.R.; Carle, R.; Yi, W.; Sellappan, S.; Felker, P. Color, betalain pattern, and antioxidant properties of cactus pear (*Opuntia* spp.) clones. *J. Agric. Food Chem.* **2005**, *53*, 442–451. [CrossRef]
21. Chougui, N.; Djerroud, N.; Naraoui, F.; Hadjal, S.; Aliane, K.; Zeroual, B.L.R. Physicochemical properties and storage stability of margarine containing *Opuntia ficus-indica* peel extract as antioxidant. *Food Chem.* **2015**, *15*, 382–390. [CrossRef]
22. Namir, M.; Elzahar, K.; Ramadan, M.F.; Allaf, K. Cactus pear peel snacks prepared by instant pressure drop texturing: Effect of process variables on bioactive compounds and functional properties. *J. Food Meas. Charact.* **2017**, *11*, 388–400. [CrossRef]
23. Abbas, E.Y.; Ezzat, M.I.; El Hefnawy, H.M.; Abdel-Sattar, E. An overview and update on the chemical composition and potential health benefits of *Opuntia ficus-indica* (L.) Miller. *J. Food Biochem.* **2022**, e14310. [CrossRef]

24. Belayneh, A.; Batu, W. Application of biosorbent derived from cactus peel for removal of colorful manganese ions from ground water. *J. Water Resour. Ocean Sci.* **2015**, *4*, 18–23. [CrossRef]
25. Liberati, A.; Altman, D.G.; Tetzlaff, J.; Mulrow, C.; Gøtzsche, P.C.; Ioannidis, J.P.; Clarke, M.; Devereaux, P.J.; Kleijnen, J.; Moher, D. The PRISMA statement for reporting systematic reviews and meta-analyses of studies that evaluate healthcare interventions: Explanation and elaboration. *BMJ* **2009**, *339*, b2700. [CrossRef]
26. Financiera Rural, Dirección General Adjunta de Planeación Estratégica y Análisis Sectorial. In Dirección Ejecutiva de Análisis Sectorial; Monografía del Nopal y la Tuna. México. 2011. Available online: https://www.yumpu.com/es/document/read/17605434/monografia-del-nopal-y-la-tuna-financiera-rural (accessed on 25 October 2022).
27. Reyes-Agüero, J.A.; Aguirre-Rivera, J.R.; Hernández, H.M. Systematyc notes and a detailed description of *Opuntia ficus-indica* (L.) Mill. (Cactaceae). *Agrociencia* **2005**, *39*, 395–408.
28. Kuti, J.O. Antioxidant compounds from four *Opuntia* cactus pear fruit varieties. *Food Chem.* **2004**, *85*, 527–533. [CrossRef]
29. Carrillo-López, A.; Cruz-Hernández, A.; Cárabez-Trejo, A.; Guevara-Lara, F.; Paredes-López, O. Hydrolytic activity and ultrastructural changes in fruit skins from two prickly pear (*Opuntia* sp.) varieties during storage. *J. Agric. Food Chem.* **2002**, *50*, 1681–1685. [CrossRef] [PubMed]
30. Sawaya, W.N.; Khatchadourian, H.A.; Safi, W.M.; Al-Muhammad, H.M. Chemical characterization of prickly pear pulp, *Opuntia ficus-indica*, and the manufacturing of prickly pear jam. *Int. J. Food Sci. Technol.* **2007**, *18*, 183–193. [CrossRef]
31. Drennan, P.; Nobel, P. Root growth dependence on soil temperature for *Opuntia ficus-indica*: Influences of air temperature and a doubled CO2 concentration. *Funct. Ecol.* **2002**, *1*, 959–964. [CrossRef]
32. Habibi, Y.; Mahrouz, M.; Vignon, M.R. Microfibrillated cellulose from the peel of prickly pear fruits. *Food Chem.* **2009**, *115*, 423–429. [CrossRef]
33. Habibi, Y.; Heyraud, A.; Mahrouz, M.; Vignon, M.R. Structural features of pectic polysaccharides from the skin of *Opuntia ficus-indica* prickly pear fruits. *Carbohydr. Res.* **2004**, *339*, 1119–1127. [CrossRef]
34. El Gharras, H.; Hasib, A.; Jaouad, A.; El Bouadili, A. Chemical and physical characterization of three cultivars of moroccan yellow prickly pears (*Opuntia ficus-indica*) at three stages of maturity. *Cienc. y Tecnol. Aliment.* **2006**, *5*, 93–99. [CrossRef]
35. Terán, Y.; Navas, D.; Petit, D.; Garrido, E.; D'Aubeterre, R. Análisis de las características físico-químicas del fruto de *Opuntia ficus–indica* (L.) Miller, cosechados en Lara, Venezuela. *Rev. Iberoam. Tecnol. Postcosecha* **2015**, *16*, 69–74.
36. Domene, M.A.; Segura, M. *Parámetros de Calidad Interna de Hortalizas y Frutas en la Industria Agroalimentaria*; Fundación Cajamar: Barcelona, Spain, 2014; pp. 1–18. Available online: https://www.cajamar.es/storage/documents/005-calidad-interna-1410512030-cc718.pdf (accessed on 25 October 2022).
37. Amaya-Cruz, D.M.; Pérez-Ramírez, I.F.; Delgado-García, J.; Mondragón-Jacobo, C.; Dector-Espinoza, A.; Reynoso-Camacho, R. An integral profile of bioactive compounds and functional properties of prickly pear (*Opuntia ficus indica* L.) peel with different tonalities. *Food Chem.* **2019**, *278*, 568–578. [CrossRef]
38. Cho, S.S. *Handbook of Dietary Fiber*; CRC Press: Boca Raton, FL, USA, 2001.
39. Gowe, C. Review on potential use of fruit and vegetables by-products as a valuable source of natural food additives. *Food Sci. Qual. Manag.* **2015**, *45*, 47–61.
40. Andreu-Coll, L.; Cano-Lamadrid, M.; Sendra, E.; Carbonell-Barrachina, Á.; Legua, P.; Hernández, F. Fatty acid profile of fruits (pulp and peel) and cladodes (young and old) of prickly pear [*Opuntia ficus-indica* (L.) Mill.] from six Spanish cultivars. *J. Food Compos. Anal.* **2019**, *84*, 103294. [CrossRef]
41. WHO. *Vitamin and Mineral Requirements in Human Nutrition*, 2nd ed.; World Health Organization: Hong Kong, China, 2004; pp. 66–69, 200–220.
42. WHO. Sodium Intake for Adults and Children. WHO Guidel. 2012. Available online: https://www.who.int/publications/i/item/9789241504836 (accessed on 25 October 2022).
43. WHO. Potassium Intake for Adults and Children. WHO Guidel. 2012. Available online: https://www.who.int/publications/i/item/9789241504829 (accessed on 25 October 2022).
44. Fairweather-Tait, S.J.; Cashman, K. Minerals and trace elements. In *Nutrition for the Primary Care Provider*; World Rev Nutr Diet; Karger: Basel, Switzerland, 2015; Volume 111, pp. 45–52. [CrossRef]
45. Bruno, E.J.; Ziegenfuss, T.N. Water-soluble vitamins: Research update. *Curr. Sports Med. Rep.* **2005**, *4*, 207–213. [CrossRef] [PubMed]
46. El-Beltagi, H.S.; Mohamed, H.I.; Elmelegy, A.A.; Eldesoky, S.E.; Safwat, G. Phytochemical screening, antimicrobial, antioxidant, anticancer activities and nutritional values of cactus (*Opuntia ficus indica*) pulp and peel. *Fresenius Environ. Bull* **2019**, *28*, 1545–1562.
47. Rodriguez, S.; Orphee, C.; Macias, S.; Generoso, S.; Gomes, G. Tuna: Propiedades físico-químicas de dos variedades. *Aliment. Latinoam.* **1996**, *210*, 34–37.
48. Michels, A.J.; Frei, B. Myths artifacts, and fatal flaws: Identifying limitations and opportunities in vitamin C research. *Nutrients* **2013**, *5*, 5161–5192. [CrossRef]
49. Cano, M.P.; Gómez-Maqueo, A.; García-Cayuela, T.; Welti-Chanes, J. Characterization of carotenoid profile of Spanish Sanguinos and Verdal prickly pear (*Opuntia ficus-indica*, spp.) tissues. *Food Chem.* **2017**, *237*, 612–622. [CrossRef]

50. De Wit, M.; Du Toit, A.; Osthoff, G.; Hugo, A. Cactus pear antioxidants: A comparison between fruit pulp, fruit peel, fruit seeds and cladodes of eight different cactus pear cultivars (*Opuntia ficus-indica* and *Opuntia robusta*). *J. Food Meas. Charact.* **2019**, *13*, 2347–2356. [CrossRef]
51. Chiva-Blanch, G.; Visioli, F. Polyphenols and health: Moving beyond antioxidants. *J. Berry Res.* **2012**, *2*, 63–71. [CrossRef]
52. Seifried, H.E.; Anderson, D.E.; Fisher, E.I.; Milner, J.A. A review of the interaction among dietary antioxidants and reactive oxygen species. *J. Nutr. Biochem.* **2007**, *18*, 567–579. [CrossRef]
53. Jiménez-Aguilar, D.M.; López-Martínez, J.M.; Hernández-Brenes, C.; Gutiérrez-Uribe, J.A.; Welti-Chanes, J. Dietary fiber, phytochemical composition and antioxidant activity of Mexican commercial varieties of cactus pear. *J. Food Compos. Anal.* **2015**, *41*, 66–73. [CrossRef]
54. Cota-Sánchez, J. Nutritional composition of the prickly pear (*Opuntia ficus-indica*) fruit. In *Nutritional Composition of Fruit Cultivars*; Academic Press: Cambridge, MA, USA, 2016; pp. 691–712. [CrossRef]
55. Chavez-Santoscoy, R.A.; Gutierrez-Uribe, J.A.; Serna-Saldívar, S.O. Phenolic composition, antioxidant capacity and in vitro cancer cell cytotoxicity of nine prickly pear (*Opuntia* spp.) juices. *Plant Foods Hum. Nutr.* **2009**, *64*, 146–152. [CrossRef] [PubMed]
56. Anwar, M.M.; Sallam, E.M. Utilization of prickly pear peels to improve quality of pan bread. *Arab. J. Nucl. Sci. Appl.* **2016**, *49*, 151–163.
57. Cardador-Martínez, A.; Jiménez-Martínez, C. Revalorization of cactus pear (*Opuntia* spp.) wastes as a source of antioxidants. *Food Sci. Technol.* **2011**, *31*, 782–788. [CrossRef]
58. Ortega-Hernández, E.; Nair, V.; Welti-Chanes, J.; Cisneros-Zevallos, L.; Jacobo-Velázquez, D.A. Wounding and UVB light synergistically induce the biosynthesis of phenolic compounds and ascorbic acid in red prickly pears (*Opuntia ficus-indica* cv. Rojo Vigor). *Int. J. Mol. Sci.* **2019**, *20*, 5327. [CrossRef] [PubMed]
59. Rahimi, P.; Abedimanesh, S.; Mesbah-Namin, S.A.; Ostadrahimi, A. Betalains, the nature-inspired pigments, in health and diseases. *Crit. Rev. food Sci. Nutr.* **2019**, *59*, 2949–2978. [CrossRef]
60. Richhariya, G.; Kumar, A.; Tekasakul, P.; Gupta, B. Natural dyes for dye sensitized solar cell: A review. *Renew. Sustain. Energy Rev.* **2017**, *69*, 705–718. [CrossRef]
61. Gengatharan, A.; Dykes, G.A.; Choo, W.S. Betalains: Natural plant pigments with potential application in functional foods. *LWT-Food Sci. Technol.* **2015**, *64*, 645–649. [CrossRef]
62. Britton, G.; Liaaen-Jensen, S.; Pfander, H. *Carotenoids: Handbook*; Birkhäuser: Basel, Switzerland, 2004. [CrossRef]
63. Melgar, B.; Dias, M.I.; Ciric, A.; Sokovic, M.; Garcia-Castello, E.M.; Rodriguez-Lopez, A.D.; Ferreira, I. By-product recovery of *Opuntia* spp. peels: Betalainic and phenolic profiles and bioactive properties. *Ind. Crops Prod.* **2017**, *107*, 353–359. [CrossRef]
64. Romeo-Donlo, M.; Martínez-Gómez, M.J.; Pizarro-Pizarro, I. Inflammatory bowel disease: The importance of early diagnosis. *Rev. Pediatr. Aten. Primaria* **2014**, *16*, 49–53. [CrossRef]
65. Lobaton, T.; Vermeire, S.; Van Assche, G.; Rutgeerts, P. Anti-adhesion therapies for inflammatory bowel disease. *Aliment. Pharmacol. Ther.* **2014**, *39*, 579–594. [CrossRef] [PubMed]
66. Makhija, D.T.; Somani, R.R.; Chavan, A.V. Synthesis and pharmacological evaluation of antiinflammatory mutual amide prodrugs. *Indian J. Pharm. Sci.* **2013**, *75*, 353–357. [CrossRef] [PubMed]
67. Beg, S.; Swain, S.; Hasan, H.; Barkat, M.A.; Hussain, M.S. Systematic review of herbals as potential anti-inflammatory agents: Recent advances, current clinical status and future perspectives. *Pharmacogn. Rev.* **2011**, *5*, 120–137. [CrossRef] [PubMed]
68. Manthey, J.A. Biological Properties of Flavonoids Pertaining to Inflammation. *Microcirculation* **2000**, *7*. S29–S34. [CrossRef] [PubMed]
69. Joo, M.; Kim, H.S.; Kwon, T.H.; Palikhe, A.; Zaw, T.S.; Jeong, J.H.; Sohn, U.D. Anti-inflammatory effects of flavonoids on TNBS-induced colitis of rats. *Korean J. Physiol. Pharmacol.* **2015**, *19*, 43–50. [CrossRef]
70. Yun-Li, Z.; Xiong-Wu, Y.; Bai-Fen, W.; Jian-Hua, S.; Ya-Ping, L.; Zhi-Dai, X. Anti-inflammatory effect of pomelo peel and its bioactive coumarins. *J. Agric. Food Chem.* **2019**, *67*, 8810–8818. [CrossRef]
71. Ismail, T.; Sestili, P.; Akhtar, S. Pomegranate peel and fruit extracts: A review of potential anti-inflammatory and anti-infective effects. *J. Ethnopharmacol.* **2012**, *143*, 397–405. [CrossRef]
72. Chen, X.M.; Tait, A.R.; Kitts, D.D. Flavonoid composition of orange peel and its association with antioxidant and anti-inflammatory activities. *Food Chem.* **2017**, *1*, 15–21. [CrossRef]
73. Elsawi, S.A.; Radwan, R.R.; Elbatanony, M.M.; El-Feky, A.M.; Sherif, N.H. Prophylactic effect of *Opuntia ficus indica* fruit peel extract against Irradiation-Induced colon injury in rats. *Planta Med.* **2020**, *86*, 61–69. [CrossRef]
74. Rana, V.; Bachheti, R.K.; Chand, T.; Barman, A. Dietary fibre and human health. *Int. J. Food Saf. Nutr. Public. Health* **2011**, *4*, 101–118. [CrossRef]
75. Valencia, F.E.R.M. Physicalchemical and functional characterization of three commercial concentrates from dietary fiber. *Rev. Vitae* **2009**, *13*, 54–60.
76. Perez-Chabela, M.L.; Cerda-Tapia, A.; Diaz-Vela, J.; Delgadillo, P.C.; Diaz, M.M.; Aleman, G. Physiological effects of agroindustrial co-products: Cactus (*Opuntia ficus*) pear peel flour and stripe apple (*Malus domestica*) marc flour on wistar rats (*Rattus norvegicus*). *Pakistan J. Nutr.* **2015**, *14*, 346–352. [CrossRef]
77. Rodríguez-Cabezas, M.E.; Camuesco, D.; Arribas, B.; Garrido-Mesa, N.; Comalada, M.; Bailón, E.; Zarzuelo, A. The combination of fructooligosaccharides and resistant starch shows prebiotic additive effects in rats. *Clin. Nutr.* **2010**, *29*, 832–839. [CrossRef] [PubMed]

78. Wafaa, M.; Mokhtar, M.; Bishrb, M.M.; Abdel-Azizc, O.M.S. Identification and isolation of anti-pneumonia bioactive compounds from *Opuntia ficus-indica* fruits waste peels. *Food Funct.* **2020**, *11*, 5275–5283. [CrossRef]
79. Silva, M.A.; Albuquerque, T.G.; Pereira, P.; Ramalho, R.; Vicente, F.; Oliveira, M.B.P.; Costa, H.S. Opuntia ficus-indica (L.) Mill.: A multi-benefit potential to be exploited. *Molecules* **2021**, *26*, 951. [CrossRef]
80. Kamel, S.; Mohamed, S.; El-Masry, S.; Alkhalifah, D.; Hozzein, W.; Aboel-Ainin, M. Phytochemical screening and characterization of the antioxidant, anti-proliferative and antibacterial effects of different extracts of Opuntia ficus-indica peel. *J. King Saud Univ. Sci.* **2022**, *34*, 102216. [CrossRef]
81. Hashem, A.H.; Selim, T.A.; Alruhaili, M.H.; Selim, S.; Alkhalifah, D.H.M.; Al Jaouni, S.K.; Salem, S.S. Unveiling Antimicrobial and Insecticidal Activities of Biosynthesized Selenium Nanoparticles Using Prickly Pear Peel Waste. *J. Funct. Biomater.* **2022**, *13*, 112. [CrossRef]
82. Canalizo-Miranda, E.; Favela-Pérez, E.A.; Salas-Anaya, J.A.; Gómez-Díaz, R.; Jara-Espino, R.; Torres-Arreola, L.; Viniegra-Osorio, A. Guía de práctica clínica. Diagnóstico y tratamiento de las dislipidemias. *Rev. Med. Inst. Mex. Seguro Soc.* **2013**, *51*, 700–709.
83. Milán-Noris, K.; Chavez-Santoscoy, A.; Olmos-Nakamura, A.; Gutiérrez-Uribe, J.; Serna-Saldívar, O. An extract from prickly pear peel (*Opuntia ficus-indica*) affects cholesterol excretion and hepatic cholesterol levels in hamsters fed hyperlipidemic diets. *Curr. Bioact. Compd.* **2016**, *12*, 10–16. [CrossRef]
84. Tesoriere, L.; Butera, D.; Pintaudi, A.M.; Allegra, M.; Livrea, M.A. Supplementation with cactus pear (*Opuntia ficus-indica*) fruit decreases oxidative stress in healthy humans: A comparative study with vitamin C. *Am. J. Clin. Nutr.* **2004**, *80*, 391–395. [CrossRef]
85. Theobald, J.; Lunn, H.E. The health effects of dietary unsaturated fatty acids. *Nutr. Bull* **2006**, *31*, 178–224. [CrossRef]
86. Hassanpour, S.H.; Dehghani, M. Review of cancer from perspective of molecular. *J. Cancer Res. Pract.* **2017**, *4*, 127–129. [CrossRef]
87. El-Beltagi, H.S.; Mohamed, H.I.; Megahed, B.M.H.; Gamal, M.; Safwat, G. Evaluation of some chemical constituents, antioxidant, antibacterial and anticancer activities of *Beta vulgaris* L. root. *Fresenius Environ. Bull* **2018**, *27*, 6369–6378.
88. Zou, D.; Brewer, M.; Garcia, F.; Feugang, J.M.; Wang, J.; Zang, R.; Liu, H.; Zou, C. Cactus pear, a natural product in cancer chemoprevention. *Nutr. J.* **2005**, *4*, 25. [CrossRef]
89. Herzog, A.; Kuntz, S.; Daniel, H.; Wenzel, U. Identification of biomarkers for the initiation of apoptosis in human preneoplastic colonocytes by proteome analysis. *Int. J. Cancer* **2004**, *109*, 220–229. [CrossRef]
90. El-Mostafa, K.; El Kharrassi, Y.; Badreddine, A.; Andreoletti, P.; Vamecq, J.; El Kebbaj, M.H.S.; Cherkaoui-Malki, M. Nopal cactus (*Opuntia ficus-indica*) as a source of bioactive compounds for nutrition, health and disease. *Molecules* **2014**, *19*, 14879–14901. [CrossRef]
91. ADA. Improving care and promoting health in populations: Standards of medical care in diabetes—2020. *Diabetes Care* **2020**, *43*, S7–S13. [CrossRef]
92. Monro, J.A.; Paturi, G.; Butts, C.A.; Young, W.; Guzman, C.E.; McLachlan, A.; Roy, N.C.; Ansell, J. Prebiotic effects of fermentable carbohydrate polymers may be modulated by faecal bulking of non-fermentable polysaccharides in the large bowel of rats. *Int. J. Food Sci. Technol.* **2012**, *47*, 968–976. [CrossRef]
93. Godard, M.P.; Ewing, B.A.; Pischel, I.; Ziegler, A.; Benedek, B.; Feistel, B. Acute blood glucose lowering effects and long-term safety of OpunDia^{TM} supplementation in pre-diabetic males and females. *J. Ethnopharmacol.* **2010**, *130*, 631–634. [CrossRef]
94. Pischel, I.; Van Proeyen, K.; Hespel, P. Effect of OpunDia^{TM} (*O. ficus-indica* extract) on oral glucose tolerance and plasma insulin before and after exercise. *J. Int. Soc. Sports Nutr.* **2011**, *8*, 17. [CrossRef]
95. Kumar, H.; Bhardwaj, K.; Sharma, R.; Nepovimova, E.; Kuča, K.; Dhanjal, D.S.; Kumar, D. Fruit and vegetable peels: Utilization of high value horticultural waste in novel industrial applications. *Molecules* **2020**, *25*, 2812. [CrossRef] [PubMed]
96. Moussa-Ayoub, T.E.; Youssef, K.; El-Samahy, S.K.; Kroh, L.W.; Rohn, S. Flavonol profile of cactus fruits (*Opuntia ficus-indica*) enriched cereal-based extrudates: Authenticity and impact of extrusion. *Food Res. Int.* **2015**, *78*, 442–447. [CrossRef] [PubMed]
97. Miranda, D.V.; Rojas, M.L.; Pagador, S.; Lescano, L.; Sanchez-Gonzalez, J.; Linares, G. Gluten-free snacks based on brown rice and amaranth flour with incorporation of cactus pear peel powder: Physical, nutritional, and sensorial properties. *Int. J. Food Sci.* **2018**, *2018*, 7120327. [CrossRef]
98. El-Shahat, M.S.; Rabie, M.A.; Ragab, M.; Siliha, H.I. Changes on physicochemical and rheological properties of biscuits substituted with the peel and alcohol-insoluble solids (AIS) from cactus pear (*Opuntia ficus-indica*). *J. Food Sci. Technol.* **2019**, *56*, 3635–3645. [CrossRef] [PubMed]
99. García-Valencia, A.; Bautista-Cano, G.; Cerón-Latorre, B.; Muñoz-Olivares, M.M.; Pérez-González, O.; Vázquez-Rodríguez, G.A. Valorización de la cáscara de tuna en un producto de panificación. *Pädi Bol. Cient. Cienc. Basic. Ing. ICBI* **2018**, *11*, 12. [CrossRef]
100. Diboune, N.; Nancib, A.; Nancib, N.; Aníbal, J.; Boudrant, J. Utilization of prickly pear waste for baker's yeast production. *Biotechnol. Appl. Biochem.* **2019**, *66*, 744–754. [CrossRef] [PubMed]
101. Todaro, M.; Alabiso, M.; Di Grigoli, A.; Scatassa, M.L.; Cardamone, C.; Mancuso, I.; Bonanno, A. Prickly pear by-product in the feeding of livestock ruminants: Preliminary investigation. *Animals* **2020**, *10*, 949. [CrossRef]
102. Amer, F.; Mobaraz, S.; Basyony, M.; Mahrose, K.; El-Medany, S. Effect of using prickly pear and its by-products as alternative feed resources on performance of growing rabbit. *Egypt. J. Rabbit Sci.* **2019**, *29*, 99–124. [CrossRef]
103. Badr, S.E.; Fattah, M.S.A.; Elsaid, A.S. Productive performance and meat quality of commercial Cobb chicken fed diets containing different levels of prickly pear fruits (*Opuntia ficus indica*) peel. *Bull Natl. Res. Cent.* **2019**, *43*, 195. [CrossRef]

104. Ragab, M.S. Effect of partially replacing of yellow corn with prickly pear peels on the growth performance of Hy-line W-36 male chicks. *Egypt J. Nutr. Feed* **2012**, *15*, 361–373.
105. Gebretsadik, G.; Animut, G.; Tegegne, F. Assessment of the potential of cactus pear (*Opuntia ficus indica*) as livestock feed in Northern Ethiopia. *Livest. Res. Rural Dev.* **2013**, *25*, 1–10.
106. Sigurdson, G.T.; Tang, P.; Giusti, M.M. Natural colorants: Food colorants from natural sources. *Annu. Rev. Food Sci. Technol.* **2017**, *8*, 261–280. [CrossRef]
107. Ciriminna, R.; Danzì, C.; Timpanaro, G.; Locatelli, M.; Carnaroglio, D.; Fidalgo, A.; Pagliaro, M. Valued bioproducts from waste *Opuntia ficus-indica* peel via microwave-assisted hydrodiffusion and hydrodistillation. *ACS Sustain Chem. Eng.* **2017**. [CrossRef]
108. Scarano, P.; Naviglio, D.; Prigioniero, A.; Tartaglia, M.; Postiglione, A.; Sciarrillo, R.; Guarino, C. Sustainability: Obtaining natural dyes from waste matrices using the prickly pear peels of *Opuntia ficus-indica* (L.) Miller. *Agronomy* **2020**, *10*, 528. [CrossRef]
109. Koubaa, M.; Barba, F.J.; Grimi, N.; Mhemdi, H.; Koubaa, W.; Boussetta, N.; Vorobiev, E. Recovery of colorants from red prickly pear peels and pulps enhanced by pulsed electric field and ultrasound. *Innov. Food Sci. Emerg. Technol.* **2016**, *37*, 336–344. [CrossRef]
110. Aparicio-Fernández, X.; Vega-Ahuatzin, A.; Ochoa-Velasco, C.E.; Cid-Pérez, S.; Hernández-Carranza, P.; Ávila-Sosa, R. Physical and antioxidant characterization of edible films added with red prickly pear (*Opuntia ficus-indica* L.) cv. San Martín peel and/or its aqueous extracts. *Food Bioprocess Technol.* **2018**, *126*, 238–245. [CrossRef]
111. Gheribi, R.; Habibi, Y.; Khwaldia, K. Prickly pear peels as a valuable resource of added-value polysaccharide: Study of structural, functional and film forming properties. *Int. J. Biol. Macromol.* **2019**, *126*, 238–245. [CrossRef]

life

Review

Unraveling the Potential of β-D-Glucans in *Poales*: From Characterization to Biosynthesis and Factors Affecting the Content

Michaela Havrlentová [1,2,*], Václav Dvořáček [3], Lucie Jurkaninová [4] and Veronika Gregusová [1]

[1] Department of Biotechnology, Faculty of Natural Sciences, University of Ss. Cyril and Methodius, Námestie J. Herdu 2, 917 01 Trnava, Slovakia; veronika.gregusova93@gmail.com
[2] National Agricultural and Food Center—Research Institute of Plant Production, Bratislavská cesta 122, 921 68 Piešťany, Slovakia
[3] Crop Research Institute, Drnovská 507, 161 06 Prague, Czech Republic; dvoracek@vurv.cz
[4] Department of Food Science, Faculty of Agrobiology, Food and Natural Resources, Czech University of Life Sciences, Kamýcká 129, 165 00 Praha, Czech Republic; jurkaninova@af.czu.cz
* Correspondence: michaela.havrlentova@ucm.sk or michaela.havrlentova@nppc.sk

Abstract: This review consolidates current knowledge on β-D-glucans in *Poales* and presents current findings and connections that expand our understanding of the characteristics, functions, and applications of this cell wall polysaccharide. By associating information from multiple disciplines, the review offers valuable insights for researchers, practitioners, and consumers interested in harnessing the benefits of β-D-glucans in various fields. The review can serve as a valuable resource for plant biology researchers, cereal breeders, and plant-based food producers, providing insights into the potential of β-D-glucans and opening new avenues for future research and innovation in the field of this bioactive and functional ingredient.

Keywords: β-D-glucans; cereals; biosynthesis; genes; breeding; environment; functions

Citation: Havrlentová, M.; Dvořáček, V.; Jurkaninová, L.; Gregusová, V. Unraveling the Potential of β-D-Glucans in *Poales*: From Characterization to Biosynthesis and Factors Affecting the Content. *Life* **2023**, *13*, 1387. https://doi.org/10.3390/life13061387

Academic Editor: Jianfeng Xu

Received: 3 May 2023
Revised: 11 June 2023
Accepted: 12 June 2023
Published: 14 June 2023

1. Basic Characterization and Localization of β-D-Glucans in the Plant

The (1,3-1,4)-β-D-glucans (hereafter referred to as β-D-glucans) are relatively small part components of the cell wall of vegetative tissues of cereals and grasses [1,2]. Mainly cereal grains such as barley (*Hordeum vulgare* L.), oat (*Avena sativa* L.), and rye (*Secale cereale* L.) are rich sources of β-D-glucans, while wheat (*Triticum aestivum* L.), rice (*Oryza sativa* L.), and maize (*Zea mays* L.) dispose of lower concentrations of this polysaccharide [3–6].

Generally, the accumulation of β-D-glucans is observed in the cell wall of endosperm cells of the developing grains and in the surrounding maternal tissues, the aleuronal and subaleuronal layer [7–9] (Figure 1). β-D-glucans are not uniformly distributed in the grain, their localization varies among plant species and different plant tissues [8,10]. This polysaccharide has also been found in vegetative organs of the plant, namely in the root, coleoptile, stem, and leaf [11–13], and variability in the content of this polysaccharide was found during plant and tissue development [13,14].

The β-D-glucans are unsubstituted, unbranched polysaccharides composed of β-D-glucopyranosyl monomers polymerized by both β-(1,3) and β-(1,4) linkages [15,16] and therefore the cereal β-D-glucans are also called "mixed-linkage glucans—MLG". From a functional point of view, the most important feature of this molecule is the arrangement of β-(1,3) and β-(1,4) linkages along the polysaccharide chain [1,2,17]. The bonds are not arranged in regularly repeating sequences, but they are also not arranged randomly [17–19]. (1,4)-β-bonds are usually more often presented in the polysaccharide than (1,3)-β-bonds. (1,3)-β-D-glucosyl residues always occur in linear β-D-glucans chain as individual parts between (1,4)-β-D-oligoglucosyl units, which are mostly found in sequences of two or three [20].

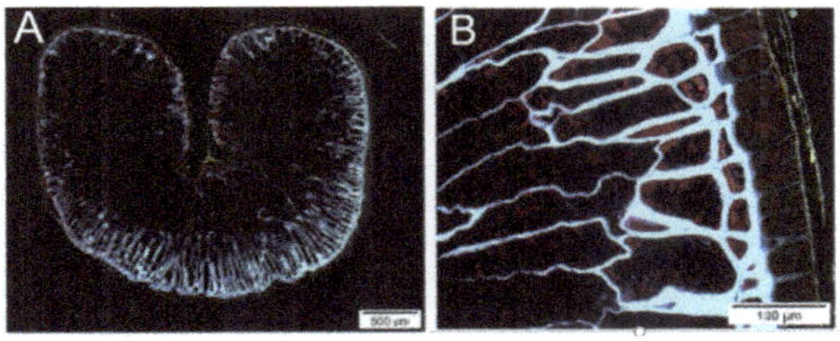

Figure 1. Localization of β-D-glucans in grains of *Avena sativa* variety SW Betania. Vertical (**A**) and longitudinal (**B**) seed sections show the presence of the polysaccharide in blue by calcofluor staining and visualized in a fluorescence microscope, adapted from [10].

The cereal β-D-glucan chain consists of glucopyranosyl monomers linked by a β-(1-4)-glycosidic bond in blocks of three or four monomers, which are called cellotriosyl and cellotetraosyl units. These units are separated by a single β-(1-3) bond, giving the chain a "staircase"-like structure [21,22] (Figure 2). Adjacent (1,3)-β-D-glucosyl residues are not present, at least not in cereal β-D-glucans [1]. Adjacent (1,4)-β-D-oligoglucosyl units located between individual (1,3)-β-D-glucosyl residues can be considered cellodextrin units, which usually consist of two or three adjacent (1,4)-β-D-glucosyl residues [1]. Most β-D-glucans in grasses have longer cellodextrin units, which consist of 5 to 20 adjacent (1,4)-β-D-oligoglucosyl units, which together make up to 10% of the polysaccharide chain [1]. Therefore, β-D-glucans in grasses can be considered (1,3)-β-linked copolymers of cellotriosyl units, cellotetraosyl units, and longer (1,4)-β-D-oligoglucosyl units in which the ratio of cellotriosyl (DP3) to cellotetraosyl (DP4) units (the ratio of β-(1-3) to β-(1-4) units) ranges from 1.5 to 4.5 depending on the source of β-D-glucans [19] with an exception in sorghum endosperm having the ratio 1.15:1 [3]. In barley, the ratio is 2.2 to 2.6:1 [1] or 1.8 to 3.5:1 [23], in wheat the ratio is 3.0 to 4.5:1, in rye it is 1.9 to 3.0:1, and in oats it is 1.5 to 2.3:1 [3,23].

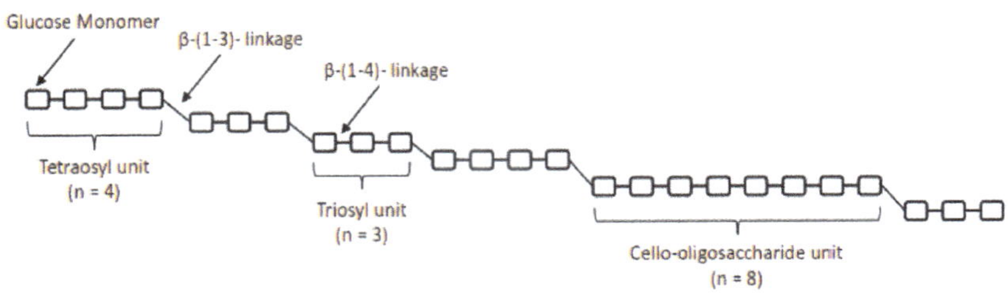

Figure 2. Basic structure of cereal β-D-glucans with β-(1-3)- and β-(1-4)- linkages and the "staircase"-like structure, adapted from [24].

The degree of polymerization (DP) of common β-D-glucans in grasses is about 1000 or more [25]. It is a unique feature of the β-D-glucans of each cereal which affects the solubility and viscosity and thus the physicochemical properties and applications of the polysaccharide in solution. For example, oat β-D-glucans with a lower molar ratio (1.5–2.3:1) are more soluble than barley and wheat β-D-glucans with a higher molar ratio (2.6:1 and 3.2:1, respectively) [26]. Differences can be observed in the same genera as genotypic variability [2]. Environmental factors such as the conditions of cultivation also affect the degree of polymerization, whereas oat varieties disposing of higher content of β-D-glucans and cultivated in drier environment show lower degree of polymerization [27].

The β-D-glucans found in cereals share the same molecular structure regardless of which source they are isolated from, but certain characteristics are specific to the source of this molecule. These characteristics are, for example, the presence and amount of long cellulose fragments, the ratio between β-(1-4) and β-(1-3) linkages, molecular size, and the ratio of cellotriosyl and cellotetraosyl units [28]. The molecular weight is approximately 31–2700 × 103 g/mol for barley, 65–3100 × 103 g/mol for oat, 21–1100 × 103 g/mol for rye, and 43–758 × 103 g/mol for wheat [23]. For sorghum, it is 36 × 103 g/mol [29]. In the case of DP3:DP4 ratio, a narrow range is, for example, observed in domestic cultivars of *Avena sativa* L. (2.05–2.11) compared to other cultivars of the genus *Avena* (1.81–2.33) [30]. The relative amount of the trisaccharide (DP3) in β-D-glucans decreases from wheat (67–72%) to barley (52–69%) and oats (53–61%), while the relative amount of tetrasaccharide (DP4) has the opposite effect trend, the growth from wheat (21–24%), through barley (25–33%) and oats (34–41%) [23]. Some structural differences between soluble and insoluble β-D-glucans show the DP3:DP4 ratio being higher for insoluble than for soluble β-D-glucans [15,21]. For example, water-soluble β-D-glucans from barley endosperm consist of about 72% of (1,4)-β-glucosyl residues and 28% of (1,3)-β-glucosyl residues [2]. However, comparing the results is difficult because the concept of insoluble β-D-glucans differs from study to study. In any case, this ratio defines the "fingerprint" of the structure of cereal β-D-glucans [26,31].

A clear and understandable result of the structural features in β-D-glucans from *Poaceae* is that polysaccharides have (1,3)-β-bonds embedded at irregular intervals along the whole β-D-glucan chain. These bonds cause irregularly distributed molecular kinks in the polysaccharide, which not only prevent the extensive intermolecular arrangement of chains into well-structured microfibrils, but also lead to the formation of polysaccharides that are able to form a gel-like matrix in cell walls and are capable of solubility in water despite its relatively high molecular weight [1,17]. Barley β-D-glucans assume an extended conformation with an axial ratio (length–width) of about 100 in aqueous media [25]. The gel-like structure allows the polysaccharide to provide some degree of structural support to the cell wall, but remains flexible, resilient, and porous enough to allow the transfer of water, nutrients, and other small molecules through the wall during plant growth and development [1].

β-D-glucans containing blocks of adjacent (1,4)-β-bonds may tend to aggregate between chains (and thus lower the solubility) through strong hydrogen bonds along the cellulose segments. On the other hand, (1,3)-β-bonds divide the regularity of the (1,4)-β-binding sequence, making the polysaccharide more soluble and flexible. The ability of a cell to change the ratio of β-cellotriosyl and β-cellotetraosyl residues provides a mechanism by which the solubility of a polysaccharide can be fine-tuned and adapted to biological requirements [32]. On the other hand, it is stated that a helix consisting of at least three cellotriosyl residues would represent a stable crystal structure in β-D-glucan molecules; it is therefore possible that a higher content of cellotriosyl fragments could cause some conformational regularity in the chain of β-D-glucans, and thus a higher degree of organization of these polymers (i.e., low solubility) [15].

The heterogeneity of the fine structure of β-D-glucans—the ratio of β-(1,4) to β-(1,3) bonds (DP3:DP4) and special distribution of bonds along the chain—obtained by chemical analysis has important implications for the physicochemical properties such as rheological behavior. The most important rheological properties of β-D-glucans include solubility in aqueous solutions and the ability to form a viscous environment [33]. Thus, moderate cellotriosyl:cellotetraosyl ratios (e.g., 1.5 to 2.5:1) would meet functional requirements on a wall such as a porous matrix, while much higher or lower ratios would characterize conformationally more regular, less soluble β-D-glucans, which would have an increased capacity for aggregation with other molecules of β-D-glucans or with cellulose and other cell wall polysaccharides, such as heteroxylans and others [2]. The solubility of β-D-glucans, an important parameter of their functional activities, is also associated with higher content of OH- groups in the structure and so high affinity to water molecules and ability to dissolve in the medium [33].

2. Content of β-D-Glucans in Grains of *Poales*

Several studies have been focused on the content of β-D-glucans in cereals such as barley and oat grains as a good natural source of this polysaccharide. Generally, barley varieties contain higher amounts of β-D-glucans compared to oat varieties; however, quality and properties of both β-D-glucans are different (Table 1). Despite this, wheat is not considered to be a good source of β-D-glucans because it has a much lower content, usually <1% on a dry basis.

Table 1. Differences in the structure and properties of the β-D-glucans among selected cereal sources.

Plant Species / β-D-Glucans	Oat (*Avena sativa* L.)	Barley (*Hordeum vulgare* L.)	Rye (*Secale cereale* L.)
Localization in the *grain*	Outer layers of the endosperm and starchy endosperm, variable among varieties [8,10,34]	Uniformly throughout the endosperm [7]	-
Level of β-D-glucans (% dwb)	2.3–8.5 [4,35–37]	2.68–8 [38–41]	1.2–1.6 [6]
Fiber structure		Extended conformation with an axial ratio of about 100 in aqueous media [25]	
DP3:DP4	1.5–2.3:1 [3,23]	1.8–3.5:1 [1,23]	1.9–30:1 [23,33]
Relative amount of DP3 in the molecule (%)	53–61 [22]	52–69 [22]	-
Solubility	Higher [26]	Lower [26]	-
Molecular size (g/mol)	65–3100 × 10^3 [22]	31–2700 × 10^3 [22]	21–1100 × 10^3 [22]
Heritability	0.55 [42]	0.75–0.84 [43]	
Breeding trends	Increase in β-D-glucan content for human nutrition	Decrease in β-D-glucan content in malting barleys [1] and barleys for poultry fattening [44]. Increase in β-D-glucan content for human nutrition [45,46]	Increase in β-D-glucan content for human nutrition
Application	Human nutrition, functional ingredient, functional foods, livestock feed	Human nutrition, livestock feed, malting, brewing [38,41,47]	Human nutrition, functional foods, functional ingredient

Barley is an important cereal grain consumed throughout the world that can be used to develop functional food products rich in β-D-glucans [38,47]. The content of β-D-glucans in barley is on average 3–4% to 8% [38,39], although barley cultivars with the content of 2–11% are observed, and the polysaccharide is in the cell wall distributed throughout the endosperm [8].

Current research of barley β-D-glucans attempts to identify suitable barley genotypes for use in a number of breeding applications such as human nutrition, livestock feed, malting, and brewing [41]. Both varietal variability in the β-D-glucans content and their degradation during germination of barley play a significant role in their application for malting and brewing. Residual malt β-D-glucan from incomplete degradation of endosperm cell walls during the malting process is associated with increased worth viscosity that can slow filtration and reduce brew house efficiency [44].

In addition to the genetic resource pools of cultivated barley, wild barley (*Hordeum spontaneum* L.) offers considerable potential as a genetic resource for barley β-D-glucans improvement. A comparative study of the β-D-glucan content between cultivated and wild barley confirmed the higher range and variability of this parameter in wild species. The β-D-glucan contents of the studied wild barley accessions ranged from 3.26% to 7.67% while the content of cultivated barley varieties ranged from 2.68% to 4.74% [41].

Another recent large-scale analysis of 117 accessions of wild barley (*Hordeum vulgare subsp. spontaneum* L.) which were selected from ICARDA's gene bank to represent 21 countries scattered along the natural geographic distribution of the species were carried out by Elouadi et al. [48]. The contents of β-D-glucans ranged from 1.44% to 11.30% in the *Hordeum spontaneum* accessions compared to 36 cultivated barley lines with contents ranging from 1.62% to 7.81%. On the other hand, a similar range (3.6–7.4%) of β-D-glucan contents was already detected by Austrian researchers in 86 hull-less forms of cultivated barely [40].

Generally, naked barley has higher starch and β-D-glucan levels than hulled barley. Additionally, mutations at the Lys3 and Lys5 loci can affect β-D-glucan content as well as other health-promoting compounds [49]. While varieties with high β-D-glucans content are preferred in human nutrition, varieties with low concentration are preferred for malting and for broiler fattening. The reliable production of low-β-D-glucan malt based on a suitable barley genotype could replace viscosity control by changing the brewing processes or adding exogenous enzymes. Therefore, novel barley β-D-glucan endohydrolase (β-glucanase) alleles with increased thermostability, e.g., from *Hordeum spontaneum* would be perspective to identify [44].

Oat is another major source of β-D-glucans among cereals. The oat usually contains 3 to 5% of this viscous and soluble fiber component [50,51]. Unlike barley grain, the main part of β-D-glucans is located in the thick cell walls in the region of the subaleural outer endosperm [8,10,34]. Therefore, core fractions consisting of subaleural layers are particularly high in this polysaccharide. The content of β-D-glucans in oat grain ranges from 2.3% to 8.5% [4,35] depending on the cultivar and other factors [4,36]. The content of β-D-glucans in diploid oat ranges from 2.85% to 6.77%, in tetraploid it ranges from 3.58% to 5.12%, and in hexaploid oat species the range is 2.88–5.90% [37].

A significantly lower variability in β-D-glucan contents was confirmed among four oat species *A. sativa* (3.60%), *A. byzantina* (3.40%), *A. abyssinica* (2.46%), and *A. strigosa* (2.97%) by the recent study of VIR oat collection by Popov et al. [52]. These results indicated that the hexaploid cultivated oat species *A. sativa* and *A. byzantina* showed a generally higher content of β-D-glucans than tetraploid and diploid wild accessions. At the same time, Loskutov and Polonkiy [53] reported a slightly higher β-D-glucan content in naked forms of *A. sativa* compared to hulled oat varieties, which was also confirmed by Havrlentova and Kraic [4]. Thus, it seems that compared to barley species, the variability of β-D-glucan content in oats is not as high and probably not sufficiently mapped.

In addition, the requirements for the specific β-D-glucan content of oat grain are not as clearly defined as in the case of malting barley processing or poultry fattening. A recent study [54] even reported that 1,3-β-D-glucan can be added to broiler feed to improve the development and integrity of the gut and enhance the immune status of birds without affecting their growth rate. However, barley β-D-glucans do not seem to have this potential and their negative effect on poultry performance is being further studied [55].

3. Biosynthesis of β-D-Glucans in *Poales*

The starting point for molecular genetic approaches to the study of cell walls of *Poales* was the sequencing of the rice genome and the subsequent identification of the superfamily of cellulose synthase genes (Cellullose Synthase A, *CesA*), while these genes are responsible for the synthesis of the hexose polysaccharide framework and cellulose itself. Cellulose Synthase-Like genes (*Csl*) were gradually discovered [56]. This family is further divided into eight subclasses: *CslA*, *CslB*, *CslC*, *CslD*, *CslE*, *CslF*, *CslG*, and *CslH*, each containing multiple genes [57]. Later, the *Csl* superfamily was supplemented by the *CslJ* family [58]. In general, *CesA* genes encode cellulose-synthesizing enzymes and *Csl* genes are responsible for the biosynthesis of hemicellulosic polysaccharides [56]. The *CslC* gene group encodes an enzyme that controls the synthesis of the xyloglucan backbone [59], while the *CslF* and *CslH* subfamily genes mediate the synthesis of β-D-glucans in *Poaceae* [60,61], with the *CslF* subfamily being key [60]. The *ClsJ* subfamily is also involved in the biosynthesis of β-D-glucans [1,58], but its phylogenetic significance is still unclear [62]. Not all *Csl* subfamilies are represented in higher plants. The *CslB* and *CslG* subfamilies are only found in dicots and gymnosperms, and *CslF* and *CslH* are only found in monocots [58], with the *CslF* branch diverging from the *CslD* subfamily, which evolved by diversification of *CslD* and *CesA* [63]. Putative functions of the candidate genes involved in the metabolism of β-D-glucans and responsible for co-regulated metabolite network of such polysaccharides in cereals such as starch and β-D-glucans are presented in Figure 3.

The biosynthesis of β-D-glucans, a unique polysaccharide of grass cell walls, is encoded by three groups of *Csl* genes [64], namely *CslF, CslH* and *CslJ* [63], while research has shown that selection pressure on *CslF* subfamily genes is related to the ratio DP3:DP4 in the molecule of polysaccharide [65]. The involvement of the *CslH* subfamily was demonstrated by genetic transformation of *Arabidopsis thaliana*, in which *CslH* genes from rice were inserted and the plant produced β-D-glucans. Similar research was also carried out with the *CslF* subfamily [66]. Likewise, research suggests that the *CslJ* subfamily encodes the synthesis of β-D-glucans in barley, and this subfamily has been observed in barley, wheat, sorghum, and maize, but not in rice [58]. All genes from the *CslF, CslH*, and *CslJ* subfamilies mediate β-D-glucan synthesis in heterologous expression systems [60,61,67], although it is unclear whether each the gene in these groups controls the synthesis of β-D-glucans in vivo [32]. Phylogenetically, the *CslJ* and *CslH* subfamilies existed before the monocot–dicot split and are now present only in monocots [63].

In barley tissues, the *CslF6* gene is the most transcribed in terms of β-D-glucan content [68–70], while it fulfils the same function in wheat, oats, and rice [68,71]. Knocking out the *CslF6* gene from its function in rice plants caused not only a significant decrease in the content of β-D-glucans in the coleoptile, but also an increased activation of mechanisms related to plant protection against bacterial infections [71].

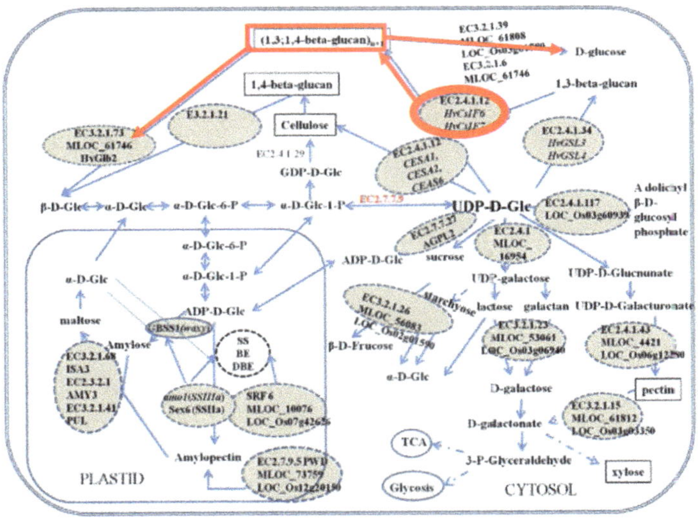

Figure 3. The candidate genes and their encoding protein are dash circled. The cell wall components are marked by black box. The dash arrows indicate there was not a single step. EC3.2.1.73:(1,3; 1,4)-β-glucanase; EC3.2.1.21: β-1,4-glucosidase; EC3.2.1.39: glucan-1,3-β-glucosidase; EC3.2.1.6: glucan-1,3-β-glucosidase; EC2.4.1.12: Cellulose synthase; EC2.4.1.34: glucan synthase; EC2.4.1.117: dilichyl-phosphate-β-glycosyltransferase; EC2.4.1: galactosyl transferase; EC2.7.7.9: UTP–glucose-1-phosphate uridylyltransferase; EC2.7.7.27: ADP-glucose pyrophosphorylase; EC3.2.1.23: β-galactosidase; EC23.2.1.26: β-fructofuranosidase; EC2.4.1.43: glucosyl transferase; EC3.2.1.15: polygalacturonase; EC2.7.9.5: phosphoglucan water dikinase; EC3.2.1.68: Isoamylase; EC3.2.2.1: amylase; EC3.2.1.41: amylopullulanase. The main points of the biosynthetic pathway of β-D-glucans are shown in red [72].

The expression levels of *CslF* subfamily genes intensively studied in different barley tissues and at different developmental stages of the grain showed significant differences in expression between samples. The *HvCslF6* gene is expressed in almost all tissues and during the entire grain development [73], but, e.g., the *HvCslF3* gene is predominantly expressed in the coleoptile and root hairs [73,74], *HvCslF9* is expressed in the coleoptile, roots, and developing seed 8–10 days after pollination, and *HvCslF7* is expressed in the stem and stalk. Based on these results, it can be concluded that the expression of *CslF* genes is regulated by tissue-specific factors. The strict regulation of *CslF* gene expression is particularly visible during seed development, where at least four genes are expressed at very specific stages. These conclusions suggest a multistep process of β-D-glucan biosynthesis, with genes early in the biosynthetic pathway expressed first, followed by other members [73].

Most of the research on the genes coding the synthesis of β-D-glucans used barley and rice as plant material. For oats, five genes from the *CslF* family are available in the GenBank database: *AsCslF3*, *AsCslF4*, *AsCslF6*, *AsCslF8*, and *AsCslF9*, with *AsCslF6*, *AsCslF8*, and *AsCslF9* showing higher similarity to barley, and *AsCslF3* and *AsCslF4* showing less conservatism. Expression analysis of genes involved in the biosynthesis of β-D-glucans in grain, leaves, and panicles of oat revealed a high level of expression of several genes in panicles, while neither the *AsCslF4* nor the *AsCslF9* gene was expressed in leaves. All genes of the *CslF* and *CslH* subfamilies were expressed during grain development. Further results revealed that all genes showed increased expression at later stages of grain development, when β-D-glucans are predicted to accumulate more prominently as grain storage polysaccharides [75]. Comparisons with published results obtained from analysis of barley showed that barley, in contrast to oats, shows increased expression of *CslF6* on the first to fourth [73] or sixth day after pollination [75] with a significant decrease on the eigth day after pollination [73], while the expression of *CslF4* and *CslF8* remains low [69].

The allohexaploid (2 n = 6 x = 42; AA, CC, and DD are subgenomes) character of oat with a large complex genome is the cause of more complex and complicated molecular–genetic experiments related to the biosynthesis of β-D-glucans [9]. In the work, Coon [75] focused on the variability of expression of *CslF6* gene homologues in species of the genus *Avena*. The genomic sequence of *CslF6* was approximately 5.2 kb, and due to the large differences between individual homologues in the intron regions, the length of the genes may vary. The A-genome ortholog was 5268 bp, the C-genome was 5162 bp, and the D-genome sequence was 5162 bp in length. The *CslF6* gene contained two introns, the first with a length of 1600 bp and the second with a size of 748 bp. The *CslF6* splice site was determined by comparison of *A. sativa* genomic sequences with *CslF6* coding sequences from *H. vulgare*. The coding sequences were highly conserved among the homologs for the analyzed samples of roots, young plants, and mature embryos at the stages of 1–3, 4–6, and 7–9 days after pollination. *CslF6* expression was significantly lower in mature embryos than at any other developmental stage of the embryo. An exception was the species *A. strigosa*, which had intermediate expression of *CslF6* in mature embryos [75].

Regarding the localization of β-D-glucan synthesis, the current dogma for cell wall polysaccharide biosynthesis is that cellulose (and callose) is synthesized at the plasma membrane, whereas matrix-phase polysaccharides are assembled in the Golgi apparatus [7]. The novel model indicates that β-D-glucan does not conform to this paradigm, and in various *Poaceae* species, the CslF6-specific antibody labelling is present in the endoplasmic reticulum, Golgi, secretory vesicles, and the plasmatic membrane and the CslH1 to the same locations apart from the membrane [76]. The updated model of the β-D-glucan synthesis shows that CslF6 is the major synthase and is proposed to act similarly to CesAs, synthesizing glucan chains *de novo* (scenario 1). A less likely but possible alternative is that *CslF6* produces cello-dextrins that are joined together at the plasma membrane by yet an unidentified protein via a single β-(1,3)-glycosidic linkage, creating the β-D-glucan chains that are then recognized by the β-D-glucan-specific antibody (scenario 2) [76] (Figure 4).

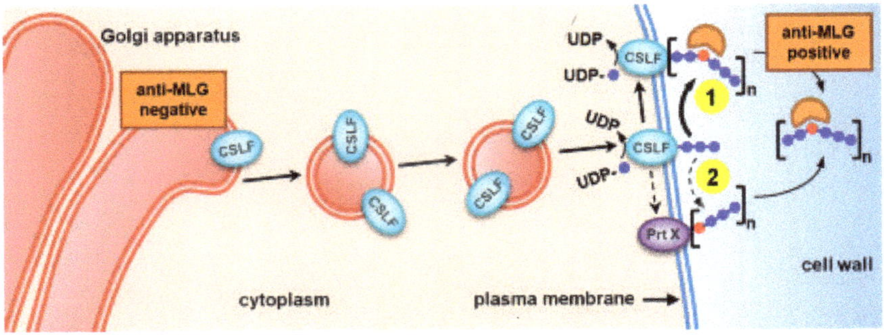

Figure 4. Novel model of the possible synthesis of β-D-glucans in cereals. CSLF6 is the major MLG synthase that is proposed to act similarly to CesAs, synthesizing *de novo* the MLG chains that are recognized by MLG-specific antibody at the cell surface (scenario 1). A less likely but possible alternative is that *CSLF6* produces cello-dextrins (shown here in blue) that are joined together at the plasma membrane by yet an unidentified protein (Prt X) via a single b-(1,3)-glycosidic linkage (shown in red), creating the MLG chains that are then recognized by the MLG-specific antibody (scenario 2) [76].

4. Environmental Factors Influencing the β-D-Glucans Content

Differences in β-D-glucan content among species such as oats, barley, wheat, and sorghum are influenced by genetic and environmental factors [77]. In addition, a mathematical model was developed that is able to pertinently assess the importance of several factors such as genetics, agronomic, and environmental factors to the content of β-D-glucans during cultivation of cereals [36]. Data collected from five locations for two years (2008, 2009) analyzing the content of β-D-glucans in 11 oat genotypes identified locality, genotype, and environmental interaction to be the factors influencing the content of this polysaccharide [37]. As summarized by Redaelli et al. [78], water is one of the most important environmental factors affecting β-D-glucans variability. However, its effect is not always unambiguous. On the one hand, Peterson et al. [79] reported an increase in the content of this cell wall polysaccharide when precipitation was insufficient. Doehlert et al. [80] also mentioned a positive correlation between β-D-glucan content and rainfall in July and August. Furthermore, results can be found in which exposure to heat stress and water deficit before harvest increases the amount of the polysaccharide in grains [47,81]. Another study further confirmed that irrigation during oat growth is responsible for the degradation of β-D-glucan content in the grain [79].

Thus, the timing of rainfall and the level of stress are likely to play a role. It can be expected that drought-stressed, e.g., "shriveled" grains show a significantly narrower ratio between starch content and other components including β-D-glucans. Excessive rainfall during the ripening stage can potentially trigger covert sprouting of grains, which could result in a considerable reduction in β-D-glucan content. It is widely recognized in barley and oat malt production [82] that germination leads to a significant breakdown of soluble dietary fiber, particularly β-D-glucans. However, there are no explicit studies demonstrating the effect of pre-harvest germination in spikes on changes in β-D-glucan content in oats or barley.

Nitrogen fertilization has been proved to be a factor influencing the level of biochemical compounds in the grain; it leads to an increase in the content of β-D-glucans [83]. Higher levels of nitrogen in the soil and the use of nitrogen fertilizers greatly increase the total content of β-D-glucans in oat and barley grains [47]. On the other hand, refs. [83,84] also confirm the influence of environment on β-D-glucan variability, but in these results, the effect of genotype dominates over that of environmental factors. This was also confirmed by the work of Dvořáček et al. [85], which compared 11 selected yield and nutritional parameters in five oat varieties. The results showed the significantly highest influence of genotype on the variability of β-D-glucans (over 40%), even in comparison with all other parameters evaluated (e.g., content of crude protein, fat, starch, thousand-grain weight, yield). Furthermore, only year was significant for β-D-glucan variability with an effect of about 21%. On the other hand, the influence of locality and conventional or organic farming management was negligible and not significant (Figure 5).

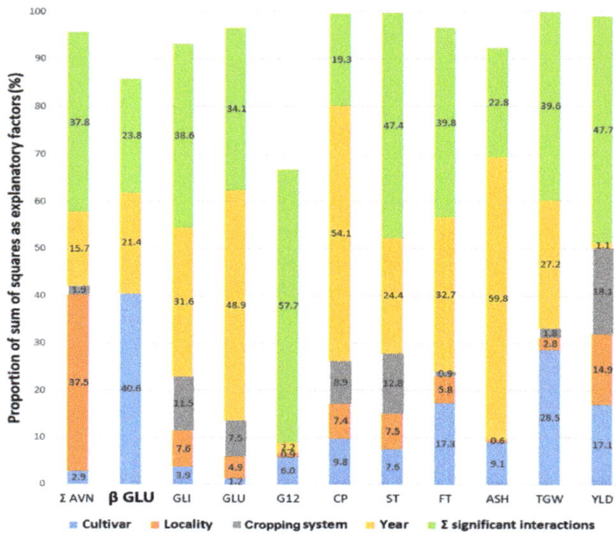

Figure 5. The effect of the main significant factors and interactions influencing the variability of oat grain β-D-glucans in comparison with other significant oat grain parameters. Calculations were performed based on the factorial proportion of the sum of squares according to the analysis of variance (ANOVA) model for selected grain parameters of five oat varieties grown at two locations and under different cropping systems (organic vs. conventional) for 3 years. Total content of avenanthramides (Σ AVN), β-D-glucans (β GLU), gliadins (GLI) and glutelin (GLU) protein fractions, immunoreactive avenin peptides (G12), crude protein (CP), starch (ST), fat (FT), ash (ASH), thousand-grain weight (TGW), and yield (YLD) was evaluated [85].

It can be further assumed that the major influence of environmental conditions on β-D-glucan level is related to the influence of the processes of synthesis transport and deposition of β-D-glucans during ontogenesis. Tiwari and Cummins [47] reported the highest levels of β-D-glucans after anthesis. During the first 15 days after flowering, 70–90% of the total content of β-D-glucans was accumulated at various nitrogen levels [14].

Comparison of the content of β-D-glucans in the panicle of hulled and naked oat samples during plant ontogenesis was also described by Hozlár et al. [13]. The accumulation of β-D-glucans in panicles during oat ontogenesis showed an increasing trend, while in all oat varieties, the content increased from 0.69% and 0.34% to 2.23% and 2.22% in both hulled and naked oat, respectively. β-D-glucan content in developing barley grains was monitored by [86]. The authors confirmed the individual time trend of β-D-glucan accumulation during grain maturation in seven tested barley cultivars.

It can thus be deduced that the variability in β-D-glucan content is strongly dependent on the genotype but also on the timing of water availability, which influences the rate of synthesis and deposition of the main grain storage substances (polysaccharides and proteins) and thus their relative proportions in the grain.

5. Function of β-D-Glucans in the Plant Organism

Interest in cereal β-D-glucans has increased after their acceptance as bioactive and functional ingredients in a healthy diet [87,88] in humans and animals. Nevertheless, β-D-glucans play an important role in the structure and functionality of cereal cell walls [26]. The architectonical function in the cell wall and the storage function in the plant seeds as a source of energy for developing sprouts has been suggested [66]; in addition, the protective role of this cell wall polysaccharide against biotic or abiotic stresses is discussed in the literature [89].

Initially, it was thought that β-D-glucans serve to store energy in elongated plant cells and seeds, because the content of this glucose cell wall polysaccharide is increased in young tissues [2,17,19]. This functionality of β-D-glucans is based on the fact that the breakdown of β-D-glucans into glucose is relatively simple and involves only two enzymatic steps that allow rapid mobilization of glucose reserves compared to a longer mobilization process of starch reserves [17]. In addition, the localization of this polysaccharide near starchy endosperm is an advantage for this molecule to serve as a source of energy for developing seedlings. It has also been shown that β-D-glucans can be metabolized as an energy source in vegetative tissues during periods of glucose deficiency. This seems to be confirmed by the fact that barley sprouts, when moved from light to dark, show increased expression of β-glucan endohydrolases and glucolases [26,90], suggesting that the plant mobilizes glucose reserves stored in β-D-glucans to compensate the decrease in photosynthetic activity in the dark [91]. In this case, plant tissues, which are not traditionally associated with photosynthetic activity, use β-D-glucans as a storage vector for glucose other than starch (or instead of the starch polysaccharide). An example is *Brachypodium distachon*, the grains of which contain up to 45% of β-D-glucans and only 6% of starch [92] compared to cereals and most wild grasses, which have 30–70% of starch as the main storage carbohydrate in the grain and generally less than 6% of β-D-glucans [93]. However, growing plants to produce seeds is very difficult.

The architectonical role of β-D-glucans stems from the localization of β-D-glucan microfibrils in the cell wall and from the physicochemical properties of this molecule. In type I cell walls found in dicots and some monocots, the cellulose is encased in a gel-like layer of pectin to provide elasticity and stability to the cell wall. Type II cell walls are found in *Poales* species and contain much less pectin, its function being taken over by β-D-glucans together with arabinoxylans [94], which are highly accumulated during cell elongation, when they can make up to 20% of dry weight of the cell wall [95].The gel-like structure allows the polysaccharide to provide structural support for the cell wall, but at the same time remain flexible, elastic, and porous enough for the transport of water, nutrients and other small molecules across the cell wall during plant tissue development [1]. In young developing plant tissues, it is important that cell walls are porous so that water molecules, nutrients and low-molecular-weight hormones can be freely transported between cells and individual tissues. In the specialized conductive tissues necessary for the long-distance transport of water molecules and nutrients through the plant, the walls, on the contrary, must be impermeable.

In most plant tissues, cell walls are also important for intercellular adhesion. These different functional requirements for cell walls are maintained by the formation of reinforced gel-like structures consisting of cellulose microfibrils that have high tensile strength and are embedded in a gel matrix phase that consists predominantly of non-cellulosic polysaccharides [26]. This matrix provides the cell with flexibility and a certain mechanical support for maintaining the functional properties of the cell wall.

In *Gramineae*, the metabolism of β-D-glucans is responsible for plant responses to environmental signals within a moderate, physiological range [96]. For example, in the work of Taketa et al. [69], barley variety producing the standard content of β-D-glucans (3.8%) was characterized by better wintering compared to a mutant with knocked enzyme responsible for producing this polysaccharide. Recent studies also point to the potential of β-D-glucans to be involved in defense mechanisms in the *Poaceae* family against selected forms of environmental stresses [89,97–99]. The gel-like layer of β-D-glucans in the cell wall can act as a defensive barrier that protects the cell from fungal invasion, but also provides a potential signaling system that indicates when such an attack is taking place. As fungi attack plant cells and release β-glucanases to digest the protective gel layer surrounding the cell, they slowly dissolve the protective layer of β-D-glucans and expose the cytoplasmic membrane itself. The antioxidant activity of β-D-glucans was also described [100], whereby antioxidant compounds are an accepted factor of the plant's defense system to cope with biotic aggressors such as fungal pathogens [101,102]. The increase or decrease in the biosynthesis and the associated content of β-D-glucans is conditioned by the expression of plant genes involved in both β-D-glucan synthesis and its degradation. After the exposure of plants to external factors such as pests or pathogens, the β-D-glucan metabolism as well as the entire β-D-glucan turnover are affected and become much more complicated. A study in rice mutants knocking out rice genes involved in the biosynthesis of β-D-glucans showed a phenotype that had a spontaneous response to the lesion, presumably suggesting that fibers of β-D-glucans work as a repressor of the signaling cascade targeting programmed cell death [71]. In addition, reduced deoxynivalenol toxin (DON) content in barley genotypes infected with *Fusarium graminearum* was observed in grains with higher content of β-D-glucans [98] and in oat grains artificially infected with *F. graminearum* and *F. culmorum*, where grains containing higher amounts of β-D-glucans showed lover content of DON and pathogenic DNA [89].

The function of β-D-glucans in *Poaceae* as well as the evolutionary preservation of this polysaccharide in monocotyledons are still discussed, as this phenomenon of β-D-glucans has not been precisely explained [1,2,17]. The conformational regularity or irregularity of β-D-glucans defines the properties and thus the physicochemical activities in the matrix in the cell wall. It is this conformational irregularity of β-D-glucans that appears to be a feature limited to *Poaceae* polysaccharides and raises the question whether this irregularity is a key feature that has led to widespread acceptance and preservation of β-D-glucans in the walls of *Poaceae* during evolution [63,67,70]. During the adaptation from aquatic to soil environment, plants had to develop a mechanism for their cells to withstand substantial expansion pressures to be able to prevent the rupture of the cytoplasmic membrane and at the same time contain a structure compatible with the growth of cells and cell tissues; the construction of the cell wall itself of photosynthetically active embryophytes is still crucial [103]. The plant cell wall is designed so that its architecture is strong, but at the same time flexible and able to withstand pressure, tension, and various adverse environmental conditions [104]. There is an assumption that the unique structure of β-D-glucans and the extensive presence of this polysaccharide in species of the genus *Poales* brings enormous evolutionary and adaptive advantages to the plant [26] and is the reason that this group of plants shows extraordinary dynamics of various pressures during evolution [62,70] and inhabits often extreme and inhospitable habitats.

6. Molecular Markers as Breeding Tools for β-D-Glucans Manipulations

In the case of β-D-glucans, the effectiveness of the standard breeding process is related to the identification of suitable genetic resources, knowledge of the responsible genes, the availability of genetic markers, and a good estimate of the heritability of the trait. Efficient phenotyping tools for screening estimation of the β-D-glucan parameters in early generations are also an advantage.

There are many studies demonstrating the moderate to high heritability of β-D-glucans found in oats and barley. The average heritability level (h2b) of β-D-glucans was estimated to be 0.55 in individual oat plants [105]. An even higher range of (h2b = 0.75–0.84) in the case of β-D-glucans was found in barley [106]. Holthaus et al. [105] also mentioned that groat β-D-glucan content in oat polygenically is controlled primarily by genes with additive effects. Genetic variation for β-D-glucans seemed adequate for effective selection, and genotype–environment interaction was minor. Swanston [107] reported that β-D-glucan content is a quantitative trait with several associated QTLs; one QTL is located on chromosome 7H and is within 5 cM of the *Nud* gene in the case of barley.

Using current molecular technologies, putative genes responsible for β-D-glucan variation in the grain are continuously studied based on the still limited information on β-D-glucan synthesis. However, gene families involved in the synthesis of these polysaccharides have been identified and include the Cellulose-synthase-like (*Csl*) genes [42], which were described in detail in the previous chapter on β-D-glucan biosynthesis. Members of the *CslF* and *CslH* gene families and Glucan synthase-like (or Callose synthases) are prime candidates for β-D-glucan synthesis. Nevertheless, the *Csl* gene families do not completely explain the variation in β-D-glucan content in cereal grains. There is accumulating evidence that several other genes, including those that hydrolyze β-D-glucans, contribute significantly to their content in mature grains [43].

New studies of genome-wide association study (GWAS) associating genetic mutations with measures of β-D-glucan content in cultivated barley have become a powerful tool in identifying candidate genes for selective breeding. Seven putative candidate genes encoding some enzymes in glucose metabolism were found to be associated with β-D-glucan content. One of the putative genes, *HORVU6Hr1G088380*, could be an important gene controlling barley β-D-glucan content [108]. Marcotuli et al. [42] identified seven genomic regions associated with β-D-glucans in a tetraploid wheat collection, located on chromosomes 1A, 2A (two), 2B, 5B, and 7A (two). An analysis of marker trait associations (MTAs) in syntenic regions of several grass species revealed putative candidate genes that might influence β-D-glucan levels in the endosperm, possibly via their participation in carbon partitioning. Walling et al. [43] applied GWAS to the Wild Barley Diversity Collection (*H. spontaneum*) and identified a total of 13 quantitative traits loci (QTL) spread across the seven barley chromosomes that explained most of the variation in β-D-glucan content. Transcriptional dynamics of two barley genotypes differing in grain β-D-glucan content during grain development was investigated by the authors of [109]. Twenty-two differentially expressed genes (DEGs) affecting β-D-glucan accumulation during late developmental stages were selected. Most of these DEGs (encoding alpha-amylase inhibitor, glucan endo-1,3-beta-glucosidase, and sugar transporter) showed different expression patterns in the two genotypes, which might explain the genotypic difference in changes in β-D-glucan content 21–28 days post-anthesis (DPA).

The hexaploid structure of the oat genome further complicates the detection of β-D-glucan-associated markers. However, Newell et al. [110] associated three DArT markers with β-D-glucan concentration in oats. These markers had sequence homology to rice; one of these DArT sequences, opt.0133, was located on rice chromosome seven and was, by our definition, adjacent to the *CslF* gene family. A recent large GWAS analysis carried out on an oat panel with 413 genotypes was evaluated for β-D-glucan content under subtropical conditions [111]. Seven quantitative trait loci (QTL) associated with β-D-glucan content were identified and located on Mrg02, Mrg06, Mrg11, Mrg12, Mrg19, and Mrg20. The QTL located on Mrg02, Mrg06, and Mrg11 seem to be genomic regions syntenic with barley.

Genome editing methods for influencing β-D-glucan content in barley were used by Garcia-Gimenez et al. [74]. The authors used CRISPR/Cas9 to generate mutations in members of the *Csl* gene superfamily that encode known (*HvCslF6* and *HvCslH1*) and putative (*HvCslF3* and *HvCslF9*) β-D-glucan synthases. Grains from *CslF6-2* (homozygous) mutants almost completely lack β-D-glucans (0.11%), whereas grains from *CslF6-2*/+ (heterozygous lines) have intermediate levels of β-D-glucans (1.45%) compared with the wild type control (5.00%). Their data further indicated that only *HvCslF6* from multiple members of the *CslF/H* family showed impact on the abundance of β-D-glucans in mature grain. Genome editing procedures in case of decrease in β-D-glucan content overcome traditional breeding practices and could be a good breeding strategy for the new malting varieties of barley.

Molecular methods in combination with effective phenotyping procedures for β-D-glucan content (e.g., modified McCleary enzymatic determination or NIRS) are key for the development of new varieties with defined content of the monitored polysaccharide in the breeding process [112,113]. NIRS models developed by Paudel et al. [113] declared high index of determination (h2 $\geq$ 0.93) and low standard error of cross-validation (SECV $\leq$ 0.23) for β-D-glucan quantification in ground and whole oat groats as well. The above findings thus hold great promise for obtaining new varieties of oats and barley with declared β-D-glucan content according to the requirements of processors.

7. The Role of β-D-Glucans in Brewing Processes

As indicated in the previous sections, barley β-D-glucans play a crucial role in beer production, and monitoring their levels is essential for optimizing the malting process and ensuring high-quality beer. Lower levels of β-D-glucan content in grains and higher levels of β-glucanase in malted barley are associated with better malting performance. If barley has a high starting β-D-glucan content, it can hinder the degradation of cell walls and the diffusion of enzymes during kernel mobilization, leading to disruptions in malt quality parameters [114].

Higher levels of β-D-glucans can contribute towards issues such as haze formation, viscous wort, and reduced wort filtration during the brewing processes. Malt extract is a complex quantitative trait that is controlled by multiple genes and its concentrations can be variable in different cultivars. It is also considered as a mega-trait which is the product of interactions between many sub-traits [115,116].

Breeders face challenges in genetically manipulating and selecting malting quality traits such as ME and β-D-glucan content due to their complex inheritance patterns. Numerous studies have identified over 250 QTLs associated with malting quality, including malt extract and β-D-glucan content. These QTLs have been found and mapped on different chromosomes of barley. Among these QTLs, QTL2 on chromosome 4H stands out as a major contributor to barley malting quality, accounting for 29% and 38% of the variation in key parameters such as β-D-glucan content and malt extract, respectively [117].

In a separate study, the telomeric region of chromosome 4H, which contains the complex associated with malting quality, underwent fine mapping. This investigation revealed the presence of 15 potential QTLs related to key malting quality parameters such as the content of β-D-glucans, ME, alpha-amylase, and diastatic power (DP) [118]. Numerous endeavors have been undertaken to develop molecular markers for the purpose of selecting barley varieties with improved malting traits, including malt extract and β-D-glucan content. For example, a study involved genotyping of approximately 1524 single-nucleotide polymorphisms (SNPs) to identify several genes associated with six malting traits, including β-D-glucan content and malt extract [119]. In recent research, QTL2, a locus that exhibits a substantial proportion of variation in malt extract and content of β-D-glucans, was selected. This locus contains a key gene, *HvTLP8*, which appears to play a role in the interaction with β-D-glucans in a redox-dependent manner [120].

The levels of β-D-glucans in the wort are influenced by the malting process itself, specifically through the action of the β-glucanase enzyme. The enzyme is predominantly synthesized in the aleurone and scutellum of germinated grains. As the malting process progresses, there is a significant decrease in β-D-glucan content due to the increased activity of this enzyme. This reduction in the concentration of the polysaccharide relies on the initial lower levels of the polysaccharide in the grain and higher quantities of the β-glucanase enzyme in the resulting malt. The activity of β-glucanase during malting leads to hydrolysis of cell walls, converting them into soluble β-dextrin with a low molecular weight. It is important to note that β-glucanase is sensitive to temperature; it is rapidly inactivated at temperatures above 50 °C during the extraction phase. However, the breakdown of β-D-glucans from intact cell walls continues, resulting in the accumulation of soluble β-D-glucan molecules in the extract. Therefore, the final characteristics of the produced malts are also influenced by factors such as temperature, humidity, and the duration of germination [121].

8. Conclusions

The review emphasizes the unique characteristics of β-D-glucans, highlighting their composition of β-D-glucopyranosyl monomers connected by β-(1,3) and β-(1,4) bonds, resulting in a distinctive "staircase" structure. It further underscores that β-D-glucans are predominantly located in cell walls in selected cereal grains such as barley (*Hordeum vulgare* L.) and oats (*Avena* ssp. L.), playing a significant role in the structural integrity, functionality, and defense mechanisms of cell walls. Moreover, the review delves into the genetic and environmental factors influencing β-D-glucan content in cereals, with particular emphasis on genotype, water availability, and nitrogen fertilization. It highlights the contemporary relevance of molecular markers and genome-wide association studies in facilitating precise breeding and genome editing approaches for manipulating β-D-glucan levels, offering promising avenues for enhancing crop traits.

Author Contributions: Conceptualization, M.H. and V.D ; methodology, M.H.; validation, M.H. and V.G.; investigation, M.H., V.D., L.J. and V.G.; writing—original draft preparation, M.H., V.D. and L.J.; writing—review and editing, M.H., V.D. and V.G.; visualization, V.G.; supervision, M.H.; project administration, M.H. and V.D.; funding acquisition, M.H. and V.D. All authors have read and agreed to the published version of the manuscript.

Funding: This research was funded by the Slovak Research and Development Agency under the Contract no. APVV-18-0154 and by the Ministry of Agriculture of the Czech Republic, institutional support MZE-RO0423.

Institutional Review Board Statement: Not applicable.

Informed Consent Statement: Not applicable.

Data Availability Statement: The data presented in this study are available on request from the corresponding author.

Conflicts of Interest: The authors declare no conflict of interest.

References

1. Fincher, G.B. Revolutionary times in our understanding of cell wall biosynthesis and remodeling in the grasses. *Plant Physiol.* **2009**, *149*, 27–37. [CrossRef] [PubMed]
2. Burton, R.A.; Gidley, M.J.; Fincher, G.B. Heterogeneity in the chemistry, structure and function of plant cell walls. *Nat. Chem. Biol.* **2010**, *6*, 724–732. [CrossRef]
3. Fincher, G.B.; Stone, B.A. Chemistry of nonstarch polysaccharides. In *Encyclopedia of Grain Science*; Wrigley, C., Ed.; Elsevier: Oxford, UK, 2004; pp. 206–223. [CrossRef]
4. Havrlentova, M.; Kraic, J. Content of β-D-glucan in cereal grains. *J. Food Nutr. Res.* **2006**, *45*, 97–103.
5. Redaelli, R.; Sgrulletta, D.; Scalfati, G.; De Stefanis, E.; Cacciatori, P. Naked oats for improving human nutrition: Genetic and agronomic variability of grain bioactive components. *Crop Sci.* **2009**, *49*, 1431–1437. [CrossRef]
6. Saastamoinen, M.; Plaami, S.; Kumpulainen, J. Pentosan and β-Glucan content of Finnish winter rye varieties as compared with rye of six other countries. *J. Cereal Sci.* **1989**, *10*, 199–207. [CrossRef]
7. Buckeridge, M.S.; Rayon, C.; Urbanowicz, B.; Tiné, M.A.S.; Carpita, N.C. Mixed linkage (1→3),(1→4)-β-D-glucans of grasses. *Cereal Chem.* **2004**, *81*, 115–127. [CrossRef]
8. Sykut-Domańska, E.; Rzedzicki, Z.; Zarzycki, P.; Sobota, A.; Błaszczak, W. Distribution of (1,3)(1,4)-beta-D-glucans in grains of polish oat cultivars and lines (*Avena sativa* L.). *Pol. J. Food Nutr. Sci.* **2016**, *66*, 51–56. [CrossRef]
9. Sikora, P.; Tosh, S.M.; Brummer, Y.; Olsson, O. Identification of high β-glucan oat lines and localization and chemical characterization of their seed kernel β-glucans. *Food Chem.* **2013**, *137*, 83–91. [CrossRef]
10. Gajdošová, A.; Petruláková, Z.; Havrlentová, M.; Červená, V.; Hozová, B.; Šturdík, E.; Kogan, G. The content of water-soluble and water-insoluble β-D-glucans in selected oats and barley varieties. *Carbohydr. Polym.* **2007**, *70*, 46–52. [CrossRef]
11. Trethewey, J.A.K.; Campbell, L.M.; Harris, P.J. (1→3),(1→4)-β-d-Glucans in the cell walls of the *Poales* (Sensu Lato): An immunogold labeling study using a monoclonal antibody. *Am. J. Bot.* **2005**, *92*, 1660–1674. [CrossRef]
12. Vega-Sanchez, M.; Verhertbruggen, Y.; Scheller, H.V.; Ronald, P. Abundance of mixed linkage glucan in mature tissues and secondary cell walls of grasses. *Plant Signal. Behav.* **2013**, *8*, e23143. [CrossRef] [PubMed]
13. Hozlár, P.; Gregusová, V.; Nemeček, P.; Šliková, S.; Havrlentová, M.; Havrlentová, M. Study of dynamic accumulation in β-D-glucan in oat (*Avena sativa* L.) during plant development. *Polymers* **2022**, *14*, 2668. [CrossRef] [PubMed]
14. Fan, M.; Zhang, Z.; Wang, F.; Li, Z.; Hu, Y. Effect of nitrogen forms and levels on β-glucan accumulation in grains of oat (*Avena Sativa* L.) plants. *Z. Pflanzenernähr. Bodenk.* **2009**, *172*, 861–866. [CrossRef]

15. Izydorczyk, M.S.; Macri, L.J.; MacGregor, A.W. Structure and physicochemical properties of barley non-starch polysaccharides—II. Alkaliextractable β-glucans and arabinoxylans. *Carbohydr. Polym.* **1998**, *35*, 259–269. [CrossRef]
16. Chang, S.-C.; Saldivar, R.K.; Liang, P.-H.; Hsieh, Y.S.Y. Structures, biosynthesis, and physiological functions of (1,3;1,4)-β-D-glucans. *Cells* **2021**, *10*, 510. [CrossRef]
17. Fincher, G.B. Exploring the evolution of (1,3;1,4)-β-D-glucans in plant cell walls: Comparative genomics can help! *Curr. Opin. Plant Biol.* **2009**, *12*, 140–147. [CrossRef]
18. Staudte, R.G.; Woodward, J.R.; Fincher, G.B.; Stone, B.A. Water-soluble (1→3), (1→4)-β-D-glucans from barley (*Hordeum vulgare*) endosperm. III. Distribution of cellotriosyl and cellotetraosyl residues. *Carbohydr. Polym.* **1983**, *3*, 299–312. [CrossRef]
19. Burton, R.A.; Fincher, G.B. (1,3;1,4)-β-D-Glucans in cell walls of the *Poaceae*, lower plants, and fungi: A tale of two linkages. *Mol. Plant.* **2009**, *2*, 873–882. [CrossRef]
20. Skendi, A.; Biliaderis, C.G.; Lazaridou, A.; Izydorczyk, M.S. Structure and rheological properties of water soluble β-glucans from oat cultivars of *Avena sativa* and *Avena bysantina*. *J. Cereal Sci.* **2003**, *38*, 15–31. [CrossRef]
21. Izydorczyk, M.S.; Biliaderis, C.G.; Macri, L.J.; MacGregor, A.W. Fractionation of oat (1→3), (1→4)-β-D-glucans and characterisation of the fractions. *J. Cereal Sci.* **1998**, *27*, 321–325. [CrossRef]
22. Tosh, S.; Brummer, Y.; Wolever, T.; Wood, P. Glycemic response to oat bran muffins treated to vary molecular weight of β-glucan. *Cereal Chem.* **2008**, *85*, 211–217. [CrossRef]
23. Lazaridou, A.; Biliaderis, C.G. Molecular aspects of cereal β-glucan functionality: Physical properties, technological applications and physiological effects. *J. Cereal Sci.* **2007**, *46*, 101–118. [CrossRef]
24. Henrion, M.; Francey, C.; Lê, K.-A.; Lamothe, L. Cereal β-glucans: The impact of processing and how it affects physiological responses. *Nutrients* **2019**, *11*, 1729. [CrossRef]
25. Woodward, J.R.; Fincher, G.B.; Stone, B.A. Water-soluble (1→3), (1→4)-β-D-glucans from barley (*Hordeum vulgare*) endosperm. II. Fine structure. *Carbohydr. Polym.* **1983**, *3*, 207–225. [CrossRef]
26. Burton, R.; Fincher, G. Current challenges in cell wall biology in the cereals and grasses. *Front. Plant Sci.* **2012**, *3*, 130. [CrossRef]
27. Doehlert, D.C.; Simsek, S. Variation in β-glucan fine structure, extractability, and flour slurry viscosity in oats due to genotype and environment. *Cereal Chem.* **2012**, *89*, 242–246. [CrossRef]
28. Izydorczyk, M.S.; Dexter, J.E. Barley β-glucans and arabinoxylans: Molecular structure, physicochemical properties, and uses in food products—A review. *Food Res. Int.* **2008**, *41*, 850–868. [CrossRef]
29. Ramesh, H.P.; Tharanathan, R.N. Carbohydrates—The renewable raw materials of high biotechnological value. *Crit. Rev. Biotechnol.* **2003**, *23*, 149–173. [CrossRef] [PubMed]
30. Miller, S.S.; Wood, P.J.; Pietrzak, L.N.; Fulcher, R.G. Mixed Linkage beta-glucan, protein content, and kernel weight in *Avena* species. *Cereal Chem.* **1993**, *70*, 231–233. Available online: https://www.cerealsgrains.org/publications/cc/backissues/1993/Documents/cc1993a47.html (accessed on 10 April 2023).
31. Cui, W.; Wood, P.J. Relationships between structural features, molecular weight and rheological properties of cereal β-D-glucans. In *Hydrocolloids*; Nishinari, K., Ed.; Elsevier Science: Amsterdam, The Netherlands, 2000; pp. 159–168. [CrossRef]
32. Bulone, V.; Schwerdt, J.G.; Fincher, G.B. Co-evolution of enzymes involved in plant cell wall metabolism in the grasses. *Front. Plant Sci.* **2019**, *10*, 1009. [CrossRef]
33. Du, B.; Meenu, M.; Liu, H.; Xu, B. A Concise review on the molecular structure and function relationship of β-glucan. *Int. J. Mol. Sci.* **2019**, *20*, 4032. [CrossRef]
34. Fulcher, R.G.; Miller, S.S. Structure of oat bran and distribution of dietary fiber components. In *Oat Bran*; Wood, P., Ed.; American Association of Cereal Chemists: St. Paul, MN, USA, 1993; pp. 1–24.
35. Welch, R.W.; Brown, J.C.W.; Leggett, J.M. Interspecific and intraspecific variation in grain and groat characteristics of wild oat (*Avena*) species: Very high groat (1→3),(1→4)-β-glucan in an *Avena atlantica* genotype. *J. Cereal Sci.* **2000**, *31*, 273–279. [CrossRef]
36. Tiwari, U.; Cummins, E. Simulation of the factors affecting β-glucan levels during the cultivation of oats. *J. Cereal Sci.* **2009**, *50*, 175–183. [CrossRef]
37. Redaelli, R.; Del Frate, V.; Bellato, S.; Terracciano, G.; Ciccoritti, R.; Germeier, C.U.; De Stefanis, E.; Sgrulletta, D. Genetic and environmental variability in total and soluble β-glucan in european oat genotypes. *J. Cereal Sci.* **2013**, *57*, 193–199. [CrossRef]
38. Markovic, S.; Djukic, N.; Knezevic, D.; Lekovic, S. Divergence of barley and oat varieties according to their content of β-glucan. *J. Serb. Chem. Soc.* **2017**, *82*, 379–388. [CrossRef]
39. MacGregor, A.W. Barley. In *Encyclopedia of Food Sciences and Nutrition*, 2nd ed.; Caballero, B., Ed.; Academic Press: Oxford, UK, 2003; pp. 379–382. [CrossRef]
40. Eticha, F.; Grausgruber, H.; Berghoffer, E. Multivariate analysis of agronomic and quality traits of hull-less spring barley (*Hordeum vulgare* L.). *J. Plant Breed.* **2010**, *2*, 81–95.
41. Nishantha, M.D.L.C.; Zhao, X.; Jeewani, D.C.; Bian, J.; Nie, X.; Weining, S. Direct comparison of β-glucan content in wild and cultivated barley. *Int. J. Food Prop.* **2018**, *21*, 2218–2228. [CrossRef]
42. Marcotuli, I.; Houston, K.; Schwerdt, J.G.; Waugh, R.; Fincher, G.B.; Burton, R.A.; Blanco, A.; Gadaleta, A. Genetic diversity and genome wide association study of β-glucan content in tetraploid wheat grains. *PLoS ONE* **2016**, *11*, e0152590. [CrossRef]
43. Walling, J.G.; Sallam, A.H.; Steffenson, B.J.; Henson, C.; Vinje, M.A.; Mahalingam, R. Quantitative trait loci impacting grain β-glucan content in wild barley (*Hordeum vulgare* ssp. *spontaneum*) reveals genes associated with cell wall modification and carbohydrate metabolism. *Crop Sci.* **2022**, *62*, 1213–1227. [CrossRef]

44. Lauer, J.C.; Cu, S.; Burton, R.A.; Eglinton, J.K. Variation in barley (1→3, 1→4)-β-glucan endohydrolases reveals novel allozymes with increased thermostability. *Theor. Appl. Genet.* **2017**, *130*, 1053–1063. [CrossRef]
45. Meints, B.; Hayes, P.M. Breeding naked barley for food, feed, and malt. In *Plant Breeding Reviews*, 1st ed.; Goldman, I., Ed.; Wiley: Mdison, WI, USA, 2019; pp. 95–119. [CrossRef]
46. Tonooka, T.; Yanagisawa, T.; Aoki, E.; Taira, M.; Yoshioka, T. Breeding of a new six-rowed waxy barley cultivar "Kihadamochi" exhibiting high levels of yield and β-glucan content. *Breed. Res.* **2022**, *24*, 146–152. [CrossRef]
47. Tiwari, U.; Cummins, E. Factors influencing β-glucan levels and molecular weight in cereal-based products. *Cereal Chem.* **2009**, *86*, 290–301. [CrossRef]
48. Elouadi, F.; Amri, A.; El-Baouchi, A.; Kehel, Z.; Salih, G.; Jilal, A.; Kilian, B.; Ibriz, M. Evaluation of a set of *Hordeum vulgare* subsp. *spontaneum* accessions for β-glucans and microelement contents. *Agriculture* **2021**, *11*, 950. [CrossRef]
49. Meints, B.; Vallejos, C.; Hayes, P. Multi-use naked barley: A New Frontier. *J. Cereal Sci.* **2021**, *102*, 103370. [CrossRef]
50. Peterson, D.M. Oat antioxidants. *J. Cereal Sci.* **2001**, *33*, 115–129. [CrossRef]
51. Mälkki, Y. Trends in dietary fibre research and development. *Acta Aliment.* **2004**, *33*, 39–62. [CrossRef]
52. Popov, V.S.; Khoreva, V.I.; Konarev, A.V.; Shelenga, T.V.; Blinova, E.V.; Malyshev, L.L.; Loskutov, I.G. Evaluating germplasm of cultivated oat species from the VIR collection under the russian northwest conditions. *Plants* **2022**, *11*, 3280. [CrossRef]
53. Loskutov, I.G.; Polonskiy, V.I. Content of β-glucans in oat grain as a perspective direction of breeding for health products and fodder. *Agric. Biol.* **2017**, *52*, 646–657. [CrossRef]
54. Amer, S.A.; Attia, G.A.; Aljahmany, A.A.; Mohamed, A.K.; Ali, A.A.; Gouda, A.; Alagmy, G.N.; Megahed, H.M.; Saber, T.; Farahat, M. Effect of 1,3-beta glucans dietary addition on the growth, intestinal histology, blood biochemical parameters, immune response, and immune expression of CD3 and CD20 in broiler chickens. *Animals* **2022**, *12*, 3197. [CrossRef]
55. Jacob, J.P.; Pescatore, A.J. Barley β-glucan in poultry diets. *Ann. Transl. Med.* **2014**, *2*, 20. [CrossRef]
56. Richmond, T.A.; Somerville, C.R. The cellulose synthase superfamily. *Plant Physiol.* **2000**, *124*, 495–498. [CrossRef]
57. Hazen, S.P.; Scott-Craig, J.S.; Walton, J.D. Cellulose synthase-like genes of rice. *Plant Physiol.* **2002**, *128*, 336–340. [CrossRef] [PubMed]
58. Farrokhi, N.; Burton, R.A.; Brownfield, L.; Hrmova, M.; Wilson, S.M.; Bacic, A.; Fincher, G.B. Plant cell wall biosynthesis: Genetic, biochemical and functional genomics approaches to the identification of key genes. *Plant Biotechnol. J.* **2006**, *4*, 145–167. [CrossRef]
59. Cocuron, J.-C.; Lerouxel, O.; Drakakaki, G.; Alonso, A.P.; Liepman, A.H.; Keegstra, K.; Raikhel, N.; Wilkerson, C.G. A gene from the cellulose synthase-like C family encodes a β-1,4 glucan synthase. *Proc. Natl. Acad. Sci. USA* **2007**, *104*, 8550–8555. [CrossRef] [PubMed]
60. Burton, R.A.; Wilson, S.M.; Hrmova, M.; Harvey, A.J.; Shirley, N.J.; Medhurst, A.; Stone, B.A.; Newbigin, E.J.; Bacic, A.; Fincher, G.B. Cellulose synthase-like *CslF* genes mediate the synthesis of cell wall (1, 3; 1, 4)-β-D-glucans. *Science* **2006**, *311*, 1940–1942. [CrossRef] [PubMed]
61. Doblin, M.S.; Pettolino, F.A.; Wilson, S.M.; Campbell, R.; Burton, R.A.; Fincher, G.B.; Newbigin, E.; Bacic, A. A barley cellulose synthase-like *CslH* gene mediates (1,3;1,4)-β-D-glucan synthesis in transgenic arabidopsis. *Proc. Natl. Acad. Sci. USA* **2009**, *106*, 5996–6001. [CrossRef]
62. Yin, Y.; Huang, J.; Xu, Y. The cellulose synthase superfamily in fully sequenced plants and algae. *BMC Plant Biol.* **2009**, *9*, 99. [CrossRef]
63. Little, A.; Lahnstein, J.; Jeffery, D.W.; Khor, S.F.; Schwerdt, J.G.; Shirley, N.J.; Hooi, M.; Xing, X.; Burton, R.A.; Bulone, V. A novel (1,4)-β-linked glucoxylan is synthesized by members of the *Cellulose Synthase-like F* Gene family in land plants. *ACS Cent. Sci.* **2019**, *5*, 73–84. [CrossRef]
64. Zhang, J.; Yan, L.; Liu, M.; Guo, G.; Wu, B. Analysis of β-d-glucan biosynthetic genes in oat reveals glucan synthesis regulation by light. *Ann. Bot.* **2021**, *127*, 371–380. [CrossRef]
65. Dimitroff, G.; Little, A.; Lahnstein, J.; Schwerdt, J.G.; Srivastava, V.; Bulone, V.; Burton, R.A.; Fincher, G.B. (1,3;1,4)-β-Glucan biosynthesis by the CSLF6 enzyme: Position and flexibility of catalytic residues influence product fine structure. *Biochemistry* **2016**, *55*, 2054–2061. [CrossRef]
66. Nemeth, C.; Freeman, J.; Jones, H.D.; Sparks, C.; Pellny, T.K.; Wilkinson, M.D.; Dunwell, J.; Andersson, A.A.M.; Åman, P.; Guillon, F.; et al. Down-regulation of the *CSLF6* gene results in decreased (1,3;1,4)-*β*-D-glucan in endosperm of wheat. *Plant Physiol.* **2010**, *152*, 1209–1218. [CrossRef]
67. Little, A.; Schwerdt, J.G.; Shirley, N.J.; Khor, S.F.; Neumann, K.; O'Donovan, L.A.; Lahnstein, J.; Collins, H.M.; Henderson, M.; Fincher, G.B.; et al. Revised phylogeny of the cellulose synthase gene superfamily: Insights into cell wall evolution. *Plant Physiol.* **2018**, *177*, 1124–1141. [CrossRef] [PubMed]
68. Burton, R.A.; Collins, H.M.; Kibble, N.A.J.; Smith, J.A.; Shirley, N.J.; Jobling, S.A.; Henderson, M.; Singh, R.R.; Pettolino, F.; Wilson, S.M.; et al. Over-expression of specific *HvCslF* cellulose synthase-like genes in transgenic barley increases the levels of cell wall (1,3;1,4)-β-d-glucans and alters their fine structure: Over-expression of *CslF* genes in barley. *Plant Biotechnol. J.* **2011**, *9*, 117–135. [CrossRef]
69. Taketa, S.; Yuo, T.; Tonooka, T.; Tsumuraya, Y.; Inagaki, Y.; Haruyama, N.; Larroque, O.; Jobling, S.A. Functional Functional characterization of barley beta-glucanless mutants demonstrates a unique role for *CslF6* in (1,3;1,4)-β-D-glucan biosynthesis. *J. Exp. Bot.* **2012**, *63*, 381–392. [CrossRef] [PubMed]

70. Schwerdt, J.G.; MacKenzie, K.; Wright, F.; Oehme, D.; Wagner, J.M.; Harvey, A.J.; Shirley, N.J.; Burton, R.A.; Schreiber, M.; Halpin, C.; et al. Evolutionary dynamics of the cellulose synthase gene superfamily in grasses. *Plant Physiol.* **2015**, *168*, 968–983. [CrossRef] [PubMed]
71. Vega-Sánchez, M.E.; Verhertbruggen, Y.; Christensen, U.; Chen, X.; Sharma, V.; Varanasi, P.; Jobling, S.A.; Talbot, M.; White, R.G.; Joo, M.; et al. Loss of *cellulose synthase—Like F6* function affects mixed-linkage glucan deposition, cell wall mechanical properties, and defense responses in vegetative tissues of rice. *Plant Physiol.* **2012**, *159*, 56–69. [CrossRef]
72. Shu, X.; Rasmussen, S.K. Quantification of amylose, amylopectin, and β-glucan in search for genes controlling the three major quality traits in barley by genome-wide association studies. *Front. Plant Sci.* **2014**, *15*, 197. [CrossRef]
73. Burton, R.A.; Jobling, S.A.; Harvey, A.J.; Shirley, N.J.; Mather, D.E.; Bacic, A.; Fincher, G.B. The genetics and transcriptional profiles of the cellulose synthase-like *HvCslF* gene family in barley. *Plant Physiol.* **2008**, *146*, 1821–1833. [CrossRef]
74. Garcia-Gimenez, G.; Barakate, A.; Smith, P.; Stephens, J.; Khor, S.F.; Doblin, M.S.; Hao, P.; Bacic, A.; Fincher, G.B.; Burton, R.A.; et al. Targeted mutation of barley (1,3;1,4)-β-glucan synthases reveals complex relationships between the storage and cell wall polysaccharide content. *Plant J.* **2020**, *104*, 1009–1022. [CrossRef]
75. Coon, M.A. Characterization and Variable Expression of the *CslF6* Homologs in Oat (*Avena* sp.). Master's Thesis, Brigham Young University, Provo, UT, USA, 2012. Available online: https://scholarsarchive.byu.edu/cgi/viewcontent.cgi?article=4749&context=etd (accessed on 2 May 2022).
76. Wilson, S.M.; Ho, Y.Y.; Lampugnani, E.R.; Van de Meene, A.M.L.; Bain, M.P.; Bacic, A.; Doblin, M.S. Determining the subcellular location of synthesis and assembly of the cell wall polysaccharide (1,3; 1,4)-β-D-glucan in grasses. *Plant Cell* **2015**, *27*, 754–771. [CrossRef]
77. Pérez-Vendrell, A.M.; Guasch, J.; Francesch, M.; Molina-Cano, J.L.; Brufau, J. Determination of β-(1–3), (1–4)-D-glucans in barley by reversed-phase high-performance liquid chromatography. *J. Chromatogr.* **1995**, *718*, 291–297. [CrossRef]
78. Redaelli, R.; Scalfati, G.; Ciccoritti, R.; Cacciatori, P.; De Stefanis, E.; Sgrulletta, D. Effects of genetic and agronomic factors on grain composition in oats. *Cereal Res. Commun.* **2015**, *43*, 144–154. [CrossRef]
79. Peterson, D.M. Genotype and environment effects on oat beta-glucan concentration. *Crop Sci.* **1991**, *31*, 1517–1520. [CrossRef]
80. Doehlert, D.C.; McMullen, M.S.; Hammond, J.J. Genotypic and environmental effects on grain yield and quality of oat grown in North Dakota. *Crop Sci.* **2001**, *41*, 1066–1072. [CrossRef]
81. Güler, M. Barley grain β-glucan content as affected by nitrogen and irrigation. *Field Crops Res.* **2003**, *84*, 335–340. [CrossRef]
82. Hübner, F.; O'Neil, T.; Cashman, K.D.; Arendt, E.K. The influence of germination conditions on beta-glucan, dietary fibre and phytate during the germination of oats and barley. *Eur. Food Res. Technol.* **2010**, *231*, 27–35. [CrossRef]
83. Dvončová, D.; Havrlentová, M.; Hlinková, A.; Hozlár, P. Effect of fertilization and variety on the β-glucan content in the grain of oats. *ZNTJ* **2010**, *17*, 108–116. [CrossRef]
84. Gorash, A.; Armonienė, R.; Mitchell Fetch, J.; Liatukas, Ž.; Danytė, V. Aspects in oat breeding: Nutrition quality, nakedness and disease resistance, challenges and perspectives: Aspects in oat breeding. *Ann. Appl. Biol.* **2017**, *171*, 281–302. [CrossRef]
85. Dvořáček, V.; Jágr, M.; Kotrbová Kozak, A.; Capouchová, I.; Konvalina, P.; Faměra, O.; Hlásná Čepková, P. Avenanthramides: Unique bioactive substances of oat grain in the context of cultivar, cropping system, weather conditions and other grain parameters. *Plants* **2021**, *10*, 2485. [CrossRef]
86. De Arcangelis, E.; Messia, M.C.; Marconi, E. Variation of polysaccharides profiles in developing kernels of different barley cultivars. *J. Cereal Sci.* **2019**, *85*, 273–278. [CrossRef]
87. Mathews, R.; Kamil, A.; Chu, Y. Global review of heart health claims for oat beta-glucan products. *Nutr. Rev.* **2020**, *78*, 78–97. [CrossRef]
88. Ahmad, A.; Anjum, F.M.; Zahoor, T.; Nawaz, H.; Dilshad, S.M.R. A valuable functional ingredient in foods. *Crit. Rev. Food Sci. Nutr.* **2012**, *52*, 201–212. [CrossRef]
89. Havrlentová, M.; Gregusová, V.; Šliková, S.; Nemeček, P.; Hudcovicová, M.; Kuzmová, D. Relationship between the content of β-D-glucans and infection with *Fusarium* pathogens in oat (*Avena sativa* L.) plants. *Plants* **2020**, *9*, 1776. [CrossRef] [PubMed]
90. Hrmova, M.; Fincher, G.B. Structure-function relationships of β-D-glucan endo- and exohydrolases from higher plants. *Plant Mol. Biol.* **2001**, *47*, 73–91. [CrossRef]
91. Roulin, S.; Buchala, A.J.; Fincher, G.B. Induction of (1→3,1→4)-β-D-glucan hydrolases in leaves of dark-incubated barley seedlings. *Planta* **2002**, *215*, 51–59. [CrossRef] [PubMed]
92. Guillon, F.; Tranquet, O.; Quillien, L.; Utille, J.-P.; Ordaz Ortiz, J.J.; Saulnier, L. Generation of polyclonal and monoclonal antibodies against arabinoxylans and their use for immunocytochemical location of arabinoxylans in cell walls of endosperm of wheat. *J. Cereal Sci.* **2004**, *40*, 167–182. [CrossRef]
93. Trafford, K.; Haleux, P.; Henderson, M.; Parker, M.; Shirley, N.J.; Tucker, M.R.; Fincher, G.B.; Burton, R.A. Grain development in *Brachypodium* and other grasses: Possible interactions between cell expansion, starch deposition, and cell-wall synthesis. *J. Exp. Bot.* **2013**, *64*, 5033–5047. [CrossRef]
94. Carpita, N.C. Structure and biogenesis of the cell walls of grasses. *Annu. Rev. Plant Biol.* **1996**, *47*, 445–476. [CrossRef] [PubMed]
95. Kim, J.-B.; Olek, A.T.; Carpita, N.C. Cell wall and membrane-associated exo-β-D-glucanases from developing maize seedlings. *Plant Physiol.* **2000**, *123*, 471–486. [CrossRef]
96. Hoson, T. Apoplast as the site of response to environmental signals. *J. Plant Res.* **1998**, *111*, 167–177. [CrossRef]

97. Havrlentová, M.; Šliková, S.; Gregusová, V.; Kovácsová, B.; Lančaričová, A.; Nemeček, P.; Hendrichová, J.; Hozlár, P. The influence of artificial *Fusarium* infection on oat grain quality. *Microorganisms* **2021**, *9*, 2108. [CrossRef] [PubMed]
98. Martin, C.; Schöneberg, T.; Vogelgsang, S.; Morisoli, R.; Bertossa, M.; Mauch-Mani, B.; Mascher, F. Resistance against *Fusarium graminearum* and the relationship to β-glucan content in barley grains. *Eur. J. Plant Pathol.* **2018**, *152*, 621–634. [CrossRef]
99. Gregusová, V.; Kaňuková, Š.; Hudcovicová, M.; Bojnanská, K.; Ondreičková, K.; Piršelová, B.; Mészáros, P.; Lengyelová, L.; Galuščáková, Ľ.; Kubová, V.; et al. The cell-wall β-d-glucan in leaves of oat (*Avena sativa* L.) affected by fungal pathogen *Blumeria graminis* f. sp. *avenae*. *Polymers* **2022**, *14*, 3416. [CrossRef]
100. Kofuji, K.; Aoki, A.; Tsubaki, K.; Konishi, M.; Isobe, T.; Murata, Y. Antioxidant activity of β-glucan. *ISRN Pharm.* **2012**, *2012*, 125864. [CrossRef]
101. Lattanzio, V.; Lattanzio, V.M.T.; Cardinali, A. Role of phenolics in the resistance mechanisms of plants against fungal pathogens and insects. *Phytochem. Adv. Res.* **2006**, *661*, 23–67.
102. Bai, Y.-P.; Zhou, H.-M.; Zhu, K.-R.; Li, Q. Effect of thermal processing on the molecular, structural, and antioxidant characteristics of highland barley β-glucan. *Carbohydr. Polym.* **2021**, *271*, 118416. [CrossRef]
103. Sørensen, I.; Willats, W.G. New insights from ancient species. *Plant Signal. Behav.* **2008**, *3*, 743–745. [CrossRef]
104. Kerstens, S.; Decraemer, W.F.; Verbelen, J.-P. Cell walls at theplant surface behave mechanically like fiber-reinforced composite materials. *Plant Physiol.* **2001**, *127*, 381–385. Available online: www.plantphysiol.org/cgi/doi/10.1104/pp.010423 (accessed on 2 May 2023). [CrossRef]
105. Holthaus, J.; Holland, J.; White, P.; Frey, K. Inheritance of B-glucan content of oat grain. *Crop Sci.* **1996**, *36*, 567–572. [CrossRef]
106. Eshghi, R.; Akhundova, E. Inheritance pattern of β-glucan and protein countents in hulless barley. *Int. J. Agric. Biol.* **2010**, *12*, 68–72.
107. Swanston, J. The barley husk: A potential barrier to future success. In *Barley: Physical Properties, Genetic Factors and Environmental Impacts on Growth*; Nova Science Publishers Inc.: New York, NY, USA, 2014; pp. 81–106.
108. Geng, L.; Li, M.; Xie, S.; Wu, D.; Ye, L.; Zhang, G. Identification of genetic loci and candidate genes related to β-glucan content in barley grain by genome-wide association study in international barley core selected collection. *Mol. Breed.* **2021**, *41*, 1–12. [CrossRef]
109. Geng, L.; He, X.; Ye, L.; Zhang, G. Identification of the genes associated with β-glucan synthesis and accumulation during grain development in barley. *Food Chem. Mol. Sci.* **2022**, *5*, 100136. [CrossRef] [PubMed]
110. Newell, M.A.; Asoro, F.G.; Scott, M.P.; White, P.J.; Beavis, W.D.; Jannink, J.-L. Genome-wide association study for oat (*Avena sativa* L.) beta-glucan concentration using germplasm of worldwide origin. *Theor. Appl. Genet.* **2012**, *125*, 1687–1696. [CrossRef] [PubMed]
111. Zimmer, C.M.; Oliveira, G.; Pacheco, M.T.; Federizzi, L.C.C. Characterization and absolute quantification of the *Cellulose synthase-like F6* homoeologs in oats. *Euphytica* **2020**, *216*, 173. [CrossRef]
112. Motilva, M.-J.; Serra, A.; Borrás, X.; Romero, M.-P.; Domínguez, A.; Labrador, A.; Peiró, L. Adaptation of the standard enzymatic protocol (Megazyme Method) to microplaque format for β-(1,3)(1,4)-D-glucan determination in cereal based samples with a wide range of β-glucan content. *J. Cereal Sci.* **2014**, *59*, 224–227. [CrossRef]
113. Paudel, D.; Caffe-Treml, M.; Krishnan, P. A single analytical platform for the rapid and simultaneous measurement of protein, oil, and beta-glucan contents of oats using near-infrared reflectance spectroscopy. *Cereal Foods World* **2018**, *63*, 17–25.
114. Habschied, K.; Lalić, A.; Horvat, D.; Mastanjević, K.; Lukinac, J.; Jukić, M.; Krstanović, V. β-Glucan degradation during malting of different purpose barley varieties. *Fermentation* **2020**, *6*, 21. [CrossRef]
115. Fang, Y.; Zhang, X.; Xue, D. Genetic analysis and molecular breeding applications of malting quality QTLs in Barley. *Front. Genet.* **2019**, *10*, 352. [CrossRef]
116. Ullrich, S.E.; Han, F.; Jones, B.L. Genetic complexity of the malt extract trait in barley suggested by QTL analysis. *J. Am. Soc. Brew. Chem.* **1997**, *55*, 1–4. [CrossRef]
117. Han, F.; Romagosa, I.; Ullrich, S.E.; Jones, B.L.; Hayes, P.M.; Wesenberg, D.M. Molecular marker-assisted selection for malting quality traits in barley. *Mol. Breed.* **1997**, *3*, 427–437. [CrossRef]
118. Iqbal, I.; Desta, Z.A.; Tripathi, R.K.; Beattie, A.; Badea, A.; Singh, J. Interaction and association analysis of malting related traits in barley. *PLoS ONE* **2023**, *18*, e0283763. [CrossRef]
119. Muñoz-Amatriaín, M.; Xiong, Y.; Schmitt, M.R.; Bilgic, H.; Budde, A.D.; Chao, S.; Smith, K.P.; Muehlbauer, G.J. Transcriptome analysis of a barley breeding program examines gene expression diversity and reveals target genes for malting quality improvement. *BMC Genom.* **2010**, *11*, 653. [CrossRef] [PubMed]
120. Singh, S.; Tripathi, R.; Lemaux, P.; Buchanan, B.; Singh, J. Redox-dependent interaction between thaumatin-like protein and β-glucan influences malting quality of barley. *Proc. Natl. Acad. Sci. USA* **2017**, *114*, 7725–7730. [CrossRef] [PubMed]
121. Farzaneh, V.; Ghodsvali, A.; Bakhshabadi, H.; Zare, Z.; Carvalho, I.S. The impact of germination time on the some selected parameters through malting process. *Int. J. Biol. Macromol.* **2017**, *94*, 663–668. [CrossRef] [PubMed]

life

Review

Plants as Biofactories for Therapeutic Proteins and Antiviral Compounds to Combat COVID-19

Corbin England [1,2], Jonathan TrejoMartinez [3], Paula PerezSanchez [3], Uddhab Karki [1,2] and Jianfeng Xu [1,4,*]

1 Arkansas Biosciences Institute, Arkansas State University, Jonesboro, AR 72401, USA
2 Molecular Biosciences Program, Arkansas State University, Jonesboro, AR 72401, USA
3 Department of Biological Sciences, Arkansas State University, Jonesboro, AR 72401, USA
4 College of Agriculture, Arkansas State University, Jonesboro, AR 72401, USA
* Correspondence: jxu@astate.edu; Tel.: +1-870-680-4812

Abstract: The outbreak of coronavirus disease 2019 (COVID-19) caused by severe acute respiratory syndrome coronavirus 2 (SARS-CoV-2) had a profound impact on the world's health and economy. Although the end of the pandemic may come in 2023, it is generally believed that the virus will not be completely eradicated. Most likely, the disease will become an endemicity. The rapid development of vaccines of different types (mRNA, subunit protein, inactivated virus, etc.) and some other antiviral drugs (Remdesivir, Olumiant, Paxlovid, etc.) has provided effectiveness in reducing COVID-19's impact worldwide. However, the circulating SARS-CoV-2 virus has been constantly mutating with the emergence of multiple variants, which makes control of COVID-19 difficult. There is still a pressing need for developing more effective antiviral drugs to fight against the disease. Plants have provided a promising production platform for both bioactive chemical compounds (small molecules) and recombinant therapeutics (big molecules). Plants naturally produce a diverse range of bioactive compounds as secondary metabolites, such as alkaloids, terpenoids/terpenes and polyphenols, which are a rich source of countless antiviral compounds. Plants can also be genetically engineered to produce valuable recombinant therapeutics. This molecular farming in plants has an unprecedented opportunity for developing vaccines, antibodies, and other biologics for pandemic diseases because of its potential advantages, such as low cost, safety, and high production volume. This review summarizes the latest advancements in plant-derived drugs used to combat COVID-19 and discusses the prospects and challenges of the plant-based production platform for antiviral agents.

Keywords: coronavirus; COVID-19; antivirals; vaccines; plant production; molecular farming; recombinant proteins; secondary metabolites

Citation: England, C.; TrejoMartinez, J.; PerezSanchez, P.; Karki, U.; Xu, J. Plants as Biofactories for Therapeutic Proteins and Antiviral Compounds to Combat COVID-19. *Life* **2023**, *13*, 617. https://doi.org/10.3390/life13030617

Academic Editor: Stefania Lamponi

Received: 17 January 2023
Revised: 14 February 2023
Accepted: 20 February 2023
Published: 23 February 2023

1. Introduction

The outbreak of coronavirus disease 2019 (COVID-19) in Wuhan, China in late December 2019 has posed a serious global public-health emergency [1]. The disease is caused by a highly transmissible and pathogenic coronavirus, named severe acute respiratory syndrome coronavirus-2 (SARS-CoV-2), which causes respiratory disease associated with high fever, difficulty breathing, and pneumonia, etc. [2,3]. As of 14 February 2023, more than 677 million people have been infected by SARS-CoV-2 globally, of which around 6.78 million lives were claimed (Worldometers.info). Equally damaging has been the global economic shutdown for fear of the threat of SARS-CoV-2 transmission. With the recent lifting of the "Zero-COVID Dynamics" policy in China, many more people will be infected, and mortality will continue to increase [4].

SARS-CoV-2 is an enveloped RNA virus with a single-stranded, positive-sense genome of ~29.9 kB in size (Figure 1) [5]. The virus consists of four major structural proteins, named spike (S), nucleocapsid (N), envelope (E), and membrane proteins (M) [1,2]. The S protein which is present as a crown-like spike on the outer surface of the virus plays a major role in viral entry into mammalian cells [6]. Specifically, the virus uses the receptor binding domain (RBD) on the S protein to interact with human angiotensin-converting enzyme 2 (ACE2) receptor as a critical initial step to enter target cells [3,7]. Therefore, both the S protein of the virus (particularly RBD) and the ACE2 of human cells have a potential target to develop therapeutics to prevent SARS-CoV-2 infection [8–11]. Since the start of the pandemic three years ago, the circulating SARS-CoV-2 virus has been constantly mutating with the emergence of multiple variants (Alpha, Beta, Gamma, Delta, Omicron) [12], which makes the control of the COVID-19 pandemic more difficult. The Omicron variant, after first being identified in South Africa in November 2021, has rapidly spread worldwide, outcompeting other variants, and becoming the predominant one for the time being [13]. The BA.5 subvariant of Omicron, which was the most prevalent coronavirus strain worldwide in 2022, has been found to escape the majority of existing SARS-CoV-2 neutralizing antibodies [12]. Fortunately, infections by Omicron were significantly less severe than those caused by Delta and other previous variants [13]. However, its immune evasiveness and high transmissibility pose a great threat to the global healthcare system [14]. Towards the end of 2022, three Omicron subvariants, BQ.1 and BQ.1.1, and then XBB.1.5 became the dominant strains in the USA, overtaking BA.5 [15]. Therefore, effective prevention and treatments of COVID-19 disease, particularly for people with risk factors for serious illness, are still essential.

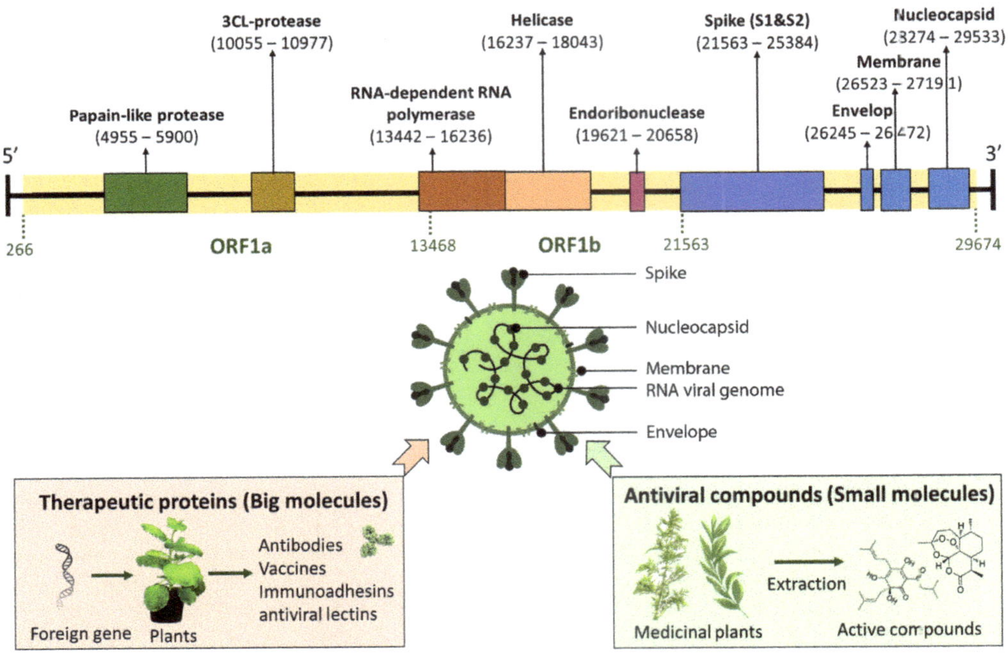

Figure 1. Schematic illustration of the single stranded RNA genome of SARS-CoV-2 (~30 kb) and pharmaceuticals (big molecules and small molecules) developed to prevent and treat COVID-19.

Great efforts have been made in the past 2 to 3 years to counteract the spread of the virus through development of vaccines [16], immune-based therapy [17], antiviral therapy [18], and natural remedies [19]. Pharmaceuticals of two general types, including biologics or "big molecules" (e.g., nucleic acids, monoclonal antibodies, therapeutic peptides, etc.) and antiviral chemical compounds or "small molecules" (e.g, Remdesivir, Olumiant, and Paxlovid, etc.) have been developed to prevent and treat COVID-19 (Figure 1). So far, COVID-19 vaccines based on messenger RNA (mRNA) (Pfizer/BioNTech, USA/Germany and Moderna, USA) [20], adenovirus vectors (Johnson & Johnson, USA; AstraZeneca, UK and Sputnik V) [21], subunit protein vaccine (Novavax, USA) [22], and inactivated virus (Sinopharm/Sinovac, China) [23] have been approved or granted Emergency Use Authorization (EUA) by the Food and Drug Administration (FDA) for vaccination in the USA and other countries. These vaccines have been effective at protecting people from getting seriously ill, being hospitalized, and even dying [24]. Almost at the same time, five anti-SARS-CoV-2 monoclonal antibodies (mAbs), including bamlanivimab plus etesevimab (Eli Lilly, USA), casirivimab plus imdevimab (Regeneron, USA), sotrovimab (GlaxoSmithKline, UK and Vir Biotechnology, USA), tocilizumab (Genentech, USA), and bebtelovimab (Eli Lilly, USA) were developed in the USA and have received EUA from the FDA for treating mild-to-moderate COVID-19 [25,26]. Among them, bebtelovimab is the only one that has shown remarkably preserved in vitro activity against all SARS-CoV-2 variants, including the Omicron subvariants BA.4 and BA.5 (but is not effective for BQ.1 and BQ.1.1. subvariants) [27]. In December 2022, tocilizumab became the first monoclonal antibody fully approved by the FDA for COVID-19 treatment. On the other hand, some traditional chemical drugs (small molecules) have been re-evaluated or re-purposed for their potential as an antiviral drug candidate against SARS-CoV-2. Remdesivir (Gilead, USA) [28] and Olumiant (baricitinib) (Eli Lilly, USA) [29] represent the first and second drugs fully approved by the FDA for treatment of hospitalized COVID-19 patients. Early in 2022, the FDA issued an EUA for the emergency use of Paxlovid (Pfizer, USA) and molnupiravir (Emory University, Ridgeback Biotherapeutics, and Merck, USA) for the treatment of mild-to-moderate COVID-19 in adults [30].

Various production platforms, mainly chemical synthesis and mammalian cell culture were utilized to manufacture the aforementioned pharmaceuticals (small and big molecules) to combat COVID-19. Plants have provided a promising alternative production platform for both natural bioactive compounds and recombinant therapeutics [31]. Plants naturally produce a diverse range of bioactive small molecules, such as alkaloids [32], flavonoids [33], terpenoids [34], and phenolic compounds [35], which are the source of countless pharmaceutical compounds for treating various diseases. Many bioactive compounds from medicinal plants, for example, those extracted from *Artemisia annua* L., *Curcuma longa* and *Tripterygium wilfordii* have been demonstrated to exhibit significant activities against SARS-CoV-2 through interfering with every step of the interaction of the virus with its host cells [36,37]. On the other hand, plants can also be genetically engineered to produce heterologous proteins (biologics) for therapeutic applications, termed "molecular farming [38]." Plants bring advantages in safety, scalability, and cost over other eukaryotic systems and have proven effective in mediating the post-translational processing required for many complex proteins [38,39]. Additionally, molecular farming in plants could facilitate rapid production of biologicals at a large scale, as demanded in the case of the COVID-19 pandemic. Extensive research has been performed to produce therapeutics against SARS-CoV-2 in plant systems, including vaccines, antibodies and other immunoadhesins. Notably, in early 2022, Canada-based biotech company Medicago announced that it had gained approval in Canada for its two-dose COVID-19 vaccine Covifenz®, an adjuvanted plant-made virus-like particles (VLP) vaccine (www.medicago.com). It represents the first COVID-19 vaccine produced by plant-based protein technology, and the promising results from a Phase III study were recently published [40].

There are many recent reviews published on the use of plant-based agents for the prevention and cure of COVID-19 [41–44]. However, these articles usually focus on one specific type of plant-produced antiviral agent, particularly bioactive natural compounds (small molecules). There is a lack of an updated review on the antiviral therapeutic proteins (big molecules) produced by plant molecular faming. In addition, most of the review articles are not up to date, because the SARS-CoV-2 virus has been mutating constantly and the therapies for COVID-19 advance rapidly. This review summarizes the latest advancements in plant-derived pharmaceuticals (both big molecules and small molecules) used to fight against SARS-CoV-2 and discusses the prospects and challenges of the plant-based production platform for antiviral agents.

2. Plant Produced Biopharmaceuticals (Biologics) against SARS-CoV-2

Plant-based expression systems, or plant molecular farming, have emerged as a promising alternative for the production of biologics. As eukaryotic organisms, plant hosts are able to perform correct post-translational modifications, such as glycosylation, allowing the development of *authentic* biologics with their efficacy being similar to those produced using other expression systems, such as mammalian or yeast-based cell cultures [38,39]. Plant-produced biologics are also regarded as safe because they do not pose the risk of introducing human and animal pathogens into biopharmaceuticals [45]. In addition, plant expression systems, particularly transient expression systems, could prompt rapid (4–8 weeks) manufacturing of target biologics on a large scale [46,47], which meet emergency demands, such as in the case of the COVID-19 pandemic. Given the aforementioned factors, plant-based expression systems have been actively adopted by pharmaceutical manufacturers. A wide range of recombinant proteins, such as vaccines, antibodies, hormones, cytokines, therapeutic enzymes, and nutritional proteins have been produced via stable and transient expression in entire plants or plant cell cultures [45,48]. The first plant-produced biologic for human use, taliglucerase alfa (Elelyso®), was approved by the FDA in 2012 for the treatment of Gaucher disease [49]. In 2019, a plant-produced influenza virus vaccine completed Phase III clinical trials with encouraging results [50]. In early 2022, the plant-made COVID-19 vaccine, Covifenz®, won first approval in Canada [40]. These successes have revived people's interest in plant-based production of biologics for human use. To combat COVID-19, plants have been used to produce vaccines [51], monoclonal antibodies [52], and other biologics that block the interactions between ACE2 and the S proteins, such as soluble ACE2 [53] and its fusion with the Fc region of human IgG1 (ACE2-Fc) [54]. In addition, plant-produced antiviral lectin has also been tested for inhibition of SARS-CoV-2 infection [55] (Figure 2).

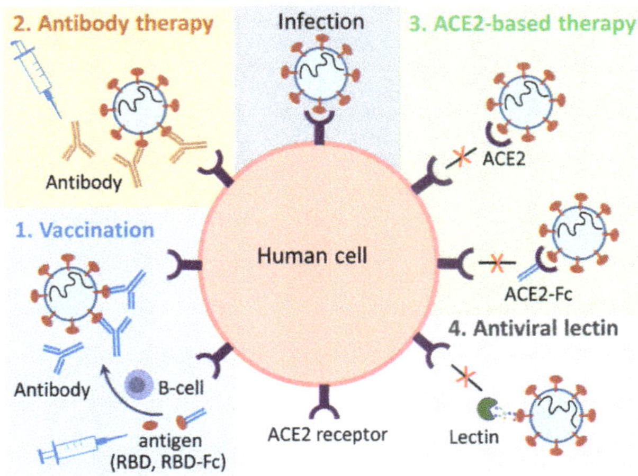

Figure 2. Schematic diagram of the plant-produced biologics functioning in preventing and treating SARS-CoV-2 infection. Plant-produced vaccines (1), antibodies (2), ACE2-based biologics (ACE2-immunoadhesin, ACE2-chewing gum) (3), and antiviral lectins (4) can be used to combat COVID-19.

2.1. Plant-Produced Vaccines

Although traditional inactivated viral vaccines and the new adenovirus vector- and mRNA-based vaccines haven been approved and widely used in the world to combat the pandemic, other types of modern vaccines, such as the protein subunit [56] or virus-like particle (VLP) varieties [57], have multiple advantages over currently used vaccines [24].

The minimum requirement for either type of vaccine is the genetic sequence of a single viral antigen rather than the genetic sequence of either virus [24]. This is safer for recipients of the vaccine, because lone antigens cannot cause or spread disease, and safer for scientists researching and manufacturing the vaccine, since no handling of live virus is required once the antigen has been sequenced [58]. SARS-CoV-2 replicates by infecting human cells via the interaction of the RBD on the viral S protein with ACE2 receptors on human cells [3], therefore, the S protein, particularly RBD, has become the focus of vaccine development efforts in the pandemic [24]. A full list of subunit vaccines and VLP vaccines that have reached or passed Phase I human trials according to the COVID-19 vaccine tracker website is available in Supplementary Table S1. Plants, either whole plants or cell suspension cultures, are suitable for producing either type of vaccine [42]. A recent comprehensive review of the use of plant-based vaccines for the prevention and cure of human viral diseases can be found in the literature [44,47,59,60]. So far, a few plant-based subunit or VLP vaccines have been developed and some of them have moved to clinical trials (Table 1).

Table 1. Plant-produced SARS-CoV-2 vaccines. Four of them have progressed to clinical trials.

Table	Trade Name	Antigen	Plant	Manufacturer	Efficacy	Commercialization Progress	Source
Virus-like particles	Covifenz	S protein	*N. benthamiana*	Medicago	69.5% to 78.8% (Phase III)	Approved: Canada Phase III Trials: Argentina, Brazil, United Kingdom, USA	[40,61,62]
	KBP-201	RBD	*N. benthamiana*	Kentucky Bioprocessing	100% (K18-hACE2 mice)	Phase I/II Trials: USA	[62–65]
	IBIO-200, IBIO-201, and IBIO-202	S protein	*N. benthamiana*	iBio, Inc.	n.d.	Pre-clinical trials	[44,66,67]
	n/a	S protein	*N. benthamiana*	n/a	n.d.	no	[68]
Subunit	Baiya SARS-CoV-2 Vax 1	RBD-Fc	*N. benthamiana*	Baiya Phytopharm	100% (K18-hACE2 mice)	Phase I Trials: Thailand	[62,69]
	Baiya SARS-CoV-2 Vax 2	RBD-Fc	*N. benthamiana*	Baiya Phytopharm	Unknown	Phase I Trials: Thailand	[62,69]
	n/a	RBD-Fc	*N. benthamiana*	n/a	n.d.	no	[51]
	n/a	RBD	*N. benthamiana*	n/a	n.d.	no	[70–72]
	n/a	S protein, RBD	Tobacco BY-2 and Medicago truncatula A17 cell	n/a	n.d.	no	[73]

n/a: not applicable; n.d.: no data.

2.1.1. Plant-Produced Subunit Vaccines

In their simplest form, subunit vaccines require only a viral protein capable of eliciting an immune response and an adjuvant. These proteins are capable of eliciting a response from B cells, helper T cells, and cytotoxic T cells, but this response is weak relative to traditional inactivated viral vaccines and necessitates the addition of an adjuvant [58]. A subunit vaccine developed by Novavax (USA) has already been granted EUA by the FDA in 2022 [22]. Studies have showed that this subunit vaccine was about 90% effective in preventing SARS-CoV-2 infections [74](Centers for Disease Control and Prevention, CDC, USA), which is similar to the efficacy of Moderna (94%) and Pfizer (95%) and better than Johnson & Johnson (66%) [75].

Plant expression platforms, mainly transient expression with *Nicotiana benthamiana*, have been used to produce subunit vaccines against SARS-CoV-2 (Table 1). The RBD alone and its fusion with the Fc region of human IgG1 (RBD-Fc) are utilized as an antigen to develop subunit vaccines. The recombinant RBD and RBD-Fc showed specific binding to human ACE2 receptor [51,70–72]. In animal tests, the plant-produced RBD and RBD-Fc antigens elicited potent neutralizing responses in mice and non-human primates [51,70]. In order to increase the immunogenicity of the antigen, RBD fused to flagellin of *Salmonella typhimurium* (Flg), known as mucosal adjuvant, was also transiently expressed with *N. benthamiana* using a self-replicating viral vector [76]. As an alternative to the transient expression platform, tobacco BY-2 and *Medicago truncatula* A17 cell suspension cultures were also used to stably express both full-length S protein and RBD [73]. The results showed that recombinant S protein and RBD could be secreted into the culture medium, which facilitated the subsequent purification and reduced the downstream processing costs. This represents the first report on the stable expression of SARS-CoV-2 antigen protein with plant cell culture system, though the bioactivity of the expressed proteins was not assessed.

Plant-produced subunit vaccines have been moved to commercial development. Of particular interest is the subunit vaccines developed by Baiya Phytopharm Co., Ltd. (Bangkok, Thailand), trade names Baiya SARS-CoV-2 Vax 1 and Baiya SARS-CoV-2 Vax 2, that utilizes a *N. benthamiana*-produced RBD as its antigen. A publication on preclinical results states that the RBD protein has been modified by fusing it with the Fc region [51]. When used with alum as adjuvant, Vax 1 induced potent immunological responses in both mice and cynomolgus monkeys [69,77,78]. Vax 1 was also reported to show 100% efficacy against infection in K18-hACE2 mice [79]. The efficacy of, and adjuvant for, the Vax 2 variant has not been revealed so far.

2.1.2. Plant-Produced VLP Vaccines

VLP vaccines make use of one or more viral structural proteins that are capable of self-assembling into nanostructures that mimic the size and shape of a virus [80]. The building blocks of the particle may serve as viral antigens capable of eliciting an immune response and the shape of the overall VLP may conform to a pathogen-associated molecular pattern recognized by the immune system [81]. Where a true virus has a cavity that contains its genetic material, VLPs have a hollow cavity that may be used to deliver small molecules to further enhance the immune response triggered by the vaccine [82].

According to the COVID-19 Vaccine Tracker website (https://covid19.trackvaccines.org/, accessed on 15 January 2023), maintained by scientists at McGill University, at the time of writing only one plant-produced SARS-CoV-2 vaccine has been approved for use. As detailed in Table 1, this VLP vaccine, with the trade name of Covifenz® (Medicago, Canada), has only been approved for use in Canada and it is in Phase III trials in several others. Three other plant-produced vaccines have reached the point of clinical trials, but none of these have yet passed the Phase III trials. Medicago's VLP vaccine utilizes the recombinant, full-length S protein from an original SARS-CoV-2 strain. These proteins associate into trimers within a lipid membrane from the cell membranes of the host *N. benthamiana* cells. The protein contains modifications made to improve the stability of the protein as well as increase the formation of VLPs [83]. According to Phase III human trials, the vaccine efficacy was 69.5% against symptomatic infection and 78.8% against infection with symptoms ranging from moderate to severe [40]. According to the Canadian government, Covifenz® is administered in two doses 21 days apart and alongside the adjuvant AS03 [61].

Kentucky Bioprocessing's VLP vaccine, trade name KBP-201, utilizes a recombinant S protein's RBD and inactivated tobacco mosaic virus. The S protein's RBD, serving as the antigen, and tobacco mosaic virus, serving as the VLP's structural component, are each expressed in *N. benthamiana,* and chemically conjugated together following purification [65]. KBP-201's RBD is fused with the Fc domain of human IgG1 to improve protein stability and an *N. benthamiana* extensin peptide to allow protein secretion and folding in the host species [65]. Preclinical trials with K18-hACE2 mice showed efficacies of 71.4% and 100%, for one and two doses effectively, against lethal infection [64]. The combined Phase I/II trial listed on ClinicalTrials.gov details a two dose regimen 21 days apart using cytosine phosphoguanine as an adjuvant [63].

In addition to Covifenz® and KBP-201, more potential SARS-CoV-2 vaccines (VLP) from iBio, Inc. (Bryan, TX, USA): IBIO-200, IBIO-201 and IBIO-202 have been reported as in pre-clinical trials in separate review publications [44,67]. However, tracking the progress of potential SARS-CoV-2 vaccines that are still in pre-clinical stages poses certain challenges that prevent us from providing accurate, current information on their status. While success in benchtop and animal models may be publicized by academic labs, corporate labs may restrict publications until entry into clinical trials. This is the case with iBio's VLP vaccines. The citations provided for the status were from iBio's website (ibioinc.com), a news media website, or another review publication over the same topic rather than a direct, peer reviewed article from scientists responsible for the research. A press release by iBio, dated after the publication of these reviews, has stated that the company will no longer continue development of IBIO-202 and none of the three vaccines appear on the company's public pipeline [84,85].

2.2. Plant-Produced Antibodies

Rather than providing the long-term protection of a vaccine, therapeutic antibodies can be used in the moment to treat people infected by a disease. mAbs targeting epitopes on the virus or infected cells may, alone or as a cocktail, reduce the viral load and thereby reduce the severity of symptoms experienced by the patient [86]. mAb-based therapeutics against the S protein have been shown to be effective treatments for SARS-CoV-2 infection, especially the original viral strain. Up to now, five kinds of FDA approved (EUA) antibodies, either alone or as a mAb cocktail, have been developed to treat COVID-19. However, the current mAbs produced in mammalian cells are expensive and might be unaffordable for many [87]. Plants may provide a low-cost and safety-friendly alternative platform to produce efficacious and affordable antibodies against SARS-CoV-2. As a new production platform, plants have already been demonstrated to have the capability of producing mAbs with quality and characteristics matching those produced in mammalian cells [88]. For example, a plant-made anti-HIV mAb has been found to meet all regulatory specifications for human application in a clinical study [89]. However, to the best of our knowledge, at the time of writing no plant-based antibodies for the treatment of SARS-CoV-2 were in clinical trials. The scope of this portion of the review has been restricted to antibodies expressed within plant cells that have been demonstrated, either in vitro or in vivo, to have a neutralizing effect on at least one variant of a SARS-CoV-2 lineage (Table 2).

Table 2. Plant-produced antibodies against the SARS-CoV-2 virus. Neutralizing capability is indicated by neutralizing titer *, meaning the dilution factor needed to reduce antibody levels below detectable limits; IC_{50} †, meaning half maximal inhibitory concentration; or NT_{100} ‡, meaning complete protection from cytotoxic effects of infection.

Antibody Name	Plant	Affected Lineages	Neutralizing Capability (Neutralizing Titer *, IC_{50} † or NT_{100} ‡)	Source
CR3022	*N. benthamiana*	Original strain	Fail to neutralize *	[71]
B38	*N. benthamiana*	Unidentified	640 at 0.492 μg/mL *	[90]
H4	*N. benthamiana*	Unidentified	40 at 5.45 μg/mL *	[90]
H4-IgG1-4	*N. benthamiana*	Unidentified	591 nM for H4-IgG3 ‡	[91]
CA1	*N. benthamiana*	Original strain, Delta	9.29 nM: Original † 89.87 nM: Delta †	[87]
CB6	*N. benthamiana*	Original strain, Delta	0.93 nM: Original † 0.75 nM: Delta †	[87]
11D7	*N. benthamiana*	Original strain, Delta, Omicron	25.37 μg/mL: Original † 59.52 μg/mL: Delta † 948.7 μg/mL: Omicron †	[46]

The first reported plant-made functional mAbs against SARS-CoV-2 were B38 and H4, which were collected from blood sera of a convalescent patient [52]. These antibodies could block binding between the RBD of the virus and the cellular receptor ACE2. Transient co-expression of heavy- and light-chain sequences of both the antibodies in *N. benthamiana* by using a geminiviral vector resulted in rapid accumulation of correctly assembled mAbs in plant leaves. Both mAbs purified from plant leaves demonstrated specific binding to RBD of SARS-CoV-2 and exhibited efficient virus neutralization activity in vitro [90]. Before this, the same research group tried to express another mAb CR3022 in *N. benthamiana*. However, this plant-produced mAb was found to bind to SARS-CoV-2 but fail to neutralize the virus in vitro [71]. These findings provide proof-of-concept for using plants as an expression system to produce SARS-CoV-2 antibodies.

Plant-made H4 was then examined in greater detail by being expressed in the four human IgG subclasses present in human serum (IgG1–4) [91]. Four constructs, each with the same variable region but different heavy chain regions, were adapted for expression in glyco-engineered *N. benthamiana*. H4-IgG3 demonstrated an up to 50-fold superior neutralization ability compared to the other three IgG against live SARS-CoV-2 virus in vivo. Complete protection from cytotoxic effects of infection (NT_{100}) using Vero cells was attained with an H4-IgG3 concentration of 5.91 nM.

Using a cocktail of mAbs that bind to complementary neutralizing epitopes represents a strategy to prevent escape of the SARS-CoV-2 mutant from mAb treatment [87]. To develop mAb cocktail-based therapeutics against SARS-CoV-2 in plants, two neutralizing mAbs, CA1 and CB6 were expressed in *N. benthamiana*. The effectiveness of plant-produced mAbs against the original SARS-CoV-2 virus and a member of the Delta lineage was tested in vitro. Both mAbs retained target epitope recognition and neutralized multiple SARS-CoV-2 variants [87]. The half maximal inhibitory concentration (IC_{50}) of CA1 was 9.29 nM for the original strain and 89.87 nM against the Delta strain. The IC_{50} of CB6 was 0.93 nM for the original strain and 0.75 nM for the Delta strain [87]. Both also demonstrated neutralizing potential against a mouse adapted strain of SARS-CoV-2 in vitro. It was also shown that one plant-made mAb has neutralizing synergy with other mAbs developed in hybridomas by the authors. A third neutralizing mAb, 11D7, which was a chimeric human IgG, was then expressed in DeltaXFT *N. benthamiana* to produce a mAb with human-like, highly homogenous *N*-linked glycans [92]. Plant-produced 11D7 was found to maintain recognition against the RBD of original, Delta and Omicron strains and neutralizing activity. Because 11D7 neutralizes SARS-CoV-2 through a mechanism not typical among currently developed mAbs, it may be useful in providing additional synergy to existing mAbs cocktails.

2.3. *Plant-Produced ACE2-Based Biologics*

2.3.1. Plant-Produced ACE2-Immunoadhesins

Although vaccines and antibodies have been developed to effectively combat COVID-19 worldwide, the rapid emergence of SARS-CoV-2 variants with altered RBD can severely affect the efficacy of such immunotherapeutic agents [14]. This problem seems to be especially pronounced with the Omicron variants that resist many of the previously isolated monoclonal antibodies [93]. Immunoadhesins, which are antibody-like molecules, make another class of immunotherapeutic agents that may complement the current therapy issue with vaccines and antibodies [94]. Immunoadhesins consist of an engineered binding domain fused to an Fc region of an antibody [95]. In the case of SARS-CoV-2, the viral cellular receptor ACE2 (extracellular domain) can serve as a binding domain for constructing such immunoadhesins, which can then function as a decoy to block the interaction of the virus with cellular ACE2 receptors [54,96]. Fusing ACE2 with the Fc region offers advantages over the treatment with ACE2 alone. This is because the Fc domain can provide effector functions, allowing the recruitment of some phagocytic immune cells and facilitating the activation of the host antiviral immune response through triggering antibody-dependent cellular cytotoxicity (ADCC) and complement-dependent cytotoxicity (CDC). Furthermore, the Fc domain can prolong the half-life, binding affinity and neutralization efficacy of the binding domain [54,97,98]. So far, more than 13 Fc fusion proteins have been approved by the FDA.

In the past 3 years, many ACE2-based immunoadhesins, including the enhanced ACE2 for binding to S protein of SARS-CoV-2 were developed [8,94,96,99–105]. These ACE2-immunoadhesins were effective in neutralizing multiple SARS-CoV-2 variants, including the Delta and the Omicron variants, suggesting that immunoadhesins-based immunotherapy is less prone to escape by the virus [94]. Again, plants can provide an economic platform to rapidly produce these biologics.

With transient expression in *N. benthamiana*, ACE2-Fc was produced at up to 100 μg/g fresh leaf. The recombinant ACE2-Fc exhibited potent anti-SARS-CoV-2 activity in vitro, and dramatically inhibited SARS-CoV-2 infectivity in Vero cells with an IC_{50} value of 0.84 μg/mL. Furthermore, treating Vero cells with ACE2-Fc at the pre-entry stage suppressed SARS-CoV-2 infection with an IC_{50} (half maximal inhibitory concentration) of 94.66 ug/mL [98].

Because ACE2 is heavily glycosylated and its glycans impact on binding to the S protein and virus infectivity, the ACE2-Fc was also expressed in glycol-engineered *N. benthamiana*. It was found that the recombinant dimeric ACE2-Fc was glycosylated with mainly complex human-type *N*-glycans and showed function in peptidase activity, binding to the RBD of the virus and neutralizing the wild-type SARS-CoV-2 virus [106].

2.3.2. Plant-Produced ACE2 and ACE2-Based Chewing Gum

Besides the ACE2-based immunoadhesins, ACE2 alone could also be developed as a therapeutic to inhibit the virus spread, though there are limitations, such as short circulating half-life [54]. Human soluble (truncated) ACE2 was reported to express in *N. benthamiana* with a high-level yield (about ~750 μg/g fresh leaf). Plant-produced ACE2 could bind to the SARS-CoV-2 S protein. Both glycosylated and deglycosylated forms of ACE2 demonstrated strong anti-SARS-CoV-2 activities in vitro, with an IC_{50} being ~1.0 and 8.48 μg/mL, respectively [53].

Of special interest is the ACE2-based chewing gum developed by Dr. Henry Daniell and his colleagues at the University of Pennsylvania [107–109]. This virus-trapping gum contains plant-made CTB-ACE2, which is ACE2 fused with non-toxic cholera toxin subunit B (CTB). CTB-ACE2 is made in chloroplasts of transgenic lettuce. The lettuce was then powdered and blended with cinnamon-flavored chewing gum. The CTB-ACE2 can efficiently bind to both GM1 and ACE2 receptors, effectively blocking binding of the S protein and viral entry into human cells. As oral epithelial cells are enriched with both receptors, this gum was designed to trap and neutralize SARS-CoV-2 in the saliva and diminish the amount of virus left in the mouth. The Phase I/II clinical trial of the chewing gum started in June 2022 (ClinicalTrials.gov Identifier: NCT05433181). If the gum proves safe and effective, it could be given to patients whose infection status is unknown or even for dental check-ups to reduce the likelihood of passing the virus to caregivers [109].

2.4. *Plant Produced Antiviral Lectins*

Lectins from plants and algae, which are carbohydrate-binding proteins of non-immune origin, were earlier found to inhibit several viral diseases, such as HIV, hepatitis C, influenza A/B, herpes, Japanese encephalitis, Ebola, and SARS coronavirus that occurred in 2003 [110–113]. Recently, some lectins have shown significant activity against SARS-CoV-2 [114–116]. For example, Griffithsin, a red algae-derived lectin of 121 amino acids, is a high mannose-specific lectin that has been recognized as a potential viral entry inhibitor [117]. Griffithsin was tested for SARS-CoV-2 entry and found that it could significantly inhibit the SARS-CoV-2 infection in a dose-dependent manner. Remarkably, the IC_{50} of griffithsin was 63 nmol/L, which is about 11-fold more potent than Remdesivir [55]. Other research demonstrated that griffithsin could block the entry of SARS-CoV-2 and its variants, Delta and Omicron, into the Vero E6 cell lines and $IFNAR^{-/-}$ mouse models by targeting the S proteins of the virus [118]. Similarly, recent molecular docking studies have shown that a banana-derived mannose-specific lectin could also neutralize SARS-CoV-2 infectivity [119]. Lectins are natural proteins which are cheap and easily accessible. They have been proven to be active against SARS-CoV-2. However, their clinical application is still hampered by several obstacles. These include the high-cost purification, short stability in the body, potential cytotoxicity and mitogenicity, and the possibility for eliciting deleterious responses in the immune system [120]. Future investigations are needed to develop plant lectins as a new antiviral agent against COVID-19.

2.5. Challenges in Commercialization of Plant-Produced Biologics against SARS-CoV-2

Numerous anti-SARS-CoV-2 biologics, including vaccines, antibodies, and other biologics against the virus have been expressed in plant systems, as mentioned above. However, compared with other production systems, such as bacterial and mammalian cell culture, plant systems suffer from a major disadvantage: low production levels of the desired proteins [48]. Additionally, isolation and purification of the recombinant proteins from plant tissues is quite expensive [38]. Although plant systems have proven effective in performing glycosylation required for complex proteins [48,121], there is a major difference in the plant and mammalian glycan structure. The N-linked glycans produced by plants carry two plant specific residues, β-1,2-xylose and core α-1,3-fucose, which are absent from mammalian cell produced proteins [122]. The immunogenicity and allergenicity of plant-specific N-glycans has been a key concern in human therapy [122]. So far, there is only one plant cell produced biopharmaceutical, taliglucerase alfa (Elelyso®), approved by FDA. Concerted research efforts based on molecular biology strategies, such as enhancing gene transcription and translation, minimizing post-translational degradation, and glycoengineering to humanize glycosylation, and engineering strategies, such as improving bioreactor design and operation and optimizing the protein purification procedure, are still needed for the commercial success of plant-based production platforms.

3. Medicinal Plant-Produced Metabolites (Small Molecules) against SARS-CoV-2

Although some vaccines and mAbs have been successfully developed in the past 2 to 3 years to combat COVID-19 disease, the lack of effective therapeutics against the virus has prompted the shift of some interests toward plant-based therapy. This is because many drugs in use are either plant materials or derived from their bioactive compounds. There is a remarkable prospect of discovering anti-COVID-19 from medicinal plants [123].

Plant-produced secondary metabolites (PSMs) are a rich source of bioactive compounds with a broad spectrum of antiviral activities [37]. Due to their high bioavailability, relatively low cost, and potential for large-scale production, PSMs represent a promising field of study to find new treatments against SARS-CoV-2 [124–126]. The potential of PSMs to treat COVID-19 is immense. They can be utilized as prophylactics, antivirals, and even adjuvants to reduce morbidity during COVID-19 treatment [127]. Natural medicines derived from PSMs are usually non-toxic, well-tolerated with minimum side effects, and highly absorptive by the human body [128]. A list of 162 PSMs found in medicinal plants that showed antiviral activity has been published earlier [37]. Among them, around 76 PSMs from different plant species are effective against COVID-19 [37]. These PSMs can be generally classified as polyphenols, alkaloids, flavonoids, coumarins and essential oils, which are able to inhibit main targets in the virus life cycle, including the viral proteins, the lipid envelope and viral nucleic acids [129]. In addition, advanced bioinformatics applications have opened a new arena in predicting PSMs as a potential COVID-19 suppressor [128,130]. In silico analysis has revealed that PSMs could be one of the most valuable drug targets against SARS-CoV-2 [37]. There are many recent review papers published on the PSMs against SARS-CoV-2 [36,37,125,129,131–136]. A comprehensive list of medicinal plants and their active compounds with inhibitory activity against SARS-CoV-2 can be seen in the recent reviews [37,131,132,137]. In this section, we will discuss the major antiviral mechanisms of PSMs against SARS-CoV-2 and summarize some of the newly published data involving the application of PSMs in the prevention and treatment of COVID-19 infections.

3.1. Antiviral Mechanisms of PSMs

Many PSMs have broad-spectrum antiviral activity. They can inhibit multiple steps in viral infection and replication and have been previously used in the treatment of SARS, MERS, influenza, and dengue virus [131,138]. Specifically, PSMs may function in inhibiting viral proteins, intercalating viral nucleic acids, blocking the ACE2 receptor, and modulating the immune system (Figure 3) [129,132,135,138,139].

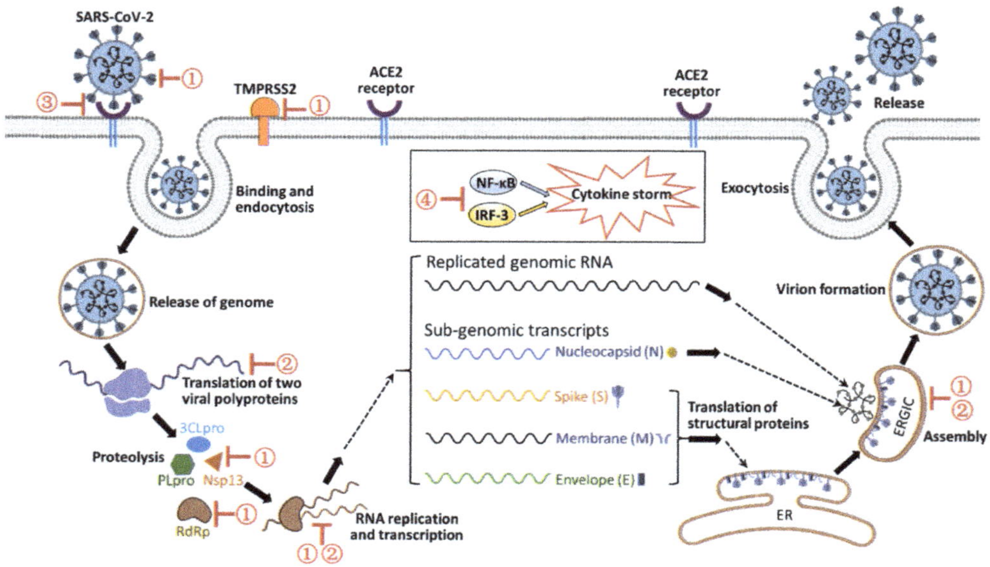

Figure 3. Schematic illustration of anti-SARS-CoV-2 mechanisms of PSMs from medicinal plants. Possible inhibition targets of PSMs during the viral life cycle are indicated by ①: inhibition of viral proteins; ②: intercalation of nucleic acids; ③: blocking of ACE2 receptor; ④: Immune modulation. TMPRSS2: Transmembrane protease, serine 2; PLpro: papain-like protease; 3CLpro: chymotrypsin-like protease; Nsp13: Nsp13 helicase; RdRp: RNA-dependent RNA polymerase; NF-κB: nuclear factor kappa-light-chain-enhancer of activated B cell; IRF3: interferon regulatory factor 3; ER: endoplasmic reticulum; ERGIC: ER-Golgi intermediate compartment.

3.1.1. Inhibition of Viral Proteins

The major drug targets that have been identified for SARS-CoV-2 through host-virus interaction studies include the SARS-CoV-2 main protease (Mpro), chymotrypsin-like protease (3CLpro), papain-like protease (PLpro), RNA-dependent RNA polymerase (RdRp), helicase Nsp13 and S proteins [36]. These viral proteins are critically involved in the viral replication and transcription process, and thus considered as the most promising targets for drug discovery against SARS-CoV-2 [135]. Since the outbreak of the COVID-19 pandemic, research has been conducted to screen for potential PSMs inhibiting the SARS-CoV-2 proteases, RdRp and other viral proteins using molecular docking analysis. Possible PSM inhibitors against major proteases and helicases of COVID-19 were summarized in recent reviews [129,140]. For example, six FDA-approved antiviral compounds, including Withaferin A, Nelfinavir, Rhein, Withanolide D, Enoxacin, and Aloe-emodin were discovered as possible COVID-19 main protease inhibitors [141]. In other research, the binding of a library of polyphenols with SARS-CoV-2 RdRp was assessed, and the study revealed that epigallocatechin gallate and three theaflavin derivatives could strongly bind to the active site of RdRp with high binding stability and with low toxicity, thus representing an effective therapy for COVID-19 [142]. In addition, in an in vivo study using hamsters as a disease model, extracts of *Perilla frutescens* and *Mentha haplocalyx* were found to be effective in inhibiting viral 3CL protease and RdRp [143], thus these compounds could be further developed as plant-derived anti–SARS-CoV-2 agents.

3.1.2. Intercalation of Nucleic Acids

Some types of PSMs, particularly alkaloids, can directly interact with DNA or RNA and hence stabilize those in single-stranded form. These DNA- or RNA-alkaloids conjugate readily, inhibit their further replication, and consequently prevent viral replication [128]. The well-known alkaloids that have been found to intercalate viral nucleic acids include berberine, emetine, sanguinarine, isoquinoline, beta-carboline, quinoline, paraquinine, dictamine, skimmianine [129]. Due to this mechanism, plant-derived alkaloids represent an important group of PSMs used to combat SARS-CoV-2.

3.1.3. Blocking of ACE2 Receptor

Because SARS-CoV-2 enters human cells through the ACE2 receptor, the simplest way to prevent viral infection is through blocking this receptor. Several PSMs, such as flavonoids, xanthones, proanthocyanidins, and secoiridoids have shown their binding affinities towards ACE2, thus becoming potential natural drugs against COVID-19 [144]. For example, quercetin was found to efficiently inhibit ACE2 [145]; essential oils isolated from lemon and Geranium could significantly downregulate the expression of the ACE2 receptor in human epithelial cells [146].

3.1.4. Immune Modulation

A growing body of clinical data has indicated that COVID-19 may cause a "cytokine storm" in patients due to an extreme inflammatory response, which is also a crucial cause of death from COVID-19. Certain PSMs have shown positive immunomodulation effects against this "cytokine storm". For example, alcoholic extract of hop (*Humulus lupulus*) and bark of cinnamon (*Cinnamum verum*) was found to inhibit NF-κB (nuclear factor kappa-light-chain-enhancer of activated B cell) which acts as a pro-inflammatory element [147].

3.2. Major Classes of PSMs against SARS-CoV-2

PSMs from three major classes: alkaloids, polyphenols and terpenoids/terpenes, have shown activities in preventing and treating COVID-19 infections [123].

3.2.1. Antiviral Alkaloids

Alkaloids represent a large class of PSMs that contain at least one nitrogen atom. According to their biosynthetic pathway alkaloids can be classified into several groups: tropanes, quinolines, indoles, purines, isoquinolines, imidazoles, pyrrolidines, pyrrolizidines and pyridines. The pharmacological effects of these alkaloid compounds include antioxidant, antifungal, antimalarial, antibacterial, and antiviral activities. The antiviral activity of some alkaloids, such as emetine, Ipecac, Macetaxime, tylophorine, and 7-methoxy cryptopleurine, is shown by inhibiting viral proteases, RNA synthesis and protein synthesis [37,124]. In silico screening analysis indicated that some alkaloid compounds, for example 10-hydroxyusambarensine and cryptoquindoline isolated from African medicinal plants, exert anti-SARS-CoV-2 activity through inhibition of 3CLpro [148]. In addition, some alkaloids, such as tetrandrine, fangchinoline, cepharanthine, and lycorine, inhibit the virus through intercalating into nucleic acid and inhibiting spike and nucleocapsid proteins [149]. So far, the alkaloid compounds showing the greatest inspiring antiviral effects against SARS-CoV-2 are papaverine, caffeine, berberine, colchicine, cryptospirolepine, deoxynortryptoquivaline, cryptomisrine, 10-hydroxyusambarensine, emetine, ergotamine, camptothecin, lycorine, nigellone, norboldine, and quinine [133,150]. These compounds could be further developed by being used alone or in combination with other drugs for treating COVID-19.

3.2.2. Antiviral Polyphenols

Polyphenols contain multiple aromatic rings and one or more hydroxyl groups. Polyphenols are broadly classified as flavonoids, lignans, stilbenes and phenolic acids according to the number of aromatic rings they contain and of the structural elements binding these rings together [151]. Numerous polyphenols are considered as antiviral agents. A comprehensive review on the phenolic compounds against SARS-CoV-2 was recently published [152]. Polyphenols show antiviral activities using diverse mechanisms, including intercalating into nucleic acid and inhibiting the activity of protease, helicase and RdRp [152,153]. This is because the hydroxyl group of polyphenols can interact with the positively charged amino groups of proteins and consequently destroy the three-dimensional structure of the protein [123]. In silico analysis revealed that polyphenols can inhibit the Mpro protease and RdRp of SARS-CoV-2 effectively [154]. Flavonoids consisting of two aromatic rings bound together by three carbon atoms comprise the most studied group of polyphenols. Flavonoid compounds were found to be able to inactivate the Mpro protease of SARS-CoV-2 [155]. In addition, the flavonoid scutellarein from the root of *Lamiacaea* was shown to inhibit the NSP13 helicase of SARS-CoV-2 by altering its ATPase activity [156]. In the past three years, many clinical trials of polyphenols as a possible treatment for patients with COVID-19 have been reported [152]. Polyphenols from different plant species have been shown to improve symptoms (fever, chills, cough, myalgia, and tachypnea), increase lymphocyte count, and decrease inflammation, etc. [152].

3.2.3. Antiviral Terpenoids/Terpenes

Terpenoids constitute a large group of PSMs with a broad spectrum of structures and effects. They are lipophilic compounds found in essential oils of many plants. Terpenoids could be used as antioxidant, anti-cancer, anti-inflammatory, antibacterial and antiviral reagents [123]. In terms of antiviral activity, lipophilic terpenoids can disturb the lipid envelope of viruses. Certain terpenes, such as celandine-B, betulinic acid, and ursolic acid have demonstrated strong antiviral effects as they can destroy the lipid layer of the virus [157]. Recent in silico screening indicated that some terpenes from African plants, such as 6-oxoisoiguesterin, 22-hydroxyhopan-3-one and 20-epi-isoiguesterino, could interact with the 3CLpro of SARS-CoV-2, and had binding affinities surpassing that of two reference compounds, lopinavir and ritonavir [148]. Recently, Glycyrrhizin, a triterpenoid saponin from licorice (*Glycyrrhyza glabra*) roots, was reported to be valuable in the treatment of COVID-19 due to its multi-target mode of action, such as binding to ACE2, downregulating proinflammatory cytokines, and stimulating endogenous interferon [158]. In addition, cannabidiol (CBD) from *Cannabis sativa* has been shown to downregulate ACE2 expression in COVID-19 target tissues, thus reducing COVID-19 severity [156].

3.3. Potential Anti-SARS-CoV-2 Compounds

Although numerous studies have focused on the inhibition of SARS-CoV and MERS-CoV with PSM compounds in the past years, there are few studies on the direct treatment of COVID-19 disease, which are limited to in silico studies [126]. A compilation of PSMs with potential inhibitory and regulatory activity against SARS-CoV-2 is listed in Table 3. Their chemical structure is shown in Figure 4. These metabolites were curated based on their potential ability to stop, prevent, and treat COVID-19 infections. Only those with potential direct activity against SARS-CoV-2 were considered. Additionally, most PSMs in Table 3 show little toxicity, as indicated by in vitro or in silico evaluations.

Given the novelty of the disease and the shift in attention to other areas, PSMs have not been studied thoroughly for treating COVID-19 and most research is still in early phases, as evidenced by the predominance of in vitro models throughout the table. Therefore, the metabolites presented are only prospects selected by their confirmed in vitro activity and/or their theorized capabilities by in silico models. More research is needed to fully corroborate their action against SARS-CoV-2 in humans. The PSM compounds introduced below are a selection of those with the most advanced stages in research.

Table 3. Compounds of medicinal plants found to be effective against SARS-CoV-2 through in vitro, in vivo or in silico analysis.

Major Active Compounds	Plant Species	Efficacy/Mechanism of Action *	References
Artemisinin, flavonoids, artesunate, artemether, nonidentified metabolites	*Artemisia annua* L.	Inhibiting viral replication	[159]
		Inhibiting replication of five virus variants including Delta	[160]
		Inhibiting viral infection	[161]
Astersaponin I (AI)	*Aster koraiensis*	Inhibiting virus entry pathways at plasma membrane and within endosomal compartments	[162]
Curcumin	*Curcuma longa*	Binding and inhibiting S protein of Omicron variant (in silico analysis)	[163]
Emetine	*Carapichea ipecacuanha*	Blocking viral entry into cells; inhibiting virus replication; anti-inflammation	[164]
Hesperidin (Hesperetin)	Various species	Reducing expression of TMPRSS2 and ACE2 (in silico analysis)	[165]
Hypericin	*Hypericum perforatum*	Binding viral envelope and reducing its infectivity	[166]
Licorice-saponin A3 (A3) and glycyrrhetinic acid (GA)	*Glycyrrhiza uralensis*	Inhibiting viral infection	[167]
Luteolin	Various species	Reducing viral replication by inhibition of RdRp	[168]
Myricetin	Various species	Inhibiting viral replication and transcription by inhibition of protease (Mpro); anti-inflammation (in vivo analysis)	[169]
Nonalkaloid compounds	*Rhazya stricta*	Binding key residues of S proteins and impeding viral infection (in silico analysis)	[170]
Panduratin A	*Boesenbergia rotunda*	Inhibiting viral replication	[171]
Persimmon-derived tannins	*Diospyros kaki*	Inhibiting virus replication; potential as a prophylactic agent	[172]
Piperine	*Piper* spp.	Inhibiting viral replication	[173]
Quercetin	Various species	Impeding viral replication by inhibition of RdRp	[168]
Thapsigargin	*Thapsia garganica*	Inhibiting viral replication by inducing stress in ER and increasing the viability of infected cells	[174]
Withaferin A, Withanone, Withanolide A	*Withania somnifera* (L.)	Inhibiting viral infection and replication; anti-inflammation and proinflammatory cytokines (in vivo analysis)	[175]

*: all those unindicated are in vitro analysis.

3.3.1. Artemisinin

Artemisinin derivatives from *Artemisia annua* L. are effective in treating malaria. They are also well documented as antiviral drugs [176]. The study by Nair et al. (2021) revealed a potent inhibitory action of *A. annua* L. leaf extracts on viral infection in Vero E6 cells with relatively low IC_{50} values [159]. The study also demonstrated that artemisinin is not the main or only metabolite with antiviral properties in the extracts. Furthermore, the metabolites present in the leaf extracts were found effective against five variants of the virus (Alpha, Beta, Gamma, Delta and Kappa) and seemed to have great stability [160]. Dry leaf samples still showed antiviral activity after staying frozen for twelve years [159]. In another study by Zhou et al. (2021), it was found that *A. annua* extracts as well as individual compounds (artemisinin, artesunate, and artemether) all showed inhibitory effects on viral infection of Vero E6 cells, human hepatoma Huh7.5 cells and human lung cancer A549-hACE2 cells. Among them, artesunate proved most potent in different cell types, and it targeted SARS-CoV-2 at the post-entry level [161].

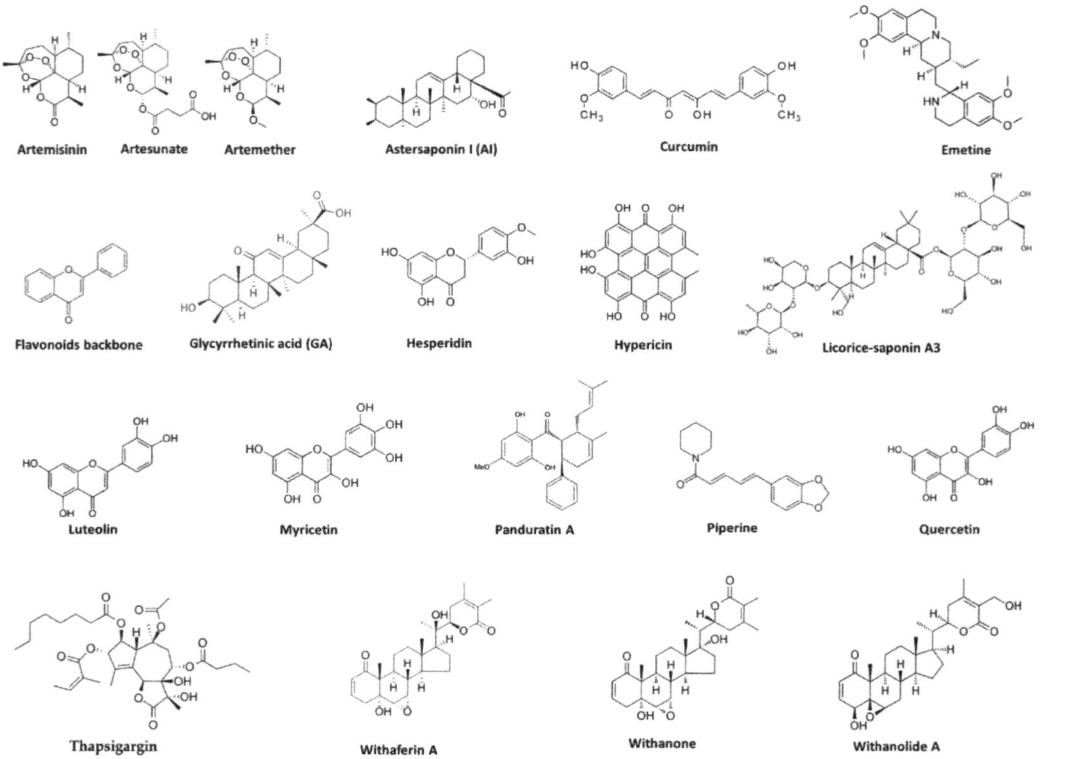

Figure 4. Chemical structures of some bioactive PSM compounds against SARS-CoV-2. Those with defined structures are shown here.

3.3.2. Hesperidin/Hesperetin

Hesperidin and its aglycone hesperetin, compounds isolated from citrus plants, are also regarded as potential antiviral drugs. Both compounds were initially studied due to their predicted interaction in molecular models with ACE2 and TMPRSS2 [165]. These molecules are crucial in the SARS-CoV-2 cell hijacking mechanism. Cheng et al. (2022) applied enzymatic activity assays and in vitro studies with Vero E6 cells and particles of pseudo-virus, demonstrating a strong disruption of ACE2-S protein interaction, but a slight inhibition of TMPRSS2. Nonetheless, their research also revealed another action mechanism for hesperidin and hesperetin—the downregulation of both ACE2 and TMPRSS2 in lung epithelial lung cancer cell lines. These metabolites' disruptive activity and little cytotoxicity make them prospective agents to prevent further cell entry in infected individuals, allowing for a quicker recovery.

3.3.3. Emetine

Emetine, an alkaloid extracted from the *ipecacuanha* plant (Brazilian root), is another compound with strong antiviral activity [177]. Its effect against other coronaviruses is well documented and SARS-CoV-2 is no exception. Several papers have documented the multifaceted approach of emetine against SARS-CoV-2, inhibiting not only viral entry, but also replication and proliferation [164]. In a study published by Wang et al. (2020) emetine swiftly reduced viral activity in SARS-CoV-2 infected Vero cells and blocked viral entry in pretreated cells [178]. The metabolite disrupts the interaction between viral mRNA and key molecules involved with translation, including ribosomes, viral polymerase RdRp and replication–transcription complex translation initiation factor eIF4E [179,180], as suggested by in silico simulations. In addition, the action of emetine remains effective even at low concentrations, making it a potent drug. However, the researchers also noted that emetine has been associated with cardiac complications and possible cytotoxicity, leaving its viability as a drug yet to be assessed.

3.3.4. Luteolin and Quercetin

Similar to emetine, luteolin and quercetin, which are ancestors of flavonoid natural compounds, exhibit in vitro inhibition against RdRp of SARS-CoV-2 [168]. Previous in vitro studies have also suggested their inhibitory activity against viral protease 3CLpro [71]. Their ability to reduce SARS-CoV-2 replication by preventing translation and post-translational processing makes both metabolites promising treatment drugs for COVID-19.

3.3.5. Panduratin A

Panduratin A, a diarylheptanoid found in *Renealmia nicolaioides* and *Boesenbergia rotunda* can not only prevent infection but also slow viral replication. As discovered by Kanjanasirirat et al. (2020), panduratin A significantly stops the activity of SARS-CoV-2 in Vero E6 and Calu-3 cell lines in pre- and post- infection stages with a performance comparable to the already approved COVID-19 treatment drugs. Additionally, it is important to highlight the minimal toxicity shown by panduratin A, as evidenced by cytotoxicity assays in five different cell lines [171].

3.3.6. Tannins

The case of persimmon-derived tannins has vital importance as it represents one of the few studies to include in vivo models. After the hamster model was treated with persimmon-derived tannins and then inoculated with SARS-CoV-2 viral particles, it remained healthy and presented low antigen and viral load levels in their lungs, as indicated by immunohistochemistry assays and qPCR [172]. By contrast, the control group presented severe lung inflammation and greater pathophysiology and a higher antigen count. Even though the researchers demonstrated the prophylactic potential of persimmon-derived tannins to prevent/regulate SARS-CoV-2 infection, further studies are needed to assess the efficacy of the compound in already infected organisms.

3.4. *Challenges in Clinical Applications of PSMs against SARS-CoV-2*

PSMs from various medicinal plants serve as a treasure of bioactive compounds that have shown promising results against SARS-CoV-2. However, due to the novelty of COVID-19, and lack of experimental evidence and safety studies, the use of most PSMs is still limited [126]. So far, none of the isolated PSM compounds from medicinal plants have been successfully used for clinically treating COVID-19. First of all, there is still a lack of sufficient in vivo and clinical studies to demonstrate the effectiveness of PSM compounds in preventing the viral infection or alleviating symptoms associated with virus infection [181]. In fact, numerous compounds showing high antiviral activity in vitro could be found to be inactive in pre-clinical or clinical trials [138]. Second, because the PSMs are small molecules, they are usually stable and can be delivered orally as plant extracts without the need to purify from other by-products. However, the treatment with plant extracts becomes complicated due to the existence of various compounds. The antiviral ability of individual compounds may be different from their functions in extracts, and they can be additive or synergistic, or even antagonistic [156]. In addition, some PSMs could be toxic at certain levels, so it is necessary to conduct in vitro and in vivo research to evaluate the safety and therapeutic levels of each compound before conducting human clinical trials [131]. Therefore, tremendous research is still needed to find the most effective PSM compounds or a combination of them that would be effective in treating COVID-19 infection.

4. Prospects

The COVID pandemic has been affecting the world for three years. It is generally accepted that SARS-CoV-2 will not be fully eradicated. Most likely, the virus will coexist with humans, and the disease will become an endemicity. There is a critical need to develop new and effective pharmaceuticals for prophylactic and therapeutic purposes. The rapid development of vaccines against SARS-CoV-2 has provided a major step forward in reducing COVID-19's impact, thus representing a scientific victory. However, many people, even those fully vaccinated lose protection over time. This problem seems to be especially pronounced with the Omicron-related variants. Therefore, there is still a pressing need for globally available vaccines that can provide more lasting immunity against current and future coronavirus variants [94]. A ferritin-based COVID-19 nanoparticle vaccine that elicits robust, durable, broad-spectrum neutralizing antisera against known variants of concern, including Omicron BQ.1 and the previous virus version SARS-CoV-1, was just reported [182]. This offers great potential for the rapid response of the emerging SARS-CoV-2 variants and provides versatility for the future development of vaccines against other emerging coronaviruses. In addition, all currently approved vaccines are administered parenterally, which may not be effective in preventing mucosal infection of respiratory pathogens like SARS-CoV-2. Therefore, mucosal COVID-19 vaccines, such as those administered orally or intranasally, would potentially be more effective in offering protection against SARS-CoV-2 infection, because they offer the dual benefit of inducing potent mucosal and systemic immunity [183]. Around 100 mucosal COVID-19 vaccines are in development globally. Among them, 20 have reached clinical trials in humans [184]. In late 2022, two mucosal COVID-19 vaccines that are delivered through the nose or mouth have been approved for use in China and India [185]. These approvals validate the need for mucosal vaccines, though their effectiveness in preventing COVID-19 infection needs be further assessed. In addition to vaccines, other antiviral agents with the potential to limit virus transmission or block infection, including both big molecules and small molecules, will continue to be explored.

Plants provide attractive bioproduction platforms for both recombinant therapeutics and natural bioactive metabolites to combat the COVID pandemic. Molecular farming in plants is an unprecedented opportunity for developing vaccines, antibodies, and other biologics for pandemic diseases because of its potential advantages, such as low cost, safety, and high production volume. As mentioned above, numerous anti-SARS-CoV-2 biologics have been expressed in plant systems. Because many plants or plant products are edible, advances in this area can lead to oral vaccines that can induce mucosal immunity, and that are low cost, easy-to-administer, and have high thermostability [186]. This could be especially applicable for vaccinating people in developing countries, such as those in Africa, where high costs and logistical problems can constrain massive vaccine programs [60]. However, compared to the major expression hosts (bacteria, yeast and mammalian cells), plants are still largely underutilized, mainly due to low productivity and non-human glycosylation [187,188]. Modern molecular biology tools, such as RNAi and the latest genome editing technology, could be exploited to modulate the genome of plant cells to create new plant lines exhibiting improved "traits" for therapeutic protein production [189].

Alternatively, medicinal plants provide a significant prospect for discovering new and effective anti-COVID-19 drugs. The PSMs from medicinal plants have been demonstrated to be powerful in fighting against SARS-CoV-2 as they can interfere with the viral life cycle, including viral entrance, replication, assembly, and virus-specific host targets [190]. Many potential antiviral PSMs against SARS-CoV-2 have been tested in vitro, in vivo or predicted by in silico analysis. However, none of the PSM compounds have been approved by the FDA for treating COVID-19 so far. More pre-clinical and clinical evaluation of the therapeutic effectiveness of these PSMs is a major concern for further development of safe and effective treatments. Because of the novelty of the virus and the disease caused, safety is still a main concern for the use of PSMs [138]. Although compared to synthetic medicines, PSMs are regarded as less toxic because of their natural origin and long-term use as traditional medicines, these compounds may have potential adverse or toxic effects at certain concentrations [36]. Therefore, further investigations, particularly pre-clinical evaluation, are necessary for determining the safe therapeutic dose of each compound before clinical application. In addition, plant metabolomics is currently used as a tool to discover novel drugs from plant resources [37]. Characterization of genes and enzymes involved in secondary metabolic pathways is also very crucial for understanding the biosynthesis of bioactive compounds [191]. This will pave the way for further genetic modifications of medicinal plants to synthesize novel PSM compounds that are most effective in treating COVID-19. Finally, in order to improve the use of PSM compounds, combination treatments, for example the treatments in combination with the FDA-approved anti-SARS-CoV-2 drugs or with the assistance of nanotechnology, may be a promising strategy to develop as they exhibit better synergistic and/or additive effects against COVID-19 [137].

5. Conclusions

The COVID-19 pandemic not only caused a public health crisis, but also severely affected the global economy. Although the epidemic has been alleviated to a great extent around the world, the virus will continue to coexist with human beings and constantly mutate. Plants provide a promising bioproduction platform for both recombinant therapeutics (big molecules) and natural bioactive compounds (small molecules) that can be used to combat the virus. "Molecular farming" in plants proposes a superior bioproduction platform for recombinant therapeutics as compared to other eukaryotic systems in terms of safety, scalability, and cost. In the future, advances in this area could also lead to oral vaccines that may be convenient and easy to deploy. Alternatively, plants represent a dramatically underutilized source of bioactive compounds with a broad spectrum of antiviral activities. In vitro, in vivo, and in silico analyses have revealed numerous plant-derived compounds with promising anti-SARS-CoV-2 activity. Therefore, these molecules will be able to develop new natural solutions for treating COVID-19. In summary, it will take the combined efforts of plant genetic engineering and natural plant medicine research to ultimately extinguish this pandemic.

Supplementary Materials: The following supporting information can be downloaded at: https://www.mdpi.com/article/10.3390/life13030617/s1, Table S1: Subunit vaccines and virus-like particle vaccines against COVID-19 that have reached or passed Phase I human trials according to the COVID-19 vaccine tracker website.

Author Contributions: Conceptualization, J.X.; writing—original draft preparation, C.E., J.X., J.T., U.K. and P.P.; writing—review and editing, J.X. and P.P.; funding acquisition, J.X. All authors have read and agreed to the published version of the manuscript.

Funding: This work was supported by the Arkansas IDeA Network of Biomedical Research Excellence—Research and Development Grant (Grant No. P20GM103429), the National Institute of Health (Grant No. R15DK128757), and the Arkansas Biosciences Institute, the major research component of the Arkansas Tobacco Settlement Proceeds Act of 2000.

Institutional Review Board Statement: Not applicable.

Informed Consent Statement: Not applicable.

Data Availability Statement: Not applicable.

Conflicts of Interest: The authors declare no conflict of interest.

References

1. Lu, R.; Zhao, X.; Li, J.; Niu, P.; Yang, B.; Wu, H.; Wang, W.; Song, H.; Huang, B.; Zhu, N.; et al. Genomic characterisation and epidemiology of 2019 novel coronavirus: Implications for virus origins and receptor binding. *Lancet* **2020**, *395*, 565–574. [CrossRef] [PubMed]
2. Jackson, C.B.; Farzan, M.; Chen, B.; Choe, H. Mechanisms of SARS-CoV-2 entry into cells. *Nat. Rev. Mol. Cell Biol.* **2022**, *23*, 3–20. [CrossRef] [PubMed]
3. Lan, J.; Ge, J.; Yu, J.; Shan, S.; Zhou, H.; Fan, S.; Zhang, Q.; Shi, X.; Wang, Q.; Zhang, L.; et al. Structure of the SARS-CoV-2 spike receptor-binding domain bound to the ACE2 receptor. *Nature* **2020**, *581*, 215–220. [CrossRef]
4. Leung, K.; Lau, E.H.Y.; Wong, C.K.H.; Leung, G.M.; Wu, J.T. Estimating the transmission dynamics of SARS-CoV-2 Omicron BF.7 in Beijing after the adjustment of zero-COVID policy in November–December 2022. *Nat. Med.* 2023, *ahead of print*. [CrossRef]
5. Pal, M.; Berhanu, G.; Desalegn, C.; Kandi, V. Severe acute respiratory syndrome coronavirus-2 (SARS-CoV-2): An update. *Cureus* **2020**, *12*, e7423. [CrossRef] [PubMed]
6. Huang, Y.; Yang, C.; Xu, X.F.; Xu, W.; Liu, S.W. Structural and functional properties of SARS-CoV-2 spike protein: Potential antivirus drug development for COVID-19. *Acta Pharmacol. Sin.* **2020**, *41*, 1141–1149. [CrossRef]
7. Zhang, H.; Penninger, J.M.; Li, Y.; Zhong, N.; Slutsky, A.S. Angiotensin-converting enzyme 2 (ACE2) as a SARS-CoV-2 receptor: Molecular mechanisms and potential therapeutic target. *Intensive Care Med.* **2020**, *46*, 586–590. [CrossRef]
8. Lei, C.; Qian, K.; Li, T.; Zhang, S.; Fu, W.; Ding, M.; Hu, S. Neutralization of SARS-CoV-2 spike pseudotyped virus by recombinant ACE2-Ig. *Nat. Commun.* **2020**, *11*, 2070. [CrossRef]
9. Martinez-Flores, D.; Zepeda-Cervantes, J.; Cruz-Resendiz, A.; Aguirre-Sampieri, S.; Sampieri, A.; Vaca, L. SARS-CoV-2 vaccines based on the spike glycoprotein and implications of new viral variants. *Front. Immunol.* **2021**, *12*, 701501. [CrossRef]
10. Premkumar, L.; Segovia-Chumbez, B.; Jadi, R.; Martinez, D.R.; Raut, R.; Markmann, A.; Cornaby, C.; Bartelt, L.; Weiss, S.; Park, Y.; et al. The receptor binding domain of the viral spike protein is an immunodominant and highly specific target of antibodies in SARS-CoV-2 patients. *Sci. Immunol.* **2020**, *5*, eabc8413. [CrossRef]
11. Abduljaleel, Z.; Shahzad, N.; Aziz, S.A.; Malik, S.M. Monoclonal antibody designed for SARS-nCoV-2 spike protein of receptor binding domain on antigenic targeted epitopes for inhibition to prevent viral entry. *Mol. Divers.* 2022, *published online ahead of print*. [CrossRef]
12. Cao, Y.; Wang, J.; Jian, F.; Xiao, T.; Song, W.; Yisimayi, A.; Huang, W.; Li, Q.; Wang, P.; An, R.; et al. Omicron escapes the majority of existing SARS-CoV-2 neutralizing antibodies. *Nature* **2022**, *602*, 657–663. [CrossRef]
13. Antonelli, M.; Pujol, J.C.; Spector, T.D.; Ourselin, S.; Steves, C.J. Risk of long COVID associated with delta versus omicron variants of SARS-CoV-2. *Lancet* **2022**, *399*, 2263–2264. [CrossRef] [PubMed]
14. Wang, Z.; Yang, L.; Song, X.Q. Oral GS-441524 derivatives: Next-generation inhibitors of SARS-CoV-2 RNA-dependent RNA polymerase. *Front. Immunol.* **2022**, *13*, 1015355. [CrossRef] [PubMed]
15. Imai, M.; Ito, M.; Kiso, M.; Yamayoshi, S.; Uraki, R.; Fukushi, S.; Watanabe, S.; Suzuki, T.; Maeda, K.; Sakai-Tagawa, Y.; et al. Efficacy of antiviral agents against omicron subvariants BQ.1.1 and XBB. N. *Engl. J. Med.* **2023**, *388*, 89–91. [CrossRef] [PubMed]
16. Tomalka, J.A.; Suthar, M.S.; Deeks, S.G.; Sekaly, R.P. Fighting the SARS-CoV-2 pandemic requires a global approach to understanding the heterogeneity of vaccine responses. *Nat. Immunol.* **2022**, *23*, 360–370. [CrossRef]
17. Galati, D.; Zanotta, S.; Capitelli, L.; Bocchino, M. A bird's eye view on the role of dendritic cells in SARS-CoV-2 infection: Perspectives for immune-based vaccines. *Allergy* **2022**, *77*, 100–110. [CrossRef]
18. Wang, Z.; Yang, L. Broad-spectrum prodrugs with anti-SARS-CoV-2 activities: Strategies, benefits, and challenges. *J. Med. Virol.* **2022**, *94*, 1373–1390. [CrossRef]
19. Jin, Y.H.; Jeon, S.; Lee, J.; Kim, S.; Jang, M.S.; Park, C.M.; Song, J.H.; Kim, H.R.; Kwon, S. Broad spectrum antiviral properties of cardiotonic steroids used as potential therapeutics for emerging coronavirus infections. *Pharmaceutics* **2021**, *13*, 1839. [CrossRef]
20. Polack, F.P.; Thomas, S.J.; Kitchin, N.; Absalon, J.; Gurtman, A.; Lockhart, S.; Perez, J.L.; Perez Marc, G.; Moreira, E.D.; Zerbini, C.; et al. Safety and efficacy of the BNT162b2 mRNA COVID-19 vaccine. *N. Engl. J. Med.* **2020**, *383*, 2603–2615. [CrossRef]
21. Mendonca, S.A.; Lorincz, R.; Boucher, P.; Curiel, D.T. Adenoviral vector vaccine platforms in the SARS-CoV-2 pandemic. *NPJ Vaccines* **2021**, *6*, 97. [CrossRef]
22. Larkin, H.D. Novavax COVID-19 vaccine booster authorized. *JAMA* **2022**, *328*, 2101. [CrossRef]
23. Tan, C.Y.; Chiew, C.J.; Lee, V.J.; Ong, B.; Lye, D.C.; Tan, K.B. Comparative effectiveness of 3 or 4 doses of mRNA and inactivated whole-virus vaccines against COVID-19 infection, hospitalization and severe outcomes among elderly in Singapore. *Lancet Reg. Health West. Pac.* **2022**, *29*, 100654. [CrossRef] [PubMed]
24. Tregoning, J.S.; Flight, K.E.; Higham, S.L.; Wang, Z.; Pierce, B.F. Progress of the COVID-19 vaccine effort: Viruses, vaccines and variants versus efficacy, effectiveness and escape. *Nat. Rev. Immunol.* **2021**, *21*, 626–636. [CrossRef] [PubMed]
25. Taylor, P.C.; Adams, A.C.; Hufford, M.M.; de la Torre, I.; Winthrop, K.; Gottlieb, R.L. Neutralizing monoclonal antibodies for treatment of COVID-19. *Nat. Rev. Immunol.* **2021**, *21*, 382–393. [CrossRef] [PubMed]
26. Jaworski, J.P. Neutralizing monoclonal antibodies for COVID-19 treatment and prevention. *Biomed. J.* **2021**, *44*, 7–17. [CrossRef]
27. Hentzien, M.; Autran, B.; Piroth, L.; Yazdanpanah, Y.; Calmy, A. A monoclonal antibody stands out against omicron subvariants: A call to action for a wider access to bebtelovimab. *Lancet Infect. Dis.* **2022**, *22*, 1278. [CrossRef]
28. Beigel, J.H.; Tomashek, K.M.; Dodd, L.E.; Mehta, A.K.; Zingman, B.S.; Kalil, A.C.; Hohmann, E.; Chu, H.Y.; Luetkemeyer, A.; Kline, S.; et al. Remdesivir for the treatment of COVID-19—Final report. *N. Engl. J. Med.* **2020**, *383*, 1813–1826. [CrossRef]

29. Anonymous. An EUA for baricitinib (Olumiant) for COVID-19. *Med. Lett. Drugs Ther.* **2020**, *62*, 202–203.
30. Wen, W.; Chen, C.; Tang, J.; Wang, C.; Zhou, M.; Cheng, Y.; Zhou, X.; Wu, Q.; Zhang, X.: Feng, Z.; et al. Efficacy and safety of three new oral antiviral treatment (molnupiravir, fluvoxamine and Paxlovid) for COVID-19: A meta-analysis. *Ann. Med.* **2022**, *54*, 516–523. [CrossRef]
31. Yao, J.; Weng, Y.; Dickey, A.; Wang, K.Y. Plants as factories for human pharmaceuticals: Applications and challenges. *Int. J. Mol. Sci.* **2015**, *16*, 28549–28565. [CrossRef]
32. El-Demerdash, A.; Hassan, A.; Abd El-Aziz, T.M.; Stockand, J.D.; Arafa, R.K. Marine brominated tyrosine alkaloids as promising inhibitors of SARS-CoV-2. *Molecules* **2021**, *26*, 6171. [CrossRef]
33. Wang, Z.; Yang, L. Chinese herbal medicine: Fighting SARS-CoV-2 infection on all fronts. *J. Ethnopharmacol.* **2021**, *270*, 113869. [CrossRef] [PubMed]
34. Sahoo, A.; Fuloria, S.; Swain, S.S.; Panda, S.K.; Sekar, M.; Subramaniyan, V.; Panda, M.; Jena, A.K.; Sathasivam, K.V.; Fuloria, N.K. Potential of marine terpenoids against SARS-CoV-2: An in silico drug development approach. *Biomedicines* **2021**, *9*, 1505. [CrossRef] [PubMed]
35. Haddad, M.; Gaudreault, R.; Sasseville, G.; Nguyen, P.T.; Wiebe, H.; Van De Ven, T.; Bourgault, S.; Mousseau, N.; Ramassamy, C. Molecular interactions of tannic acid with proteins associated with SARS-CoV-2 infectivity. *Int. J. Mol. Sci.* **2022**, *23*, 2643. [CrossRef] [PubMed]
36. Merarchi, M.; Dudha, N.; Das, B.C.; Garg, M. Natural products and phytochemicals as potential anti-SARS-CoV-2 drugs. *Phytother. Res.* **2021**, *35*, 5384–5396. [CrossRef]
37. Bhuiyan, F.R.; Howlader, S.; Raihan, T.; Hasan, M. Plants metabolites: Possibility of natural therapeutics against the COVID-19 pandemic. *Front. Med.* **2020**, *7*, 444. [CrossRef]
38. Schillberg, S.; Finnern, R. Plant molecular farming for the production of valuable proteins—Critical evaluation of achievements and future challenges. *J. Plant Physiol.* **2021**, *258–259*, 153359. [CrossRef]
39. Fischer, R.; Buyel, J.F. Molecular farming—The slope of enlightenment. *Biotechnol Adv.* **2020**, *40*, 107519. [CrossRef]
40. Hager, K.J.; Perez Marc, G.; Gobeil, P.; Diaz, R.S.; Heizer, G.; Llapur, C.; Makarkov, A.I.; Vasconcellos, E.; Pillet, S.; Riera, F.; et al. Efficacy and safety of a recombinant plant-based adjuvanted COVID-19 vaccine. *N. Engl. J. Med.* **2022**, *386*, 2084–2096. [CrossRef]
41. Tuse, D.; Nandi, S.; McDonald, K.A.; Buyel, J.F. The Emergency Response Capacity of Plant-based biopharmaceutical manufacturing-what it is and what it could be. *Front. Plant Sci.* **2020**, *11*, 594019. [CrossRef]
42. Capell, T.; Twyman, R.M.; Armario-Najera, V.; Ma, J.K.; Schillberg, S.; Christou, P. Potential applications of plant biotechnology against SARS-CoV-2. *Trends Plant Sci.* **2020**, *25*, 635–643. [CrossRef]
43. Mahmood, N.; Nasir, S.B.; Hefferon, K. Plant-based drugs and vaccines for COVID-19. *Vaccines* **2020**, *9*, 15. [CrossRef] [PubMed]
44. Ortega-Berlanga, B.; Pniewski, T. Plant-based vaccines in combat against coronavirus diseases. *Vaccines* **2022**, *10*, 138. [CrossRef] [PubMed]
45. Xu, J.; Towler, M.; Weathers, P.J. Platforms for plant-based protein production. In *Bioprocessing of Plant In Vitro Systems*, 1st ed.; Pavlov, A., Bley, T., Eds.; Springer International Publishing AG: Midtown Manhattan, NY, USA, 2016; pp. 1–40. [CrossRef]
46. Jugler, C.; Sun, H.; Nguyen, K.; Palt, R.; Felder, M.; Steinkellner, H.; Chen, Q. A novel plant-made monoclonal antibody enhances the synergetic potency of an antibody cocktail against the SARS-CoV-2 Omicron variant. *Plant Biotechnol. J.* 2022, *published online ahead of print*. [CrossRef]
47. Ruocco, V.; Strasser, R. Transient expression of glycosylated SARS-CoV-2 antigens in Nicotiana benthamiana. *Plants* **2022**, *11*, 1093. [CrossRef] [PubMed]
48. Schillberg, S.; Raven, N.; Spiegel, H.; Rasche, S.; Buntru, M. Critical Analysis of the Commercial potential of plants for the production of recombinant proteins. *Front. Plant Sci.* **2019**, *10*, 720. [CrossRef]
49. Tekoah, Y.; Shulman, A.; Kizhner, T.; Ruderfer, I.; Fux, L.; Nataf, Y.; Bartfeld, D.; Ariel, T.; Gingis-Velitski, S.; Hanania, U.; et al. Large-scale production of pharmaceutical proteins in plant cell culture-the protalix experience. *Plant Biotechnol. J.* **2015**, *13*, 1199–1208. [CrossRef]
50. Ward, B.J.; Makarkov, A.; Seguin, A.; Pillet, S.; Trepanier, S.; Dhaliwall, J.; Libman, M.D.; Vesikari, T.; Landry, N. Efficacy, immunogenicity, and safety of a plant-derived, quadrivalent, virus-like particle influenza vaccine in adults (18–64 years) and older adults (>/=65 years): Two multicentre, randomised phase 3 trials. *Lancet* **2020**, *396*, 1491–1503. [CrossRef]
51. Siriwattananon, K.; Manopwisedjaroen, S.; Shanmugaraj, B.; Rattanapisit, K.; Phumiamorn, S.; Sapsutthipas, S.; Trisiriwanich, S.; Prompetchara, E.; Ketloy, C.; Buranapraditkun, S.; et al. Plant-produced receptor-binding domain of SARS-CoV-2 elicits potent neutralizing responses in mice and non-human primates. *Front. Plant Sci.* **2021**, *12*, 682953. [CrossRef]
52. Wu, Y.; Wang, F.; Shen, C.; Peng, W.; Li, D.; Zhao, C.; Li, Z.; Li, S.; Bi, Y.; Yang, Y.; et al. A noncompeting pair of human neutralizing antibodies block COVID-19 virus binding to its receptor ACE2. *Science* **2020**, *368*, 1274–1278. [CrossRef]
53. Mamedov, T.; Gurbuzaslan, I.; Yuksel, D.; Ilgin, M.; Mammadova, G.; Ozkul, A.; Hasanova, G. Soluble human angiotensin-converting enzyme 2 as a potential therapeutic tool for COVID-19 is produced at high levels in *Nicotiana benthamiana* plant with potent anti-SARS-CoV-2 activity. *Front. Plant Sci.* **2021**, *12*, 742875. [CrossRef]
54. Alfaleh, M.A.; Zawawi, A.; Al-Amri, S.S.; Hashem, A.M. David versus goliath: ACE2-Fc receptor traps as potential SARS-CoV-2 inhibitors. *MAbs* **2022**, *14*, 2057832. [CrossRef]

55. Cai, Y.; Xu, W.; Gu, C.; Cai, X.; Qu, D.; Lu, L.; Xie, Y.; Jiang, S. Griffithsin with a broad-spectrum antiviral activity by binding glycans in viral glycoprotein exhibits strong synergistic effect in combination with a pan-coronavirus fusion inhibitor targeting SARS-CoV-2 spike S2 subunit. *Virol. Sin.* **2020**, *35*, 857–860. [CrossRef]
56. Heidary, M.; Kaviar, V.H.; Shirani, M.; Ghanavati, R.; Motahar, M.; Sholeh, M.; Ghahramanpour, H.; Khoshnood, S. A comprehensive review of the protein subunit vaccines against COVID-19. *Front. Microbiol.* **2022**, *13*, 927306. [CrossRef]
57. Tariq, H.; Batool, S.; Asif, S.; Ali, M.; Abbasi, B.H. Virus-like particles: Revolutionary platforms for developing vaccines against emerging infectious diseases. *Front. Microbiol.* **2021**, *12*, 790121. [CrossRef]
58. Moyle, P.M.; Toth, I. Modern subunit vaccines: Development, components, and research opportunities. *ChemMedChem* **2013**, *8*, 360–376. [CrossRef] [PubMed]
59. Peyret, H.; Steele, J.F.C.; Jung, J.W.; Thuenemann, E.C.; Meshcheriakova, Y.; Lomonossoff, G.P. Producing vaccines against enveloped viruses in plants: Making the impossible, difficult. *Vaccines* **2021**, *9*, 780. [CrossRef] [PubMed]
60. El Jaddaoui, I.; Al Idrissi, N.; Hamdi, S.; Wakrim, L.; Nejjari, C.; Amzazi, S.; Elouahabi, A.; Bakri, Y.; Ghazal, H. Plant-based vaccines against COVID-19 for massive vaccination in Africa. *Front. Drug Deliv.* **2022**, *2*, 909958. [CrossRef]
61. Medicago Covifenz COVID-19 Vaccine. Available online: https://www.canada.ca/en/health-canada/services/drugs-health-products/covid19-industry/drugs-vaccines-treatments/vaccines/medicago.html#a4 (accessed on 1 December 2022).
62. COVID-19 Vaccine Development and Approvals Tracker. Available online: https://covid19.trackvaccines.org (accessed on 2 December 2022).
63. KBP-201 COVID-19 Vaccine Trial in Healthy Volunteers. Available online: https://clinicaltrials.gov/ct2/show/NCT04473690 (accessed on 1 December 2022).
64. Demarco, J.K.; Royal, J.M.; Severson, W.E.; Gabbard, J.D.; Hume, S.; Morton, J.; Swope, K.; Simpson, C.A.; Shepherd, J.W.; Bratcher, B.; et al. CoV-RBD121-NP vaccine candidate protects against symptomatic disease following SARS-CoV-2 challenge in K18-hACE2 mice and induces protective responses that prevent COVID-19-associated immunopathology. *Vaccines* **2021**, *9*, 1346. [CrossRef] [PubMed]
65. Royal, J.M.; Simpson, C.A.; McCormick, A.A.; Phillips, A.; Hume, S.; Morton, J.; Shepherd, J.; Oh, Y.; Swope, K.; De Beauchamp, J.L.; et al. Development of a SARS-CoV-2 vaccine candidate using plant-based manufacturing and a tobacco mosaic virus-like nano-particle. *Vaccines* **2021**, *9*, 1347. [CrossRef] [PubMed]
66. Maharjan, P.M.; Choe, S. Plant-based COVID-19 vaccines: Current status, design, and development strategies of candidate vaccines. *Vaccines* **2021**, *9*, 992. [CrossRef]
67. Uthaya Kumar, A.; Kadiresen, K.; Gan, W.C.; Ling, A.P.K. Current updates and research on plant-based vaccines for coronavirus disease 2019. *Clin. Exp. Vaccine Res.* **2021**, *10*, 13. [CrossRef]
68. Balieu, J.; Jung, J.W.; Chan, P.; Lomonossoff, G.P.; Lerouge, P.; Bardor, M. Investigation of the N-glycosylation of the SARS-CoV-2 S protein contained in VLPs produced in Nicotiana benthamiana. *Molecules* **2022**, *27*, 5119. [CrossRef]
69. Shanmugaraj, B.; Khorattanakulchai, N.; Panapitakkul, C.; Malla, A.; Im-Erbsin, R.; Inthawong, M.; Sunyakumthorn, P.; Hunsawong, T.; Klungthong, C.; Reed, M.C.; et al. Preclinical evaluation of a plant-derived SARS-CoV-2 subunit vaccine: Protective efficacy, immunogenicity, safety, and toxicity. *Vaccine* **2022**, *40*, 4440–4452. [CrossRef]
70. Mamedov, T.; Yuksel, D.; Ilgin, M.; Gurbuzaslan, I.; Gulec, B.; Yetiskin, H.; Uygut, M.A.; Islam Pavel, S.T.; Ozdarendeli, A.; Mammadova, G.; et al. Plant-produced glycosylated and in vivo deglycosylated receptor binding domain proteins of SARS-CoV-2 Induce potent neutralizing responses in mice. *Viruses* **2021**, *13*, 1595. [CrossRef]
71. Rattanapisit, K.; Shanmugaraj, B.; Manopwisedjaroen, S.; Purwono, P.B.; Siriwattananon, K.; Khorattanakulchai, N.; Hanittinan, O.; Boonyayothin, W.; Thitithanyanont, A.; Smith, D.R.; et al. Rapid production of SARS-CoV-2 receptor binding domain (RBD) and spike specific monoclonal antibody CR3022 in Nicotiana benthamiana. *Sci. Rep.* **2020**, *10*, 17698. [CrossRef]
72. Demone, J.; Maltseva, M.; Nourimand, M.; Nasr-Sharif, M.; Galipeau, Y.; Alarcon, E.I.; Langlois, M.A.; MacLean, A.M. Scalable agroinfiltration-based production of SARS-CoV-2 antigens for use in diagnostic assays and subunit vaccines. *PLoS ONE* **2022**, *17*, e0277668. [CrossRef] [PubMed]
73. Rebelo, B.A.; Folgado, A.; Ferreira, A.C.; Abranches, R. Production of the SARS-CoV-2 spike protein and its receptor binding domain in plant cell suspension cultures. *Front. Plant Sci.* **2022**, *13*, 995429. [CrossRef] [PubMed]
74. Granwehr, B.P. In adults who had not had COVID-19, Novavax vaccine had 90% efficacy at >/=7 d after the second dose. *Ann. Intern. Med.* **2022**, *175*, JC52. [CrossRef] [PubMed]
75. Marabotti, C. Efficacy and effectiveness of COVID-19 vaccine—Absolute vs. relative risk reduction. Expert Rev. *Vaccines* **2022**, *21*, 873–875. [CrossRef]
76. Mardanova, E.S.; Kotlyarov, R.Y.; Ravin, N.V. High-yield production of receptor binding domain of SARS-CoV-2 linked to bacterial flagellin in plants using self-replicating viral vector pEff. *Plants* **2021**, *10*, 2682. [CrossRef] [PubMed]
77. Khorattanakulchai, N.; Srisutthisamphan, K.; Shanmugaraj, B.; Manopwisedjaroen, S.; Rattanapisit, K.; Panapitakkul, C.; Kemthong, T.; Suttisan, N.; Malaivijitnond, S.; Thitithanyanont, A.; et al. A recombinant subunit vaccine candidate produced in plants elicits neutralizing antibodies against SARS-CoV-2 variants in macaques. *Front. Plant Sci.* **2022**, *13*, 901978. [CrossRef] [PubMed]

78. Khorattanakulchai, N.; Manopwisedjaroen, S.; Rattanapisit, K.; Panapitakkul, C.; Kemthong, T.; Suttisan, N.; Srisutthisamphan, K.; Malaivijitnond, S.; Thitithanyanont, A.; Jongkaewwattana, A.; et al. Receptor binding domain proteins of SARS-CoV-2 variants produced in Nicotiana benthamiana elicit neutralizing antibodies against variants of concern. *J. Med. Virol.* **2022**, *94*, 4265–4276. [CrossRef]
79. Phoolcharoen, W.; (Chulalongkorn University, Bangkok, Thailand). Personal communication, 2022.
80. Nooraei, S.; Bahrulolum, H.; Hoseini, Z.S.; Katalani, C.; Hajizade, A.; Easton, A.J.; Ahmadian, G. Virus-like particles: Preparation, immunogenicity and their roles as nanovaccines and drug nanocarriers. *J. Nanobiotechnol.* **2021**, *19*, 59. [CrossRef]
81. Kushnir, N.; Streatfield, S.J.; Yusibov, V. Virus-like particles as a highly efficient vaccine platform: Diversity of targets and production systems and advances in clinical development. *Vaccine* **2012**, *31*, 58–83. [CrossRef]
82. Schwarz, B.; Uchida, M.; Douglas, T. Biomedical and Catalytic Opportunities of Virus-Like Particles in Nanotechnology. *Adv. Virus Res.* **2017**, *97*, 1–60. [CrossRef]
83. Ward, B.J.; Gobeil, P.; Séguin, A.; Atkins, J.; Boulay, I.; Charbonneau, P.Y.; Couture, M.; D'Aoust, M.A.; Dhaliwall, J.; Finkle, C.; et al. Phase 1 randomized trial of a plant-derived virus-like particle vaccine for COVID-19. *Nat. Med.* **2021**, *27*, 1071–1078. [CrossRef]
84. iBio Reports Preliminary Unaudited Fiscal Year 2022 Financial Results and Provides Corporate Update. Available online: https://www.globenewswire.com/news-release/2022/09/27/2523812/0/en/iBio-Reports-Preliminary-Unaudited-Fiscal-Year-2022-Financial-Results-and-Provides-Corporate-Update.html#:~:text=Preliminary%20Unaudited%20Financial%20Results%3A,comparable%20period%20in%20fiscal%202021 (accessed on 1 December 2022).
85. Pipeline Therapeutic Candidates. Available online: https://ibioinc.com/pipeline/ (accessed on 2 December 2022).
86. Lu, R.M.; Hwang, Y.C.; Liu, I.J.; Lee, C.C.; Tsai, H.Z.; Li, H.J.; Wu, H.C. Development of therapeutic antibodies for the treatment of diseases. *J. Biomed. Sci.* **2020**, *27*, 1–30. [CrossRef] [PubMed]
87. Jugler, C.; Sun, H.; Grill, F.; Kibler, K.; Esqueda, A.; Lai, H.; Li, Y.; Lake, D.; Chen, Q. Potential for a plant-made SARS-CoV-2 neutralizing monoclonal antibody as a synergetic cocktail component. *Vaccines* **2022**, *10*, 772. [CrossRef] [PubMed]
88. Chen, Q. Development of plant-made monoclonal antibodies against viral infections. *Curr. Opin. Virol.* **2022**, *52*, 148–160. [CrossRef] [PubMed]
89. Ma, J.K.; Drossard, J.; Lewis, D.; Altmann, F.; Boyle, J.; Christou, P.; Cole, T.; Dale, P.; van Dolleweerd, C.J.; Isitt, V.; et al. Regulatory approval and a first-in-human phase I clinical trial of a monoclonal antibody produced in transgenic tobacco plants. *Plant Biotechnol. J.* **2015**, *13*, 1106–1120. [CrossRef] [PubMed]
90. Shanmugaraj, B.; Rattanapisit, K.; Manopwisedjaroen, S.; Thitithanyanont, A.; Phoolcharoen, W. Monoclonal antibodies B38 and H4 produced in Nicotiana benthamiana neutralize SARS-CoV-2 in vitro. *Front. Plant Sci.* **2020**, *11*, 589995. [CrossRef] [PubMed]
91. Kallolimath, S.; Sun, L.; Palt, R.; Stiasny, K.; Mayrhofer, P.; Gruber, C.; Kogelmann, B.; Chen, Q.; Steinkellner, H. Highly active engineered IgG3 antibodies against SARS-CoV-2. *Proc. Natl. Acad. Sci. USA* **2021**, *118*, e2107249118. [CrossRef] [PubMed]
92. Jung, J.W.; Zahmanova, G.; Minkov, I.; Lomonossoff, G.P. Plant-based expression and characterization of SARS-CoV-2 virus-like particles presenting a native spike protein. *Plant Biotechnol. J.* **2022**, *20*, 1363–1372. [CrossRef] [PubMed]
93. VanBlargan, L.A.; Errico, J.M.; Halfmann, P.J.; Zost, S.J.; Crowe, J.E., Jr.; Purcell, L.A.; Kawaoka, Y.; Corti, D.; Fremont, D.H.; Diamond, M.S. An infectious SARS-CoV-2 B.1.1.529 Omicron virus escapes neutralization by therapeutic monoclonal antibodies. *Nat. Med.* **2022**, *28*, 490–495. [CrossRef] [PubMed]
94. Cohen-Dvashi, H.; Weinstein, J.; Katz, M.; Eilon-Ashkenazy, M.; Mor, Y.; Shimon, A.; Achdout, H.; Tamir, H.; Israely, T.; Strobelt, R.; et al. Anti-SARS-CoV-2 immunoadhesin remains effective against Omicron and other emerging variants of concern. *iScience* **2022**, *25*, 105193. [CrossRef] [PubMed]
95. Chamow, S.M.; Ashkenazi, A. Immunoadhesins: Principles and applications. *Trends Biotechnol.* **1996**, *14*, 52–60. [CrossRef]
96. Chen, Y.; Sun, L.; Ullah, I.; Beaudoin-Bussieres, G.; Anand, S.P.; Hederman, A.P.; Tolbert, W.D.; Sherburn, R.; Nguyen, D.N.; Marchitto, L.; et al. Engineered ACE2-Fc counters murine lethal SARS-CoV-2 infection through direct neutralization and Fc-effector activities. *Sci. Adv.* **2022**, *8*, eabn4188. [CrossRef]
97. Yasui, F.; Kohara, M.; Kitabatake, M.; Nishiwaki, T.; Fujii, H.; Tateno, C.; Yoneda, M.; Morita, K.; Matsushima, K.; Koyasu, S.; et al. Phagocytic cells contribute to the antibody-mediated elimination of pulmonary-infected SARS coronavirus. *Virology* **2014**, *454–455*, 157–168. [CrossRef]
98. Siriwattananon, K.; Manopwisedjaroen, S.; Kanjanasirirat, P.; Budi Purwono, P.; Rattanapisit, K.; Shanmugaraj, B.; Smith, D.R.; Borwornpinyo, S.; Thitithanyanont, A.; Phoolcharoen, W. Development of plant-produced recombinant ACE2-Fc fusion protein as a potential therapeutic agent against SARS-CoV-2. *Front. Plant Sci.* **2021**, *11*, 604663. [CrossRef]
99. Glasgow, A.; Glasgow, J.; Limonta, D.; Solomon, P.; Lui, I.; Zhang, Y.; Nix, M.A.; Rettko, N.J.; Zha, S.; Yamin, R.; et al. Engineered ACE2 receptor traps potently neutralize SARS-CoV-2. *Proc. Natl. Acad. Sci. USA* **2020**, *117*, 28046–28055. [CrossRef]
100. Higuchi, Y.; Suzuki, T.; Arimori, T.; Ikemura, N.; Mihara, E.; Kirita, Y.; Ohgitani, E.; Mazda, O.; Motooka, D.; Nakamura, S.; et al. Engineered ACE2 receptor therapy overcomes mutational escape of SARS-CoV-2. *Nat. Commun.* **2021**, *12*, 3802. [CrossRef]
101. Bernardi, A.; Huang, Y.; Harris, B.; Xiong, Y.; Nandi, S.; McDonald, K.A.; Faller, R. Development and simulation of fully glycosylated molecular models of ACE2-Fc fusion proteins and their interaction with the SARS-CoV-2 spike protein binding domain. *PLoS ONE* **2020**, *15*, e0237295. [CrossRef]

102. Mou, H.; Quinlan, B.D.; Peng, H.; Liu, G.; Guo, Y.; Peng, S.; Zhang, L.; Davis-Gardner, M.E.; Gardner, M.R.; Crynen, G.; et al. Mutations derived from horseshoe bat ACE2 orthologs enhance ACE2-Fc neutralization of SARS-CoV-2. *PLoS Pathog.* **2021**, *17*, e1009501. [CrossRef] [PubMed]
103. Chan, K.K.; Dorosky, D.; Sharma, P.; Abbasi, S.A.; Dye, J.M.; Kranz, D.M.; Herbert, A.S.; Procko, E. Engineering human ACE2 to optimize binding to the spike protein of SARS coronavirus 2. *Science* **2020**, *369*, 1261–1265. [CrossRef]
104. Tada, T.; Fan, C.; Chen, J.S.; Kaur, R.; Stapleford, K.A.; Gristick, H.; Dcosta, B.M.; Wilen, C.B.; Nimigean, C.M.; Landau, N.R. An ACE2 microbody containing a single immunoglobulin Fc domain is a potent inhibitor of SARS-CoV-2. *Cell Rep.* **2020**, *33*, 108528. [CrossRef] [PubMed]
105. Tsai, T.I.; Khalili, J.S.; Gilchrist, M.; Waight, A.B.; Cohen, D.; Zhuo, S.; Zhang, Y.; Ding, M.; Zhu, H.; Mak, A.N.; et al. ACE2-Fc fusion protein overcomes viral escape by potently neutralizing SARS-CoV-2 variants of concern. *Antivir. Res.* **2022**, *199*, 105271. [CrossRef] [PubMed]
106. Castilho, A.; Schwestka, J.; Kienzl, N.F.; Vavra, U.; Grunwald-Gruber, C.; Izadi, S.; Hiremath, C.; Niederhofer, J.; Laurent, E.; Monteil, V.; et al. Generation of enzymatically competent SARS-CoV-2 decoy receptor ACE2-Fc in glycoengineered Nicotiana benthamiana. *Biotechnol. J.* **2021**, *16*, e2000566. [CrossRef]
107. Daniell, H.; Nair, S.K.; Esmaeili, N.; Wakade, G.; Shahid, N.; Ganesan, P.K.; Islam, M.R.; Shepley-McTaggart, A.; Feng, S.; Gary, E.N. Debulking SARS-CoV-2 in saliva using angiotensin converting enzyme 2 in chewing gum to decrease oral virus transmission and infection. *Mol. Ther.* **2022**, *30*, 1966–1978. [CrossRef]
108. Daniell, H.; Nair, S.K.; Guan, H.; Guo, Y.; Kulchar, R.J.; Torres, M.D.; Shahed-Al-Mahmud, M.; Wakade, G.; Liu, Y.M.; Marques, A.D. Debulking different Corona (SARS-CoV-2 delta, omicron, OC43) and influenza (H1N1, H3N2) virus strains by plant viral trap proteins in chewing gums to decrease infection and transmission. *Biomaterials* **2022**, *288*, 121671. [CrossRef]
109. Ganesan, P.K.; Kulchar, R.J.; Kaznica, P.; Montoya-Lopez, R.; Green, B.J.; Streatfield, S.J.; Daniell, H. Optimization of biomass and target protein yield for Phase III clinical trial to evaluate Angiotensin Converting Enzyme 2 expressed in lettuce chloroplasts to reduce SARS-CoV-2 infection and transmission. *Plant Biotechnol. J.* 2022, *published online ahead of print*. [CrossRef]
110. Ghafoor, D.; Ahmed, S.; Zaman, N. Lectins; a hope of treatment for COVID-19. *Am. J. Biomed. Sci. Res.* **2021**, *12*, 280–282. [CrossRef]
111. Ahmed, M.N.; Jahan, R.; Nissapatorn, V.; Wilairatana, P.; Rahmatullah, M. Plant lectins as prospective antiviral biomolecules in the search for COVID-19 eradication strategies. *Biomed. Pharmacother.* **2022**, *146*, 112507. [CrossRef]
112. Naik, S.; Kumar, S. Lectins from plants and algae act as anti-viral against HIV, influenza and coronaviruses. *Mol. Biol. Rep.* **2022**, *49*, 12239–12246. [CrossRef] [PubMed]
113. Barre, A.; Van Damme, E.J.M.; Simplicien, M.; Le Poder, S.; Klonjkowski, B.; Benoist, H.; Peyrade, D.; Rouge, P. Man-Specific lectins from plants, fungi, algae and cyanobacteria, as potential blockers for SARS-CoV, MERS-CoV and SARS-CoV-2 (COVID-19) Coronaviruses: Biomedical perspectives. *Cells* **2021**, *10*, 1619. [CrossRef] [PubMed]
114. Nascimento da Silva, L.C.; Mendonca, J.S.P.; de Oliveira, W.F.; Batista, K.L.R.; Zagmignan, A.; Viana, I.F.T.; Dos Santos Correia, M.T. Exploring lectin-glycan interactions to combat COVID-19: Lessons acquired from other enveloped viruses. *Glycobiology* **2021**, *31*, 358–371. [CrossRef] [PubMed]
115. Chan, J.F.; Oh, Y.J.; Yuan, S.; Chu, H.; Yeung, M.L.; Canena, D.; Chan, C.C.; Poon, V.K.; Chan, C.C.; Zhang, A.J.; et al. A molecularly engineered, broad-spectrum anti-coronavirus lectin inhibits SARS-CoV-2 and MERS-CoV infection in vivo. *Cell Rep. Med.* **2022**, *3*, 100774. [CrossRef]
116. Liu, Y.M.; Shahed-Al-Mahmud, M.; Chen, X.; Chen, T.H.; Liao, K.S.; Lo, J.M.; Wu, Y.M.; Ho, M.C.; Wu, C.Y.; Wong, C.H.; et al. A carbohydrate-binding protein from the edible lablab beans effectively blocks the infections of influenza viruses and SARS-CoV-2. *Cell Rep.* **2020**, *32*, 108016. [CrossRef]
117. Lee, C. Griffithsin, a highly potent broad-spectrum antiviral lectin from red algae: From discovery to clinical application. *Mar. Drugs* **2019**, *17*, 567. [CrossRef]
118. Ahan, R.E.; Hanifehnezhad, A.; Kehribar, E.S.; Oguzoglu, T.C.; Foldes, K.; Ozcelik, C.E.; Filazi, N.; Oztop, S.; Palaz, F.; Onder, S.; et al. A highly potent SARS-CoV-2 blocking lectin protein. *ACS Infect. Dis.* **2022**, *8*, 1253–1264. [CrossRef]
119. Lokhande, K.B.; Apte, G.R.; Shrivastava, A.; Singh, A.; Pal, J.K.; Swamy, K.V.; Gupta, R.K. Sensing the interactions between carbohydrate-binding agents and N-linked glycans of SARS-CoV-2 spike glycoprotein using molecular docking and simulation studies. *J. Biomol. Struct. Dyn.* **2022**, *40*, 3880–3898. [CrossRef]
120. Nabi-Afjadi, M.; Heydari, M.; Zalpoor, H.; Arman, I.; Sadoughi, A.; Sahami, P.; Aghazadeh, S. Lectins and lectibodies: Potential promising antiviral agents. *Cell Mol. Biol. Lett.* **2022**, *27*, 37. [CrossRef]
121. Gecchele, E.; Merlin, M.; Brozzetti, A.; Falorni, A.; Pezzotti, M.; Avesani, L. A comparative analysis of recombinant protein expression in different biofactories: Bacteria, insect cells and plant systems. *J. Vis. Exp.* **2015**, *97*, 52459. [CrossRef]
122. Gomord, V.; Fitchette, A.C.; Menu-Bouaouiche, L.; Saint-Jore-Dupas, C.; Plasson, C.; Michaud, D.; Faye, L. Plant-specific glycosylation patterns in the context of therapeutic protein production. *Plant Biotechnol. J.* **2010**, *8*, 564–587. [CrossRef] [PubMed]
123. Al-Kuraishy, H.M.; Al-Fakhrany, O.M.; Elekhnawy, E.; Al-Gareeb, A.I.; Alorabi, M.; De Waard, M.; Albogami, S.M.; Batiha, G.E. Traditional herbs against COVID-19: Back to old weapons to combat the new pandemic. *Eur. J. Med. Res.* **2022**, *27*, 186. [CrossRef] [PubMed]
124. Islam, M.T.; Sarkar, C.; El-Kersh, D.M.; Jamaddar, S.; Uddin, S.J.; Shilpi, J.A.; Mubarak, M.S. Natural products and their derivatives against coronavirus: A review of the non-clinical and pre-clinical data. *Phytother. Res.* **2020**, *34*, 2471–2492. [CrossRef] [PubMed]

125. Bagde, H.; Dhopte, A. Effects of plant metabolites on the growth of COVID-19 (Coronavirus Disease-19) including omicron strain. *Cureus* **2022**, *14*, e26549. [CrossRef]
126. Bhattacharya, R.; Dev, K.; Sourirajan, A. Antiviral activity of bioactive phytocompounds against coronavirus: An update. *J. Virol. Methods* **2021**, *290*, 114070. [CrossRef]
127. Pawar, K.S.; Mastud, R.N.; Pawar, S.K.; Pawar, S.S.; Bhoite, R.R.; Bhoite, R.R.; Kulkarni, M.V.; Deshpande, A.R. Oral curcumin with piperine as adjuvant therapy for the treatment of COVID-19: A randomized clinical trial. *Front. Pharmacol.* **2021**, *12*, 669362. [CrossRef]
128. Bhar, A.; Jain, A.; Das, S. Natural therapeutics against SARS-CoV-2: The potentiality and challenges. *Vegetos*, 2022, *published online ahead of print*. [CrossRef]
129. Holghoomi, R.; Abdolmaleki, A.; Asadi, A.; Kajkolah, M.; Gurushankar, K.; Bidarlord, M. Plant-derived metabolites as potent inhibitor and treatment for COVID-19. *J. Pharm. Care* **2022**, *10*, 84–93. [CrossRef]
130. Lopes, A.J.O.; Calado, G.P.; Froes, Y.N.; Araujo, S.A.; Franca, L.M.; Paes, A.M.A.; Morais, S.V.; Rocha, C.Q.D.; Vasconcelos, C.C. Plant metabolites as SARS-CoV-2 inhibitors candidates: In silico and in vitro studies. *Pharmaceuticals* **2022**, *15*, 1045. [CrossRef]
131. Mukherjee, P.K.; Efferth, T.; Das, B.; Kar, A.; Ghosh, S.; Singha, S.; Debnath, P.; Sharma, N.; Bhardwaj, P.K.; Haldar, P.K. Role of medicinal plants in inhibiting SARS-CoV-2 and in the management of post-COVID-19 complications. *Phytomedicine* **2022**, *98*, 153930. [CrossRef]
132. Llivisaca-Contreras, S.A.; Naranjo-Moran, J.; Pino-Acosta, A.; Pieters, L.; Vanden Berghe, W.; Manzano, P.; Vargas-Perez, J.; Leon-Tamariz, F.; Cevallos-Cevallos, J.M. Plants and natural products with activity against various types of coronaviruses: A review with focus on SARS-CoV-2. *Molecules* **2021**, *26*, 4099. [CrossRef]
133. Al-Harrasi, A.; Behl, T.; Upadhyay, T.; Chigurupati, S.; Bhatt, S.; Sehgal, A.; Bhatia, S.; Singh, S.; Sharma, N.; Vijayabalan, S.; et al. Targeting natural products against SARS-CoV-2. *Environ. Sci Pollut. Res. Int.* **2022**, *29*, 42404–42432. [CrossRef] [PubMed]
134. Jamshidnia, M.; Sewell, R.D.E.; Rafieian-Kopaei, M. An update on promising agents against COVID-19: Secondary metabolites and mechanistic aspects. *Curr. Pharm. Des.* **2022**, *28*, 2415–2425. [CrossRef] [PubMed]
135. Zrig, A. The effect of phytocompounds of medicinal plants on coronavirus (2019-NCOV) infection. *Pharm. Chem. J.* **2022**, *55*, 1080–1084. [CrossRef] [PubMed]
136. Trivedi, P.; Abbas, A.; Lehmann, C.; Rupasinghe, H.P.V. Antiviral and anti-inflammatory plant-derived bioactive compounds and their potential use in the treatment of COVID-19-related pathologies. *J. Xenobiot.* **2022**, *12*, 289–306. [CrossRef] [PubMed]
137. Yang, L.; Wang, Z. Natural products, alone or in combination with FDA-Approved drugs, to treat COVID-19 and lung cancer. *Biomedicines* **2021**, *9*, 689. [CrossRef] [PubMed]
138. Khan, T.; Khan, M.A.; Mashwani, Z.U.; Ullah, N.; Nadhman, A. Therapeutic potential of medicinal plants against COVID-19: The role of antiviral medicinal metabolites. *Biocatal. Agric. Biotechnol.* **2021**, *31*, 101890. [CrossRef] [PubMed]
139. Remali, J.; Aizat, W.M. A Review on Plant Bioactive Compounds and Their Modes of Action Against Coronavirus Infection. *Front. Pharmacol.* **2020**, *11*, 589044. [CrossRef]
140. Park, J.; Park, R.; Jang, M.; Park, Y.I.; Park, Y. Coronavirus enzyme inhibitors-experimentally proven natural compounds from plants. *J. Microbiol.* **2022**, *60*, 347–354. [CrossRef] [PubMed]
141. Kumar, D.; Chandel, V.; Raj, S.; Rathi, B. In silico identification of potent FDA approved drugs against Coronavirus COVID-19 main protease: A drug repurposing approach. *Chem. Biol. Lett.* **2020**, *7*, 166–175. [CrossRef]
142. Singh, S.; Sk, M.F.; Sonawane, A.; Kar, P.; Sadhukhan, S. Plant-derived natural polyphenols as potential antiviral drugs against SARS-CoV-2 via RNA-dependent RNA polymerase (RdRp) inhibition: An in-silico analysis. *J. Biomol. Struct. Dyn.* **2021**, *39*, 6249–6264. [CrossRef]
143. Jan, J.T.; Cheng, T.R.; Juang, Y.P.; Ma, H.H.; Wu, Y.T.; Yang, W.B.; Cheng, C.W.; Chen, X.; Chou, T.H.; Shie, J.J.; et al. Identification of existing pharmaceuticals and herbal medicines as inhibitors of SARS-CoV-2 infection. *Proc. Natl. Acad. Sci. USA* **2021**, *118*, e2021579118. [CrossRef] [PubMed]
144. Muchtaridi, M.; Fauzi, M.; Khairul Ikram, N.K.; Mohd Gazzali, A.; Wahab, H.A. Natural flavonoids as potential angiotensin-converting enzyme 2 inhibitors for anti-SARS-CoV-2. *Molecules* **2020**, *25*, 3980. [CrossRef]
145. Liu, X.; Raghuvanshi, R.; Ceylan, F.D.; Bolling, B.W. Quercetin and Its Metabolites Inhibit Recombinant Human Angiotensin-Converting Enzyme 2 (ACE2) Activity. *J. Agric. Food Chem.* **2020**, *68*, 13982–13989. [CrossRef] [PubMed]
146. Senthil Kumar, K.J.; Gokila Vani, M.; Wang, C.S.; Chen, C.C.; Chen, Y.C.; Lu, L.P.; Huang, C.H.; Lai, C.S.; Wang, S.Y. Geranium and lemon essential oils and their active compounds downregulate angiotensin-converting enzyme 2 (ACE2), a SARS-CoV-2 spike receptor-binding domain, in epithelial cells. *Plants* **2020**, *9*, 770. [CrossRef]
147. Lucas, K.; Frohlich-Nowoisky, J.; Oppitz, N.; Ackermann, M. Cinnamon and hop extracts as potential immunomodulators for severe COVID-19 cases. *Front. Plant Sci.* **2021**, *12*, 589783. [CrossRef] [PubMed]
148. Gyebi, G.A.; Ogunro, O.B.; Adegunloye, A.P.; Ogunyemi, O.M.; Afolabi, S.O. Potential inhibitors of coronavirus 3-chymotrypsin-like protease (3CL(pro)): An in silico screening of alkaloids and terpenoids from African medicinal plants. *J. Biomol. Struct. Dyn.* **2021**, *39*, 3396–3408. [CrossRef]
149. Wink, M. Potential of DNA intercalating alkaloids and other plant secondary metabolites against SARS-CoV-2 causing COVID-19. *Diversity* **2020**, *12*, 175. [CrossRef]

150. Gonzalez, B.L.; de Oliveira, N.C.; Ritter, M.R.; Tonin, F.S.; Melo, E.B.; Sanches, A.C.C.; Fernandez-Llimos, F.; Petruco, M.V.; de Mello, J.C.P.; Chierrito, D.; et al. The naturally-derived alkaloids as a potential treatment for COVID-19: A scoping review. *Phytother. Res.* **2022**, *36*, 2686–2709. [CrossRef]
151. Truzzi, F.; Tibaldi, C.; Zhang, Y.; Dinelli, G.; D'Ammen, E. An overview on dietary polyphenols and their biopharmaceutical classification system (BCS). *Int. J. Mol. Sci.* **2021**, *22*, 5514. [CrossRef]
152. Tirado-Kulieva, V.A.; Hernandez-Martinez, E.; Choque-Rivera, T.J. Phenolic compounds versus SARS-CoV-2: An update on the main findings against COVID-19. *Heliyon* **2022**, *8*, e10702. [CrossRef]
153. Annunziata, G.; Sanduzzi Zamparelli, M.; Santoro, C.; Ciampaglia, R.; Stornaiuolo, M.; Tenore, G.C.; Sanduzzi, A.; Novellino, E. May polyphenols have a role against coronavirus infection? An overview of in vitro evidence. *Front. Med.* **2020**, *7*, 240. [CrossRef]
154. Abdelmageed, M.I.; Abdelmoneim, A.H.; Mustafa, M.I.; Elfadol, N.M.; Murshed, N.S.; Shantier, S.W.; Makhawi, A.M. Design of a multiepitope-based peptide vaccine against the e protein of human COVID-19: An immunoinformatics approach. *Biomed. Res. Int.* **2020**, *2020*, 2683286. [CrossRef] [PubMed]
155. Owis, A.I.; El-Hawary, M.S.; El Amir, D.; Aly, O.M.; Abdelmohsen, U.R.; Kamel, M.S. Molecular docking reveals the potential of Salvadora persica flavonoids to inhibit COVID-19 virus main protease. *RSC Adv.* **2020**, *10*, 19570–19575. [CrossRef] [PubMed]
156. Sytar, O.; Brestic, M.; Hajihashemi, S.; Skalicky, M.; Kubes, J.; Lamilla-Tamayo, L.; Ibrahimova, U.; Ibadullayeva, S.; Landi, M. COVID-19 prophylaxis efforts based on natural antiviral plant extracts and their compounds. *Molecules* **2021**, *26*, 727. [CrossRef] [PubMed]
157. Reichling, J. Plant-microbe interactions and secondary metabolites with antibacterial, antifungal and antiviral properties. In *Annual Plant Reviews: Functions and Biotechnology of Plant Secondary Metabolites*, 2nd ed.; Wink, M., Ed.; Wiley-Blackwell: Hoboken, NJ, USA, 2010; Volume 39, pp. 214–347. [CrossRef]
158. Gomaa, A.A.; Mohamed, H.S.; Abd-Ellatief, R.B.; Gomaa, M.A.; Hammam, D.S. Advancing combination treatment with glycyrrhizin and boswellic acids for hospitalized patients with moderate COVID-19 infection: A randomized clinical trial. *Inflammopharmacology* **2022**, *30*, 477–486. [CrossRef] [PubMed]
159. Nair, M.S.; Huang, Y.; Fidock, D.A.; Polyak, S.J.; Wagoner, J.; Towler, M.J.; Weathers, P.J. Artemisia annua L. extracts inhibit the in vitro replication of SARS-CoV-2 and two of its variants. *J. Ethnopharmacol.* **2021**, *274*, 114016. [CrossRef]
160. Nair, M.S.; Huang, Y.; Fidock, D.A.; Towler, M.J.; Weathers, P.J. Artemisia annua L. hot-water extracts show potent activity in vitro against COVID-19 variants including delta. *J. Ethnopharmacol.* **2022**, *284*, 114797. [CrossRef]
161. Zhou, Y.; Gilmore, K.; Ramirez, S.; Settels, E.; Gammeltoft, K.A.; Pham, L.V.; Fahnoe, U.; Feng, S.; Offersgaard, A.; Trimpert, J.; et al. In vitro efficacy of artemisinin-based treatments against SARS-CoV-2. *Sci. Rep.* **2021**, *11*, 14571. [CrossRef]
162. Kim, T.Y.; Kim, J.Y.; Kwon, H.C.; Jeon, S.; Lee, S.J.; Jung, H.; Kim, S.; Jang, D.S.; Lee, C.J. Astersaponin I from Aster koraiensis is a natural viral fusion blocker that inhibits the infection of SARS-CoV-2 variants and syncytium formation. *Antivir. Res.* **2022**, *208*, 105428. [CrossRef]
163. Nag, A.; Banerjee, R.; Paul, S.; Kundu, R. Curcumin inhibits spike protein of new SARS-CoV-2 variant of concern (VOC) Omicron, an in silico study. *Comput. Biol. Med.* **2022**, *146*, 105552. [CrossRef]
164. Valipour, M. Different aspects of emetine's capabilities as a highly potent SARS-CoV-2 inhibitor against COVID-19. ACS Pharmacol. *Transl. Sci.* **2022**, *5*, 387–399. [CrossRef]
165. Cheng, F.J.; Huynh, T.K.; Yang, C.S.; Hu, D.W.; Shen, Y.C.; Tu, C.Y.; Wu, Y.C.; Tang, C.H.; Huang, W.C.; Chen, Y.; et al. Hesperidin is a potential inhibitor against SARS-CoV-2 infection. *Nutrients* **2021**, *13*, 2800. [CrossRef]
166. Delcanale, P.; Uriati, E.; Mariangeli, M.; Mussini, A.; Moreno, A.; Lelli, D.; Cavanna, L.; Bianchini, P.; Diaspro, A.; Abbruzzetti, S.; et al. The interaction of hypericin with SARS-CoV-2 reveals a multimodal antiviral activity. *ACS Appl. Mater. Interfaces* **2022**, *14*, 14025–14032. [CrossRef] [PubMed]
167. Yi, Y.; Li, J.; Lai, X.; Zhang, M.; Kuang, Y.; Bao, Y.O.; Yu, R.; Hong, W.; Muturi, E.; Xue, H.; et al. Natural triterpenoids from licorice potently inhibit SARS-CoV-2 infection. *J. Adv. Res.* **2022**, *36*, 201–210. [CrossRef] [PubMed]
168. Munafo, F.; Donati, E.; Brindani, N.; Ottonello, G.; Armirotti, A.; De Vivo, M. Quercetin and luteolin are single-digit micromolar inhibitors of the SARS-CoV-2 RNA-dependent RNA polymerase. *Sci. Rep.* **2022**, *12*, 10571. [CrossRef] [PubMed]
169. Xiao, T.; Cui, M.; Zheng, C.; Wang, M.; Sun, R.; Gao, D.; Bao, J.; Ren, S.; Yang, B.; Lin, J.; et al. Myricetin inhibits SARS-CoV-2 viral replication by targeting M(pro) and ameliorates pulmonary inflammation. *Front. Pharmacol.* **2021**, *12*, 669642. [CrossRef]
170. Baeshen, N.A.; Albeshri, A.O.; Baeshen, N.N.; Attar, R.; Karkashan, A.; Abbas, B.; Bouback, T.A.; Aljaddawi, A.A.; Refai, M.Y.; Abdelkader, H.S.; et al. In silico screening of some compounds derived from the desert medicinal plant Rhazya stricta for the potential treatment of COVID-19. *Sci. Rep.* **2022**, *12*, 11120. [CrossRef]
171. Kanjanasirirat, P.; Suksatu, A.; Manopwisedjaroen, S.; Munyoo, B.; Tuchinda, P.; Jearawuttanakul, K.; Seemakhan, S.; Charoensutthivarakul, S.; Wongtrakoongate, P.; Rangkasenee, N.; et al. High-content screening of Thai medicinal plants reveals Boesenbergia rotunda extract and its component Panduratin A as anti-SARS-CoV-2 agents. *Sci. Rep.* **2020**, *10*, 19963. [CrossRef]
172. Furukawa, R.; Kitabatake, M.; Ouji-Sageshima, N.; Suzuki, Y.; Nakano, A.; Matsumura, Y.; Nakano, R.; Kasahara, K.; Kubo, K.; Kayano, S.I.; et al. Persimmon-derived tannin has antiviral effects and reduces the severity of infection and transmission of SARS-CoV-2 in a Syrian hamster model. *Sci. Rep.* **2021**, *11*, 23695. [CrossRef]
173. Shekunov, E.V.; Efimova, S.S.; Yudintceva, N.M.; Muryleva, A.A.; Zarubaev, V.V.; Slita, A.V.; Ostroumova, O.S. Plant alkaloids inhibit membrane fusion mediated by calcium and fragments of MERS-CoV and SARS-CoV/SARS-CoV-2 fusion peptides. *Biomedicines* **2021**, *9*, 1434. [CrossRef]

174. Shaban, M.S.; Mayr-Buro, C.; Meier-Soelch, J.; Albert, B.V.; Schmitz, M.L.; Ziebuhr, J.; Kracht, M. Thapsigargin: Key to new host-directed coronavirus antivirals? *Trends Pharmacol. Sci.* **2022**, *43*, 557–568. [CrossRef] [PubMed]
175. Saggam, A.; Limgaokar, K.; Borse, S.; Chavan-Gautam, P.; Dixit, S.; Tillu, G.; Patwardhan, B. Withania somnifera (L.) Dunal: Opportunity for Clinical Repurposing in COVID-19 Management. *Front. Pharmacol.* **2021**, *12*, 623795. [CrossRef]
176. Farmanpour-Kalalagh, K.; Beyraghdar Kashkooli, A.; Babaei, A.; Rezaei, A.; van der Krol, A.R. Artemisinins in combating viral infections like SARS-CoV-2, inflammation and cancers and options to meet increased global demand. *Front. Plant Sci.* **2022**, *13*, 780257. [CrossRef]
177. Wang, Z.; Yang, L. Turning the tide: Natural products and natural-product-inspired chemicals as potential counters to SARS-CoV-2 infection. *Front. Pharmacol.* **2020**, *11*, 1013. [CrossRef]
178. Wang, A.; Sun, Y.; Liu, Q.; Wu, H.; Liu, J.; He, J.; Yu, J.; Chen, Q.Q.; Ge, Y.; Zhang, Z.; et al. Low dose of emetine as potential anti-SARS-CoV-2 virus therapy: Preclinical in vitro inhibition and in vivo pharmacokinetic evidences. *Mol. Biomed.* **2020**, *1*, 14. [CrossRef] [PubMed]
179. Ren, P.X.; Shang, W.J.; Yin, W.C.; Ge, H.; Wang, L.; Zhang, X.L.; Li, B.Q.; Li, H.L.; Xu, Y.C.; Xu, E.H.; et al. A multi-targeting drug design strategy for identifying potent anti-SARS-CoV-2 inhibitors. *Acta Pharmacol. Sin.* **2022**, *43*, 483–493. [CrossRef] [PubMed]
180. Kumar, R.; Afsar, M.; Khandelwal, N.; Chander, Y.; Riyesh, T.; Dedar, R.K.; Gulati, B.R.; Pal, Y.; Barua, S.; Tripathi, B.N.; et al. Emetine suppresses SARS-CoV-2 replication by inhibiting interaction of viral mRNA with eIF4E. *Antivir. Res.* **2021**, *189*, 105056. [CrossRef] [PubMed]
181. Morales-Paredes, C.A.; Rodriguez-Diaz, J.M.; Boluda-Botella, N. Pharmaceutical compounds used in the COVID-19 pandemic: A review of their presence in water and treatment techniques for their elimination. *Sci. Total Environ.* **2022**, *814*, 152691. [CrossRef]
182. Weidenbacher, P.A.; Sanyal, M.; Friedland, N.; Tang, S.; Arunachalam, P.S.; Hu, M.; Kumru, O.S.; Morris, M.K.; Fontenot, J.; Shirreff, L.; et al. A ferritin-based COVID-19 nanoparticle vaccine that elicits robust, durable, broad-spectrum neutralizing antisera in non-human primates. *bioRxiv* **2022**. *preprint*. [CrossRef]
183. Hameed, S.A.; Paul, S.; Dellosa, G.K.Y.; Jaraquemada, D.; Bello, M.B. Towards the future exploration of mucosal mRNA vaccines against emerging viral diseases; lessons from existing next-generation mucosal vaccine strategies. *NPJ Vaccines* **2022**, *7*, 71. [CrossRef]
184. Waltz, E. How nasal-spray vaccines could change the pandemic. *Nature* **2022**, *609*, 240–242. [CrossRef] [PubMed]
185. Waltz, E. China and India approve nasal COVID vaccines—Are they a game changer? *Nature* **2022**, *609*, 450. [CrossRef]
186. Chan, H.T.; Daniell, H. Plant-made oral vaccines against human infectious diseases-Are we there yet? *Plant Biotechnol. J.* **2015**, *13*, 1056–1070. [CrossRef] [PubMed]
187. Xu, J.; Dolan, M.C.; Medrano, G.; Cramer, C.L.; Weathers, P.J. Green factory: Plants as bioproduction platforms for recombinant proteins. *Biotechnol. Adv.* **2012**, *30*, 1171–1184. [CrossRef] [PubMed]
188. Xu, J.; Zhang, N. On the way to commercializing plant cell culture platform for biopharmaceuticals: Present status and prospect. *Pharm. Bioprocess.* **2014**, *2*, 499–518. [CrossRef] [PubMed]
189. Karki, U.; Fang, H.; Guo, W.; Unnold-Cofre, C.; Xu, J. Cellular engineering of plant cells for improved therapeutic protein production. *Plant Cell Rep.* **2021**, *40*, 1087–1099. [CrossRef]
190. Mlozi, S.H. The role of natural products from medicinal plants against COVID-19: Traditional medicine practice in Tanzania. *Heliyon* **2022**, *8*, e09739. [CrossRef] [PubMed]
191. Li, D.; Halitschke, R.; Baldwin, I.T.; Gaquerel, E. Information theory tests critical predictions of plant defense theory for specialized metabolism. *Sci. Adv.* **2020**, *6*, eaaz0381. [CrossRef]

MDPI
St. Alban-Anlage 66
4052 Basel
Switzerland
www.mdpi.com

Life Editorial Office
E-mail: life@mdpi.com
www.mdpi.com/journal/life

www.ingramcontent.com/pod-product-compliance
Lightning Source LLC
Chambersburg PA
CBHW061815210926
51850CB00006B/711
* 9 7 8 3 0 3 6 5 9 4 6 5 1 *